Nonlinear Image Processing

Academic Press Series in Communications, Networking, and Multimedia

EDITOR-IN-CHIEF

Jerry D. Gibson
Southern Methodist University

This series has been established to bring together a variety of publications that represent the latest in cutting-edge research, theory, and applications of modern communication systems. All traditional and modern aspects of communications as well as all methods of computer communications are to be included. The series will include professional handbooks, books on communication methods and standards, and research books for engineers and managers in the world-wide communications industry.

Books in the Series:
Handbook of Image and Video Processing, Al Bovik, editor
The E-Commerce Book, Steffano Korper and Juanita Ellis
Multimedia Communications, Jerry Gibson, editor
Nonlinear Image Processing, Sanjit K. Mitra and Giovanni L. Sicuranza, editors

Nonlinear Image Processing

EDITORS

SANJIT K. MITRA
University of California
Santa Barbara, California, USA

GIOVANNI L. SICURANZA
University of Trieste
Trieste, Italy

ACADEMIC PRESS
A Harcourt Science and Technology Company
SAN DIEGO / SAN FRANCISCO / NEW YORK / BOSTON / LONDON / SYDNEY / TOKYO

This book is printed on acid-free paper. 

ACADEMIC PRESS
A Harcourt Science and Technology Company
525 B Street, Suite 1900, San Diego, CA 92101-4495, USA
http://www.academicpress.com

Academic Press
Harcourt Place, 32 Jamestown Road, London, NW1 7BY, UK

Library of Congress Catalog Number: 00-104376

ISBN: 0-12-500451-6

Printed in the United States of America

00 01 02 03 04 05 HP 9 8 7 6 5 4 3 2 1

Contents

Preface

In recent years, nonlinear methods and techniques have emerged as intensive research topics in the fields of signal and image processing. The reason for this increased interest in nonlinear methods of image processing is mainly due to the following observations. First, the human visual system (HVS) includes some nonlinear effects that need to be considered in order to develop effective image processing algorithms. Therefore, to comply with the characteristics of the HVS and, thus, obtain better visual results, nonlinear algorithms are necessary. Moreover, the nonlinear behavior of optical imaging systems and their related image formation systems must be taken into account. Finally, images are signals that in general do not satisfy the widely used hypotheses of Gaussianity and stationarity that are usually assumed to validate linear models and filtering techniques. In this respect, it is well known, for example, that linear filters are not able to remove an impulsive noise superimposed on an image without blurring its edges and small details. Other situations in which linear filters perform poorly are those cases where signal-dependent or multiplicative noises are present in the images.

Although linear filters continue to play an important role in signal processing because they are inherently simple to implement, the advances of computers and digital signal processors, in terms of speed, size, and cost, make the implementation of more sophisticated algorithms practical and effective.

These considerations are the basis for the increased interest in the development of new nonlinear techniques for image processing, with particular emphasis on the applications that benefit greatly from a nonlinear approach, such as edge preserving smoothing, edge enhancement, noise filtering, image segmentation, and feature extraction.

An interesting aspect of the recent studies on nonlinear image processing is the fact that an attempt has been made to organize the previously scattered contributions in a few homogeneous sectors. While a common framework is far from being derived (or it is simply out of reach since nonlinearity is defined as the lack of a property, that is, linearity), suitable classes of nonlinear operators have been introduced. A (not exhaustive) list of these classes includes:

- Homomorphic filters, relying on a generalized superposition principle;
- Nonlinear mean filters, using nonlinear definitions of means;
- Morphological filters, based on geometrical rather than analytical properties;
- Order statistics filters, based on ordering properties of the input samples;
- Polynomial filters, using polynomial expressions in the input and output samples;
- Fuzzy filters, applying fuzzy reasoning to model the uncertainty that is typical of some image processing issues; and
- Nonlinear operators modeled in terms of nonlinear partial differential equations (PDEs).

All of these filter families are considered in this book, but with a different stress according to their popularity and impact on image processing tasks.

Another relevant aspect that constitutes at present a trend in the area of nonlinear filters is the search for relationships and hierarchies among the above-mentioned classes. For example, interrelations have been pointed out between order statistic filters and PDE models; and between these two classes and morphological filters, some forms of polynomial filters can be expressed as PDEs, etc. Moreover, the generalization efforts permit the well-known filters to be considered members of broader classes. In this respect, homomorphic filters can be viewed as belonging to the class of nonlinear mean filters, the ubiquitous median filter can be described as an element of more general categories of order statistic nonlinear filters, and so on. Such aspects are considered in the appropriate chapters of this book and the relevant interrelations and hierarchies are referenced and illustrated.

Finally, an emerging research line in the field of nonlinear filtering is the joint exploitation of different information and features typical of different filter classes. The aim of the approaches based on this methodology is clearly to exploit the advantages offered by the various classes of nonlinear operators while reducing their drawbacks. This result can be achieved by combining different information to define new filter classes, as shown for example in some of the contributions contained in this book by the joint use of spatial and rank ordering information. An alternative approach, especially useful for the solution of well-defined image processing tasks, is based on the successive use of different kinds of filters depending on the specific application considered. In our opinion the material presented in this book aids this goal and thus permits the realizations of actual application-oriented algorithms and systems.

The first three chapters of the book deal with variations of order statistic filters. Chapter 1 introduces stack and weighted order statistics filters. The interrelations between different kinds of filters that can be viewed as cases of the general threshold decomposition and Boolean logic treatment are shown. The most important tools available for the analysis and optimization of these filters are presented. Both deterministic and statistical properties are considered in order to give a comprehensive understanding of the reasons why these filters work so well in certain applications.

Chapter 2 deals with an extended class of nonlinear filters derived from the median operator, that is, the local weighted median filter. After reviewing the principles of weighted medians, smoothers allowing positive as well as negative weights are introduced. These nonlinear tools are applied to image enhancement and analysis, with specific applications to image denoising and sharpening, zooming, and edge detection. Methods for designing optimal frequency-selective weighted median filters are also described.

Chapter 3 explores the joint use of spatial and rank ordering information on the input samples in the framework of the so-called selection filters. In such filters, spatial ordering is used to exploit correlations between neighboring samples while rank order is used to isolate outliers and ensure robust behavior. The chapter theoretically motivates selection filters and develops several class subsets and extensions that utilize partial/full/extended spatial and rank ordering information. The developed filters are applied to various image processing tasks, such as noise smoothing, interpolation, and image restoration.

Chapter 4 contains another example of a combination of different kinds of operations to derive new sets of nonlinear filters. In fact, the signal-dependent rank-ordered-mean filters exploit the rank ordering properties of the input samples together with operations such as mean and differences acting on the input samples ordered by rank. In particular, the rank-ordered differences provide information about the likelihood of corruption for the current pixel. The resulting nonlinear algorithms are particularly efficient to remove impulse noises from highly corrupted images while preserving details and features.

Nonlinear mean filters, described in Chapter 5, can be considered as another alternative to median filters and their extension to remove impulse noises effectively, especially when the impulses occur with a high probability. They have a very simple structure and thus are suitable for real-time processing applications. From a statistical point of view, they rely on the nonlinear means that are well-known location estimators. This approach produces a general filter structure that encompasses homomorphic, order statistics, and morphological filters. Effective edge detectors and edge preserving filters are demonstrated, together with soft gray-scale morphological filters, which are shown to be useful for removal of both Rayleigh and signal-dependent Gaussian speckle noises that usually affect ultrasonic images.

Chapters 6 and 7 deal with polynomial filters and their application to image processing tasks. The interest in such filters is mainly due to the fact that they can be considered to be the most natural extension of linear operators. In Chapter 6, the Teager filter is described in the framework of the most general class of quadratic Volterra filters. This filter has the property that sinusoidal inputs generate constant outputs that are approximately proportional to the square of the input frequency. Its properties are presented and appropriate two-dimensional versions are derived. Efficient design techniques are proposed and applications in image enhancement are demonstrated.

Chapter 7 provides an overview of polynomial filters based on the discrete Volterra series and of their extensions to two dimensions. Then, rational filters are introduced. These nonlinear filters, whose input–output relationship is given in the form of a ratio of two polynomials with respect to the input samples, are universal approximators, as polynomial functions, but they can achieve the desired level of accuracy with lower complexity and better extrapolation capabilities. Applications of polynomial filters for contrast enhancement, texture segmentation, and edge extraction are considered. Applications of rational filters to detail-preserving noise smoothing and interpolation with accurate edge reproduction are presented, together with a contrast enhancement technique that provides results comparable to those obtained with the previously described polynomial technique.

Using partial differential equations (PDEs) and curve/surface flows leads to model images in a continuous domain. The understanding of discrete local nonlinear filters is facilitated when one lets the grid mesh tend to zero and thus rewrites the discrete filter, thanks to an asymptotic expansion, as a partial differential operator. An advantage of such an approach is the possibility of achieving high accuracy and stability according to the extensive available research on numerical analysis. This emerging research area is considered in Chapter 8. After a general presentation, the first part of the chapter deals with the use of PDEs for image segmentation, while the second part discusses the use of PDEs to process multivalued data defined on nonflat manifolds, for example, directional data.

In Chapter 9 the basic concepts of morphological filtering and the corresponding operators are introduced and described by examples. Then, the basic notions related to a recent set of morphological filtering tools, called connected operators, are presented. Connected operators are essentially region-based filtering tools since they do not modify individual pixel values, but instead act directly on the connected components of the space where the image is constant. The two most successful strategies to define connected operators, based on reconstruction processes and tree representations, are discussed. The interest for morphological region-based tools is related to the recent developments in the new area of multimedia applications and services, where content-based compression and indexing of image and video signals are typical examples of situations where new modeling strategies are necessary.

Differential morphology is the topic presented in Chapter 10. Morphological image processing has traditionally been based on modeling images as sets or as points in a complete lattice of functions and viewing morphological image transformations as set or lattice operations. In parallel, there is a recently growing part of morphological image processing that is based on ideas from differential calculus and dynamic systems. Therefore, the unifying theme that defines the differential morphology is a collection of nonlinear differential/difference equations modeling the scale or space dynamics of morphological systems. In this chapter a unified view of the various interrelated ideas in this area is presented. Some system analysis tools in both space and transform domains are developed. Moreover, the connections between nonlinear PDEs and multiscale morphological filtering are fully discussed.

Chapter 11 presents the fundamental definitions and properties of the coordinate logic filters that constitute a tool for processing gray-level images as a set of binary images. In fact, these filters are coincident with morphological filters for binary images, while maintaining a similar functionality for gray-level images. The remarkable advantage of coordinate logic filters is that their simplicity allows for very fast implementations because sorting operations are not required. Typical applications for image enhancement and analysis are presented. Another relevant property of these filters is their direct relation with fractal structures. In the last part of the chapter, examples and simple rules for designing fractal forms and cellular automata are given.

Since its introduction in 1965 as a mathematical tool able to model the concept of partial membership, the theory of fuzzy sets has been used in many fields of engineering. Recently, this approach has been extended to cover also image processing applications because fuzzy filters are well suited to address the uncertainty that typically occurs when opposite needs have to be guaranteed, for example, noise cancellation and detail preservation. After a brief introduction to fuzzy models, the principal families of nonlinear filters based on fuzzy systems are described in detail in Chapter 12. Both indirect approaches, which typically adopt the basic structure of a weighted mean filter and use fuzzy models to evaluate the corresponding weights, and direct approaches, which adopt special fuzzy systems for directly yielding the output values, are presented. Relevant applications to noise removal are also shown.

Chapter 13 deals with digital halftoning, that is, the procedure that allows the reproduction of original continuous-tone photographs with binary patterns. After a brief review of the major techniques used in the past, current solutions for high print resolution and accurate color reproduction are described in detail. Several metrics for the characterization of stochastic dither patterns in both the spatial and spectral domain are introduced. Two approaches, based on blue-noise and green-noise models, are discussed. The blue-noise model, which is just the high-frequency component of a white noise, constitutes at present the basis of techniques widely applied in the printing industry. In contrast, the green-noise model, which is essentially the mid-frequency component of a white noise, represents a new approach to stochastic halftoning that provides higher resolutions.

Finally, in Chapter 14 the concept of intrinsic dimensionality is introduced as a relevant property of images. In fact, most local areas of natural images are nearly constant, and thus are classified as intrinsic zero-dimensional structures, while some other areas, such as straight lines and edges, are intrinsically one dimensional and only a minority of zones are intrinsically two dimensional, such as junctions and corners. Chapter 14 shows that, while the separation of intrinsic zero-dimensional signals from other signals requires only some kind of linear filtering and a subsequent threshold operation, the selective processing of intrinsic two-dimensional signals requires specific Volterra operators. The derivation of the necessary and sufficient conditions for the definition of suitable quadratic operators is provided, together with actual examples of different types of such operators. Some further extensions related to higher order statistics and an analysis of the relations of local intrinsic dimensionality to basic neurophysiological and psychophysical aspects of biological image processing conclude the chapter.

As might be clear from the discussion of its contents, our objective in editing this book has been to present both an overview of the state of the art and an exposition of some recent advances in the area of nonlinear image processing. We have attempted to present a comprehensive description of the most relevant classes of nonlinear filters, even though some other contributions could have been inserted. An example of these contributions is the evolutionary and learning-based nonlinear operators, including models that exploit training methods and algorithms based on machine learning paradigms, often copied from biological structures, such as neural networks and intelligent agents. In consideration of the vast number of contributions in this area that are still evolving, and with interest mainly oriented toward the methodologies rather than specific image processing tasks, our decision was not to include a report on these approaches in this book.

According to the choice of topics included here and the style of their presentation, this book is suitable, in our opinion, both as an introductory text to nonlinear image processing and as an updating report for a few specific, advanced areas. In fact, tutorial aspects have been preferred in some chapters or in some sections, while more specific techniques and applications have been considered in other chapters or sections. For this reason parts of the book can be usefully adopted as textbook material for graduate studies, whereas other parts can be used as an up-to-date reference for practicing engineers.

We believe the field of nonlinear image processing has matured sufficiently to justify bringing out a more up-to-date book on the subject. Because of the diversity of topics in the field, it would be difficult for one or two authors to write such a book. This is the reason for publishing an edited book. We would like to point out that the authors contributing chapters to this book are leading experts in their respective fields. We would also like to express to each of them our gratitude for their timely contributions of high-quality texts.

We have made every attempt to ensure the accuracy of all materials in this book. However, we would very much appreciate readers bringing to our attention any errors that may have appeared in the book due to reasons beyond our control and that of the publisher. These errors and any other comments can be communicated to either of us by email addressed to `mitra@ece.ucsb.edu` or `sicuranza@gnbts.univ.trieste.it`.

We thank Dr. Jayanta Mukhopadhyay of the Indian Institute of Technology, Kharagpur, India, for his critical review of all chapters. We also thank Patricia Monohon for her assistance in the preparation of the LaTeX files of this book.

SANJIT K. MITRA
GIOVANNI L. SICURANZA

1

Analysis and Optimization of Weighted Order Statistic and Stack Filters

Sari Peltonen, Pauli Kuosmanen, Karen Egiazarian,
Moncef Gabbouj, and Jaakko Astola

Department of Information Technology
Tampere University of Technology
Tampere, Finland

1.1 Introduction

In this chapter we consider stack and weighted order statistic filters and the most important tools available for their analysis and optimization. Both deterministic and statistical properties are covered to give a comprehensive understanding of why these filters work so well in certain applications.

1.2 Median and Order Statistic Filters

The median filter was introduced in the 1970s by Tukey under the name "running median" for smoothing of discrete data [Tuk74]. Since median filters attenuate impulsive noise effectively and preserve signal edges well, these filters have been studied and used widely in the field of signal processing (e.g., [Ast97]). Edge preservation is especially essential in image processing due to the nature of visual perception.

The references in this chapter follow the literature published in English but the filters considered also were studied extensively at the same time in the former Soviet Union (see [Gil76] and references therein).

At time instant n let the samples $X(n-k), X(n-k+1), \ldots, X(n+k)$ be in the filter window. For simplicity we denote these samples by $X_1 = X(n-k), X_2 = X(n-k+1), \ldots, X_N = X(n+k)$. Let $X_{(1)}, X_{(2)}, \ldots, X_{(N)}$ be the samples in increasing order; we call element $X_{(t)}$ the tth order statistic. Now, the output of the median filter is the sample $X_{(k+1)}$, i.e., the middle sample. If instead of the median sample the tth order statistic of the values inside the window is chosen to be the output of the filter, the filter is called the tth order statistic filter or ranked order filter.

The median filter, although offering clear advantages over linear filters, also has its shortcomings, such as streaking [Bov87b], edge jittering [Bov87a], and loss of small details from the images [Arc89, Nie87]. The main reason for the loss of details is that the median filter uses only rank order information, discarding the temporal order (spatial order) information. Thus, a natural extension of the median filter is a weighted median (WM) filter, where more emphasis can be given to the samples that are assumed to be more reliable, i.e., the samples near the center sample of the window. The filtering procedure is similar to the one for median filtering with the exception that the samples are duplicated to the number of the corresponding weight. The WM filters are considered in the next chapter and have been surveyed thoroughly elsewhere [Yli91, Yin96].

In the same way that the median filter has a weighted version, the order statistic filter also has one, called the *weighted order statistic* (WOS) filter. We have adopted the notation of Yli-Harja et al. [Yli91], where the weights, separated by commas, and the threshold, separated by a semicolon, are listed in angle brackets, that is, $\langle w_1, w_2, \ldots, w_N; T\rangle$. The output $Y(\mathbf{x})$ of this WOS filter with input vector $\mathbf{x} = (X_1, X_2, \ldots, X_N)$ is given by

$$Y(\mathbf{x}) = T\text{th largest value of multiset } \{w_1 \blacklozenge X_1, w_2 \blacklozenge X_2, \ldots, w_N \blacklozenge X_N\},$$

where $\blacklozenge$ denotes the repetition (duplication) operation, that is, $r \blacklozenge x = \underbrace{x, x, \ldots, x}_{r \text{ times}}$.

The recursive version of the median filter is defined as

$$Y(\mathbf{x}) = \text{MEDIAN}\{Y_1, Y_2, \ldots, Y_k, X_{k+1}, X_{k+2}, \ldots, X_N\},$$

where $Y_1, Y_2, \ldots, Y_k$ are outputs already computed. The recursive counterpart of every filter can be obtained in the same way.

1.3 Stack Filters

Motivated by the success of the median filter Wendt et al. [Wen86] developed a new filter class that shares two important properties of the median filter, namely, threshold decomposition and stacking properties. The former is a limited superposition property giving a new filter architecture and the latter is an ordering property.

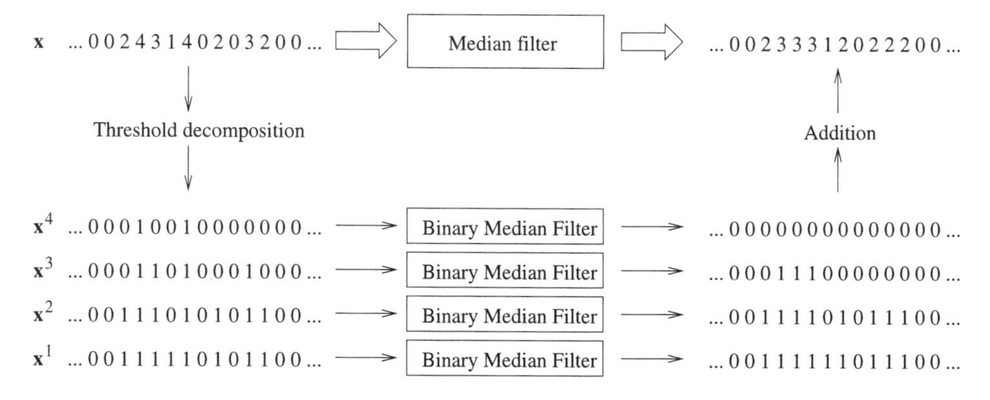

Figure 1.1: Illustration of stack filtering operation using threshold decomposition. The broad arrows show the overall filtering operation. The slender arrows show the same operation in the threshold decomposition architecture. The Boolean function used in the illustration is $f(\mathbf{x}) = x_1x_2 + x_1x_3 + x_2x_3$, which corresponds to the three-point median filter.

1.3.1 Definition

Consider an M-valued vector $\mathbf{x} = [X_1, X_2, \ldots, X_N]$, where $X_i \in \{0, 1, \ldots, M-1\}$. The threshold decomposition of $\mathbf{x}$ means the slicing of $\mathbf{x}$ into $M-1$ binary vectors $\mathbf{x}^1, \mathbf{x}^2, \ldots, \mathbf{x}^{M-1}$, obtained by the following thresholding rule:

$$x_n^m = T_m(X_n) = \begin{cases} 1 & \text{if } X_n \geq m, \\ 0 & \text{otherwise.} \end{cases} \tag{1.1}$$

In other words an element x_n^k of the binary vector $\mathbf{x}^k$ is equal to 1 whenever the element X_n of the input signal is greater than or equal to k but its value is zero otherwise. Thresholding does not lose any information about the signal; it only changes this information into simpler binary form. The original multivalued signal can be reconstructed from its thresholded binary vectors simply by adding them together:

$$X_n = \sum_{m=1}^{M-1} x_n^m.$$

What happens when we filter each binary slice $\mathbf{x}^k$ separately by the median filter? First, the ordering operation for the binary signals reduces to additions, and second, the filtered binary slices form the threshold decomposition of the output of the median filter applied to the original signal, as shown in Fig. 1.1. If instead of the binary median function the filtering of binary slices is done by any binary function that possesses the property of commuting with threshold decomposition, we obtain a stack filter. Now we should find out which binary functions possess this property in order to formally define the stack filter. First some definitions are needed.

Let $\mathbf{x}$ and $\mathbf{y}$ be binary vectors (signals) of fixed length. Define

$$\mathbf{x} \leq \mathbf{y} \quad \text{if and only if} \quad X_n \leq Y_n \quad \text{for all } n. \tag{1.2}$$

Since the relation defined by Eq. (1.2) is reflexive, antisymmetric, and transitive, it defines a partial ordering on the set of binary vectors of fixed length. Now consider a signal $\mathbf{x}$ and its thresholded binary signals $\mathbf{x}^1, \mathbf{x}^2, \ldots, \mathbf{x}^{M-1}$. Clearly, $\mathbf{x}^i \leq \mathbf{x}^j$ if $i \geq j$ and the binary signals $\mathbf{x}^1, \mathbf{x}^2, \ldots, \mathbf{x}^{M-1}$ form a nonincreasing sequence.

A Boolean function $f(\cdot)$ is called a positive Boolean function (PBF) if it can be written as a Boolean expression that contains only uncomplemented input variables. For a PBF $f(\cdot)$ it holds that

$$f(\mathbf{x}) \geq f(\mathbf{y}) \quad \text{if } \mathbf{x} \geq \mathbf{y}. \tag{1.3}$$

The property given by Eq. (1.3) is called the *stacking property*. In practice, the stacking property means that when the binary output signals are piled on top of each other, as in Fig. 1.1, there can be only zeros on top of a zero. So in the reconstruction phase the binary signals do not have to be added together but a simple binary search can be used to find the levels just before the transitions from 1 to 0 take place. Now we can define the stack filter.

Definition 1.1. A stack filter $S_f(\cdot)$ is defined by a positive Boolean function $f(\cdot)$ as follows:

$$S_f(\mathbf{x}) = \sum_{m=1}^{M-1} f(\mathbf{x}^m). \tag{1.4}$$

Thus, filtering a vector $\mathbf{x}$ with a stack filter $S_f(\cdot)$ based on the PBF $f(\cdot)$ is equivalent to decomposing $\mathbf{x}$ to binary vectors $\mathbf{x}^m$, $1 \leq m \leq M-1$, by thresholding, filtering each threshold level with the binary filter $f(\cdot)$, and reconstructing the output vector as the sum Eq. (1.4).

By using Eq. (1.4) stack filters are completely characterized by their operation on binary vectors. Thus, all of their properties can be deduced from their action on binary signals.

A stack filter defined by a PBF is a WOS filter if and only if the PBF is linearly separable, that is, it can be represented in the form

$$f(x_1, x_2, \ldots, x_N) = \begin{cases} 1 & \text{if } \sum_{i=1}^{N} w_i x_i \geq T, \\ 0 & \text{otherwise,} \end{cases}$$

where x_i are binary variables and the weights w_i and threshold T are constants. It should be noted that different weights and thresholds can lead to the same PBF and thus the stack filter defined by a PBF is not unique.

Definition 1.2. The dual $f^{\mathrm{D}}(\mathbf{x})$ of a Boolean function $f(\mathbf{x})$ is defined by $f^{\mathrm{D}}(\mathbf{x}) = \overline{f(\overline{\mathbf{x}})}$, where $\overline{\mathbf{x}}$ is the complement of $\mathbf{x}$. A Boolean function $f(\mathbf{x})$ is self-dual if and only if $f(\mathbf{x}) = f^{\mathrm{D}}(\mathbf{x})$.

If in addition to being linearly separable the PBF is also self-dual, the stack filter defined by this PBF is a WM filter.

Remark 1.1. For real-valued signals the stack filter (continuous amplitude) defined by a positive Boolean function $f(x_1, x_2, \ldots, x_N)$ with input vector $\mathbf{x} = [X_1, X_2, \ldots,$

$X_N]$ can be defined as, for example, [Yli91],

$$S_f(\mathbf{x}) = \max \{\beta \in \mathbf{R} : f(T_\beta(X_1), T_\beta(X_2), \ldots, T_\beta(X_N)) = 1\},$$

where the thresholding function is defined by Eq. (1.1), or it can be defined by the following connection with the PBF. The PBF

$$f(x_1, x_2, \ldots, x_N) = \sum_{i=1}^{K} \prod_{j \in P_i} x_j, \tag{1.5}$$

where the P_i are subsets of $\{1, 2, \ldots, N\}$, if and only if the stack filter $S_f(\cdot)$ corresponding to $f(x_1, x_2, \ldots, x_N)$ is

$$S_f(\mathbf{x}) = \max\left\{\min\{X_j : j \in P_1\}, \min\{X_j : j \in P_2\}, \ldots, \min\{X_j : j \in P_K\}\right\}. \tag{1.6}$$

The first formulation reflects the threshold expression of the original definition of the discrete stack filter and the second tells that the real domain stack filter corresponding to a PBF can be expressed by replacing "and" and "or" with "min" and "max", respectively. For example, the three-point median filter over real variables X_1, X_2, and X_3 (see also Fig. 1.1) is a stack filter defined by the PBF $f(x_1, x_2, x_3) = x_1x_2 + x_1x_3 + x_2x_3$, that is,

$$\text{med}\{X_1, X_2, X_3\} = \max\left\{\min\{X_1, X_2\}, \min\{X_1, X_3\}, \min\{X_2, X_3\}\right\}.$$

There is a close connection between stack filters and morphological filtering. For binary signals one can view morphological erosion as a stack filter defined by a single monomial. Thus, stack filters are essentially a union of erosions with flat structuring elements [Dou87].

1.3.2 Impulse and Step Responses

Usually when nonlinear filters are considered, their impulse removal and edge preservation capabilities are mentioned as important properties; whether a filter has these properties can be found by studying its impulse and step responses.

The impulse response of a filter is studied by filtering an input signal in which one sample is an impulse having the value -1 or 1 and the rest of the signal values are zero. When can this positive or negative impulse be the output of a stack filter in the case of a nontrivial Boolean function, that is, a function not identically equal to zero or one? This question can be answered easily by examining the subsets P_i of the max-min representation of Eq. (1.6) of a stack filter. For a positive impulse to be the output, at least one of the minima must be equal to the value of the impulse; this can happen only if there is j such that $|P_j| = 1$. If there is a negative impulse in the input, all of the minima must be equal to this value for it to be the maximum of them, that is,

$$\bigcap_{j=1}^{K} P_j \neq \varnothing,$$

where $\emptyset$ denotes the null set. For a WOS filter the impulses are completely removed if each weight

$$w_i < \min\left\{T, \sum_{j=1}^{N} w_j - T + 1\right\},$$

and for the median filter the impulse response is zero.

The class of stack filters includes filters with varying detail and edge preservation properties. With weighting, the detail preservation properties of median and order statistic filters can be improved, but only at the expense of lower noise suppression. Weights also can be chosen in such a way that certain structures, for example, lines, in the signals are preserved.

A step signal has two constant areas of different values, between which there is an edge. Stack filters can only translate binary edges, and because of threshold decomposition edges, can be translated but not blurred.

1.3.3 Root Signals

Analysis of root signals is important for an understanding of the operation of stack filters. Root signals that pass through the filter unaltered give valuable information about it, so filters can be designed such that certain image patterns are root signals and thus are not disturbed by the filtering operation.

To be able to filter the outmost input samples of a finite signal when parts of the filter window fall outside the input signal we need to append samples to the ends of the signal. A common appending strategy is to replicate the outmost input samples as many times as needed. The roots of a stack filter are all of the appended signals that are invariant under filtering.

The median filter has the very nice property of converging to a root in a finite number of passes of the filter. This and the structure of the root signals is important for determining which filtering problems can be solved by the median filter. In a similar manner the root signals and convergence of stack filters has been analyzed to better understand which filtering problems are solvable by stack filters. We state here few simple roots of stack filters [Wen86] but encourage interested reader to consult a more profound presentation of the convergence and root signals of stack filters [Gab92].

We denote by 0^m and 1^m, a 0 and 1 repeated m times. A stack filter defined by a nontrivial PBF preserves all constant signals. An increasing signal is preserved by a stack filter defined by a nontrivial PBF with window width $N = 2k + 1$ if and only if the output of the PBF with input $0^k 1^{k+1}$ is equal to 1 and with input $0^{k+1} 1^k$ is equal to 0. By interchanging 0 and 1 we obtain similar result for decreasing signals.

1.3.4 Output Distributions and Moments of Stack Filters

Since nonlinear filters are all of those filters that are not linear filters, there is a wide variety of different nonlinear filters and no common theory for such a heterogeneous filter class. However, within this filter class stack filters form a very specific subclass of signal smoothers with the possibility to derive analytical results for their statistical properties. A basic statistical descriptor that can be used to study the noise attenuation properties of stack filters is the output distribution. It can also be used for determining biasedness or unbiasedness of the estimator and in the optimization of the filter.

We give here the output distribution of a stack filter only for the case of independent and identically distributed (i.i.d.) input values that can be generalized to the case of nonidentically distributed samples [Yli91].

Proposition 1.1. *Let the input values $X_1, X_2, \ldots, X_N$, in the window of the stack filter $S_f(\cdot)$ defined by a positive Boolean function $f(\cdot)$ be i.i.d. random variables having a common distribution function $\Phi(x)$. The distribution function of the output $\Psi(x)$ of the stack filter $S_f(\cdot)$ is*

$$\Psi(x) = \sum_{i=0}^{N} A_i [1 - \Phi(x)]^i \Phi(x)^{N-i}, \tag{1.7}$$

where the numbers A_i are defined by

$$A_i = \left| \{\mathbf{x} : f(\mathbf{x}) = 0, w_{\mathrm{H}}(\mathbf{x}) = i\} \right|, \tag{1.8}$$

with $w_{\mathrm{H}}(\mathbf{x})$ denoting the number of 1s in $\mathbf{x}$, that is, its Hamming weight.

Example 1.1. Let the input values X_1, X_2, and X_3, in the window B of a stack filter $S_f(\cdot)$ defined by a positive Boolean function $f(x_1, x_2, x_3) = x_1x_2 + x_1x_3$ be i.i.d. random variables having a common distribution function $\Phi(x)$. From Proposition 1.1 the output distribution function $\Psi(x)$ of the stack filter $S_f(\cdot)$ is

$$\Psi(x) = \Phi^3(x) + 3\Phi^2(x)[1 - \Phi(x)] + \Phi(x)[1 - \Phi(x)]^2 = \Phi(x) + \Phi^2(x) - \Phi^3(x).$$

For the numbers A_i we have limits $0 \le A_i \le \binom{N}{i}$, $i = 1, 2, \ldots, N$. To guarantee that the stack filter is not defined by the trivial PBFs $f(\mathbf{x}) \equiv 0$ or $f(\mathbf{x}) \equiv 1$, we require that $A_0 = 1$ and $A_N = 0$, and thus we can omit $i = N$ from the sum (1.7). For self-dual stack filters we have the following useful property:

$$A_i + A_{N-i} = \binom{N}{i}, \quad i = 1, 2, \ldots, N. \tag{1.9}$$

For recursive stack filters a method relying on finite automata and Markov chain theory has been provided for deriving the output distribution function [Shm99]. Just as for one stack filter, we similarly can give the joint output distribution of two or more stack filters [Aga95].

The kth order moment about the origin of the output Y of a stack filter can be expressed as

$$\mu_k = E\{Y^k\} = \sum_{i=0}^{N-1} A_i M(\Phi, k, N, i), \tag{1.10}$$

where

$$M(\Phi,k,N,i) = \int_{-\infty}^{\infty} x^k \frac{\mathrm{d}}{\mathrm{d}x} \left\{[1-\Phi(x)]^i \Phi(x)^{N-i}\right\} \mathrm{d}x, \quad i = 0,1,\ldots,N-1. \quad (1.11)$$

By using the output moments about the origin we easily obtain output central moments, denoted by $\mu_k = E\left\{(Y - E\{Y\})^k\right\}$. For example, the second order central output moment is

$$\mu_2 = \sum_{i=0}^{N-1} A_i M(\Phi,2,N,i) - \left[\sum_{i=0}^{N-1} A_i M(\Phi,1,N,i)\right]^2. \quad (1.12)$$

The output variance is a measure of the noise attenuation capability of a filter, quantifying the spread of the output samples with respect to their mean value. Equation (1.12) gives an expression for the output variance in terms of the quantities $M(\Phi,1,N,i)$, and $M(\Phi,2,N,i)$ and the coefficients A_i. The important factor is that the $M(\Phi,k,N,i)$ depend only on the input distribution Φ and the window size N but not on the stack filter in question. The coefficients A_i, on the other hand, depend only on the stack filter and not on the input distribution.

To avoid technical difficulties that might obscure the main ideas, we assume that the input distribution $\Phi(t)$ is very smooth. Specifically, we assume that the density function $\phi(t)$ is positive for all t, and symmetric with respect to the origin and that $\lim_{t\to-\infty} t^2\Phi(t) = \lim_{t\to\infty} t^2[1-\Phi(t)] = 0$. It follows from these assumptions that the expectation of the input is zero, that is, $\mu_1 = 0$, and that $\Phi(t) = 1 - \Phi(-t)$. Even though the numbers $M(\Phi,k,N,i)$ are defined by integrals, they are quite easy to compute numerically because they satisfy the following recurrence formula:

$$M(\Phi,k,N,i) = M(\Phi,k,N-1,i-1) - M(\Phi,k,N,i-1), \quad 1 \le i \le N, \quad (1.13)$$

with initial values

$$M(\Phi,k,N,0) = \int_{-\infty}^{\infty} x^k \frac{\mathrm{d}}{\mathrm{d}x}\left[\Phi(x)^N\right] \mathrm{d}x.$$

The quantities $M(\Phi,2,N,i)$ satisfy

$$M(\Phi,2,N,i) \begin{cases} > 0 & i = 0 \text{ or } N/2 < i < N, \\ < 0 & \text{otherwise.} \end{cases} \quad (1.14)$$

Calculation of A_i for WOS Filters

The numbers A_i can be found by listing for every i those vectors $\mathbf{x}$ having a Hamming weight i and calculating those that give $f(\mathbf{x}) = 0$. However, this is a time-consuming procedure and for WOS filters a more efficient method utilizing generating functions has been given [Ast94]. First, we form the product

$$P(\xi,\eta) = \prod_{i=1}^{N}(1+\xi\eta^{w_i}) = \sum_{\mathbf{x}\in\{0,1\}^N} \prod_{i=1}^{N} (\xi\eta^{w_i})^{x_i} = \sum_{\mathbf{x}\in\{0,1\}^N} \xi^{\sum x_i}\eta^{\sum w_i x_i}.$$

Now the coefficients A_i can be calculated by using the following algorithm:

1. Form $P(\xi, \eta) = \prod_{i=1}^{N} (1 + \xi\eta^{w_i})$.
2. Expand $P(\xi, \eta) = \sum \xi^{\sum x_i} \eta^{\sum w_i x_i}$.
3. Collect the powers of η: $P(\xi, \eta) = \sum_{k=0}^{\sum w_i} S_k(\xi)\eta^k$.
4. Truncate with respect to η at $T - 1$: $Q(\xi, \eta) = \sum_{k=0}^{T-1} S_k(\xi)\eta^k$.
5. Now, $A(\xi) = Q(\xi, 1) = \sum_{i=0}^{N} A_i \xi^i$.

1.3.5 Selection Probabilities

Selection probabilities give a very intuitive way of understanding the role of different samples in the filter window. They provide a general view of how important different samples are with respect to each other for a given filter. The sample selection probabilities can be used for determining the detail preservation properties of a filter and the rank selection probabilities give a useful way of expressing the output distribution and can be used as optimization constraints.

The output of a stack filter is always one of the input samples, which can clearly be noticed from the max-min representation in Eq. (1.6). This gave Prasad et al. [Pra90] the idea to determine the probability that the ith smallest sample $X_{(i)}$ or the jth sample X_j is the output of the filter. These probabilities are called rank and sample selection probabilities, respectively.

Definition 1.3. The ith rank selection probability is denoted by $P[Y = X_{(i)}]$, $1 \le i \le N$, and is the probability that the output $Y = X_{(i)}$. The jth sample selection probability is denoted by $P[Y = X_j]$, $1 \le j \le N$, and is the probability that the output $Y = X_j$. The rank selection probability vector is the row vector $\mathbf{r} = (r_1, r_2, \dots, r_N)$, where $r_i = P[Y = X_{(i)}]$, $1 \le i \le N$, and the sample selection probability vector is the row vector $\mathbf{s} = (s_1, s_2, \dots, s_N)$, where $s_j = P[Y = X_j]$, $1 \le j \le N$.

Example 1.2. Let $S_f(X_1, X_2, X_3) = \max\left\{\min\{X_1, X_2\}, \min\{X_1, X_3\}\right\}$. Below is a table presenting all six possible orderings of the inputs and the corresponding outputs of the filter.

Ordering	$S_f(X_1, X_2, X_3)$
$X_1 \le X_2 \le X_3$	$X_1 = X_{(1)}$
$X_1 \le X_3 \le X_2$	$X_1 = X_{(1)}$
$X_2 \le X_1 \le X_3$	$X_1 = X_{(2)}$
$X_3 \le X_1 \le X_2$	$X_1 = X_{(2)}$
$X_2 \le X_3 \le X_1$	$X_3 = X_{(2)}$
$X_3 \le X_2 \le X_1$	$X_2 = X_{(2)}$

Now the rank selection probability vector is $\mathbf{r} = [1/3, 2/3, 0]$ and the sample selection probability vector is $\mathbf{s} = [2/3, 1/6, 1/6]$.

For the five point median filter the rank and sample selection probability vectors are $\mathbf{r} = [0, 0, 1, 0, 0]$ and $\mathbf{s} = [1/5, 1/5, 1/5, 1/5, 1/5]$. For detail preservation it would be important that the center sample of the window be more probable than the other samples in the window, but for the median every position of the samples is equally probable. If we increase the weight of the center sample from 1 to 3, the rank and sample selection probabilities of the obtained center WM (CWM) filter are $\mathbf{r} = [0, 2/5, 1/5, 2/5, 0]$ and $\mathbf{s} = [1/10, 1/10, 3/5, 1/10, 1/10]$. Now the center sample clearly has higher probability of becoming the output than other samples in the window, and thus the detail preservation properties were improved by weighting, but the robustness of the filter deteriorated simultaneously as can be observed from vector $\mathbf{r}$.

The distribution and density functions of the ith order statistics are denoted by $\Phi_i(\cdot)$ and $\phi_i(\cdot)$, respectively, and are given as follows [Roh76]:

$$\Phi_i(x) = \sum_{k=i}^{N} \binom{N}{k} \Phi^k(x)[1 - \Phi(x)]^{N-k} \tag{1.15}$$

and

$$\phi_i(x) = i\binom{N}{i} \Phi^{i-1}(x)[1 - \Phi(x)]^{N-i}\phi(x). \tag{1.16}$$

The output distribution function of a stack filter can be given as a weighted sum of the distribution functions of the order statistics with rank selection probabilities as weights [Pra90].

Proposition 1.2. *The output distribution function* $\Psi(\cdot)$ *and the density function* $\psi(\cdot)$ *of a stack filter having its rank selection probability vector* $\mathbf{r} = [r_1, r_2, \ldots, r_N]$ *and continuous i.i.d. inputs are given by*

$$\Psi(x) = \sum_{i=1}^{N} r_i \Phi_i(x)$$

and

$$\psi(x) = \sum_{i=1}^{N} r_i \phi_i(x),$$

where $\Phi_i(\cdot)$ *and* $\phi_i(\cdot)$ *are the distribution function and the density function, respectively, of the* i*th order statistics for i.i.d. inputs given by Eqs. (1.15) and (1.16).*

1.3.6 Finite-Sample Breakdown Points

The global reliability of the filter is an important aspect of robustness. The breakdown point, a simple quantitative global robustness measure, can be used to measure this directly. It describes up to what distance from the model distribution the information given by the estimator is relevant.

Roughly speaking, the breakdown point gives the minimum fraction of outliers that can carry the value of the estimator over all bounds [Ham74, Ham86]. For example, the arithmetic mean and the median have breakdown points $1/N$ and $(k+1)/N$, respectively. This means that a single outlier can carry the mean to infinity, but for the median, $k+1$ same sided impulses are needed for the output to be one of them. Often the breakdown points of arithmetic mean and median are given to equal 0 and $1/2$, which are obtained by letting $N \to \infty$. A positive breakdown point is closely related to, but not identical with, the robustness of the estimator.

Since we have a finite number of samples in the input window of the filter, we provide here a finite-sample definition of the breakdown point. Slightly different definitions for the finite-sample breakdown point also can be given (e.g., [Don83, Ham86]).

Definition 1.4. The finite-sample breakdown point ε_N^* of the estimator $T_N(\cdot)$ at the sample $\{X_1, X_2, \ldots, X_N\}$, $X_i \neq \pm\infty$, is given by

$$\varepsilon_N^* = \frac{1}{N} \min \left\{ m : \sup_{y_1, y_2, \ldots, y_m} |T_N(z_1, z_2, \ldots, z_N)| = \infty \right\}, \tag{1.17}$$

where the sample $\{z_1, z_2, \ldots, z_N\}$ is obtained by replacing any m sample points $X_{i_1}, X_{i_2}, \ldots, X_{i_m}$ by the arbitrary values $y_1, y_2, \ldots, y_m$.

Note that the finite-sample breakdown point of Eq. (1.17) usually is independent of the sample $\{X_1, X_2, \ldots, X_N\}$. For simplicity we use "breakdown point" to mean the finite-sample breakdown point in the following.

The breakdown points of stack filters can be found by using the following proposition [Kuo96].

Proposition 1.3. *Let $\mathbf{r} = [r_1, r_2, \ldots, r_N]$ be the rank selection vector of a stack filter $S_f(\cdot)$ with a window size N. Then the breakdown point of $S_f(\cdot)$ for i.i.d. inputs is given by*

$$\varepsilon_N^* = \frac{1}{N} \min \left\{ \min\{i : r_i > 0\}, N + 1 - \max\{i : r_i > 0\} \right\}, \tag{1.18}$$

and it is independent of the input samples.

A fast method for finding the breakdown point of a stack filter of window size N defined by a positive Boolean function $f(\cdot)$ in its minimum sum of products form is given by the following corollary [Kuo96].

Corollary 1.1. *Let $S_f(\cdot)$ be the stack filter of window size N defined by the positive Boolean function represented in the minimum sum-of-products form of Eq. (1.5) where P_i are subsets of $\{1, 2, \ldots, N\}$. Then the breakdown point of $S_f(\cdot)$ is given by*

$$\begin{aligned} \varepsilon_N^* = \frac{1}{N} \min \Big(&\left\{ |P_i| : i \in \{1, 2, \ldots, K\} \right\} \\ &\cup \left\{ |A| : A \cap P_i \neq \emptyset, i \in \{1, 2, \ldots, K\} \right\} \Big). \end{aligned} \tag{1.19}$$

Example 1.3. Consider stack filtering with a filter defined by $f(x_1,x_2,x_3,x_4,x_5) = x_1x_3x_5+x_2x_3x_4+x_2x_4$. Now $P_1 = \{1,3,5\}, P_2 = \{2,3,4\}$ and $P_3 = \{2,4\}$. Therefore, $m_1 = \min\{|P_i| : i \in \{1,2,3\}\} = 2$ and $m_2 = \min\{|A| : A \cap P_i \neq \emptyset \ i \in \{1,2,3\}\} = |\{2,3\}| = 2$. (Another choice for the minimal set A would be $\{3,4\}$.) Thus, by using Corollary 1.1 we obtain $\varepsilon_N^* = 2/5$.

If there are no constraints for the stack filter to satisfy, the stack filter that maximizes the breakdown point is the median filter.

1.3.7 Breakdown Probabilities

Breakdown points describe well the number of outliers that can break down the statistics of a filter; that is, the output of a filter is also an outlier. The concept of breakdown points has been extented into impulsive noise breakdown probabilities [Mal80]. The breakdown probability is defined to be the probability for an impulse to be the output when the probability of impulses is given, so it gives the average fraction of impulses remaining after filtering but does not give any information about the number of impulses that the filter can handle. The breakdown probability clearly gives much more information about the robustness of an estimator than the breakdown point does.

Consider the case of an i.i.d. input signal corrupted by independent impulsive (salt and pepper) noise, where a signal sample is replaced by a negative impulse $-\infty$ with probability p, and a positive impulse ∞ with probability q but is otherwise unaltered. The breakdown probability of an estimator $T_N(\cdot)$ is denoted by $\vartheta_N(p,q)$. We trivially obtain $\vartheta_N(0,0) = 0$ and $\vartheta_N(p,1-p) = \vartheta_N(1-q,q) = 1$, so in the following we assume that $p+q \neq 0,1$.

The following proposition gives the breakdown probability of a stack filter with a fixed rank selection vector $\mathbf{r}$.

Proposition 1.4. *The breakdown probability of a stack filter* $S_f(\cdot)$ *with rank selection vector* $\mathbf{r} = [r_1, r_2, \ldots, r_N]$ *is given by*

$$\vartheta_N(p,q) = \sum_{i=1}^{N} r_i \left[\sum_{j=i}^{N} \binom{N}{j} p^j (1-p)^{N-j} + \sum_{j=N-i+1}^{N} \binom{N}{j} q^j (1-q)^{N-j} \right]. \tag{1.20}$$

The breakdown probability of a stack filter $S_f(\cdot)$ has an even more simple form when it is expressed in terms of the coefficients A_i:

Proposition 1.5. *The breakdown probability of a stack filter* $S_f(\cdot)$ *with coefficients* $A_0, A_1, \ldots, A_{N-1}$ *is given by*

$$\vartheta_N(p,q) = \begin{cases} 1 + \sum_{i=0}^{N-1} A_i[p^{N-i}(1-p)^i - q^i(1-q)^{N-i}] & p,q \neq 0, \\ p^N + \sum_{i=1}^{N-1} A_i p^{N-i}(1-p)^i & q = 0, \\ 1-(1-q)^N - \sum_{i=1}^{N-1} A_i q^i (1-q)^{N-i} & p = 0. \end{cases} \tag{1.21}$$

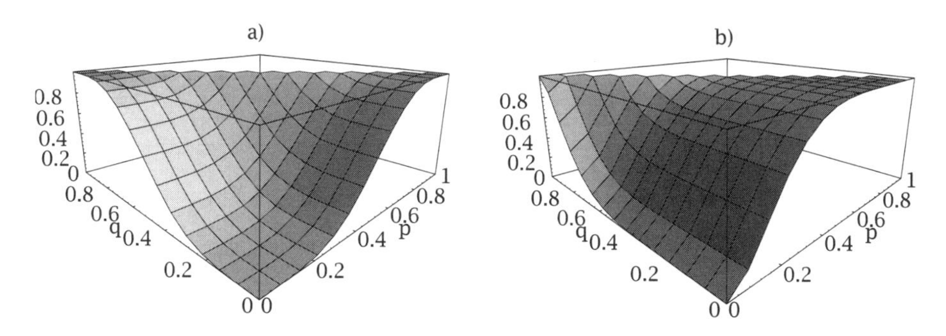

Figure 1.2: Breakdown probability functions of (a) 9-point median and (b) second order statistic filters.

Now, the breakdown probability $\vartheta_N(p, q)$ is a function of p, q and rank selection vector $\mathbf{r}$ in Eq. (1.20) [or coefficients $A_0, A_1, \ldots, A_{N-1}$ in Eq. (1.21)]. The first partial derivatives are shown to be positive [Kuo94], and thus $\vartheta_N(p, q)$ is an increasing function.

In Fig. 1.2 (also see color insert), two selected breakdown probability functions have been plotted.

The following propositions give the stack filters that minimize the breakdown probability when p and q are fixed:

Proposition 1.6. *Let the probabilities of negative and positive impulses p and q be given. Then the stack filters that minimize the breakdown probability are*

$$S_{\text{opt}} = \begin{cases} \text{1st order statistic filter} & \text{for } q = 0, \\ N\text{th order statistic filter} & \text{for } p = 0, \\ \left(N - \left\lfloor \dfrac{N \ln\left(\frac{1-q}{p}\right)}{\ln\left(\frac{(1-p)(1-q)}{pq}\right)} \right\rfloor\right) \text{th order statistic filter} & \text{for } p, q \neq 0, \end{cases}$$

where $\lfloor \cdot \rfloor$ is the floor operation.

This proposition has the following interesting corollary.

Corollary 1.2. *Assume that $p = q \neq 0$ and that N is odd. Then the stack filter that minimizes the breakdown probability $\vartheta_N(p, q)$ is the median filter.*

1.3.8 Finite-Sample Influence Function

The influence function (IF) is a useful heuristic tool of robust statistics introduced by Hampel [Ham68, Ham74] under the name influence curve (IC) for studying the performance of estimators T given as functionals under noisy conditions.

Definition 1.5. The IF of estimator T at underlying probability distribution Φ is given by

$$\text{IF}(x; T, \Phi) = \lim_{t \to 0^+} \frac{T\left((1-t)\Phi + t\Delta_x\right) - T(\Phi)}{t}$$

for those x where this limit exists.

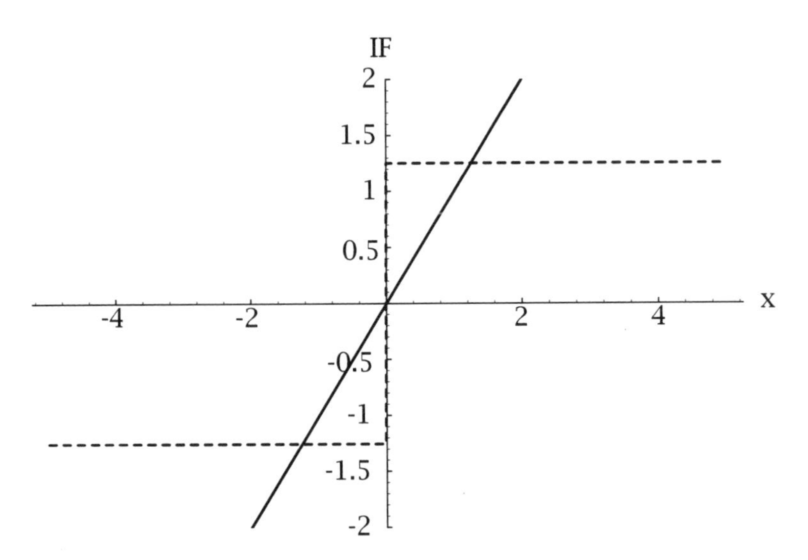

Figure 1.3: The IFs of the mean (solid line) and the median (dashed line) at the standard normal distribution.

In this definition Δ_x is the probability measure that puts mass 1 at the point x. The IF gives the effect that an infinitesimal contamination at point x has on the estimator T when divided by the mass of the contamination, so it gives the asymptotic bias caused by the contamination and thus characterizes the properties of the estimator as the number of observations approaches infinity.

The influence functions for the mean and the median are shown in Fig. 1.3 where the underlying distribution is the standard normal distribution. For the mean, the gross error sensitivity, that is, the worst influence a small amount of contamination of fixed size can have on the value of the estimator, is equal to infinity, and for the median it is finite and equals $\sqrt{\pi/2} \approx 1.253$. So for the mean a single outlier can carry the estimate over all bounds, but for the median an outlier has a fixed influence.

Since the IF is an asymptotic measure, it describes properties of infinite length filters, which may differ from those of finite length filters used in real world filtering applications. It would be more useful and more interesting to examine properties of these finite length filters rather than the asymptotic properties. For this purpose some finite sample versions of an IF based on an actual sample $(X_1, X_2, \ldots, X_{N-1})$ have been proposed. Stylized versions of them [And72] can be obtained by using an artificial sample, instead of an actual sample, which can be obtained, for example, by taking the $N - 1$ expected order statistics from a random sample of $N - 1$ or by setting $X_i = \Phi^{-1}(i/N)$.

We present here briefly the three well known finite-sample versions of the IF: the empirical IF, the sensitivity curve (SC), and a version using jackknifing. The first of these, the empirical IF [Ham82] of the estimator T_N, $N \geq 1$, at sample $(X_1, X_2, \ldots, X_{N-1})$ is a plot of

$$T_N(X_1, X_2, \ldots, X_{N-1}, x)$$

as a function of x. The SC [Tuk77] is defined as

$$\mathrm{SC}_N(x) = N\left[T_N(X_1, X_2, \ldots, X_{N-1}, x) - T_{N-1}(X_1, X_2, \ldots, X_{N-1})\right]$$

or when the estimator is a functional, that is, $T_N(X_1, X_2, \ldots, X_N) = T(\Phi_N)$, as

$$\mathrm{SC}_N(x) = \frac{T\left(\left(1 - \frac{1}{N}\right)\Phi_{N-1} + \frac{1}{N}\Delta_x\right) - T\left(\Phi_{N-1}\right)}{\frac{1}{N}},$$

where Φ_{N-1} is the empirical distribution function of $(X_1, X_2, \ldots, X_{N-1})$. The third version is based on the ith jackknifed pseudovalue [Que56]:

$$T_{Ni}^{*} = NT_N(X_1, X_2, \ldots, X_N) - (N-1)T_{N-1}(X_1, X_2, \ldots, X_{i-1}, X_{i+1}, \ldots, X_N).$$

Now the finite-sample IF using jackknifing is defined as

$$T_{Ni}^{*} - T_N(X_1, X_2, \ldots, X_N).$$

1.3.9 Output Distributional Influence Function

For the finite-sample influence functions either a real sample $(X_1, X_2, \ldots, X_{N-1})$ or an artificial sample generated from the distribution Φ of the input samples is needed, and this sample itself, or the way it is derived from the distribution Φ, affects the result. What we would like to have is a general method that uses the distribution function Φ of the input sample itself and not any artificial sample derived from Φ. In the case where the output distribution of a filter can be expressed in a closed form as a function of the distribution functions of the input samples, the output distributional influence function (ODIF) has been introduced [Pel99a] for analyzing the robustness of the finite length filters.

We assume here that the input samples are i.i.d. random variables. First we need a way to denote the output distribution function of a filter when a fraction ε of the input samples has different distribution than the rest of the samples. We denote by $\Psi_{(1-\varepsilon)\Phi+\varepsilon G_y}(\cdot)$ the output distribution $\Psi(\cdot)$ of the filter, where every occurrence of the common distribution function Φ of the input samples is replaced by $(1-\varepsilon)\Phi+\varepsilon G_y$, with G_y being a distribution function with mean y. The following definition was given for the ODIF for the distribution function [Pel99a]:

Definition 1.6. Let the output distribution function of a filter be $\Psi(\cdot)$, the common distribution function of the input samples be $\Phi(\cdot)$, and $G_y(\cdot)$ be a distribution function having mean y. Then the ODIF for the distribution function $\Omega(\cdot)$ is

$$\Omega(x, y) = \lim_{\varepsilon \to 0^+} \frac{\Psi_{(1-\varepsilon)\Phi+\varepsilon G_y}(x) - \Psi(x)}{\varepsilon}$$

for those x and y where this limit exists.

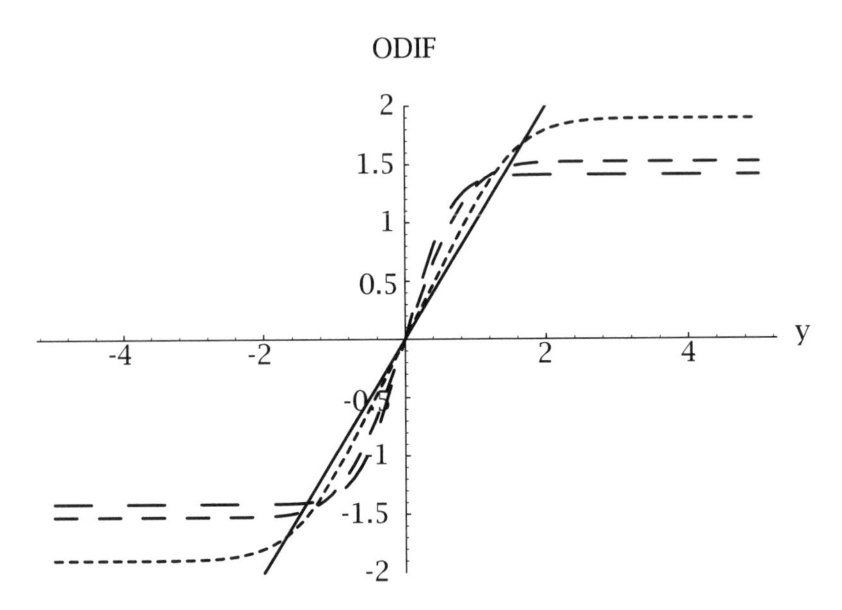

Figure 1.4: The ODIFs for the expectation of the CWM filter of length 7 having center weights 1 (long dashes), 3 (medium dashes), 5 (short dashes), and 7 (solid line) at the standard normal distribution and $G_y = \Delta_y$.

In the same way as for the distribution function in Definition 1.6 the ODIF was defined for the density function and moments [Pel99a].

The following proposition gives the ODIF for the distribution function of a stack filter by using the coefficients A_i [Pel99b].

Proposition 1.7. *Let the distribution function of the input samples be $\Phi(\cdot)$ and let $G_y(\cdot)$ be a distribution function having mean y. Then the ODIF for the distribution function of the stack filter of length N is given by*

$$\Omega(x,y) = \sum_{i=0}^{N-1} A_i \left(1 - \Phi(x)\right)^{i-1} \Phi(x)^{N-i-1} \left(N - i - N\Phi(x)\right) \left(G_y(x) - \Phi(x)\right),$$

where the coefficients A_i are defined by Eq. (1.8).

The simplest of the WM filters are the CWM filters having the weight of only the center sample different from 1 and all the other weights equal to 1. As an example we consider the CWM filter of length 7 with four different weights of the center sample. Figure 1.4 shows the ODIFs for the expectation $\omega_\mu(y) = \int_{-\infty}^{\infty} x\,[(d/dx)\Omega(x,y)]\,dx$ for the four different center weights at the standard normal distribution and $G_y = \Delta_y$. For the center weights 1, 3, and 5 the graphs are bounded; the supremum of the absolute value of each of the graphs is limited for these filters and has a smaller value for a smaller center weight. For center weight 7 the filter becomes the identity filter and the ODIF for the expectation is the solid straight line, which is not bounded.

Since for smaller center weights small fluctuations in the observations create large fluctuations in the average output of the filter, the CWM filters with a larger center weight behave better in the area close to the origin than those with a smaller center weight do. When the contamination is farther from zero a good behavior is the boundedness of the graph, which makes the filter robust against outliers. Since the CWM filter with the smallest center weight has the smallest absolute value in the constant area of the graph, it is the most robust against outliers.

1.3.10 Optimization

There are several different approaches to optimize stack filters. An adaptive filtering approach was developed by Lin et al. [Lin90]. Gabbouj and Coyle [Gab90] first developed an approach called optimal filtering under structural constraints for finding the optimal filter attenuating noise and preserving certain signal features. One approach would be training based on a representative training set containing the ideal signal and the corrupted signal. A training framework for the optimal nonlinear filter design problem and fast procedures for obtaining optimal solutions have been developed for stack filters [Tab96]. We consider here model-based optimization with a constant ideal signal and a known noise distribution.

Assume that we wish to find the self-dual stack filter of length N (N odd) that best attenuates i.i.d. symmetric noise. Because the expectation of the output is zero, the output variance is, according to Eq. (1.12),

$$\mu_2 = \sum_{i=0}^{N-1} A_i M(\Phi, 2, N, i).$$

From the above it follows that always $A_0 = 1$, $A_N = 0$, and $M(\Phi, 2, N, i) < 0$ for $1 \le i \le (N-1)/2$ and > 0 for $(N+1)/2 \le i \le N-1$. This means that we must make A_i as large as possible for $1 \le i \le (N-1)/2$ and as small as possible for $(N+1)/2 \le i \le N-1$. This obviously happens if we choose $A_i = \binom{N}{i}$ for $1 \le i \le (N-1)/2$ and $A_i = 0$ for $(N+1)/2 \le i \le N-1$. This choice gives the median filter!

In a more meaningful situation we have additional constraints on the coefficients A_i that arise, for example, from requirements that the filter have a certain degree of robustness and also be able to preserve details of a prescribed type. The constraints that give detail preservation can be given, for example, by fixing predetermined values of the defining Boolean function.

The coefficients A_i of a stack filter $S_f(\cdot)$ of window size N and the rank selection vector $\mathbf{r} = [r_1, r_2, \ldots, r_N]$ satisfy [Kuo94]

$$r_j = \frac{A_{N-j}}{\binom{N}{j}} - \frac{A_{N-j+1}}{\binom{N}{j-1}}, \quad j = 1, 2, \ldots, N. \tag{1.22}$$

The rank selection probabilities give an intuitively appealing way of constraining a stack filter. For instance, a certain amount of robustness is guaranteed if we require that $r_1 = r_2 = \cdots = r_k = 0$ and $r_l = r_{l+1} = \cdots = r_N = 0$. This will give a stack filter that is "trimmed" in the same way as an L-filter with the coefficients corresponding to a number of the largest and smallest coefficients equal to zero. Constraints on the rank selection probabilities translate immediately into constraints on A_i because of the above relation.

From the breakdown points we achieve new constraints for the coefficients A_i. Consider, for example, the design of a stack filter for which $\varepsilon_N^* > a$ for some $a \in [0,1]$. Then

$$A_{N-1} = A_{N-2} = \ldots = A_{N-\lfloor aN \rfloor} = 0 \quad \text{and} \quad A_j = \binom{N}{j}, \quad j = 1, 2, \ldots, \lfloor aN \rfloor. \tag{1.23}$$

Similarly, we can have breakdown probability constraints. For example, if we wish to have $\vartheta_N(p,q) < a$ for some $a \in [0,1]$ then

$$\sum_{i=0}^{N-1} A_i \left(p^{N-i}(1-p)^i - q^i(1-q)^{N-i} \right) < a - 1.$$

The relations given above make it possible to optimize stack filters in the mean square sense without performing a full search over all stack filters, which otherwise would be impossible, except for small window sizes, because of the very large number of stack filters. For instance, for the window size N the number of different stack filters is greater than $2^{2^N/N}$ [Aga95, Shm95]. The optimization consists of finding a solution of the integer linear programming task

$$\text{minimize} \quad \sum_{i=0}^{N-1} A_i M(\Phi, 2, N, i), \tag{1.24}$$

under the constraints for A_i and then determining a stack filter with the above coefficients A_i if it exists. It must be emphasized that usually there is no guarantee that such a stack filter exists. However, once we have the target coefficients A_i the search for the optimal stack filter is simpler. We also can take a stack filter that has coefficients A_i close to the solution of the optimization problem Eq. (1.24) and then check to see if its filtering behavior is satisfactory.

In the above, we considered the rank selection probabilities and used them to constrain the filter in the statistical sense. In many image processing problems it is not enough to know the "average" behavior of the filter; we need to be sure that it will handle certain signal segments in a prescribed way. This can be achieved using so-called structural constraints, the goal of which is to preserve some desired signal details, for example, pulses in 1-D signals, or lines in images, and to remove undesired signal patterns. The structural constraints consist of a list of different structures to be preserved, deleted, or modified. Since stack filters obey the threshold decomposition, the structural constraints need to be considered only in the context of binary signals. That is, they can be specified by a set of binary vectors and their outputs. The binary vectors are divided into two subsets, type 1 constraints and type 0 constraints [Yin95].

A binary vector that is specified by the structural constraints is called a type 1 constraint if its output is 1; otherwise, it is called a type 0 constraint. Denote the set of all type 1 constraints by $\Gamma_1 = \{\mathbf{x}_1, \mathbf{x}_2, \ldots, \mathbf{x}_p\}$ and the set of all type 0 constraints by $\Gamma_0 = \{\mathbf{y}_1, \mathbf{y}_2, \ldots, \mathbf{y}_q\}$.

Structural constraints induce two new constraints for the coefficients A_i: Let the number of vectors $\mathbf{x} \in \Gamma_1$ with $w_H(\mathbf{x}) = i$ be $y_i^{(1)}$ for all $1 \le i \le N-1$. Then

$$A_i \le \binom{N}{i} - y_i^{(1)}, \qquad 1 \le i \le N-1.$$

Let the number of vectors $\mathbf{x} \in \Gamma_0$ with $w_H(\mathbf{x}) = i$ be $y_i^{(0)}$ for all $1 \le i \le N-1$. Then

$$A_i \ge y_i^{(0)}, \qquad 1 \le i \le N-1.$$

Since many stack filters can have the same coefficients A_i, the filter has to be chosen from among them such that the structural constraints are satisfied.

Example 1.4. We find the coefficients A_i of the stack filters of window length $N = 5$ that, for white Gaussian N(0,1) noise, minimize the second order central output moment under constraints. The breakdown point of the filter is greater than $1/4$ and the probability that the filter outputs the median of the values in the window is at most 0.5.

Constraint of Eq. (1.23) gives equalities $A_4 = 0$ and $A_1 = 5$ and from Eq. (1.22) we obtain

$$\frac{A_2}{\binom{5}{3}} - \frac{A_3}{\binom{5}{2}} = r_3 \le 0.5,$$
$$A_3 \ge A_2 - 5.$$

Because rank selection probabilities are nonnegative we obtain from Eq. (1.22) the inequality

$$A_{i+1} \le \frac{N-i}{i+1} A_i, \qquad i = 0, 1, \ldots, N-1. \tag{1.25}$$

Now, using the formula of Eq. (1.12) we see that the optimal coefficients A_i are found by minimizing the expression

$$(1,5,A_2,A_3,0)\begin{pmatrix}1.80002\\-0.24869\\-0.02697\\0.02697\\0.24869\end{pmatrix} - \left[(1,5,A_2,A_3,0)\begin{pmatrix}1.16296\\-0.13359\\-0.04950\\-0.04950\\-0.13359\end{pmatrix}\right]^2$$
$$= 0.55657 + 0.02697(A_3 - A_2) - (0.49501 - 0.04950(A_2 + A_3))^2$$

subject to

$$0 \le A_2 \le 10, \qquad A_2 - 5 \le A_3 \le A_2.$$

The coefficients are computed from the recurrence Eq. (1.13). Because in this example we do not require the filter to be self-dual, the first moment also appears in the above minimization. By using nonlinear programming the solutions are found to be

$$\mathbf{a}_1 = (1,5,10,10,0) \quad \text{and} \quad \mathbf{a}_2 = (1,5,0,0,0),$$

which corresponds to the rank selection probability vectors

$$\mathbf{r}_1 = (0, 1, 0, 0, 0) \quad \text{and} \quad \mathbf{r}_2 = (0, 0, 0, 1, 0).$$

Thus, the optimal filters are simply the second and fourth order statistic filters. If we further require that the filter must be self-dual, the solution is given by a weighted median filter.

It generally is difficult to find a stack filter corresponding to the optimal coefficients A_i even if such a filter exists. For self-dual stack filters, the optimal filter usually can be obtained in closed form, and the following simple optimization method [Kuo94] is given for the cases with only structural constraints.

Proposition 1.8. *Assume that there are only structural constraints. Then the Boolean function corresponding to the optimal self-dual stack filter is*

$$f(\mathbf{x}) = \left[\sum_{i=1}^{p} f_i(\mathbf{x}) + f_{\mathrm{med}}(\mathbf{x})\right] \prod_{j=1}^{p} f_j^{\mathrm{D}}(\mathbf{x}),$$

where $f_{\mathrm{med}}(\cdot)$ is the Boolean function of the N point standard median filter, $f_i(\cdot)$ is the elementary conjunction of an element of Γ_1, and p is the number of elements in the set Γ_1.

Example 1.5. Consider a 2-D 3×3 window in which the structures to be preserved are the four "basic" straight lines. The structuring constraints are presented before indexing the vectors:

$$\Gamma_1 = \left\{ \begin{pmatrix} 0 & 1 & 0 \\ 0 & 1 & 0 \\ 0 & 1 & 0 \end{pmatrix}, \begin{pmatrix} 0 & 0 & 0 \\ 1 & 1 & 1 \\ 0 & 0 & 0 \end{pmatrix}, \begin{pmatrix} 0 & 0 & 1 \\ 0 & 1 & 0 \\ 1 & 0 & 0 \end{pmatrix}, \begin{pmatrix} 1 & 0 & 0 \\ 0 & 1 & 0 \\ 0 & 0 & 1 \end{pmatrix} \right\}$$

and

$$\Gamma_0 = \left\{ \begin{pmatrix} 1 & 0 & 1 \\ 1 & 0 & 1 \\ 1 & 0 & 1 \end{pmatrix}, \begin{pmatrix} 1 & 1 & 1 \\ 0 & 0 & 0 \\ 1 & 1 & 1 \end{pmatrix}, \begin{pmatrix} 1 & 1 & 0 \\ 1 & 0 & 1 \\ 0 & 1 & 1 \end{pmatrix}, \begin{pmatrix} 0 & 1 & 1 \\ 1 & 0 & 1 \\ 1 & 1 & 0 \end{pmatrix} \right\}.$$

When we index the variables in the following manner,

$$\begin{pmatrix} 1 & 2 & 3 \\ 4 & 5 & 6 \\ 7 & 8 & 9 \end{pmatrix},$$

we obtain the Boolean function defining the optimal filter using Proposition 1.8 as

$$\begin{aligned} f(\mathbf{x}) = {} & [f_{\mathrm{med}}(\mathbf{x}) + x_2x_5x_8 + x_4x_5x_6 + x_7x_5x_3 + x_1x_5x_9](x_2 + x_5 + x_8) \\ & \times (x_4 + x_5 + x_6)(x_7 + x_5 + x_3)(x_1 + x_5 + x_9). \end{aligned}$$

Optimal WM Filters under Structural Constraints

An adaptive approach to optimal WOS filtering using neural networks and an algorithm similar to the classical backpropagation algorithm has been used for solving the problem of optimal WOS filter design [Tab93]. A new optimality theory was developed [Yan95] for WM filters with a new expression utilizing L-vector and M-vector parameters for the output moments of WM filtered data, and we formulate this approach here by using coefficients A_i in a case where we have only structural constraints.

For WM filters we can express the coefficients A_i as functions of the filter weights as

$$A_i = \sum_{\mathbf{x} \in S_i} U\left(\frac{1}{2}\sum_{j=1}^{N} w_j - \sum_{j=1}^{N} w_j x_j\right),$$

where $U(\cdot)$ is the unit step function and S_i is the set $S_i = \{\mathbf{x} \in \{0,1\}^N : w_{\mathrm{H}}(\mathbf{x}) = i\}$. Thus, for WM filters the optimization problem of minimizing the sum

$$\sum_{i=0}^{N-1} A_i M(\Phi, 2, N, i)$$

becomes a problem of finding the optimal weights w_i instead of the optimal coefficients A_i. Thus, we do not have the problem of the possible nonexistence of a stack filter with optimal coefficients A_i since for WM filters we obtain directly the weights defining the optimal filter. However, the constraints of the optimization problem also have to be given as functions of weights, which is simple for structural constraints but may be harder for other types of constraints. The optimization problem is nonlinear since there is a nondifferentiable unit step function in the objective function. However, we can approximate the unit step function by a sigmoidal function $U_s(z) = 1/(1 + e^{-\beta z})$, that for large β approximates it well and is differentiable.

Example 1.6. Given a WM filter with weights $\mathbf{w} = [w_1, w_2, w_3, w_4, w_5]$, suppose that constant pulses of length 2 shall be preserved. Then the constraints on the weights should be

$$w_2 + w_3 \geq T,$$
$$w_3 + w_4 \geq T.$$

Replacing T by $1/2 \sum_{i=1}^{5} w_i$, we can rewrite the above inequalities as follows:

$$w_2 + w_3 \geq w_1 + w_4 + w_5,$$
$$w_3 + w_4 \geq w_1 + w_2 + w_5.$$

The optimal symmetric filter found by minimizing the sum $\sum_{i=0}^{N-1} A_i M(\Phi, 2, N, i)$ under the above constraints has weights $\mathbf{w} = [1, 2, 3, 2, 1]$.

1.4 Image Processing Applications

Impulsive noise or noise having a heavy-tailed distribution can produce image pixels whose values differ significantly from those of their neighbors. This type of noise is much more disturbing to the eye than, for example, Gaussian noise and thus image processing filters should be able to remove this type of noise well.

Detail preservation is another important property of the image processing filters. By using the filters studied in this chapter good compromises between these two contradictory goals can be obtained, see, e.g., [Ast97], [Tab96] and [Yin96]. The filters studied in this chapter are good not only in image restoration but can also be used in a variety of other image processing tasks, e.g., in skeletonization [Pet94], edge detection [Pet95] and lossless compression [Pet99].

From the properties of the human vision system it follows that edges often carry the main information in an image. Thus edge preservation is extremely important in image processing. Since any stack filter preserves an ideal edge (but may cause edge shift), the filters studied in this chapter are suitable in image processing also in this respect.

Figure 1.5a shows left half of the test image "Baboon." We have added to this image impulsive noise generated by introducing random bit errors (with probability 0.05) to obtain a more realistic case than the often used "pure" salt-and-pepper noise with amplitudes 0 and 255. Figure 1.5b shows the noisy image. The outputs of 5×5 median filter and 5×5 CWM filter with center weight 13 are shown in Figs. 1.5c and d, respectively. The original image has very many details and the median filter in addition to removing the noise very well also destroys the details of the image. The CWM filter preserves the details and removes the noise sufficiently well, giving much sharper and visually more pleasing results. Also the edges of the image are not blurred but remain very sharp.

1.5 Summary

In this chapter we considered stack and weighted order statistic filters. We studied the main deterministic properties of these filters including impulse and step responses as well as root signals. Statistical properties were also investigated by using output distributions and moments. The concepts of selection probability, finite-sample breakdown point, breakdown probability, and output distributional influence function were defined and their basic properties as possible tools for analyzing and optimizing these filters were introduced. Also, illustrative optimization examples were given. Finally, the application of these filters to image processing was considered.

References

[Aga95] S. Agaian, J. Astola, and K. Egiazarian. *Binary Polynomial Transforms and Nonlinear Digital Filters.* Marcel Dekker, New York (1995).

[And72] D. F. Andrews, P. J. Bickel, F. R. Hampel, P. J. Huber, W. H. Rogers, and J. W. Tukey. *Robust Estimates of Location: Survey and Advances*, Princeton University Press, Princeton, NJ (1972).

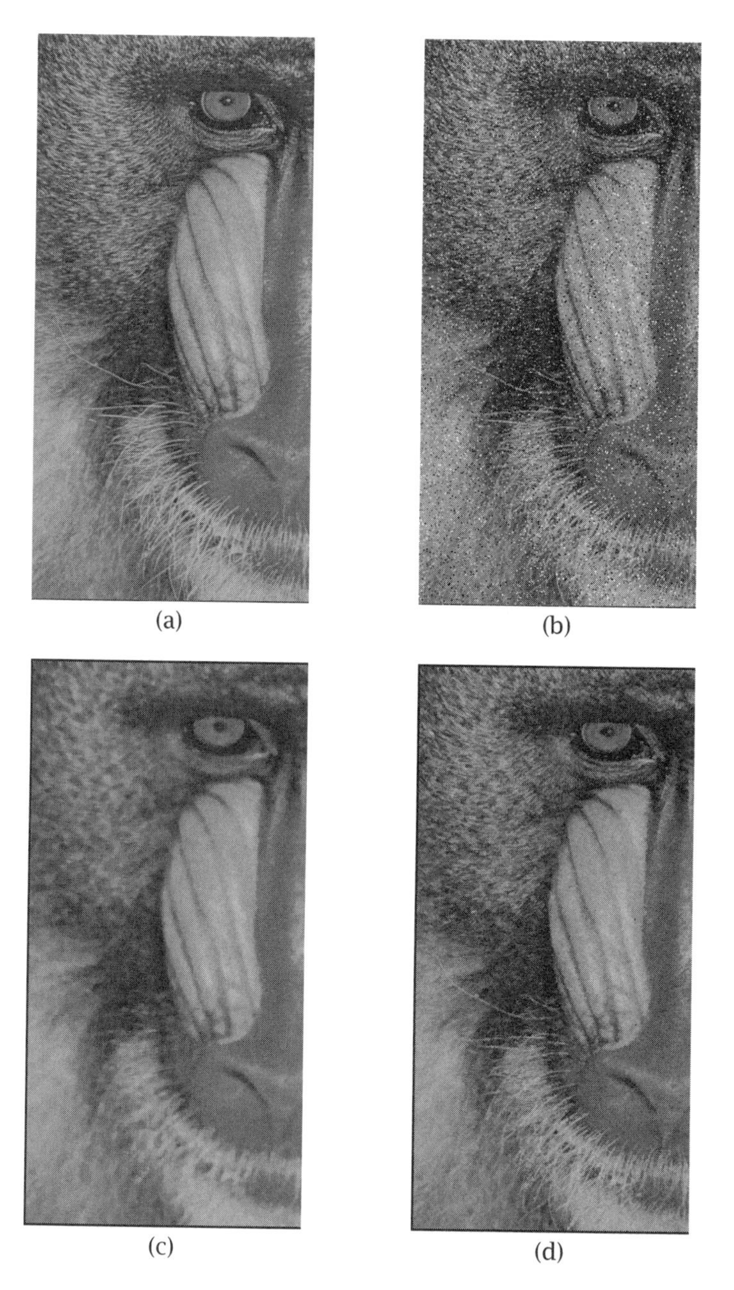

(a) (b)

(c) (d)

Figure 1.5: (a) Original image, (b) noisy image, (c) 5×5 median filtered image, and (d) 5×5 CWM filtered image.

[Arc89] G. R. Arce and R. E. Foster. Detail preserving ranked-order based filters for image processing. *IEEE Trans. Acoust. Speech Signal Process.* **37**(1), 83–98 (January 1989).

[Ast94] J. Astola and Y. Neuvo. An efficient tool for analyzing weighted median filters. *IEEE Trans. Circ. Syst.* **41**(7), 487–489 (July 1994).

[Ast97] J. Astola and P. Kuosmanen. *Fundamentals of Nonlinear Digital Filtering.* CRC Press, Boca Raton, FL (1997).

[Bov87a] A. C. Bovik, T. S. Huang, and D. C. Munson, Jr. The effect of median filtering on edge estimation and detection. *IEEE Trans. Patt. Anal. Machine Intell.* **9**(2), 181–194 (March 1987).

[Bov87b] A. C. Bovik. Streaking in median filtered images. *IEEE Trans. Acoust. Speech Signal Process.* **35**(4), 493–503 (April 1987).

[Don83] D. L. Donoho and P. J. Huber. The notion of breakdown point. In *A festschrift for Erich L. Lehmann* (P. J. Bickel et al., eds.), pp. 157–184. Wadsworth, Belmont, CA (1983).

[Dou87] E. R. Dougherty and C. R. Giardina. *Image Processing: Continuous to Discrete*, Vol. I, Prentice-Hall, Englewood Cliffs, NJ (1987).

[Gab90] M. Gabbouj and E. J. Coyle. Minimum mean absolute error stack filtering with structuring constraints and goals. *IEEE Trans. Acoust. Speech Signal Process.* **38**(6), 955–968 (June 1990).

[Gab92] M. Gabbouj, P.-T. Yu, and E. J. Coyle. Convergence behavior and root signal sets of stack filters. *Circ. Syst. Signal Process.*, Special Issue on Median and Morphological Filtering, **11**(1), 171–194 (1992).

[Gil76] E. P. Gilbo and I. B. Chelpanov. *Obrabotka Signalov na Osnove Uporyadochennogo Vybora (Signal Processing Based on Ordering)*, Izd. Sovetskoe Radio, Moscow (1976). In Russian.

[Ham68] F. R. Hampel. *Contribution to the Theory of Robust Estimation.* Ph.D. thesis, University of California, Berkeley (1968).

[Ham74] F. R. Hampel. The influence curve and its role in robust estimation. *J. Amer. Statist. Assoc.* **69**(346), 383–393 (1974).

[Ham82] F. R. Hampel, A. Marazzi, E. Ronchetti, P. J. Rousseeuw, W. A. Stahel, and R. E. Welsch. Handouts for the instructional meeting on robust statistics, Palermo, Italy. Fachgruppe für Statistik, ETH, Zürich (1982).

[Ham86] F. R. Hampel, P. J. Rousseeuw, E. M. Ronchetti, and W. A. Stahel. *Robust Statistics: The Approach Based on Influence Functions.* Wiley, New York (1986).

[Kuo94] P. Kuosmanen. *Statistical Analysis and Optimization of Stack Filters.* Ph.D. thesis, Acta Polytechnica Scandinavica, Electrical Engineering Series No. 77, Helsinki, Finland (1994).

[Kuo96] P. Kuosmanen and J. Astola. Breakdown points, breakdown probabilities, midpoint sensitivity curves and optimization of stack filters. *Circ. Syst. Signal Process.* **15**(2), 165–211 (February 1996).

[Lin90] J. H. Lin, T. M. Sellke, and E. J. Coyle. Adaptive stack filtering under the mean absolute error criterion. *IEEE Trans. Acoust. Speech Signal Process.* **38**(6), 938–954 (June 1990).

[Mal80] C. L. Mallows. Some theory of nonlinear smoothers. *Ann. Statist.* **8**(4), 695–715 (1980).

[Nie87] A. Nieminen, P. Heinonen, and Y. Neuvo. A new class of detail-preserving filters for image processing. *IEEE Trans. Patt. Anal. Machine Intell.* **9**(1), 74–90 (January 1987).

[Pel99a] S. Peltonen, P. Kuosmanen, and J. Astola. Output distributional influence function. In *Proc. IEEE-EURASIP Workshop on Nonlinear Signal and Image Processing*, Vol. 1, pp. 33–37 (Antalya, Turkey, June 1999).

[Pel99b] S. Peltonen and P. Kuosmanen. ODIF for weighted median filters. In *Proc. 6th IEEE Int. Conf. on Electronics, Circuits and Systems*, pp. 245–248 (Pafos, Cyprus, Sept. 1999).

[Pet94] D. Petrescu, I. Tabus, and M. Gabbouj. Adaptive skeletonization using multistage boolean and stack filtering, In *Proc. 1994 European Signal Processing Conf.*, pp. 951–954 (Edinborough, Sept. 1994).

[Pet95] D. Petrescu, I. Tabus, and M. Gabbouj. Edge detection based on optimal stack filtering under given noise distribution, In *Proc. 1995 European Conf. on Circuit Theory and Design*, pp. 1023–1026 (Istanbul, Turkey, Aug. 27-31, 1995).

[Pet99] D. Petrescu, I. Tabus, and M. Gabbouj. Prediction capabilities of boolean and stack filters for lossless image compression. *Multidimensional Syst. Signal Proc.* **10**(2), 161–187 (April 1999).

[Pra90] M. K. Prasad and Y. H. Lee. Stack filters and selection probabilities. In *Proc. IEEE Int. Sympos. on Circuits and Systems*, pp. 1747–1750 (May 1990).

[Que56] M. H. Quenouille. Notes on bias in estimation. *Biometrika* **43**, 353–360 (1956).

[Roh76] V. K. Rohatgi. *An Introduction to Probability Theory and Mathematical Statistics.* Wiley, New York (1976).

[Shm95] I. Shmulevich, T. M. Sellke, M. Gabbouj, and E. J. Coyle. Stack filters and free distributive lattices. In *Proc. IEEE Workshop on Nonlinear Signal and Image Processing*, pp. 927–930 (Halkidiki, Greece, 1995).

[Shm99] I. Shmulevich, O. Yli-Harja, K. Egiazarian, and J. Astola. Output distributions of recursive stack filters. *IEEE Signal Process. Lett.* **6**(7), 175–178 (July 1999).

[Tab93] I. Tabus, M. Gabbouj, and L. Yin. Real domain WOS filtering using neural network approximations. In *Proc. IEEE Winter Workshop on Nonlinear Digital Signal Processing*, pp. 7.2–1.1–6 (Tampere, Finland, January 1996).

[Tab96] I. Tabus, D. Petrescu, and M. Gabbouj. A training framework for stack and Boolean filtering—fast optimal design procedures and robustness case study. *IEEE Trans. Image Process.* **5**(6), 809–826 (June 1996).

[Tuk74] J. W. Tukey. Nonlinear (nonsuperposable) methods for smoothing data. In Congr. Rec. EASCON, p. 673 (1974). (Abstract only.)

[Tuk77] J. W. Tukey. *Exploratory Data Analysis.* Addison-Wesley, Reading, MA (1977; preliminary edition 1970/71).

[Wen86] P. Wendt, E. Coyle, and N. Gallagher. Stack filters. *IEEE Trans. Acoust. Speech Signal Process.* **34**(4), 898–911 (August 1986).

[Yan95] R. Yang, L. Yin, M. Gabbouj, J. Astola, and Y. Neuvo. Optimal weighted median filtering under structural constraints. *IEEE Trans. Signal Process.* **43**(3), 591–604 (March 1995).

[Yin95] L. Yin. Stack filter design: A structural approach. *IEEE Trans. Signal Process.* **43**(4), 831–840 (April 1995).

[Yin96] L. Yin, R. Yang, M. Gabbouj, and Y. Neuvo. Weighted median filters: A tutorial. *IEEE Trans. Circ. Syst. II: Analog and Digital Signal Processing.* **43**(3), 157–192 (March 1996).

[Yli91] O. Yli-Harja, J. Astola, and Y. Neuvo. Analysis of the properties of median and weighted median filters using threshold logic and stack filter representation. *IEEE Trans. Signal Process.* **39**(2), 395–410 (February 1991).

2

Image Enhancement and Analysis with Weighted Medians

Gonzalo R. Arce and Jose L. Paredes*

Department of Electrical and Computer Engineering
University of Delaware
Newark, Delaware

2.1 Introduction

Digital image enhancement and analysis have played and will continue to play an important role in scientific, industrial, and military applications. In addition to these, image enhancement and analysis are being used increasingly in consumer electronics. World Wide Web users, for instance, not only rely on built-in image processing protocols such as JPEG and interpolation but also have become image processing users equipped with powerful yet inexpensive software such as PhotoShop. Users not only retrieve digital images from the Web but now are able to acquire their own by use of digital cameras or through digitization services of standard 35-mm analog film. The end result is that consumers are beginning to use home computers to enhance and manipulate their own digital pictures.

Image enhancement tools are often classified into (a) point operations and (b) spatial operators. Point operations include contrast stretching, histogram modification, and pseudo-coloring. Point operations are, in general, simple nonlinear

*José L. Paredes is also with the Electrical Engineering Department, University of Los Andes, Mérida-Venezuela.

operations that are well known in the image processing literature and are covered elsewhere. Spatial operations used in image processing today, on the other hand, typically are linear operations. The reason for this is that spatial linear operations are simple and easily implemented. Although linear image enhancement tools are often adequate in many applications, significant advantages in image enhancement can be attained if nonlinear techniques are applied [Ast97, Lee85, Pit90, Yin96]. Nonlinear methods effectively preserve edges and details of images, while methods using linear operators tend to blur and distort them. Additionally, nonlinear image enhancement tools are less susceptible to noise, which is always present due to the physical randomness of image acquisition systems. For example, underexposure and low-light conditions in analog photography, which together with the image signal itself are captured during the digitization process, lead to images with film-grain noise.

This chapter focuses on nonlinear and spatial image enhancement and analysis. The nonlinear tools described are easily implemented on currently available computers. Rather than using linear combinations of pixel values within a local window, these tools use the local weighted median. In Sec. 2.2, the principles of weighted median (WM) operators are presented. WM smoothers allowing only positive weights are reviewed, and WM filters admitting positive as well as negative weights are introduced. Weighted medians have striking analogies with traditional linear FIR filters, yet their behavior is often markedly different. In Sec. 2.3, we show how WM filters can be easily used for image denoising. In particular, the center WM filter and a simple permutation WM filter are described as tunable smoothers highly effective in impulse noise. Sec. 2.4 focuses on image enlargement, or zooming, using WM smoother structures, which unlike standard linear interpolation methods cause little edge degradation. Section 2.5 describes image sharpening algorithms based on WM filters. These methods offer significant advantages over traditional linear sharpening tools whenever noise is present in the underlying images. Sec. 2.6 descibes methods for designing optimal frequency selective WM filters. Sec. 2.7 goes beyond image enhancement and focuses on the analysis of images. In particular, edge detection methods based on WM filters are described, as well as their advantages over traditional edge-detection algorithms.

2.2 Weighted Median Smoothers and Filters

2.2.1 Weighted Median Smoothers[1]

The running median was first suggested as a nonlinear smoother for time series data by Tukey in 1974 [Tuk74]. To define the running median smoother, let $\{x(\cdot)\}$

[1]Weighted median smoothers admitting only positive weights have traditionally been referred to in the literature as WM filters. Smoothers, due to their nonnegative weights, are limited to low-pass operators. In this chapter, we denote these structures as "smoothers" so as to differentiate them from the more powerful WM filter structures admitting both positive and negative weights, which can synthesize general frequency selecting filtering.

be a discrete time sequence. The running median passes a window over the sequence $\{x(\cdot)\}$ that selects, at each instant n, a set of samples to comprise the observation vector $\mathbf{x}(n)$. The observation window is centered at n, resulting in

$$\mathbf{x}(n) = [x(n - N_L), \dots, x(n), \dots, x(n + N_R)]^T, \tag{2.1}$$

where N_L and N_R may range in value over the nonnegative integers, with $N = N_L + N_R + 1$ being the window size. The median smoother operating on the input sequence $\{x(\cdot)\}$ produces the output sequence $\{y(\cdot)\}$, where at time index n

$$y(n) = \text{MEDIAN}[x(n - N_L), \dots, x(n), \dots, x(n + N_R)] \tag{2.2}$$

$$= \text{MEDIAN}[x_1(n), \dots, x_N(n)], \tag{2.3}$$

where $x_i(n) = x(n - N_L + i - 1)$ for $i = 1, 2, \dots, N$. That is, the samples in the observation window are sorted and the middle, or median, value is taken as the output. If $x_{(1)}, x_{(2)}, \dots, x_{(N)}$ are the sorted samples in the observation window, the median smoother outputs

$$y(n) = \begin{cases} x_{\left(\frac{N+1}{2}\right)} & \text{for } N \text{ odd,} \\ \frac{x_{\left(\frac{N}{2}\right)} + x_{\left(\frac{N}{2}+1\right)}}{2} & \text{otherwise.} \end{cases} \tag{2.4}$$

In most cases, the window is symmetric about $x(n)$ and $N_L = N_R$.

The input sequence $\{x(\cdot)\}$ may be either finite or infinite in extent. For the finite case, the samples of $\{x(\cdot)\}$ can be indexed as $x(1), x(2), \dots, x(L)$, where L is the length of the sequence. Due to the symmetric nature of the observation window, the window extends beyond a finite extent input sequence at both the beginning and end. These end effects are generally accounted for by appending N_L samples at the beginning and N_R samples at the end of $\{x(\cdot)\}$. Although the appended samples can be arbitrarily chosen, typically they are selected so that those appended at the beginning of the sequence have the same value as the first signal sample and those appended at the end have the value of the last signal sample.

Figure 2.1 illustrates the appending of an input sequence and the median smoother operation. In this example, the input signal $\{x(\cdot)\}$ consists of 20 observations from a six-level process, $\{x : x(n) \in \{0, 1, \dots, 5\}, n = 1, 2, \dots, 20\}$. The figure shows the input sequence and the resulting output sequence for a median smoother of window size 5. Note that to account for end effects, two samples have been appended to both the beginning and end of the sequence. The median smoother output at the window location shown in the figure is

$$\begin{aligned} y(9) &= \text{MEDIAN}[x(7), x(8), x(9), x(10), x(11)] \\ &= \text{MEDIAN}[1, 1, 4, 3, 3] \\ &= 3. \end{aligned}$$

The running one-dimensional or two-dimensional median, at each instant in time, computes the sample median, which in many respects resembles the sample mean.

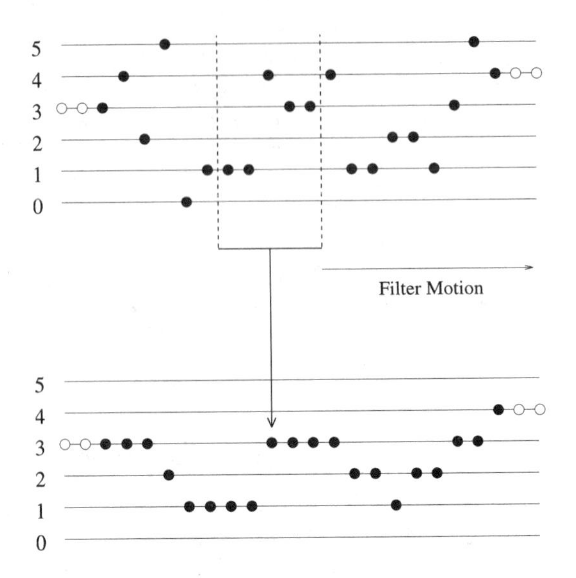

Figure 2.1: The operation of the window size 5 median smoother; the open circles are the appended points.

That is, given N samples $x_1, \ldots, x_N$, the sample mean $\bar{x}$ and sample median $\tilde{x}$ minimize the expression

$$G(\beta) = \sum_{i=1}^{N} |x_i - \beta|^p, \tag{2.5}$$

with respect to β, for $p = 2$ and $p = 1$, respectively. Thus, the median of an odd number of samples emerges as the sample whose sum of absolute distances to all other samples in the set is the smallest. Likewise, the sample mean is given by the value β whose square distance to all samples in the set is the smallest possible. The analogy between the sample mean and median extends into the statistical domain of parameter estimation, where it can be shown that the sample median is the maximum likelihood (ML) estimator of location of a constant parameter in Laplacian noise. Likewise, the sample mean is the ML estimator of location of a constant parameter in Gaussian noise [Leh83]. This result has profound implications in signal processing, as most tasks where non-Gaussian noise is present will benefit from signal processing structures using medians, particularly when the noise statistics can be characterized by probability densities having heavier than Gaussian tails (which leads to noise with impulse characteristics) [Arn92, Bov83, Dav82].

Although the median is a robust estimator that possesses many optimality properties, the performance of a running median is limited by the fact that it is temporally blind. That is, all observation samples are treated equally, regardless of their location within the observation window. Much like weights can be incorporated into the sample mean to form a weighted mean, a weighted median can be defined as the sample that minimizes the weighted cost function

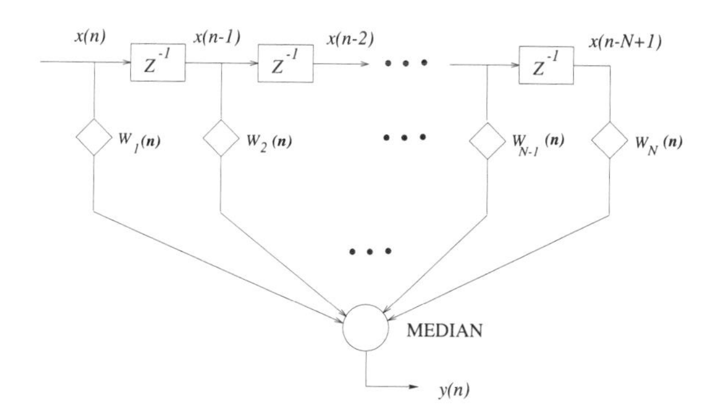

Figure 2.2: The weighted median smoothing operation.

$$G_p(\beta) = \sum_{i=1}^{N} W_i |x_i - \beta|^p, \tag{2.6}$$

with respect to β, where W_i are the weights and $p = 1$. For $p = 2$, the cost function $G_p(\beta)$ in Eq. (2.6) is a quadratic function, and the value β minimizing it is the normalized weighted mean:

$$\hat{\beta} = \arg\min_{\beta} \sum_{i=1}^{N} W_i (x_i - \beta)^2 = \frac{\sum_{i=1}^{N} W_i x_i}{\sum_{i=1}^{N} W_i}, \tag{2.7}$$

with $W_i > 0$. For $p = 1$, $G_1(\beta)$ is piecewise linear and convex for $W_i \geq 0$. The value β minimizing Eq. (2.6) is thus guaranteed to be one of the samples $x_1, x_2, \ldots, x_N$ and is referred to as the weighted median, originally introduced over a hundred years ago by Edgemore [Edg87]. After some algebraic manipulations, it can be shown that the running weighted median output is defined as follows:

Weighted Median Smoothers. Given a set of N positive weights $\langle W_1, W_2, \ldots, W_N \rangle$ and the observation vector $\mathbf{x} = [x_1, x_2, \ldots, x_N]^T$, the weighted median smoother output is

$$y(n) = \text{MEDIAN}\,[W_1 \blacklozenge x_1(n),\; W_2 \blacklozenge x_2(n), \ldots, W_N \blacklozenge x_N(n)], \tag{2.8}$$

where $W_i > 0$ and is the replication operator defined as $W_i \blacklozenge x_i = \underbrace{x_i, x_i, \ldots, x_i}_{W_i \text{ times}}$.

Weighted median smoothers were introduced in the signal processing literature by Brownrigg in 1984 and have since received considerable attention [Bro84, Ko91, Yin96]. The WM smoothing operation can be schematically described as in Fig. 2.2. Weighted medians admitting only positive weights are low-pass filters by nature and consequently these signal processing structures are here refered to as "smoothers."

The computation of weighted median smoothers is simple. Consider the WM smoother of window size 5 defined by the symmetric weight vector $\mathbf{W} = [1, 2, 3, 2, 1]$. For the observation $\mathbf{x}(n) = [12, 6, 4, 1, 9]$, the weighted median smoother output is found as

$$\begin{aligned} y(n) &= \text{MEDIAN}[1\blacklozenge 12,\ 2\blacklozenge 6,\ 3\blacklozenge 4,\ 2\blacklozenge 1,\ 1\blacklozenge 9] \\ &= \text{MEDIAN}[12, 6, 6, 4, 4, 4, 1, 1, 9] \\ &= \text{MEDIAN}[1, 1, 4, 4, \underline{4}, 6, 6, 9, 12] \\ &= 4. \end{aligned} \tag{2.9}$$

The large weighting on the center input sample results in this sample being taken as the output. As a comparison, the standard median output for the given input is $y(n) = 6$. More on the computation of WM smoothers will be described later.

The Center Weighted Median Smoother

The weighting mechanism of WM smoothers allows for great flexibility in emphasizing or deemphasizing specific input samples. In most applications, not all samples are equally important. Due to the symmetric nature of the observation window, the sample most correlated with the desired estimate is, in general, the center observation sample. This observation leads to the center weighted median (CWM) smoother, which is a relatively simple subset of WM smoother that has proven useful in many applications [Ko91].

The CWM smoother is realized by allowing only the center observation sample to be weighted. Thus, the output of the CWM smoother is given by

$$y(n) = \text{MEDIAN}[x_1, \ldots, x_{c-1},\ W_c \blacklozenge x_c,\ x_{c+1}, \ldots, x_N], \tag{2.10}$$

where W_c is an odd positive integer and $c = (N+1)/2 = N_1 + 1$ is the index of the center sample. When $W_c = 1$, the operator is a median smoother, and for $W_c \geq N$, the CWM reduces to an identity operation.

The effect of varying the center sample weight is perhaps best seen by way of an example. Consider a segment of recorded speech. The voiced waveform "a" is shown at the top of Fig. 2.3. This speech signal is taken as the input of a CWM smoother of size 9. The outputs of the CWM, as the weight parameter $W_c = 2w + 1$ for $w = 0, \ldots, 3$, are shown in the figure. Clearly, as W_c is increased less smoothing occurs.

The CWM smoother has an intuitive interpretation. It turns out that the output of a CWM smoother is equivalent to computing

$$y(n) = \text{MEDIAN}\left[x_{(k)}, x_c, x_{(N-k+1)}\right], \tag{2.11}$$

where $k = (N+2-W_c)/2$ for $1 \leq W_c \leq N$ and $k = 1$ for $W_c > N$; $x_{(i)}$ is the ith order statistics, defined as $x_{(1)} < x_{(2)} \ldots < x_{(N)}$. Since $x(n)$ is the center sample in the observation window, that is, $x_c = x(n)$, the output of the smoother is identical to

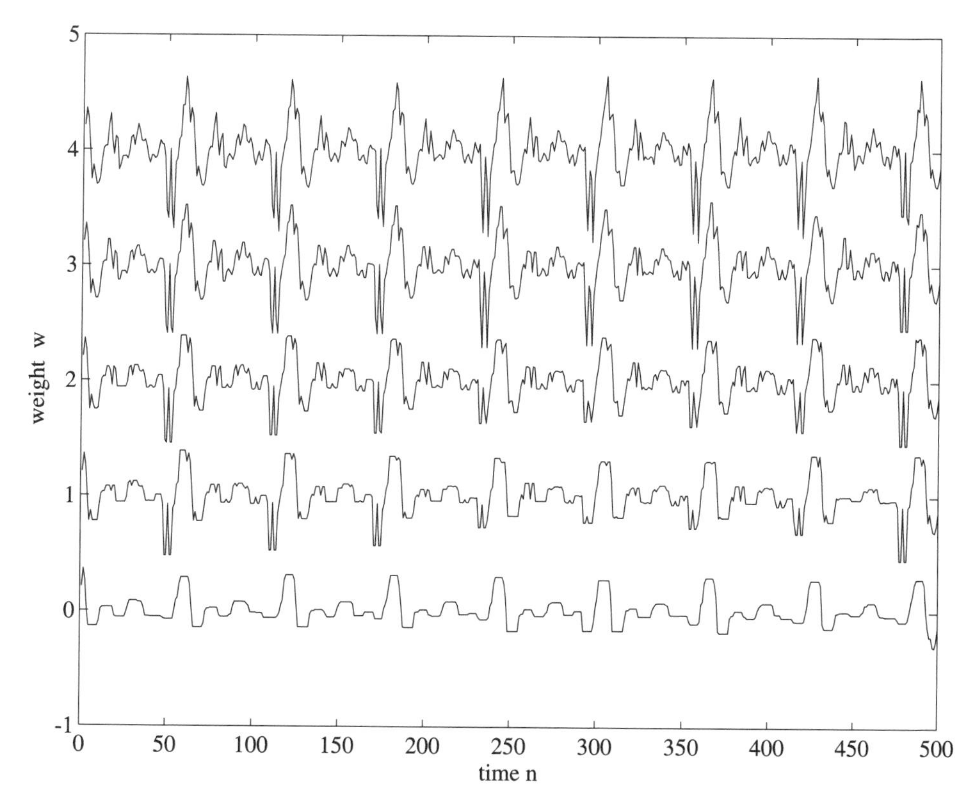

Figure 2.3: Effects of increasing the center weight of a CWM smoother of size $N = 9$ operating on the voiced speech "a". The CWM smoother output is shown for $W_c = 2w + 1$, with $w = 0, 1, 2, 3$. Note that for $W_c = 1$ the CWM reduces to median smoothing, and for $W_c = 9$ it becomes the identity operator.

the input as long as the $x(n)$ lies in the interval $[x_{(k)}, x_{(N+1-k)}]$. If the center input sample is greater than $x_{(N+1-k)}$, the CWM smoother outputs $x_{(N+1-k)}$, guarding against a high rank order (large) aberrant data point being taken as the output. Similarly, the smoother's output is $x_{(k)}$ if the sample $x(n)$ is smaller than this order statistic. This CWM smoother performance characteristic is illustrated in Fig. 2.4, which shows how the input sample is left unaltered if it is between the order statistics $x_{(k)}$ and $x_{(N+1-k)}$ but is mapped to one of these order statistics if it is outside this range.

2.2.2 Weighted Median Filters Admitting Negative Weights

Weighted median smoothers admit only positive weights. This is a limitation as WM smoothers are, in essence, limited to having "low-pass" type filtering characteristics. Although WM smoothers have some analogies with linear FIR filters, they are equivalent to the normalized weighted average with only nonnegative weights—a severely constrained subset of linear FIR filter. A large number of

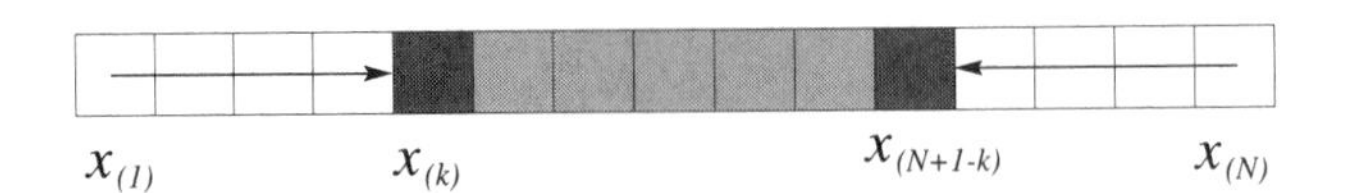

Figure 2.4: The center weighted median smoothing operation. The center observation sample is mapped to the order statistic $x_{(k)}$ ($x_{(N+1-k)}$) if the center sample is less (greater) than $x_{(k)}$ ($x_{(N+1-k)}$), but is left unaltered otherwise.

image processing applications require "band-pass" or "high-pass" frequency filtering characteristics. Linear FIR equalizers admitting only positive filter weights, for instance, would lead to unacceptable results in many signal processing tasks. Thus, it is not surprising that weighted median smoothers admitting only positive weights lead to inadequate results in a number of applications.

In order to formulate the general weighted median filter structure, it is logical to ask how linear FIR filters arise within the location estimation problem; the answer could provide the key to the formulation of the general WM filter. To this end, consider N samples $x_1, x_2, \ldots, x_N$ obeying a multivariate Gaussian distribution

$$f(\mathbf{x}) = \frac{1}{(2\pi)^{N/2}[\det(\mathbf{C})]^{1/2}} \exp\left[-\frac{1}{2}(\mathbf{x}-\mathbf{e}\beta)^T \mathbf{C}^{-1}(\mathbf{x}-\mathbf{e}\beta)\right], \tag{2.12}$$

where $\mathbf{x} = [x_1, x_2, \ldots, x_N]^T$ is the observation vector, $\mathbf{e} = [1, 1, \ldots, 1]^T$, β is the location parameter, $\mathbf{C}$ is the covariance matrix, and $\det(\mathbf{C})$ is the determinant of $\mathbf{C}$. The maximum likelihood estimate of the location parameter β results in

$$\hat{\beta} = \frac{\mathbf{e}^T\mathbf{C}^T\mathbf{x}}{\mathbf{e}^T\mathbf{C}\mathbf{e}} = \mathbf{W}^T\mathbf{x}, \tag{2.13}$$

where $\mathbf{e}^T\mathbf{C}\mathbf{e} > 0$ due to the positive definite nature of the covariance matrix and where elements in the vector $\mathbf{e}^T\mathbf{C}^T$ can take on positive as well as negative values. Thus, Eq. (2.13) takes on the structure of a linear FIR filter whose weights W_i may take on negative values, depending on the mutual correlation of the observation samples.

The extension of the above to the case of Laplacian distributed samples, and in general to other non-Gaussian distributions, unfortunately becomes too cumbersome. The multivariate Laplacian distribution, and in general all non-Gaussian multivariate distributions, do not lead to simple ML location estimates. The complexity in these solutions has hindered the development of nonlinear filters having attributes comparable to those of linear FIR filters. Notably, however, a novel approach was discovered that can overcome these limitations [Arc98]. In this approach, a generalization of the sample mean leads to the class of linear FIR filters. This generalization is, in turn, applied to the sample median, forming the class of weighted median filters that admits both positive and negative weights. The extension, not only turns out to be natural, leading to a significantly richer filter class, but it is simple as well.

The sample mean $\text{MEAN}(x_1, x_2, \ldots, x_N)$ can be generalized to the class of linear FIR filters as

$$\bar{\beta} = \text{MEAN}(W_1 \cdot x_1, W_2 \cdot x_2, \ldots, W_N \cdot x_N), \tag{2.14}$$

where $W_i \in R$, with R denoting the set of real numbers. For us to apply the analogy to the median filter structure, Eq. (2.14) must be written as

$$\bar{\beta} = \text{MEAN}\,[|W_1| \blacklozenge \operatorname{sgn}(W_1)x_1, |W_2| \blacklozenge \operatorname{sgn}(W_2)x_2, \ldots, |W_N| \blacklozenge \operatorname{sgn}(W_N)x_N], \quad (2.15)$$

where the sign of the weight affects the corresponding input sample and the weighting is constrained to be nonnegative. By analogy, the class of weighted median filters admitting real-valued weights emerges as defined next.

Weighted Median Filters. Given a set of N real-valued weights $\langle W_1, W_2, \ldots, W_N \rangle$ and the observation vector $\mathbf{x} = [x_1, x_2, \ldots, x_N]^T$, the weighted median filter output is defined as

$$\tilde{\beta} = \text{MEDIAN}\,[|W_1| \blacklozenge \operatorname{sgn}(W_1)x_1, |W_2| \blacklozenge \operatorname{sgn}(W_2)x_2, \ldots, |W_N| \blacklozenge \operatorname{sgn}(W_N)x_N], \quad (2.16)$$

with $W_i \in R$ for $i = 1, 2, \ldots, N$. Note that the weight signs are uncoupled from the weight magnitude values and are merged with the observation samples. The weight magnitudes play the equivalent role of positive weights in the framework of weighted median smoothers.

Weighted Median Filter Computation

The computation of the WM filter is best illustrated by means of an example. Consider first the case where the weights are integer-valued and where these add up to an odd integer number. Let the window size be 5 defined by the symmetric weight vector $\mathbf{W} = \langle 1, -2, 3, -2, 1 \rangle$. For the observation vector $\mathbf{x}(n) = [2, -6, 9, 1, 12]$, the weighted median filter output is found as

$$\begin{aligned} y(n) &= \text{MEDIAN}[1\blacklozenge 2,\ -2\blacklozenge -6,\ 3\blacklozenge 9,\ -2\blacklozenge 1,\ 1\blacklozenge 12] \\ &= \text{MEDIAN}[1\blacklozenge 2,\ 2\blacklozenge 6,\ 3\blacklozenge 9,\ 2\blacklozenge -1,\ 1\blacklozenge 12] \\ &= \text{MEDIAN}[2, 6, 6, 9, 9, 9, -1, -1, 12] \\ &= \text{MEDIAN}[-1, -1, 2, 6, \underline{6}, 9, 9, 9, 12] \\ &= 6, \end{aligned} \quad (2.17)$$

where the median filter output value is underlined.

Next consider the case where the WM filter weights add up to an even integer with $\mathbf{W} = \langle 1, -2, 2, -2, 1 \rangle$. Furthermore, assume that the observation vector consists of a set of constant-valued samples $\mathbf{x}(n) = [5, 5, 5, 5, 5]$. The weighted median filter output in this case is found as

$$\begin{aligned} y(n) &= \text{MEDIAN}[1\blacklozenge 5,\ -2\blacklozenge 5,\ 2\blacklozenge 5,\ -2\blacklozenge 5,\ 1\blacklozenge 5] \\ &= \text{MEDIAN}[1\blacklozenge 5,\ 2\blacklozenge -5,\ 2\blacklozenge 5,\ 2\blacklozenge -5,\ 1\blacklozenge 5] \\ &= \text{MEDIAN}[5, -5, -5, 5, 5, -5, -5, 5] \\ &= \text{MEDIAN}[-5, -5, -5, \underline{-5, 5}, 5, 5, 5] \\ &= 0, \end{aligned} \quad (2.18)$$

where the median filter output is the average of the underlined samples. Note that in order for the WM filter to have band- or high-pass frequency characteristics, where constant signals are annihilated, the weights' absolute values must add to an even number so that averaging of the middle rank samples can occur. When the WM filter's absolute-value weights add to an odd number, the output is one of the signed input samples, and consequently the filter is unable to suppress constant-valued signals.

In general, the WM filter output can be computed without replicating the sample data according to the corresponding weights, since this increases the computational complexity. A more efficient method to find the WM is shown next, which not only is attractive from a computational perspective but also admits real-valued weights.

The weighted median filter output for non-integer weights can be determined as follows:

1. Calculate the threshold $T_0 = 1/2 \sum_{i=1}^{N} |W_i|$.
2. Sort the "signed" observation samples $\operatorname{sgn}(W_i)x_i$.
3. Sum the magnitude of the weights corresponding to the sorted "signed" samples beginning with the maximum and continuing down in order.
4. The output is the signed sample whose magnitude weight causes the sum to become greater than or equal to T_0. For band- and high-pass characteristics, the output is the average of the signed sample whose weight magnitude causes the sum to become greater than or equal to T_0 and the next smaller signed sample.

The following example illustrates this procedure. Consider the window size 5 WM filter defined by the real valued weights $\langle W_1, W_2, W_3, W_4, W_5 \rangle = \langle 0.1, 0.2, 0.3, -0.2, 0.1 \rangle$. The output for this filter operating on the observation set $[x_1, x_2, x_3, x_4, x_5] = [-2, 2, -1, 3, 6]$ is found as follows: Summing the absolute weights gives the threshold $T_0 = 1/2 \sum_{i=1}^{5} |W_i| = 0.45$. The "signed" observation samples, sorted observation samples, their corresponding weights, and the partial sum of weights (from each ordered sample to the maximum) are

observation samples	−2,	2,	−1,	3,	6
corresponding weights	0.1,	0.2,	0.3,	−0.2,	0.1
sorted signed observation samples	−3,	−2,	−1,	2,	6
corresponding absolute weights	0.2,	0.1,	0.3,	0.2,	0.1
partial weight sums	0.9,	0.7,	$\underline{0.6}$,	0.3,	0.1

Thus, the output is −1 since when starting from the right (maximum sample) and summing the weights, the threshold $T_0 = 0.45$ is not reached until the weight associated with −1 is added. The underlined sum value indicates that this is the first sum that meets or exceeds the threshold. To warrant high- or band-pass characteristics, the WM filter output would be modified so as to compute the average between -1 and -2, leading to -1.5 as the output value.

It should be noted that as a result of the negative weights, the computation of the weighted median filter is not shift invariant. Consider the previous example and add a shift of 2 on the samples of $\mathbf{x}$ such that $x_i' = x_i + 2$. The weighted median filtering of $\mathbf{x}' = [4, -4, 11, 3, 15]$ with the weight vector $\mathbf{W} = \langle 1, -2, 3, -2, 1 \rangle$ leads to the output $y'(n) = 4$, which does not equal the previous output in Eq. (2.17) of 6 plus the appropriate shift.

Permutation Weighted Median Filters

The principle behind the CWM smoother lies in its ability to emphasize or deemphasize the center sample of the window by adjusting the center weight while keeping the weight values of all other samples at unity. In essence, the value given to the center weight indicates the "reliability" of the center sample. If that sample does not contain an impulse (high reliability), it would be desirable to make the center weight large such that no smoothing takes place (identity filter). On the other hand, if an impulse were present in the center of the window (low reliability), no emphasis should be given to the center sample (impulse), and the center weight should be given the smallest possible weight, that is, $W_c = 1$, reducing the CWM smoother structure to a simple median. Notably, this adaptation of the center weight can be easily achieved by considering the center sample's rank among all pixels in the window [Arc95, Har94]. More precisely, denoting the rank of the center sample of the window at a given location as $R_c(n)$, then the simplest *permutation* WM smoother is defined by the following modification of the CWM smoothing operation:

$$W_c(n) = \begin{cases} N & \text{for } T_L \le R_c(n) \le T_U, \\ 1 & \text{otherwise,} \end{cases} \tag{2.19}$$

where N is the window size and $1 \le T_L \le T_U \le N$ are two adjustable threshold parameters that determine the degree of smoothing. Note that the weight in Eq. (2.19) is data adaptive and may change between two values with n. The smaller (larger) the threshold parameter T_L (T_U) is set to, the better the detail preservation is. Generally, T_L and T_U are set symmetrically around the median. If the underlying noise distribution was not symmetric about the origin, a nonsymmetric assignment of the thresholds would be appropriate. An image denoising example is provided in Sec. 2.3.2, and review of permutation filters and related topics are given in Chapter 3.

The data-adaptive structure of the smoother in Eq. (2.19) can be extended so that the center weight is not switched between two possible values but can take on N different values:

$$W_c(n) = \begin{cases} W_{c(j)}(n) & \text{for } R_c(n) = j, \quad j \in \{1, 2, \ldots, N\}, \\ 0 & \text{otherwise.} \end{cases} \tag{2.20}$$

Thus, the weight assigned to x_c is drawn from the center weight set $\{W_{c(1)}, W_{c(2)}, \ldots, W_{c(N)}\}$. With an increased number of weights, the smoother in Eq. (2.20) can perform better, although the design of the weights is no longer trivial and optimization algorithms are needed [Arc95, Har94]. A further generalization of Eq. (2.20) is feasible in which weights are given to all samples in the window but their values are data dependent, real valued, and determined by the ranks of all the samples in the observation window.

Permutation WM Filters. Let $\langle W_{1(\mathfrak{R}_r)}, W_{2(\mathfrak{R}_r)}, \ldots, W_{N(\mathfrak{R}_r)} \rangle$ be rank order dependent weights assigned to the input observation samples. The output of the permutation WM filter is found as

$$y = \text{MEDIAN}[|W_{1(\mathfrak{R}_r)}| \blacklozenge \text{sgn}(W_{1(\mathfrak{R}_r)})x_1, \ldots, |W_{N(\mathfrak{R}_r)}| \blacklozenge \text{sgn}(W_{N(\mathfrak{R}_r)})x_N], \quad (2.21)$$

where $W_{i(\mathfrak{R}_r)}$ is the weight assigned to x_i, selected according to the samples' ranks, $\mathfrak{R}_r = [R_1, R_2, \ldots, R_N]$.

Note that the sign of the weights is decoupled from the replication operator and applied to the data sample. The weight assigned to x_i is drawn from the weight set $\{W_{i(1)}, W_{i(2)}, \ldots, W_{i(N!)}\}$. With $(N!)$ weights per sample, a total of $N(N!)$ values need to be stored in the computation of Eq. (2.21). In general, an optimization algorithm is needed to design the set of weights, although in some cases only a few rank order dependent weights are required and their design is simple. Permutation WM filters can provide significant improvement in performance at the higher cost of memory cells.

Optimal WM Filtering: The Least Mean Absolute Algorithm

We now consider the important issue of the design of optimal weighted median smoothers (see Chapter 1 for additional approaches to WM optimization). Much like linear filters can be optimized using the Wiener filter theory, weighted median filters enjoy an equivalent theory for optimization. To develop the various optimization algorithms, we introduce the concept of threshold decomposition of real-valued inputs. Consider the set of real-valued samples $x_1, x_2, \ldots, x_N$ with $x_i \in \mathrm{R}$ and define a weighted median filter by the corresponding real-valued weights $W_1, W_2, \ldots, W_N$. Decompose each sample x_i as $X_i^q = \text{sgn}(x_i - q)$, where $-\infty < q < \infty$ and

$$\text{sgn}(x_i - q) = \begin{cases} +1 & \text{for } x_i \geq q, \\ -1 & \text{for } x_i < q. \end{cases} \quad (2.22)$$

Thus, each sample x_i is decomposed into an infinite set of binary points taking values in $[-1, 1]$. Figure 2.5 depicts the decomposition of x_i as a function of q.

Threshold decomposition is reversible; the original real-valued sample x_i can be perfectly reconstructed from the infinite set of thresholded signals as

$$\begin{aligned} x_i &= \frac{1}{2}\int_{-\infty}^{\infty} X_i^q \, dq \\ &= \frac{1}{2}\int_{-\infty}^{\infty} \text{sgn}\,(x_i - q)\, dq. \end{aligned} \quad (2.23)$$

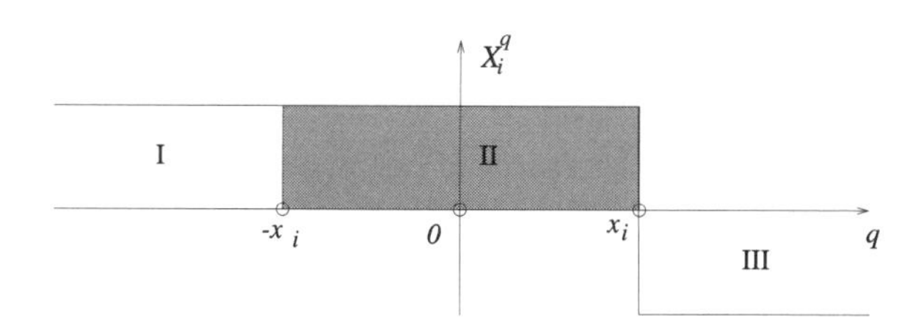

Figure 2.5: Decomposition of x_i into the three-ary X_i^q signal.

The sample x_i can be reconstructed from its corresponding set of decomposed signals, and consequently x_i has a unique threshold signal representation, and vice versa:

$$x_i \overset{\text{T.D.}}{\longleftrightarrow} \{X_i^q\}, \tag{2.24}$$

where $\overset{\text{T.D.}}{\longleftrightarrow}$ denotes the one-to-one mapping provided by the threshold decomposition operation.

Threshold decomposition in the real-valued sample domain also allows the order of the median and threshold decomposition operations to be interchanged without affecting the end result. Given N samples $x_1, x_2, \ldots, x_N$ and their corresponding threshold decomposition representations $X_1^q, X_2^q, \ldots, X_N^q$, the median of the decomposed signals at a fixed value of q is

$$Y^q = \text{MEDIAN}\left(X_1^q, X_2^q, \ldots, X_N^q\right) = \begin{cases} +1 & \text{for } x_{\left(\frac{N+1}{2}\right)} \geq q, \\ -1 & \text{for } x_{\left(\frac{N+1}{2}\right)} < q. \end{cases} \tag{2.25}$$

Reversing the threshold decomposition, y is obtained as

$$\begin{aligned} y &= \frac{1}{2}\int_{-\infty}^{\infty} Y^q \, dq \\ &= \frac{1}{2}\int_{-\infty}^{\infty} \text{sgn}\left(x_{\left(\frac{N+1}{2}\right)} - q\right) dq \\ &= x_{\left(\frac{N+1}{2}\right)}. \end{aligned} \tag{2.26}$$

Thus, applying the median operation on a set of samples and applying the median operation on a set threshold decomposed set of samples and reversing the decomposition give exactly the same result. With threshold decomposition, the weighted median filter operation can be implemented as

$$\begin{aligned} \hat{\beta} &= \text{MEDIAN}\left[|W_i| \blacklozenge \text{sgn}(W_i)x_i|_{i=1}^N\right] \\ &= \text{MEDIAN}\left(|W_i| \diamond \frac{1}{2}\int_{-\infty}^{\infty} \text{sgn}\left[\text{sgn}(W_i)x_i - q\right] dq|_{i=1}^N\right), \end{aligned} \tag{2.27}$$

where $|W_i| \diamond \text{sgn}(W_i)x_i|_{i=1}^N = |W_1| \blacklozenge \text{sgn}(W_1)x_1, |W_2| \blacklozenge \text{sgn}(W_2)x_2, \ldots, |W_N| \blacklozenge \text{sgn}(W_N)x_N$. The expression in Eq. (2.27) represents the median operation of a set of weighted integrals, each synthesizing a signed sample. Note that the same result is obtained if the weighted median of these functions, at each value of q, is taken first and the resultant signal is integrated over its domain. Thus, the order of the integral and the median operator can be interchanged without affecting the result, leading to

$$\hat{\beta} = \frac{1}{2}\int_{-\infty}^{\infty} \text{MEDIAN}\left[|W_i| \blacklozenge \text{sgn}[\text{sgn}(W_i)x_i - q]|_{i=1}^{N}\right] dq. \tag{2.28}$$

In this representation, the "signed" samples play a fundamental role; thus, we define the "signed" observation vector $\mathbf{s}$ as

$$\begin{aligned}\mathbf{s} &= [\text{sgn}(W_1)x_1, \text{sgn}(W_2)x_2, \ldots, \text{sgn}(W_N)x_N]^T \\ &= [s_1, s_2, \ldots, s_N]^T.\end{aligned} \tag{2.29}$$

The threshold decomposed signed samples, in turn, form the vector $\mathbf{S}^q$, defined as

$$\begin{aligned}\mathbf{S}^q &= [\text{sgn}[\text{sgn}(W_1)x_1 - q], \text{sgn}[\text{sgn}(W_2)x_2 - q], \ldots, \text{sgn}[\text{sgn}(W_N)x_N - q]]^T \\ &= [S_1^q, S_2^q, \ldots, S_N^q]^T.\end{aligned} \tag{2.30}$$

Letting $\mathbf{W}_a$ be the vector whose elements are the magnitude weights, $\mathbf{W}_a = [|W_1|, |W_2|, \ldots, |W_N|]^T$, the WM filter operation can be expressed as

$$\hat{\beta} = \frac{1}{2}\int_{-\infty}^{\infty} \text{sgn}\left(\mathbf{W}_a^T \mathbf{S}^q\right) dq. \tag{2.31}$$

The WM filter representation using threshold decomposition is compact even though it may seem like the integral term might be difficult to implement in practice. Equation (2.31), however, is used for the purposes of analysis, not implementation.

Next, the threshold decomposition architecture is used to develop an optimization algorithm for WM filters—namely, the *least mean absolute* (LMA) adaptive algorithm, which shares many of the desirable attributes of the LMS algorithm, including simplicity and efficiency.

Assume that the observed process $\{x(n)\}$ is statistically related to some desired process $\{d(n)\}$ of interest; $\{x(n)\}$ is typically a transformed or corrupted version of $\{d(n)\}$. Furthermore, it is assumed that these processes are jointly stationary. A window of width N slides across the input process pointwise, estimating the desired sequence. The vector containing the N samples in the window at time n is

$$\begin{aligned}\mathbf{x}(n) &= [x(n - N_1), \ldots, x(n), \ldots, x(n + N_2)]^T \\ &= [x_1(n), x_2(n), \ldots, x_N(n)]^T,\end{aligned} \tag{2.32}$$

with $N = N_1 + N_2 + 1$. The running weighted median filter output estimates the desired signal as

$$\hat{d}(n) = \text{MEDIAN}\left[|W_i| \blacklozenge \text{sgn}(W_i)x_i(n)|_{i=1}^{N}\right],$$

where both the weights W_i and samples $x_i(n)$ take on real values. The goal is to determine the weight values in $\mathbf{W} = [W_1, W_2, \ldots, W_N]^T$ that will minimize the estimation error. Under the mean absolute error (MAE) criterion, the cost to minimize is given by

$$J(\mathbf{W}) = E\left\{|d(n) - \hat{d}(n)|\right\}$$
$$= E\left\{\frac{1}{2}\left|\int_{-\infty}^{\infty} \operatorname{sgn}(d-q) - \operatorname{sgn}\left(\mathbf{W}_a^T \mathbf{S}^q\right) dq\right|\right\}, \tag{2.33}$$

where the threshold decomposition representation of the signals is used. The absolute value and integral operators in Eq. (2.33) can be interchanged since the integral acts on a strictly positive or a strictly negative function. This results in

$$J(\mathbf{W}) = \frac{1}{2}\int_{-\infty}^{\infty} E\left\{\left|\operatorname{sgn}(d-q) - \operatorname{sgn}\left(\mathbf{W}_a^T \mathbf{S}^q\right)\right|\right\} dq. \tag{2.34}$$

Furthermore, since the argument inside the absolute value operator in Eq. (2.34) can take on only values in the set $\{-2, 0, 2\}$, the absolute value operator can be replaced by a properly scaled second power operator. Thus,

$$J(\mathbf{W}) = \frac{1}{4}\int_{-\infty}^{\infty} E\left\{\left[\operatorname{sgn}(d-q) - \operatorname{sgn}\left(\mathbf{W}_a^T \mathbf{S}^q\right)\right]^2\right\} dq. \tag{2.35}$$

Evaluation of the gradient of the above results in

$$\frac{\partial}{\partial \mathbf{W}} J(\mathbf{W}) = -\frac{1}{2}\int_{-\infty}^{\infty} E\left\{e^q(n)\frac{\partial}{\partial \mathbf{W}} \operatorname{sgn}\left(\mathbf{W}_a^T \mathbf{S}^q\right)\right\} dq, \tag{2.36}$$

where $e^q(n) = \operatorname{sgn}(d-q) - \operatorname{sgn}(\mathbf{W}_a^T \mathbf{S}^q)$. Since the sign function is discontinuous at the origin, its derivative introduces Dirac impulse terms that are inconvenient for further analysis. To overcome this difficulty, the signum function in Eq. (2.36) is approximated by a smoother differentiable function. A simple approximation is given by the hyperbolic tangent function $\operatorname{sgn}(x) \approx \tanh(x) = (e^x - e^{-x})/(e^x + e^{-x})$. Since $(\partial/\partial x)\tanh(x) = \operatorname{sech}^2(x) = 2/(e^x + e^{-x})$, it follows that

$$\frac{\partial}{\partial \mathbf{W}} \operatorname{sgn}\left(\mathbf{W}_a^T \mathbf{S}^q\right) \approx \operatorname{sech}^2\left(\mathbf{W}_a^T \mathbf{S}^q\right)\frac{\partial}{\partial \mathbf{W}}\left(\mathbf{W}_a^T \mathbf{S}^q\right). \tag{2.37}$$

Evaluating the derivative in Eq. (2.37) and simplifying leads to

$$\frac{\partial}{\partial \mathbf{W}} \operatorname{sgn}\left(\mathbf{W}_a^T \mathbf{S}^q\right) \approx \operatorname{sech}^2\left(\mathbf{W}_a^T \mathbf{S}^q\right)\begin{bmatrix} \operatorname{sgn}(W_1)S_1^q \\ \operatorname{sgn}(W_2)S_2^q \\ \vdots \\ \operatorname{sgn}(W_N)S_N^q \end{bmatrix}. \tag{2.38}$$

Using Eq. (2.38) in Eq. (2.36) yields

$$\frac{\partial}{\partial W_j} J(\mathbf{W}) = -\frac{1}{2}\int_{-\infty}^{\infty} E\left\{e^q(n)\operatorname{sech}^2\left(\mathbf{W}_a^T \mathbf{S}^q\right)\operatorname{sgn}(W_j)S_j^q\right\} dq. \tag{2.39}$$

Using the gradient, the optimal coefficients can be found through the steepest descent recursive update:

$$\begin{aligned} W_j(n+1) &= W_j(n) + 2\mu\left[-\frac{\partial}{\partial W_j}J(\mathbf{W})\right] \\ &= W_j(n) + \mu\left[\int_{-\infty}^{\infty} E\left\{e^q(n)\mathrm{sech}^2\left(\mathbf{W}_a^T(n)\mathbf{S}^q(n)\right)\right.\right. \\ &\quad \left.\left.\times\ \mathrm{sgn}\left(W_j(n)\right)S_j^q(n)\right\}\, dq\right]. \end{aligned} \tag{2.40}$$

Using the instantaneous estimate for the gradient and simplifying, the fast LMA WM adaptive algorithm is derived as

$$W_j(n+1) = W_j(n)+\mu\left(d(n)-\hat{d}(n)\right)\mathrm{sgn}\left(W_j(n)\right)\mathrm{sgn}\left(\mathrm{sgn}\left(W_j(n)\right)x_j(n)-\hat{d}(n)\right), \tag{2.41}$$

for $j = 1, 2, \ldots, N$.

A convergence analysis is not available for the fast LMA WM adaptive algorithm and therefore exact bounds on the step size μ are not available. In practice, a reliable guideline for this algorithm is to select a step size of the same order as that required for the standard LMS algorithm. It can then be further tuned according to the user's requirements and by evaluating the response given by the initial step size choice.

2.2.3 Recursive Weighted Median Filters

It is natural now to extend the weighted median filters to other more general signal processing structures. Here, the class of recursive weighted median (RWM) filters admitting real-valued weights is defined. These filters are analogous to the class of infinite impulse response (IIR) linear filters, which often lead to reduced computational complexity. Much like IIR linear filters provide this advantage over linear FIR filters, RWM filters also exhibit characteristics superior to nonrecursive WM filters. RWM filters can synthesize nonrecursive WM filters of much larger window sizes, and in terms of noise attenuation, recursive median smoothers have far superior characteristics to their nonrecursive counterparts [Arc86, Arc88].

The general structure of linear IIR filters is defined by the difference equation

$$y(n) = \sum_{\ell=1}^{N} A_\ell\, y(n-\ell) + \sum_{k=-M_1}^{M_2} B_k\, x(n-k), \tag{2.42}$$

where the output is formed not only from the input but also from previously computed outputs. The filter weights consist of two sets: the feedback coefficients $\{A_\ell\}$ and the feed-forward coefficients $\{B_k\}$. In all, $N + M_1 + M_2 + 1$ coefficients are needed to define the recursive difference equation. The generalization of Eq. (2.42) to an RWM filter structure is straightforward. Following an approach similar to that used in the definition of nonrecursive WM filters, the summation operation is replaced with the *median* operation, and the *multiplication* weighting is replaced by weighting through *signed replication*:

Recursive Weighted Median Filters. Given a set of N real-valued feedback coefficients $A_i|_{i=1}^{N}$ and a set of $M+1$ real-valued feed-forward coefficients $B_i|_{i=0}^{M}$, the noncausal RWM filter output is defined as [Arc00]

$$y(n) = \text{MEDIAN}\,[|A_N| \blacklozenge \operatorname{sgn}(A_N)y(n-N), \ldots, |A_1| \blacklozenge \operatorname{sgn}(A_1)y(n-1), |B_0| \blacklozenge \operatorname{sgn}(B_0)x(n), \ldots, |B_M| \blacklozenge \operatorname{sgn}(B_M)x(n+M)]. \tag{2.43}$$

For notational convenience, recursive WM filters are denoted with double angle brackets and with the first feed-forward coefficient underlined; the RWM filter in Eq. (2.43), for example, is denoted by $\langle\langle A_N, \ldots, A_1, \underline{B_0}, B_1, \ldots, B_M \rangle\rangle$.

The recursive WM filter output for non-integer weights can be determined as follows:

1. Calculate the threshold $T_0 = 1/2\left(\sum_{\ell=1}^{N} |A_\ell| + \sum_{k=0}^{M} |B_k|\right)$.
2. Jointly sort the "signed" past output samples $\operatorname{sgn}(A_\ell)y(n-\ell)$ and the "signed" input observations $\operatorname{sgn}(B_k)x(n+k)$.
3. Sum the magnitudes of the weights corresponding to the sorted "signed" samples, beginning with the maximum and continuing down in order.
4. Choose the output as the average of the signed sample whose weight magnitude causes the sum to become greater than or equal to T_0 and the next smaller signed sample. For "selection" type filtering, choose the output as the signed sample whose weight magnitude causes the sum to become greater than or equal to T_0.

The following example illustrates this procedure. Consider the window size 6 RWM filter defined by the real-valued weights $\langle A_2, A_1, \underline{B_0}, B_1, B_2, B_3 \rangle = \langle 0.2, 0.4, \underline{0.6}, -0.4, 0.2, 0.2 \rangle$. The output for this filter operating on the observation set $[y(n-2), y(n-1), x(n), x(n+1), x(n+2), x(n+3)]^T = [-2, 2, -1, 3, 6, 8]^T$ is found as follows: Summing the absolute weights gives the threshold $T_0 = (1/2)(|A_1| + |A_2| + |B_0| + |B_1| + |B_2| + |B_3|) = 1$. The "signed" set of samples spanned by the filter's window, the sorted set, their corresponding weights, and the partial sum of weights (from each ordered sample to the maximum) are:

sample set in the window	-2,	2,	-1,	3,	6,	8
corresponding weights	0.2,	0.4,	$\underline{0.6}$,	-0.4,	0.2,	0.2
sorted signed samples	-3,	-2,	-1,	2,	6,	8
corresponding absolute weights	0.4,	0.2,	0.6,	0.4,	0.2	0.2
partial weight sums	2.0,	1.6,	$\underline{1.4}$,	0.8,	0.4	0.2

Thus, the output is $(-1-2)/2 = -1.5$ since when starting summing from the right (maximum sample), the threshold $T_0 = 1$ is not reached until the weight associated with -1 is added. The underlined sum value indicates that this is the first sum that meets or exceeds the threshold.

Stability of Recursive WM Filters. One of the main problems in the design of linear IIR filters is the stability under the bounded-input bounded-output (BIBO) criterion, which establishes certain constraints on the feedback filter coefficient values. To guarantee the BIBO stability of a linear IIR filter, the poles of its transfer function must lie within the unit circle in the complex plane [Mit00]. Unlike linear IIR filters, it has been shown that recursive WM filters are always stable [Par99].

> *Recursive weighted median filters, as defined in Eq. (2.43), are stable under the bounded-input bounded-output criterion, regardless of the values taken by the feedback coefficients* $\{A_\ell\}$ *for* $\ell = 1, 2, \ldots, N$.

Optimal Recursive Weighted Median Filtering

Under the MAE criterion the goal is to determine the weights $\{A_\ell\}|_{\ell=1}^{N}$ and $\{B_k\}|_{k=0}^{M}$ so as to minimize the cost function

$$J(A_1, \ldots, A_N, B_0, \ldots, B_M) = E\{|d(n) - y(n)|\}, \tag{2.44}$$

where $E\{\cdot\}$ denotes the statistical expectation, $y(n)$ is the output of the recursive WM filter given in Eq. (2.43), and $\{d(n)\}$ is a desired process.

To form an iterative optimization algorithm, the steepest descent algorithm is used, in which the filter coefficients are updated according to

$$\begin{aligned} A_\ell(n+1) &= A_\ell(n) + 2\mu\left[-\frac{\partial}{\partial A_\ell}J(A_1, \ldots, A_N, B_0, \ldots, B_M)\right] \\ &\quad \ell = 1, \ldots, N; \\ B_k(n+1) &= B_k(n) + 2\mu\left[-\frac{\partial}{\partial B_k}J(A_1, \ldots, A_N, B_0, \ldots, B_M)\right] \\ &\quad k = 0, \ldots, M. \end{aligned} \tag{2.45}$$

The gradient (∇J) of the cost function in Eq. (2.45), has to be computed to update the filter weights. Due to the feedback operation inherent in the recursive WM filter, however, the computation of ∇J becomes intractable.

To overcome this problem, we use the optimization framework referred to as *equation error formulation* [Arc00], which is used in the design of linear IIR filters and is based on the fact that ideally the filter's output is close to the desired response [Shy89]. The lagged values of $y(n)$ in Eq. (2.43) can thus be replaced with the corresponding lagged values $d(n)$. Hence, the previous outputs $y(n-\ell)|_{\ell=1}^{N}$ are replaced with the previous desired outputs $d(n-\ell)|_{\ell=1}^{N}$ to obtain a two-input, single-output filter that depends on the input samples $x(n+k)|_{k=0}^{M}$ and on delay samples of the desired response $d(n-l)|_{l=1}^{N}$, namely,

$$\begin{aligned} \hat{y}(n) = \text{MEDIAN}\,[&|A_N| \blacklozenge \operatorname{sgn}(A_N)d(n-N), \ldots, |A_1| \blacklozenge \operatorname{sgn}(A_1)d(n-1), \\ &|B_0| \blacklozenge \operatorname{sgn}(B_0)x(n), \ldots, |B_M| \blacklozenge \operatorname{sgn}(B_M)x(n+M)]. \end{aligned} \tag{2.46}$$

The approximation leads to an output $\hat{y}(n)$ that does not depend on delayed output samples; therefore, the filter no longer introduces feedback, reducing the output to a two-input, single-output nonrecursive system. This "recursive decoupling" optimization approach provides the key to a gradient-based optimization algorithm for recursive WM filters.

According to the approximate filtering structure, the cost function to be minimized is

$$\hat{J}(A_1, \ldots, A_N, B_0, \ldots, B_M) = E\{|d(n) - \hat{y}(n)|\}, \tag{2.47}$$

where $\hat{y}(n)$ is the nonrecursive filter output given by Eq. (2.46). Since $d(n)$ and $x(n)$ are not functions of the feedback coefficients, the derivative of $\hat{J}(A_1, \ldots, A_N, B_0, \ldots, B_M)$ with respect to the filter weights is nonrecursive and its computation is straightforward. The adaptive optimization algorithm using the steepest descent method of Eq. (2.45) but with $J(\cdot)$ replaced by $\hat{J}(\cdot)$ is derived following steps similar to those used in the derivation of the adaptive algorithms for nonrecursive WM filters. The recursion for the fast LMA RWM filter algorithm is found as

$$\begin{aligned} A_\ell(n+1) &= A_\ell(n) + \mu\,[d(n) - \hat{y}(n)]\,\mathrm{sgn}\,(A_\ell(n))\,\mathrm{sgn}\left(s_{d_\ell} - \hat{y}(n)\right), \\ B_k(n+1) &= B_k(n) + \mu\,[d(n) - \hat{y}(n)]\,\mathrm{sgn}\,(B_k(n))\,\mathrm{sgn}\,(s_{x_k} - \hat{y}(n)), \end{aligned} \tag{2.48}$$

for $\ell = 1, 2, \ldots, N$ and $k = 0, 1, \ldots, M$ and where $s_{x_k} = \mathrm{sgn}(B_k)x(n+k)$ and $s_{d_\ell} = \mathrm{sgn}(A_\ell)[d(n-\ell)]$.

Due to the nonlinear nature of the adaptive algorithm, a convergence analysis cannot be derived; thus, exact bounds on the step size μ are not available. We have observed in practice that a reliable guideline for this algorithm is to select a step size of the same order as that required for the standard LMS algorithm.

2.3 Image Denoising

2.3.1 Denoising with CWM Smoothers

Median smoothers are widely used in image processing to restore images corrupted by impulse noise. Median filters are particularly effective at removing outliers. Although a weighted median smoother can be designed to "best" remove the noise, CWM smoothers often provide similar results at a much lower complexity [Ko91]. By simply selecting the center weight a user can obtain the desired level of smoothing. Of course, as the center weight is decreased to attain the desired level of impulse suppression, the output image suffers increased distortion, particularly around fine details. Nonetheless, CWM smoothers can be highly effective in removing salt-and-pepper noise while preserving the fine image details.

Figures 2.6a and 2.6b depict a noise-free image and the corresponding image with salt-and-pepper noise. Each pixel in the image has a 10% probability of being contaminated with an impulse. The impulses occur randomly and have been generated by MATLAB's `imnoise` funtion. Figures 2.6c and 2.6d depict the noisy image processed with a 5×5 window CWM smoother with center weights 15 and

5, respectively. The tradeoff between impulse rejection and detail preservation in CWM smoothing is illustrated in these figures.

At the extreme, for $W_c = 1$, the CWM smoother reduces to the median smoother, which is effective at removing impulse noise while preserving edges; however, it is unable to preserve the image's fine details. Figure 2.7 shows zoomed sections of the noise-free image (left) and of the noisy image after the median smoother has been applied (center); as can be seen, severe blurring is introduced by the median smoother. As a reference, the output of a running mean of the same size is shown in the right-hand image; it is severely degraded since each impulse is smeared to neighboring pixels by the averaging operation.

Figures 2.6 and 2.7 show that CWM smoothers can be effective at removing impulse noise. If increased detail preservation is sought and the center weight is increased, CWM smoothers begin to break down and impulses appear on the output. One simple way to ameliorate this limitation is to employ a recursive mode of operation. In essence, past inputs are replaced by previous outputs, with the difference being that only the center sample is weighted; all the other samples in the window are weighted by one. Figures 2.8a and 2.8b show zoomed sections of the nonrecursive CWM filter output (left) and the corresponding recursive CWM smoother output (center), both with the same center weight ($W_c = 15$). These figures illustrate the increased noise attenuation provided by recursive filtering without the loss of image resolution.

Both recursive and nonrecursive CWM smoothers can produce outputs with disturbing artifacts, particularly when the center weights are increased to improve the detail preservation characteristics of the smoothers. The artifacts are most apparent around the image's edges and details; edges at the output appear jagged, and impulse noise can break through next to the image detail features. The distinct responses of the CWM smoother in different regions of the image are due to the fact that images are nonstationary in nature; abrupt changes in the image's local mean and texture carry most of the visual information content. CWM smoothers process the entire image with fixed weights and are inherently limited in this sense by their static nature. Although some improvement is attained by introducing recursion or by using more weights in a properly design WM smoother structure, these approaches are also static and do not properly address the nonstationary nature of images.

2.3.2 Denoising with Permutation CWM Smoothers

We now apply the simple permutation WM filter of Eq. (2.19) to impulse noise removal from images. The permutation WM smoother we utilize is defined by the smoothing operation

$$W_c(n) = \begin{cases} N_0 & \text{if } T_L \leq R_c(n) \leq T_U, \\ 1 & \text{otherwise,} \end{cases} \tag{2.49}$$

where $R_c(n)$ denotes the rank of the center sample of the window at location n, N_0 is a parameter to be adjusted, and $1 \leq T_L \leq T_U \leq N$ are two adjustable threshold

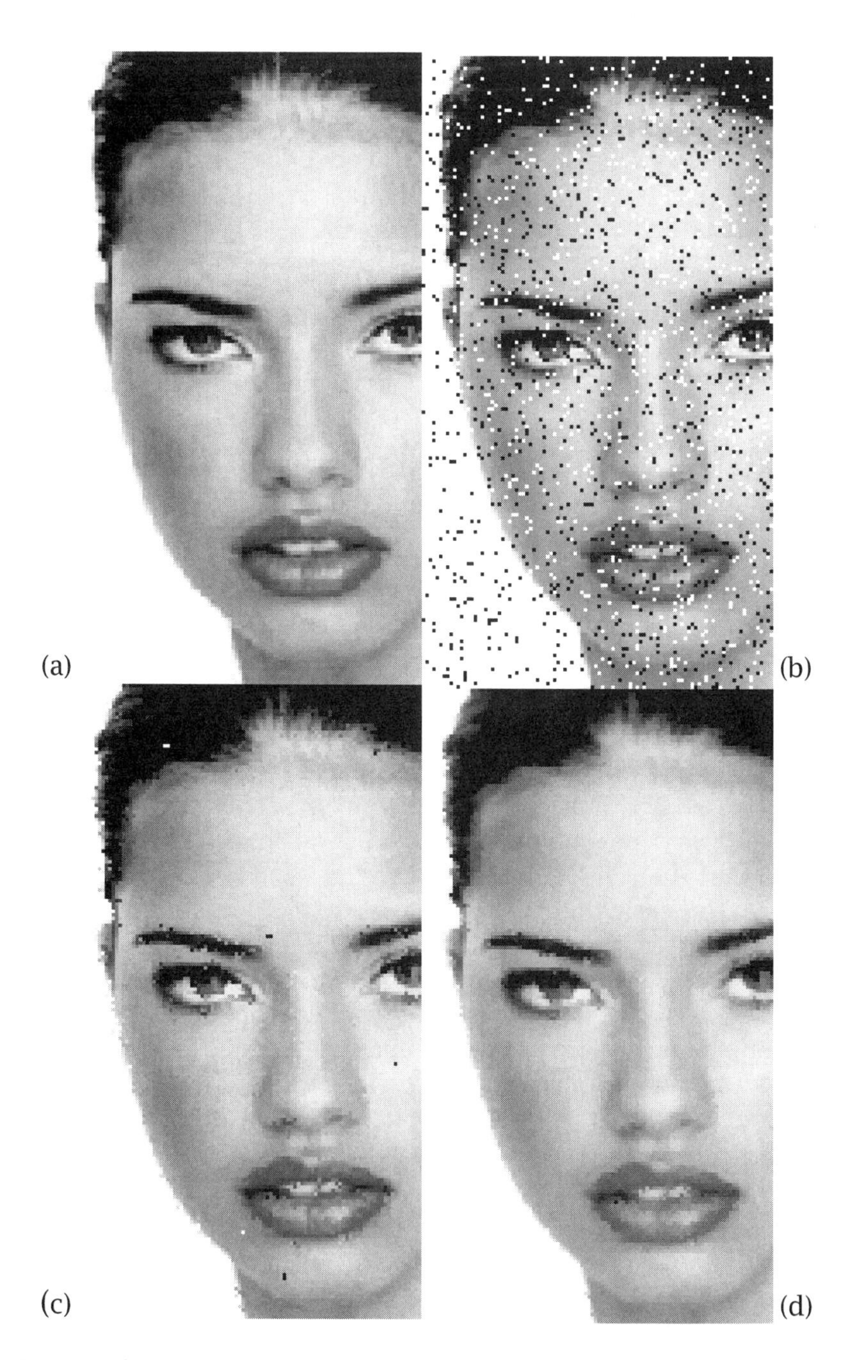

Figure 2.6: Impulse noise cleaning with a 5×5 CWM smoother: (a) original "portrait" image, (b) image with salt-and-pepper noise, (c) CWM smoother with $W_c = 15$, (d) CWM smoother with $W_c = 5$.

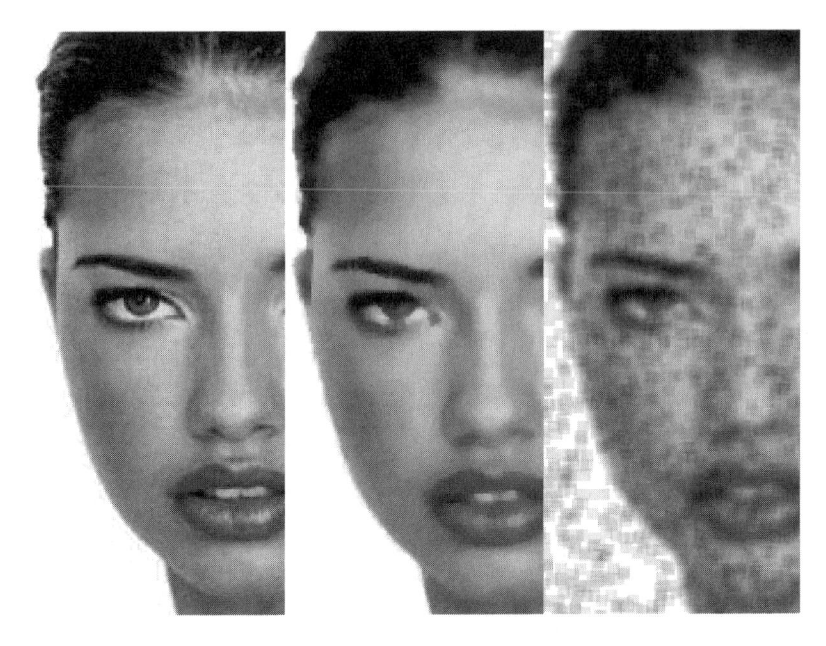

Figure 2.7: Noise-free image (left), 5×5 median smoother output (center), and 5×5 mean smoother output (right).

parameters that determine the degree of smoothing. Note that the weight in Eq. (2.49) is data adaptive and may change between two values with n. The smaller (larger) the threshold parameter T_L (T_U) is set to, the better the detail preservation.

Figure 2.8 depicts the outputs of the 5×5 CWM smoother, the recursive CWM smoother, and the permutation CWM smoother. The image on the right shows the output of the permutation CWM filter of Eq. (2.19) when the salt-and-pepper degraded "portrait" image is given as an input. The parameters used are the values $N_0 = 15$, $T_L = 6$, and $T_U = 20$. The improvement achieved by switching W_c between just two different values is significant; the impulses are deleted without exception, the details are preserved, and the jagged artifacts typical of CWM smoothers are not present in the output.

2.3.3 Optimal WM Filter Denoising

Next, consider the WM filter image denoising when the weight parameters are optimized through the use of the LMA algorithm. We use 3×3 windows in all examples. A recursive center WM filter and a nonrecursive center WM filter with the same set of weights are included [Ko91]. Figures 2.9d and 2.9c show their respective filter outputs with a center weight $W_c = 5$. The recursive WM filter is more effective than its nonrecursive counterpart.

A small 60×60 pixel area in the upper left part of the original and noisy images were used to train the recursive WM filter using the LMA algorithm and to train a nonrecursive WM filter. The initial conditions for the weights for both algorithms

Figure 2.8: CWM smoother output (left), recursive CWM smoother output (center), and permutation CWM smoother output (right). Window size is 5×5.

were the filter coefficients of the center WM filters described above, and the step size was 10^{-3}. The optimal weights found by the adaptive algorithms are

$$\left\langle \begin{array}{ccc} 1.38 & 1.64 & 1.32 \\ 1.50 & \underline{5.87} & 2.17 \\ 0.63 & 1.36 & 2.24 \end{array} \right\rangle \quad \text{and} \quad \left\langle \begin{array}{ccc} 1.24 & 1.52 & 2.34 \\ 2.07 & \underline{4.89} & 1.45 \\ 1.95 & 0.78 & 2.46 \end{array} \right\rangle$$

for the nonrecursive and recursive WM filters, respectively, where the underlined weight is associated with the center sample of the 3×3 window. The optimal filters determined by the training algorithms were used to filter the entire image. Figures 2.9f and 2.9e show the outputs of the optimal RWM filter and the optimal nonrecursive WM filter, respectively. The normalized mean square and mean absolute errors produced by each of the filters are listed in Table 2.1. As can be seen by a visual comparison of the various images and by the error values, recursive WM filters outperform nonrecursive WM filters. Before we end this section, it should be mentioned that there exist other nonlinear methods to suppress impulse noise; in particular, in Chapter 4, the signal-dependent rank ordered mean (SD-ROM) filters are described and their performances are compared with those of the median smoothers.

2.4 Image Zooming

Zooming is used in many imaging applications. It is implemented by inserting zero-valued pixels into the image to expand its size and interpolating the new

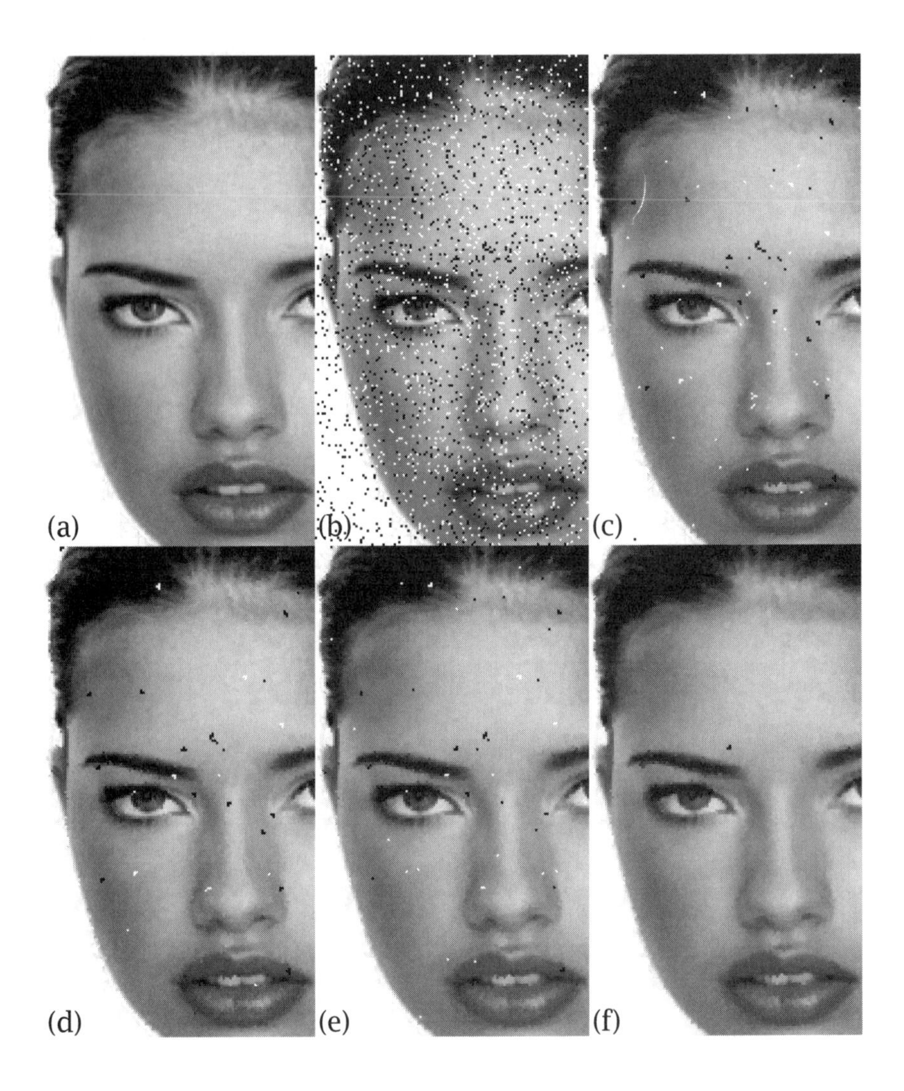

Figure 2.9: Image denoising using 3×3 recursive and nonrecursive WM filters: (a) original, (b) image with salt-and-pepper noise, (c) nonrecursive center WM filter output, (d) recursive center WM filter output, (e) optimal nonrecursive WM filter output, and (f) optimal RWM filter output.

pixels from the surrounding original pixels. Consider the zooming of an image by a factor of powers of 2. General zooming with non-integer factors is also possible with simple modifications of the method described next.

To double the size of an image, first an empty array is constructed with twice the number of rows and columns as the original and the original pixels are placed in alternating rows and columns (the "00" pixels in Fig. 2.10a). To interpolate the remaining pixels, the method known as polyphase interpolation is used. In the method, each new pixel that has original pixels at its four corners (the "11" pixels

Table 2.1: Results for Impulse Noise Removal

Image	Normalized mean square error	Normalized mean absolute error
Noisy image	2545.20	12.98
Recursive center WM filter	189.44	1.69
Nonrecursive center WM filter	243.83	1.92
Optimal nonrecursive WM filter	156.30	1.66
Optimal RWM filter	88.13	1.57

00	01	00	01	00	01
10	11	10	11	10	11
00	01	00	01	00	01
10	11	10	11	10	11
00	01	00	01	00	01
10	11	10	11	10	11

(a)

•	01	•	01	•	01
10	11	10	11	10	11
•	01	•	01	•	01
10	11	10	11	10	11
•	01	•	01	•	01
10	11	10	11	10	11

(b)

•	01	•	01	•	01
10	■	10	■	10	■
•	01	•	01	•	01
10	■	10	■	10	■
•	01	•	01	•	01
10	■	10	■	10	■

(c)

•	□	•	□	•	□
◊	■	◊	■	◊	■
•	□	•	□	•	□
◊	■	◊	■	◊	■
•	□	•	□	•	□
◊	■	◊	■	◊	■

(d)

Figure 2.10: The steps of polyphase interpolation.

in Fig. 2.10b) is interpolated first by using the weighted median of the four nearest original pixels as the value for that pixel. Since all original pixels are equally trustworthy and the same distance from the pixel being interpolated, a weight of 1 is used for the four nearest original pixels, the resulting array is shown in Fig. 2.10c. The remaining pixels are determined by taking a weighted median of the four closest pixels. Thus, each of the "01" pixels in Fig. 2.10c is interpolated using the two original pixels to the left and right and the two previously interpolated pixels above and below. Similarly, the "10" pixels are interpolated with original pixels above and below and interpolated pixels ("11" pixels) to the right and left.

Since the "11" pixels were interpolated, they are less reliable than the original pixels and should be given lower weights in determining the "01" and "10" pixels. Therefore, the "11" pixels are given weights of 0.5 in the median, while the "00" original pixels have weights of 1 associated with them. The weight 0.5 is used because it implies that when both "11" pixels have values that are not between the two "00" pixel values, one of the "00" pixels or their average will be used. Thus, "11" pixels differing from the "00" pixels do not greatly affect the result of the weighted median; only when the "11" pixels lie between the two "00" pixels will they have a direct effect on the interpolation. The choice of 0.5 for the weight is arbitrary since any weight greater than 0 and less than 1 will produce the same result. When implementing the polyphase method, the "01" and "10" pixels must be treated differently due to the fact that the orientation of the two closest original pixels is different for the two types of pixels. Figure 2.10d shows the final result of doubling the size of the original array.

To illustrate the process, consider an expansion of the grayscale image represented by an array of pixels, the pixel in the ith row and jth column having brightness $a_{i,j}$. The array $a_{i,j}$ is interpolated into the array $x_{i,j}^{pq}$, with p and q taking values 0 or 1, indicating in the same way as above the type of interpolation required:

$$\begin{bmatrix} a_{1,1} & a_{1,2} & a_{1,3} \\ a_{2,1} & a_{2,2} & a_{2,3} \\ a_{3,1} & a_{3,2} & a_{3,3} \end{bmatrix} \Rightarrow \begin{bmatrix} x_{1,1}^{00} & x_{1,1}^{01} & x_{1,2}^{00} & x_{1,2}^{01} & x_{1,3}^{00} & x_{1,3}^{01} \\ x_{1,1}^{10} & x_{1,1}^{11} & x_{1,2}^{10} & x_{1,2}^{11} & x_{1,3}^{10} & x_{1,3}^{11} \\ x_{2,1}^{00} & x_{2,1}^{01} & x_{2,2}^{00} & x_{2,2}^{01} & x_{2,3}^{00} & x_{2,3}^{01} \\ x_{2,1}^{10} & x_{2,1}^{11} & x_{2,2}^{10} & x_{2,2}^{11} & x_{2,3}^{10} & x_{2,3}^{11} \\ x_{3,1}^{00} & x_{3,1}^{01} & x_{3,2}^{00} & x_{3,2}^{01} & x_{3,3}^{00} & x_{3,3}^{01} \\ x_{3,1}^{10} & x_{3,1}^{11} & x_{3,2}^{10} & x_{3,2}^{11} & x_{3,3}^{10} & x_{3,3}^{11} \end{bmatrix}.$$

The pixels are interpolated as follows:

$$\begin{aligned} x_{i,j}^{00} &= a_{i,j}, \\ x_{i,j}^{11} &= \text{MEDIAN}[a_{i,j},\ a_{i+1,j},\ a_{i,j+1},\ a_{i+1,j+1}], \\ x_{i,j}^{01} &= \text{MEDIAN}[a_{i,j},\ a_{i,j+1},\ 0.5\blacklozenge x_{i-1,j}^{11},\ 0.5\blacklozenge x_{i+1,j}^{11}], \\ x_{i,j}^{10} &= \text{MEDIAN}[a_{i,j}, a_{i+1,j},\ 0.5\blacklozenge x_{i,j-1}^{11},\ 0.5\blacklozenge x_{i,j+1}^{11}]. \end{aligned}$$

An example comparing median interpolation with bilinear interpolation is given in Fig. 2.11. The zooming factor of 4 was obtained by two consecutive interpolations, each doubling the size of the input. Bilinear interpolation uses the average of the nearest two original pixels to interpolate the "01" and "10" pixels in Fig. 2.10b and the average of the nearest four original pixels for the "11" pixels. The edge preserving advantage of the weighted median interpolation is readily seen in this figure. Further work in nonlinear interpolation can be found elsewhere [Bin92].

In closing, it should be pointed out that WM interpolation has been applied to video de-interlacing and image compression [Que95, Yin96].

2.5 Image Sharpening

In principle, image sharpening consists of adding to the original image a signal that is proportional to a high-pass-filtered version of the original image. Figure 2.12 illustrates this procedure, often referred to as unsharp masking on a

Figure 2.11: Zooming by 4: (top) original image with area of interest outlined, (bottom left) bilinear interpolation of the area, and (bottom right) weighted median interpolation.

one-dimensional signal [Jai89]. Other methods of image sharpening are described in Chapters 6 and 7 where polynomial filters are used.

As shown in Fig. 2.12, the original image is first filtered by a high-pass filter, which extracts the high-frequency components, and then a scaled version of the high-pass-filter output is added to the original image, thus producing a sharpened version of the original image. Note that the homogeneous regions of the signal, that is, those where the signal is constant, remain unchanged. The sharpening operation can be represented by

$$y(m,n) = x(m,n) + \lambda \mathcal{F}\left(x(m,n)\right), \tag{2.50}$$

where $x(m,n)$ is the original pixel value at the coordinate (m,n), $\mathcal{F}(\cdot)$ is the output of the high-pass filter, λ is a tuning parameter greater than or equal to zero, and $y(m,n)$ is the sharpened pixel at the coordinate (m,n). The value taken by λ depends on the grade of sharpness desired, increasing λ yields a more sharpened image, but if background noise is present, it will rapidly amplify the noise.

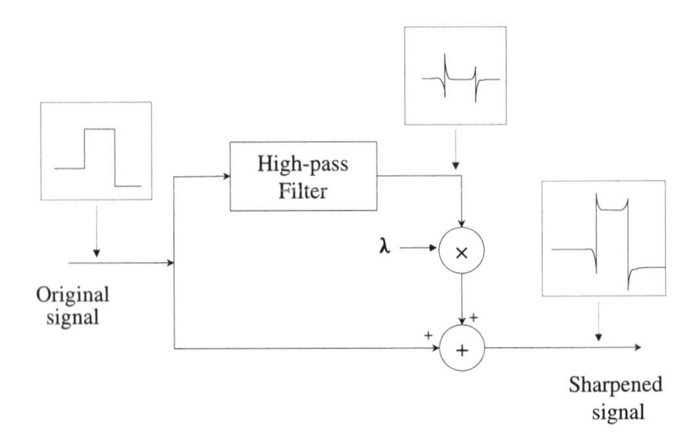

Figure 2.12: Image sharpening by high-frequency emphasis.

2.5.1 Sharpening with WM Filters

The key point in the effective sharpening process lies in the choice of the high-pass-filtering operation. Traditionally, linear filters have been used to implement the high-pass filter; however, linear techniques can lead to rapid performance degradation should the input image be corrupted with noise. A tradeoff between noise attenuation and edge highlighting can be obtained if a weighted median filter with appropriate weights is used. To illustrate this, we consider a WM filter applied to a gray-scale image where the following filter mask is used:

$$W = \left\langle \begin{array}{ccc} -1 & -1 & -1 \\ -1 & 8 & -1 \\ -1 & -1 & -1 \end{array} \right\rangle . \tag{2.51}$$

Due to the weight coefficients in Eq. (2.51), for each position of the moving window the output is proportional to the difference between the center pixel and the smallest pixel around it. Thus, the filter output takes relatively large values for prominent edges in an image, but small values in regions that are fairly smooth, being zero only in regions that have a constant gray level.

Although this filter can effectively extract the edges contained in an image, the effect that this filtering operation has on negative-slope edges is different from that obtained for positive-slope edges.[2] Since the filter output is proportional to the difference between the center pixel and the smallest pixel around the center, for negative-slope edges the center pixel takes small values, producing small values at the filter output. Moreover, the filter output is zero if the center pixel and smallest pixel around it have the same values. This implies that negative-slope edges are not extracted in the same way as positive-slope edges. To overcome this limitation we must modify the basic image sharpening structure shown in Fig. 2.12 such that

[2] A change from one gray level to a lower gray level is referred to as a negative-slope edge, whereas a change to a higher gray level is referred to as a positive-slope edge.

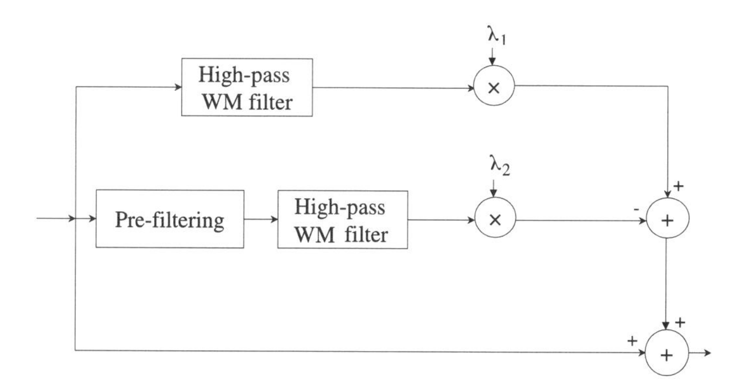

Figure 2.13: Image sharpening based on the weighted median filter.

positive-slope edges as well as negative-slope edges are highlighted in the same proportion. A simple way to accomplish that is as follows:

1. Extract the positive-slope edges by filtering the original image with the filter mask described above.
2. Extract the negative-slope edges by first preprocessing the original image such that the negative-slope edges become positive slopes, and then filter the preprocessed image with the filter described above.
3. Combine appropriately the original image and the filtered versions of the original image and the preprocessed image to form the sharpened image.

Thus, both positive- and negative-slope edges are equally highlighted. This procedure is illustrated in Fig. 2.13, in which the top branch extracts the positive-slope edges and the middle branch extracts the negative ones. To understand the effects of edge sharpening, we plot a row of a test image in Fig. 2.14, together with the row from the sharpened image when only the positive-slope (2.14a), and negative-slope (2.14b), edges are highlighted and when both are jointly highlighted (2.14c).

The λ_1 and λ_2 in Fig. 2.13, are tuning parameters that control the amount of sharpness desired in the positive- and negative-slope directions, respectively. Their values are generally selected to be equal. The output of the prefiltering operation is defined as

$$x(m,n)' = M - x(m,n), \tag{2.52}$$

with M equal to the maximum pixel value of the original image. This prefiltering operation can be thought of as a flipping and a shifting operation of the values of the original image such that the negative-slope edges are converted to positive-slope edges. Since the original and the prefiltered images are filtered by the same WM filter, the positive- and negative-slope edges are sharpened in the same way.

In Fig. 2.15, the performance of the WM filter image sharpening is compared with that of traditional image sharpening based on linear FIR filters. For the linear sharpener, the scheme shown in Fig. 2.12 was used and the parameter λ was set

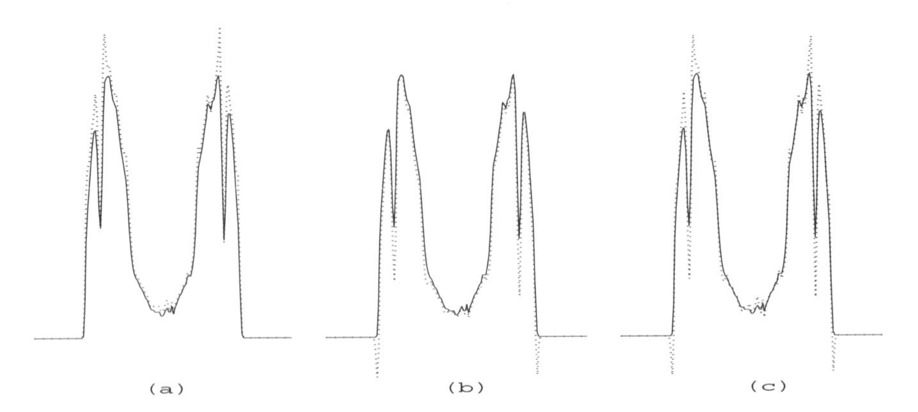

Figure 2.14: Original row of a test image (solid lines) and row sharpened (dotted lines) with (a) only positive-slope edges, (b) only negative-slope edges, and (c) both positive- and negative-slope edges.

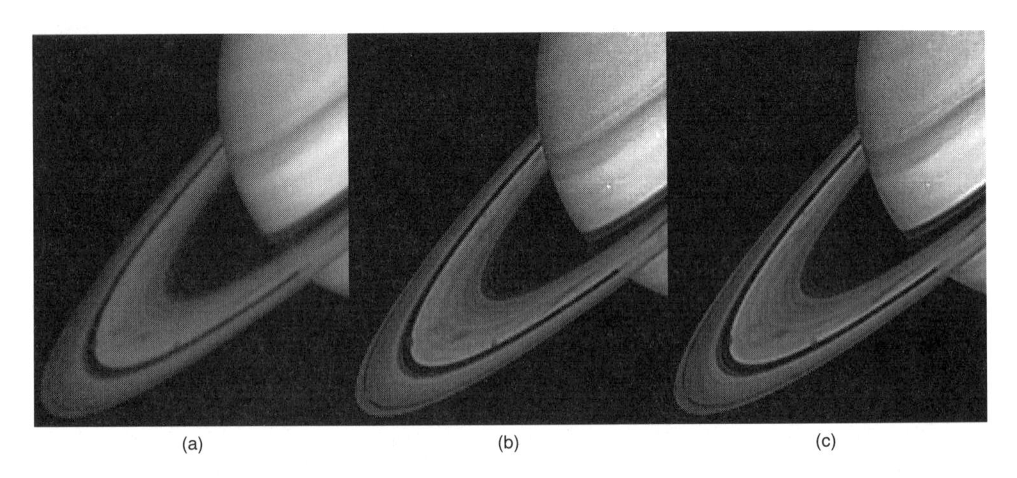

Figure 2.15: (a) Original image, sharpened with (b) the FIR-sharpener and (c) the WM sharpener.

to 1. For the WM sharpener, the scheme of Fig. 2.13 was used with $\lambda_1 = \lambda_2 = 2$. The filter mask given by Eq. (2.51) was used in median image sharpening, whereas the filter mask for the linear image sharpening is $1/3W$, where W is given by Eq. (2.51). Sharpening with WM filters does not introduce as much noise amplification as sharpeners equipped with FIR filters do.

2.5.2 Sharpening with Permutation WM filters

Linear high-pass filters are inadequate in unsharp masking whenever background noise is present. Although WM high-pass filters ameliorate the problem, the goal is to improve on their performance by allowing the WM filter weights to take on rank-dependent values.

The unsharp WM filter structure shown in Fig. 2.13 is used with the exception that permutation WM filters are now used to synthesize the high-pass-filter operation. The weight mask for the permutation WM high-pass filter is

$$W = \left\langle \begin{array}{ccc} W_{1(R_1,R_c)} & W_{2(R_2,R_c)} & W_{3(R_3,R_c)} \\ W_{4(R_4,R_c)} & \underline{W_{c(R_c)}} & W_{6(R_6,R_c)} \\ W_{7(R_7,R_c)} & W_{8(R_8,R_c)} & W_{9(R_9,R_c)} \end{array} \right\rangle, \tag{2.53}$$

where $W_{i(R_i,R_c)}$ depends only on the rank of the ith sample and the rank of the center sample. $W_{i(R_i,R_c)} = -1$, for $i = 1,\ldots,9$, $i \neq 5$, $R_c = 1,\ldots,9$, with the following exceptions: The center weight is given the value according to

$$W_{c(R_c)} = \begin{cases} 8 & \text{for } R_c = 2,3,\ldots,8, \\ -1 & \text{otherwise.} \end{cases} \tag{2.54}$$

That is, the value of the center weight is 8 if the center sample is not the smallest or largest in the observation window. If it happens to be the smallest or largest, its reliability is low, and the weighting strategy must be altered such that the center weight is set to -1 and the weight of 8 is given to the sample closest in rank to the center sample, leading to

$$\begin{aligned} W_{\ell_{(8)}(8,9)} &= \begin{cases} 8 & \text{for } x_c = x_{(9)}, \\ -1 & \text{otherwise;} \end{cases} \\ W_{\ell_{(2)}(2,1)} &= \begin{cases} 8 & \text{for } x_c = x_{(1)}, \\ -1 & \text{otherwise.} \end{cases} \end{aligned} \tag{2.55}$$

Here $\ell_{(i)}$ refers to the location of the ith smallest sample in the observation window and $W_{\ell_{(i)}}$ refers to its weight.

This weighting strategy can be extended to the case where the L smallest and L largest samples in the window are considered unreliable, and the weighting strategy applied in Eq. (2.55) now applies to the weights $W_{\ell_{(L+1)}(L+1,L)}$ and $W_{\ell_{(N-L)}(N-L,N-L+1)}$.

Figure 2.16 illustrates the image sharpening performance when permutation WM filters are used. A Saturn image with added Gaussian background noise is shown in Fig. 2.16a. The other images show this image sharpened with (b) a Lower-Upper-Middle (LUM) sharpener [Har93], (c) a linear FIR filter sharpener, (d) the WM filter sharpener, and the permutation WM filter sharpener with (e) $L = 1$ and (f) $L = 2$. The λ parameters were given a value of 1.5 for all weighted median type sharpeners a value of 1 for the linear sharpener. The linear sharpener introduces background noise amplification. The LUM sharpener does not amplify the background noise; however, it introduces severe edge distortion artifacts. The WM filter sharpener ameliorates the noise amplification and does not introduce edge artifacts. The permutation WM filter sharpeners perform best, with higher robustness attributes as L increases.

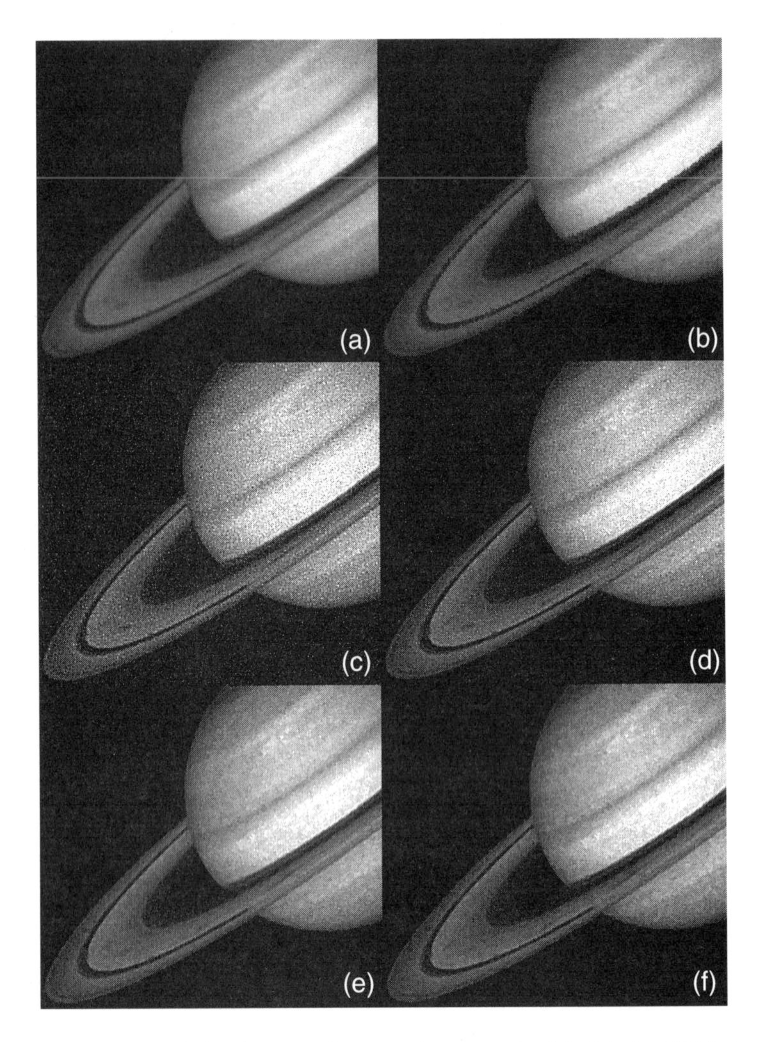

Figure 2.16: (a) Image with background noise sharpened with (b) the LUM sharpener, (c) the FIR sharpener, (d) the WM sharpener, and the permutation WM sharpener with (e) $L = 1$ and (f) $L = 2$.

2.6 Optimal Frequency Selection WM Filtering

We now consider the design of a robust band-pass recursive WM filter using the LMA adaptive optimization algorithm. The performance of the optimal recursive WM filter is compared with the performances of a linear FIR filter, a linear IIR filter, and a nonrecursive WM filter all designed for the same task. Moreover, to show the noise attenuation capability of the recursive WM filter and compare it with those of the other filters, we used an impulse-noise-corrupted test signal. Examples are shown in one-dimensional signals for illustration purposes but the extension to two-dimensional signals is straightforward.

The application at hand is the design of a 62-tap band-pass RWM filter with passband $0.075 \leq \omega \leq 0.125$ (normalized Nyquist frequency = 1). We used white Gaussian noise with zero mean and variance equal to one as input training signals. The desired signal was provided by the output of a large FIR filter (122-tap linear FIR filter) designed by MATLAB's M-file fir1 function. The 31 feedback filter coefficients were initialized to small random numbers (on the order of 10^{-3}). The feed-forward filter coefficients were initialized to the values output by MATLAB's fir1 with 31 taps and the same passband of interest. A variable step size $\mu(n)$ was used in both adaptive optimizations, where the step size $\mu(n)$ changes according to $\mu_0 e^{-n/100}$ with $\mu_0 = 10^{-2}$.

A signal that spanned the range of frequencies of interest was used as a test signal. Figure 2.17a depicts a linear swept-frequency signal spanning instantaneous frequencies from 0 to 400 Hz, with a sampling rate of 2 kHz. Figure 2.17b shows the chirp signal filtered by the 122-tap linear FIR filter used to produce the desired signal during the training stage. Figure 2.17c shows the output of a 62-tap linear FIR filter used for comparison purposes.

The adaptive optimization algorithm described in Section 2.2 was used to optimize a 62-tap nonrecursive WM filter admitting negative weights; the filtered signal attained is shown in Fig. 2.17d. Note that the nonrecursive WM filter tracks the frequencies of interest but fails to attenuate completely the frequencies out of the desired passband. MATLAB's yulewalk function was used to design a 62-tap linear IIR filter with passband $0.075 \leq \omega \leq 0.125$; Fig. 2.17e depicts its output. Finally, Fig. 2.17f shows the output of the optimal recursive WM filter determined by the LMA training algorithm described in Sec. 2.2.2. Note that the frequency components of the test signal that are not in the passband are attenuated completely. Moreover, the RWM filter generalizes very well on signals that were not used during the training stage.

Comparing the different filtered signals in Fig. 2.17, we see that the recursive filtering operation performs much better than its nonrecursive counterpart having the same number of coefficients. Likewise, to achieve a specified level of performance, a recursive WM filter generally requires considerably fewer filter coefficients than the corresponding nonrecursive WM filter.

To test the robustness of the different filters, we next contaminated the test signal with additive α-stable noise (Fig. 2.18a); the impulse noise was generated using the parameter α set to 1.4. (Fig. 2.18a is truncated so that the same scale is used in all plots.) Figures 2.18b and 2.18d show the filter outputs of the linear FIR and IIR filters, respectively; both outputs are severely affected by the noise. On the other hand, the nonrecursive and recursive WM filters' outputs, Fig. 2.18c and 2.18e, remain practically unaltered. Figure 2.18 clearly depicts the robust characteristics of median-based filters.

To better evaluate the frequency response of the various filters, we performed a frequency domain analysis. Due to the nonlinearity inherent in the median operation, traditional linear tools, such as transfer-function-based analysis, cannot be applied. However, if the nonlinear filters are treated as a single-input, single-

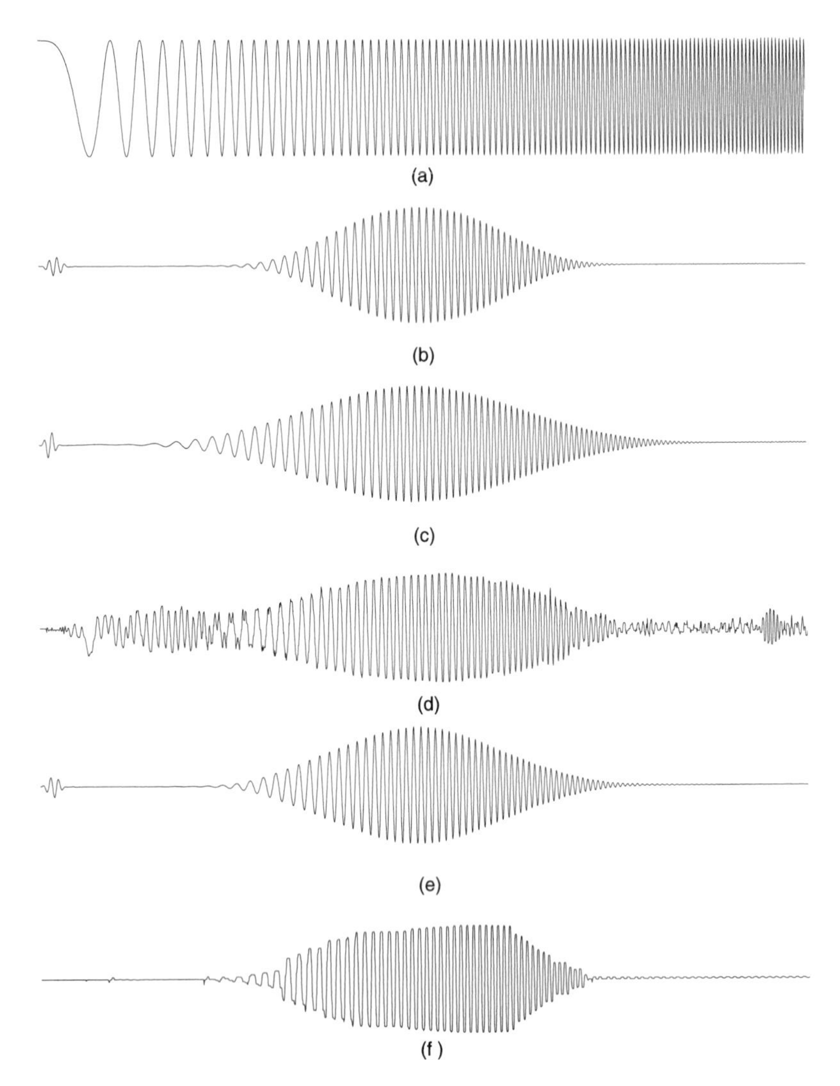

Figure 2.17: Band-pass filter design: (a) input test signal, (b) desired signal, (c) linear FIR filter output, (d) nonrecursive WM filter output, (e) linear IIR filter output, and (f) RWM filter output. (Reproduced with permission from [Arc00]. © 2000 IEEE.)

output system, the magnitude of the frequency response can be experimentally obtained as follows: A single-tone sinusoidal signal $\sin(2\pi f t)$ was given as the input to each filter, with f spanning the complete range of possible frequencies. A sufficiently large number of frequencies spanning the interval $[0, 1]$ was chosen. For each frequency value, the mean power of each filter's output was computed. Figure 2.19a shows a plot of the normalized mean power versus frequency attained by the different filters. Upon closer examination of Fig. 2.19a, it can be seen that the recursive WM filter yields the flattest response in the passband of interest. A similar conclusion can be drawn from the time domain plots shown in Fig. 2.17.

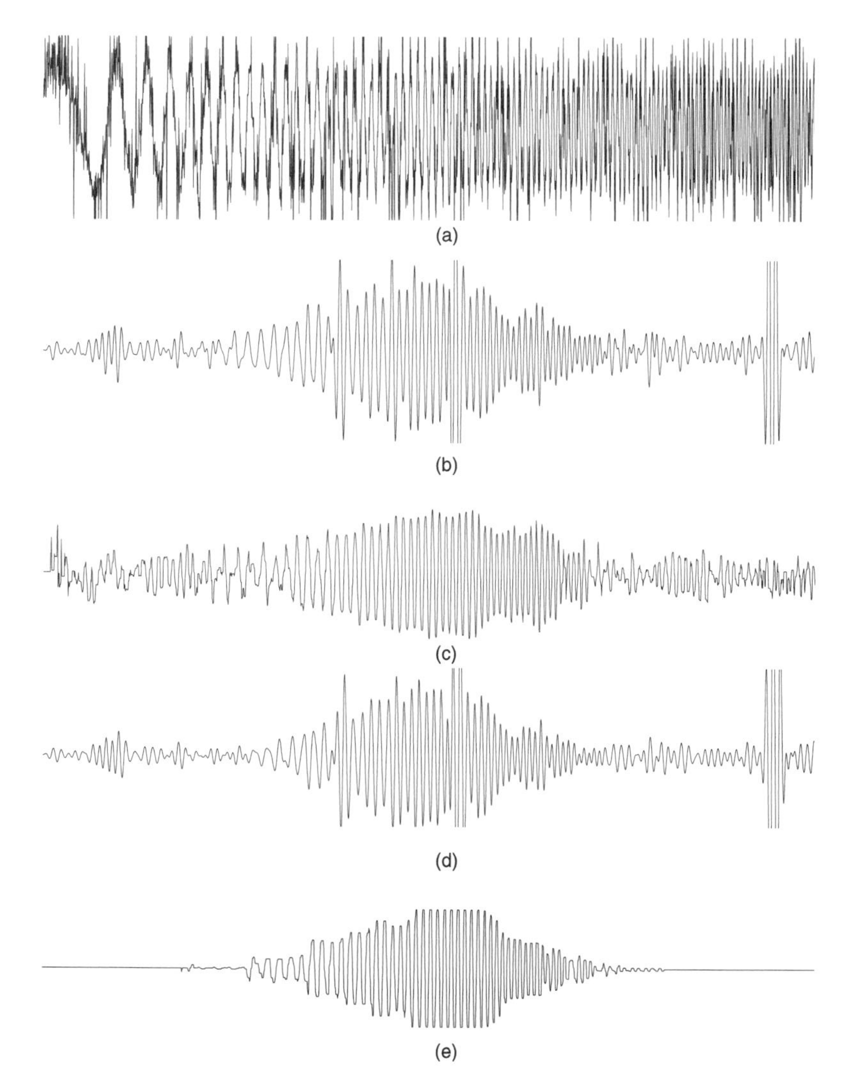

Figure 2.18: Performance of the band-pass filter in noise: (a) chirp test signal in stable noise, (b) linear FIR filter output, (c) nonrecursive WM filter output, (d) linear IIR filter output, and (e) RWM filter output. (Reproduced with permission from [Arc00]. © 2000 IEEE.)

To see the effects that impulse noise has over the magnitude of the frequency response, we input to each filter a contaminated sinusoidal signal $\sin(2\pi f t) + \eta$, where η is α-stable noise with parameter $\alpha = 1.4$. Following the same procedure described above, the mean power versus frequency diagram was obtained and is shown in Fig. 2.19b. As expected, the magnitudes of the frequency responses for the linear filters are highly distorted, whereas those for the median-based filters do not change significantly with noise.

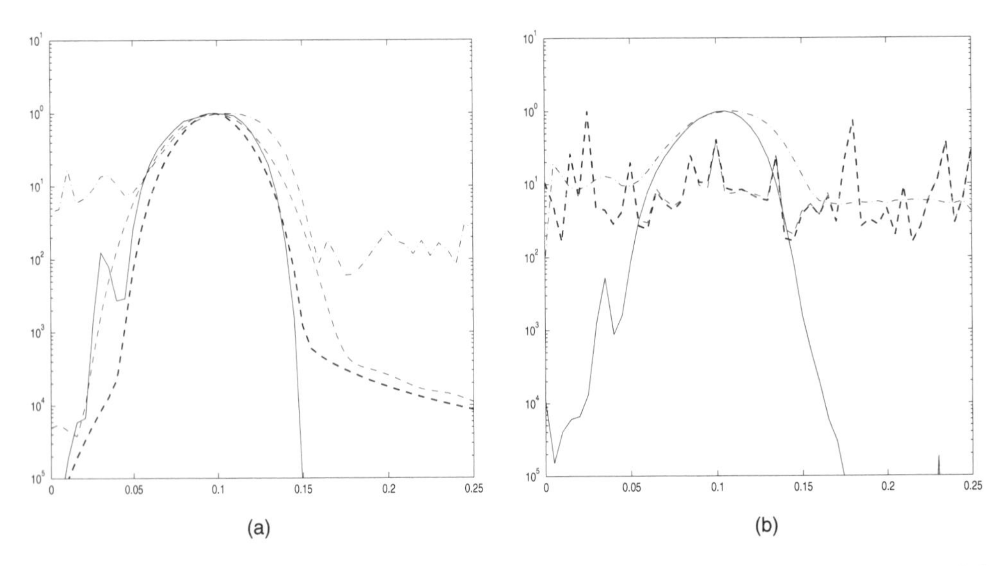

Figure 2.19: Frequency response to (a) a noiseless and (b) a noisy sinusoidal signal: (solid lines) RWM, (dotted-dashed lines) nonrecursive WM filter, (thin dashes) linear FIR filter, and (thick dashes) linear IIR filter. (Reproduced with permission from [Arc00]. © 2000 IEEE.)

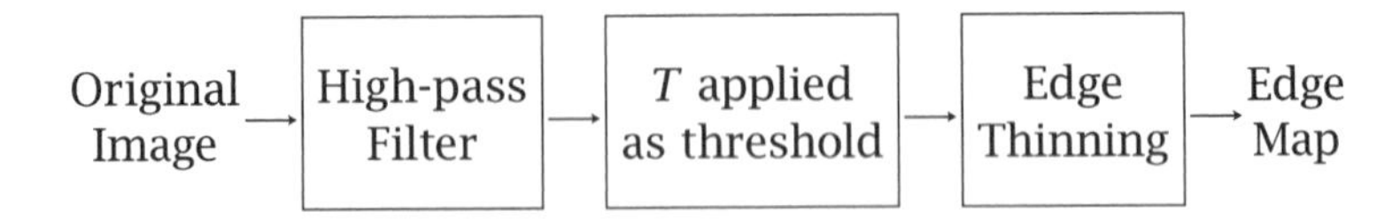

Figure 2.20: The process of edge detection

2.7 Edge Detection

Edge detection is an important tool in image analysis and is necessary for applications of computer vision in which objects need to be recognized by their outlines. An edge-detection algorithm should show the locations of major edges in the image while ignoring false edges caused by noise. The most common approach used for edge detection is illustrated in Fig. 2.20. A high-pass filter is applied to the image to obtain the amount of change present in the image at every pixel. The output of the filter is thresholded to determine those pixels that have a rate of change high enough to be considered as lying on an edge; that is, all pixels with filter output greater than some value T are taken as edge pixels. The value of T can be adjusted to give the best visual results. High thresholds lose some of the real edges, while low values may result in many false edges; thus, a tradeoff is needed to get the best results. Other techniques such as edge thinning are often applied to further pinpoint the location of the edges in an image.

The most common linear filter used for the initial high-pass filtering is the Sobel operator, which uses the following 3×3 masks:

$$\left\langle \begin{array}{ccc} -1 & -2 & -1 \\ 0 & 0 & 0 \\ 1 & 2 & 1 \end{array} \right\rangle \quad \text{and} \quad \left\langle \begin{array}{ccc} -1 & 0 & 1 \\ -2 & 0 & 2 \\ -1 & 0 & 1 \end{array} \right\rangle .$$

These two masks are convolved with the image separately to measure the strength of the horizontal and vertical edges, respectively, present at each pixel. Thus, if the amount to which a horizontal edge is present at the pixel in the ith row and jth column is represented as $E^{\mathrm{h}}_{i,j}$, and if the vertical edge indicator is $E^{\mathrm{v}}_{i,j}$, then the values are

$$\begin{aligned} E^{\mathrm{h}}_{i,j} &= -x_{i-1,j-1} - 2x_{i-1,j} - x_{i-1,j+1} + x_{i+1,j-1} + 2x_{i+1,j} + x_{i+1,j+1}, \\ E^{\mathrm{v}}_{i,j} &= -x_{i-1,j-1} - 2x_{i,j-1} - x_{i+1,j-1} + x_{i-1,j+1} + 2x_{i,j+1} + x_{i+1,j+1}, \end{aligned}$$

where $x_{i,j}$ is the pixel located at the ith row and jth column. The two strengths are combined to find the total amount to which any edge exists at a pixel: $E^{\mathrm{total}}_{i,j} = ((E^{\mathrm{h}}_{i,j})^2 + (E^{\mathrm{v}}_{i,j})^2)^{1/2}$. This value is then compared to the threshold T to determine the existence of an edge.

In place of using linear high-pass filters, WM filters with the weights from the Sobel masks can be used. The Sobel linear high-pass filters take a weighted difference between the pixels on either side of $x_{i,j}$. On the other hand, if the same weights are used in a weighted median filter, the value returned is the difference between the lowest-valued pixels on either side of $x_{i,j}$. If the pixel values are then flipped about some middle value, the difference between the *highest*-valued pixels on either side can also be obtained. The flipping can be achieved by finding some maximum pixel value M and using $x'_{i,j} = M - x_{i,j}$ as the "flipped" value of $x_{i,j}$, thus causing the highest values to become the lowest. The lower of the two differences across the pixel can then be used as the indicator of the presence of an edge. If a true edge is present, then both differences should be high in magnitude, while if noise causes one of the differences to be too high, the other difference is not necessarily affected. Thus, the horizontal and vertical edge indicators are

$$E^{\mathrm{h}}_{i,j} = \min \left(\begin{array}{l} \mathrm{MEDIAN} \left[\begin{array}{lll} -1\blacklozenge x_{i-1,j-1}, & -2\blacklozenge x_{i-1,j}, & -1\blacklozenge x_{i-1,j+1} \\ 1\blacklozenge x_{i+1,j-1}, & 2\blacklozenge x_{i+1,j}, & 1\blacklozenge x_{i+1,j+1} \end{array} \right], \\ \mathrm{MEDIAN} \left[\begin{array}{lll} -1\blacklozenge x'_{i-1,j-1}, & -2\blacklozenge x'_{i-1,j}, & -1\blacklozenge x'_{i-1,j+1}, \\ 1\blacklozenge x'_{i+1,j-1}, & 2\blacklozenge x'_{i+1,j}, & 1\blacklozenge x'_{i+1,j+1} \end{array} \right] \end{array} \right),$$

$$E^{\mathrm{v}}_{i,j} = \min \left(\begin{array}{l} \mathrm{MEDIAN} \left[\begin{array}{ll} -1\blacklozenge x_{i-1,j-1}, & 1\blacklozenge x_{i-1,j+1}, \\ -2\blacklozenge x_{i,j-1}, & 2\blacklozenge x_{i,j+1}, \\ -1\blacklozenge x_{i+1,j-1}, & 1\blacklozenge x_{i+1,j+1} \end{array} \right], \\ \mathrm{MEDIAN} \left[\begin{array}{ll} -1\blacklozenge x'_{i-1,j-1}, & 1\blacklozenge x'_{i-1,j+1}, \\ -2\blacklozenge x'_{i,j-1}, & 2\blacklozenge x'_{i,j+1}, \\ -1\blacklozenge x'_{i+1,j-1}, & 1\blacklozenge x'_{i+1,j+1} \end{array} \right] \end{array} \right),$$

and the strength of the horizontal and vertical edges $E^{\text{h,v}}_{(i,j)}$ is determined in the same way as in the linear case:

$$E^{\text{h,v}}_{i,j} = ((E^{\text{h}}_{i,j})^2 + (E^{\text{v}}_{i,j})^2)^{1/2}.$$

Horizontal and vertical indicators are not sufficient to register diagonal edges, so the following two masks must also be used as weights for the WM filter.

$$\left\langle \begin{array}{ccc} -2 & -1 & 0 \\ -1 & 0 & 1 \\ 0 & 1 & 2 \end{array} \right\rangle \quad \text{and} \quad \left\langle \begin{array}{ccc} 0 & 1 & 2 \\ -1 & 0 & 1 \\ -2 & -1 & 0 \end{array} \right\rangle.$$

Thus the strengths of the two types of diagonal edges in an image are $E^{\text{d1}}_{i,j}$ for those going from the bottom left to the top right (left mask) and $E^{\text{d2}}_{i,j}$ for those from top left to bottom right (right mask). The values are given by

$$E^{\text{d1}}_{i,j} = \min \left(\begin{array}{l} \text{MEDIAN} \left[\begin{array}{ll} -2\blacklozenge x_{i-1,j-1}, & -1\blacklozenge x_{i-1,j}, \\ -1\blacklozenge x_{i,j-1}, & 1\blacklozenge x_{i,j+1}, \\ 1\blacklozenge x_{i+1,j}, & 2\blacklozenge x_{i+1,j+1} \end{array} \right], \\ \text{MEDIAN} \left[\begin{array}{ll} -2\blacklozenge x'_{i-1,j-1}, & -1\blacklozenge x'_{i-1,j}, \\ -1\blacklozenge x'_{i,j-1}, & 1\blacklozenge x'_{i,j+1}, \\ 1\blacklozenge x'_{i+1,j}, & 2\blacklozenge x'_{i+1,j+1} \end{array} \right] \end{array} \right),$$

$$E^{\text{d2}}_{i,j} = \min \left(\begin{array}{l} \text{MEDIAN} \left[\begin{array}{ll} 1\blacklozenge x_{i-1,j}, & 2\blacklozenge x_{i-1,j+1}, \\ -1\blacklozenge x_{i,j-1}, & 1\blacklozenge x_{i,j+1}, \\ -2\blacklozenge x_{i+1,j-1}, & -1\blacklozenge x_{i+1,j} \end{array} \right], \\ \text{MEDIAN} \left[\begin{array}{ll} 1\blacklozenge x'_{i-1,j}, & 2\blacklozenge x'_{i-1,j+1}, \\ -1\blacklozenge x'_{i,j-1}, & 1\blacklozenge x'_{i,j+1}, \\ -2\blacklozenge x'_{i+1,j-1}, & -1\blacklozenge x'_{i+1,j} \end{array} \right] \end{array} \right).$$

A diagonal edge strength is determined in the same way as the horizontal and vertical edge strengths above:

$$E^{\text{d1,d2}}_{i,j} = ((E^{\text{d1}}_{i,j})^2 + (E^{\text{d2}}_{i,j})^2)^{1/2}.$$

The indicator of all edges in any direction is the maximum of the two strengths $E^{\text{h,v}}_{i,j}$ and $E^{\text{d1,d2}}_{i,j}$:

$$E^{\text{total}}_{i,j} = \max(E^{\text{h,v}}_{i,j}, E^{\text{d1,d2}}_{i,j}).$$

As in the linear case, this value is compared to the threshold T to determine whether a pixel lies on an edge. Figure 2.21 shows the results of calculating $E^{\text{total}}_{i,j}$ for an image. The results of the median edge detection are similar to the results of using the Sobel linear operator. Other approaches for edge detector based on the median filter can be found elsewhere [Bov86, Pit86].

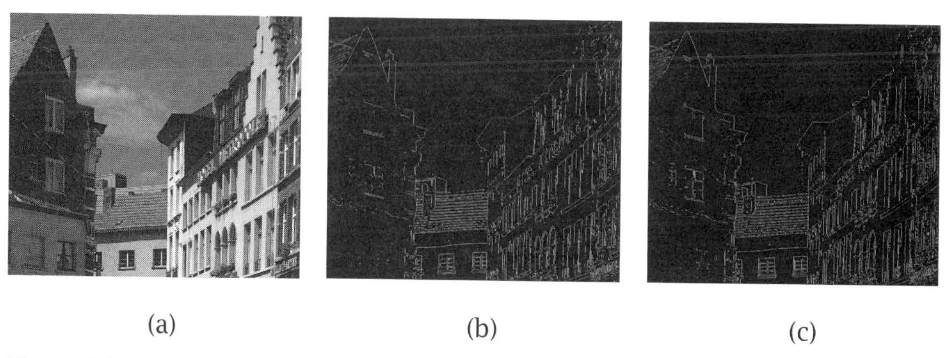

(a) (b) (c)

Figure 2.21: (a) Original image, and edge detection using (b) the linear method and (c) the median method .

2.8 Conclusion

The principles behind WM smoothers and WM filters have been presented in this chapter, as well as some of the applications of these nonlinear methods to image processing. It should be apparent to the reader that many similarities exist between linear and median filters. As illustrated here, there are several applications in image enhancement where WM filters provide significant advantages over traditional methods using linear filters. The methods presented here, and other image enhancement methods that can be easily developed using WM filters, are computationally simple and provide significant advantages. Consequently, they can be used in emerging consumer electronic products, PC and internet imaging tools, and medical and biomedical imaging systems, as well as in military applications.

Acknowledgments

This research has been supported through collaborative participation in the Advanced Telecommunications/Information Distribution Research Program (ATIRP) Consortium sponsored by the U.S. Army Research Laboratory under the Federated Laboratory Program, Cooperative Agreement DAAL01-96-2-0002, and by the NSF under grants MIP-9530923 and CDA-9703088.

References

[Arc86] G. R. Arce. Statistical threshold decomposition for recursive and nonrecursive median filters. *IEEE Trans. Inf. Theory* IT-**32**(2), 243–253 (March 1986).

[Arc88] G. R. Arce and N. C. Gallagher. Stochastic analysis of the recursive median filter process. *IEEE Trans. Inf. Theory* IT-**34**(4), 669–679 (July 1988).

[Arc95] G. R. Arce, T. A. Hall, and K. E. Barner. Permutation weighted order statistic filters. *IEEE Trans. Image Process.* **4**, 1070–1083 (August 1995).

[Arc98] G. R. Arce. A general weighted median filter structure admitting negative weights. *IEEE Trans. Signal Process.* SP-**46**(12), 3195–3205 (December 1998).

[Arc00] G. R. Arce and J. L. Paredes. Recursive weighted median filters admitting negative weights and their optimization. *IEEE Trans. Signal Process.* **48**(3), 768–799 (2000).

[Arn92] B. C. Arnold, N. Balakrishnan, and H. N. Nagaraja. *A First Course in Order Statistics.* Wiley, New York (1992).

[Ast97] J. Astola and P. Kuosmanen. *Fundamentals of Nonlinear Digital Filtering.* CRC Press, Boca Raton, FL (1997).

[Bin92] Z. Bing and A. N. Venetsanopoulos. Comparative study of several nonlinear image interpolation schemes, *Proc. SPIE*, pp. 21–29 (November 1992).

[Bov83] A. C. Bovik, T. S. Huang, and D. C. Munson, Jr. A generalization of median filtering using linear combinations of order statistics. *IEEE Trans. Acoust. Speech, Signal Process.* ASSP-**31**(6), 1342–1350 (December 1983).

[Bov86] A. C. Bovik and D. C. Munson. Edge detection using median comparisons. *Comput. Vision Graph. Image Process.* **33**(3), 377–389 (March 1986).

[Bro84] D. R. K. Brownrigg. The weighted median filter. *Commun. Assoc. Comput. Machin.* **27**(8), 807–818 (August 1984).

[Dav82] H. A. David. *Order Statistics*, Wiley Interscience, New York (1982).

[Edg87] F. Y. Edgeworth. A new method of reducing observations relating to several quantities. *Philos. Mag. (Fifth Series)*, **24**, 222–223 (1887).

[Har93] R. C. Hardie and C. G. Boncelet, Jr. LUM filters: A class rank order based filter for smoothing and sharpening. *IEEE Trans. Signal Process.* **41**(3), 1061–1076 (March 1993).

[Har94] R. C. Hardie and K. E. Barner. Rank conditioned rank selection filters for signal restoration. *IEEE Trans. Image Process.* **3**(2), 192–206 (March 1994).

[Jai89] A. K. Jain. *Fundamentals of Digital Image Processing.* Prentice Hall, Englewood Cliffs, NJ (1989).

[Ko91] S.-J. Ko and Y. H. Lee. Center weighted median filters and their applications to image enhancement. *IEEE Trans. Circ. Syst.* **38**(9), 984–993 (September 1991).

[Lee85] Y. H. Lee and S. A. Kassam. Generalized median filtering and related nonlinear filtering techniques. *IEEE Trans. Acoust. Speech Signal Process.*, ASSP-**33**(3), 672–683 (June 1985).

[Leh83] E. L. Lehmann. *Theory of Point Estimation*, Wiley, New York (1983).

[Mit00] S. K. Mitra. *Digital Signal Processing: A Computer-Based Approach*, 2nd ed. McGraw-Hill, Burr Ridge, IL (2000).

[Par99] J. L. Paredes and G. R. Arce. Stack filters, stack smoothers, and mirrored threshold decomposition. *IEEE Trans. Signal Process.* **47**(10), 2757–2767 (October 1999).

[Pit86] I. Pitas and A. N. Venetsanopoulos. Nonlinear order statistic filters for image filtering and edge detection. *Signal Process.* **10**(4), 395–413 (April 1986).

[Pit90] I. Pitas and A. N. Venetsanopoulos. *Nonlinear Digital Filters: Principles and Applications.* Kluwer, Boston (1990).

[Que95] R. Queiroz, D. Florencio, and R. Schafer. Nonexpansive pyramid for image coding using a nonlinear filterbank. *IEEE Trans. Image Process.* **7**(2), 246–252 (February 1995).

[Shy89] J. Shynk. Adaptive IIR filtering. *IEEE ASSP Mag.* **6**(2), 4–21 (April 1989).

[Tuk74] J. W. Tukey. Nonlinear (nonsuperimposable) methods for smoothing data. In *Conf. Rec. Eascon*, p. 673 (1974).

[Yin96] L. Yin, R. Yang, M. Gabbouj, and Y. Neuvo. Weighted median filters: a tutorial. *IEEE Trans. Circ. Syst. II*, **43**(3), 157–192 (March 1996).

3

Spatial–Rank Order Selection Filters

Kenneth E. Barner

Department of Electrical and Computer Engineering
University of Delaware
Newark, Delaware

Russell C. Hardie

Department of Electrical and Computer Engineering
University of Dayton
Dayton, Ohio

3.1 Introduction

Many image processing applications demand the use of nonlinear methods. Image nonstationarities, in the form of edges, and commonly occurring heavy-tailed noise result in image statistics that are decidedly non-Gaussian. Linear methods often perform poorly on non-Gaussian signals and tend to excessively smooth visually important image cues, such as edges and fine detail. Through the consideration of appropriate statistical signal and interference models, more effective image processing algorithms can be developed. Indeed, an analysis based on maximum likelihood estimation carried out in Sec. 3.2, indicates that rank order based processing of signals such as images is more appropriate than linear processing. Strict rank order methods, however, are spatially blind and cannot exploit the rich spatial correlations generally present in images.

This chapter explores the joint use of spatial and rank (SR) ordering information in a selection filter framework and applies the methods developed to several common image processing problems. Each of the marginal orderings of observed samples yields information that can be used in the design of filtering algorithms: spatial ordering can be used to exploit correlations between neighboring samples, while rank order can be used to isolate outliers and ensure robust behavior. By operating jointly on the SR orderings, sophisticated algorithms can be developed that exploit spatial correlations while producing robust outputs that appropriately process abrupt signal transitions (edges) and are immune to sample outliers.

Numerous filtering algorithms have been developed that exploit, in some fashion, spatial and rank order information in the filtering operation. A large class of such filters can be categorized as selection type, in that their output is always one of the samples from the local observation window. Restricting a filter to be selection type is rarely limiting and, as shown in Sec. 3.2, often holds advantages in image processing applications. Thus, we focus on developing the broad class of SR selection filters that operates on the SR ordering information of observed samples.

To illustrate the advantages of SR selection filters over traditional linear schemes, consider the smoothing of a noisy image. Figure 3.1 shows the results of processing a noisy image with a weighted sum filter that operates strictly on spatial order, along with the results of a selection filter that operates jointly on SR ordering information. These results indicate that weighted sum filters tend to smooth edges and obliterate fine detail. In contrast, the selection filter, by operating jointly on the SR ordering information, is able to suppress the noise while simultaneously preserving edges and fine detail.

The remainder of this chapter theoretically motivates SR selection filters, develops several filter class subsets and extensions that utilize partial, full, or extended SR ordering information, and applies the filters developed to several image processing problems. Section 3.2 begins with a theoretical discussion of maximum likelihood (ML) estimation that motivates the use of rank order in the processing of signals with heavy-tailed distributions. The ML development leads naturally to several rank order selection filters, including the median filter, which are then extended to the general class of selection filters. Additionally, a general framework for relating the spatial and rank orderings of samples is introduced in the section.

The broad class of SR selection filters is discussed in Sec. 3.3, beginning with permutation filters, which utilize the full SR ordering information. The factorial growth (with window size) in the number of SR ordering limits the size of permutation filter window that can be utilized in practice. To efficiently utilize partial SR information in the filtering process, we develop M permutation and colored permutation filters. These methods utilize the rank order information of specific spatial samples and allow ordering equivalences to be established in order to efficiently utilize the most important SR information in a given application. We extend the SR selection filtering framework to include augmented observation sets that may include functions of the observed samples.

Figure 3.1: Image Aerial broken into four quadrants: original (upper left), noisy (upper right), output of weighted sum filter operating on the sample spatial order (lower left), and output of selection filter operating jointly on the sample SR orderings (lower right).

Each of the filtering methods discussed operates under the same basic principle, selecting an input sample to be the output based on partial, full, or extended SR ordering information. A unified optimization procedure is therefore developed in Sec. 3.4. Results of applying the developed SR selection filters to several image processing problems, including noise removal in single frames and video sequences, edge sharpening, and interpolation, are presented in Sec. 3.5. Finally, extensions based on fuzzy logic, which lead to fuzzy SR relations and fuzzy order statistics, are given in Sec. 3.6, and possible future research directions are discussed.

3.2 Selection Filters and Spatial-Rank Ordering

3.2.1 ML Estimation

To motivate the development of theoretically sound signal processing methods, consider first the modeling of observation samples. In all but trivial cases, nondeterministic methods must be used. Since most signals have random components, probability based models form a powerful set of modeling methods. Accordingly, signal processing methods have deep roots in statistical estimation theory.

Consider a set of N observation samples. In most image processing applications, these are the pixel values observed from a moving window centered at some position $\mathbf{n} = [n_1, n_2]$ in the image. Such samples will be denoted as $\mathbf{x}(\mathbf{n}) = [x_1(\mathbf{n}), x_2(\mathbf{n}), \ldots, x_N(\mathbf{n})]^T$. For notational convenience, we will drop the index $\mathbf{n}$ unless necessary for clarity.

Assume now that we model these samples as independent and identically distributed (i.i.d.) random variables. Each observation sample is then characterized by the common probability density function (pdf) $f_\beta(x)$, where β is the mean, or location, of the distribution. Often β is information carrying and unknown, and thus must be estimated. The maximum likelihood estimate of the location is achieved by maximizing, with respect to β, the probability of observing $x_1, x_2, \ldots, x_N$. For i.i.d. samples, this results in

$$\hat{\beta} = \arg\max_{\beta} \prod_{i=1}^{N} f_\beta(x_i). \tag{3.1}$$

Thus, the value of β that maximizes the product of the pdfs constitutes the ML estimate.

The degree to which the ML estimate accurately represents the location is dependent, to a large extent, on how accurately the model distribution represents the true distribution of the observation process. To allow for a wide range of sample distributions, we can generalize the commonly assumed Gaussian distribution by allowing the exponential rate of tail decay to be a free parameter. This results in the *generalized Gaussian* density function,

$$f_\beta(x) = c e^{-(|x-\beta|/\sigma)^p}, \tag{3.2}$$

where p governs the rate of tail decay, $c = p/[2\sigma\Gamma(1/p)]$, and $\Gamma(\cdot)$ is the gamma function. This includes the standard Gaussian distribution as a special case ($p = 2$). For $p < 2$, the tails decay more slowly than in the Gaussian case, resulting in a heavier-tailed distribution. Of particular interest is the case $p = 1$, which yields the double exponential, or Laplacian, distribution,

$$f_\beta(x) = \frac{1}{2\sigma} e^{-|x-\beta|/\sigma}. \tag{3.3}$$

To illustrate the effect of p, consider the modeling of image samples within a local window. Figure 3.2 shows the distribution of samples about the 3×3 neighborhood mean for the image Lena (Fig. 3.4a), along with the Gaussian ($p = 2$) and

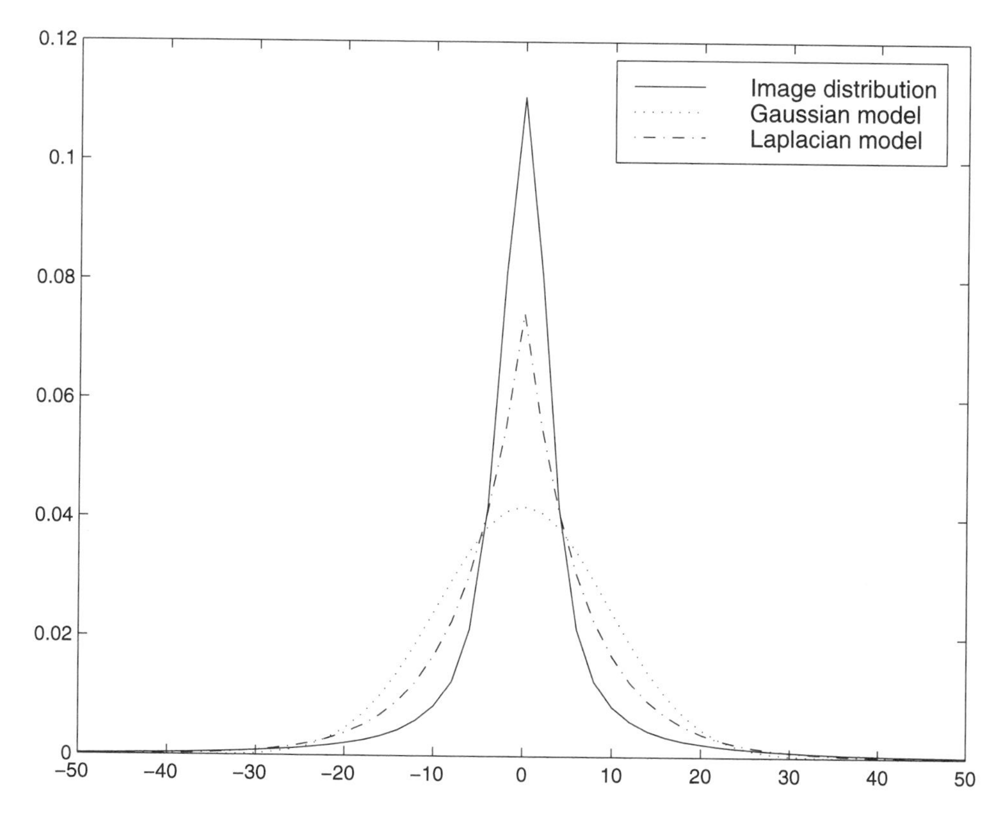

Figure 3.2: Distribution of local samples in the image Lena (Fig. 3.4a) and the generalized Gaussian distribution models for $p = 2$ (standard Gaussian distribution) and $p = 1$ (Laplacian distribution).

Laplacian ($p = 1$) approximations. As the figure shows, the Laplacian distribution models the image samples more accurately than the Gaussian distribution. Moreover, the heavy tails of the Laplacian distribution are well suited to modeling the impulse noise often observed in images.

The ML criteria can be applied to optimally estimate the location of a set of N samples distributed according to the generalized Gaussian distribution, yielding

$$\hat{\beta} = \arg\max_{\beta} \prod_{i=1}^{N} c e^{-(|x-\beta|/\sigma)^p} = \arg\min_{\beta} \sum_{i=1}^{N} |x_i - \beta|^p. \tag{3.4}$$

Determining the ML estimate is thus equivalent to minimizing

$$G_p(\beta) = \sum_{i=1}^{N} |x_i - \beta|^p \tag{3.5}$$

with respect to β. For the Gaussian case ($p = 2$), this reduces to the sample mean, or average:

$$\hat{\beta} = \arg\min_{\beta} G_2(\beta) = \frac{1}{N} \sum_{i=1}^{N} x_i. \tag{3.6}$$

A much more robust estimator is realized if the underlying sample distribution is taken to be the heavy-tailed Laplacian distribution ($p = 1$). In this case, the ML estimator of location is given by the value β that minimizes the sum of least absolute deviations,

$$G_1(\beta) = \sum_{i=1}^{N} |x_i - \beta|, \tag{3.7}$$

which can easily be shown to be the sample median:

$$\hat{\beta} = \arg\min_{\beta} G_1(\beta) = \text{median}[x_1, x_2, \ldots, x_N]. \tag{3.8}$$

The sample mean and median thus play analogous roles in location estimation: While the mean is associated with the Gaussian distribution, the median is related to the Laplacian distribution, which has heavier tails and provides a better image and impulse process model.

Although the median is a robust estimator that possesses many optimality properties, the performance of the median filter is limited by the fact that it is spatially blind. That is, all observation samples are treated equally regardless of their location within the observation window. This limitation is a direct result of the i.i.d. assumption made in the filter development. A much richer class of filters is realized if this assumption is relaxed to the case of independent but not identically distributed samples.

Consider the generalized Gaussian distribution case in which the observation samples have a common location parameter β but where each x_i has a (possibly) unique scale parameter σ_i. Incorporating the unique scale parameters into the ML criteria yields a location estimate given by the value of β that minimizes

$$G_p(\beta) = \sum_{i=1}^{N} \frac{1}{\sigma_i^p} |x_i - \beta|^p. \tag{3.9}$$

In the special case of the standard Gaussian distribution ($p = 2$), the ML estimate reduces to the normalized weighted average

$$\hat{\beta} = \arg\min_{\beta} \sum_{i=1}^{N} \frac{1}{\sigma_i^2}(x_i - \beta)^2 = \frac{\sum_{i=1}^{N} w_i \cdot x_i}{\sum_{i=1}^{N} w_i}, \tag{3.10}$$

where $w_i = 1/\sigma_i^2 > 0$. In the heavier-tailed Laplacian distribution special case ($p = 1$), the ML estimate reduces to the weighted median (WM), originally introduced over a hundred years ago by Edgemore [Edg87] and defined as

$$\hat{\beta} = \arg\min_{\beta} \sum_{i=1}^{N} \frac{1}{\sigma_i} |x_i - \beta| = \text{MEDIAN}[w_1 \blacklozenge x_1, w_2 \blacklozenge x_2, \cdots, w_N \blacklozenge x_N], \tag{3.11}$$

where $w_i = 1/\sigma_i > 0$ and the diamond $\blacklozenge$ is the replication operator, defined as

$$w_i \blacklozenge x_i = \underbrace{x_i, x_i, \cdots, x_i}_{w_i \text{ times}}.$$

More complete discussions of weighted median filters and the related class of weighted order statistic filters are given in Chapters 1 and 2.

Two important observations can be made about the median filter and the more general weighted median filter:

1. *Selection type*—The cost functions leading to the median and WM filters are piecewise linear and convex. Their output is thus guaranteed to be one of the observation samples $x_1, x_2, \ldots, x_N$. That is, the filters are selection type, choosing one of the observed samples as the output.

2. *Partial spatial-rank order use*—The median filter is spatial order blind. The WM filter, in contrast, utilizes partial spatial order information by weighting samples based on their spatial location in a process that can be interpreted as attempting to exploit spatial correlations between samples. Both filters utilize partial rank order information by selecting the central ranked sample from the observation set, or from the expanded observation set in the case of the WM filter.

These concepts are extended in the following subsections to yield a general class of filters that can be employed in a wide array of applications. Specifically, we define the general class of selection filters. This is an extremely broad class of filters, whose only restriction is that the filter output at each instant be one of the observed samples. The decision as to which input sample to select as the output is generally based on some feature that is a function of the observation samples. We show that the spatial and rank ordering information of the observed samples is a particularly useful feature that can be used in the selection rule. This leads to the broad class of spatial-rank ordering selection filters, which are the focus of this chapter.

3.2.2 Selection Filters

Selection filters constitute a large filtering class in which the only restriction placed upon a filter is that its output, at each pixel location, be one of the current observation samples. A selection filter F can thus be thought of as a mapping from the observation set to an output that belongs to the observation set. Let the observation samples from a moving window be denoted, as defined previously, by $\mathbf{x}$. The selection filter function is then given by

$$y = F(\mathbf{x}) = x_{S(\mathbf{z})}, \tag{3.12}$$

where $\mathbf{z}$ is a feature vector, derived from the observation set and lying in the feature space Ω, and $S(\mathbf{z})$ is the selection function. The selection function determines which sample to select as the output at each window location, based strictly on $\mathbf{z} \in \Omega$. Thus, the selection filter function can be expressed as

$$F : \{x_1, x_2, \ldots, x_N\} \mapsto y \in \{x_1, x_2, \ldots, x_N\} \tag{3.13}$$

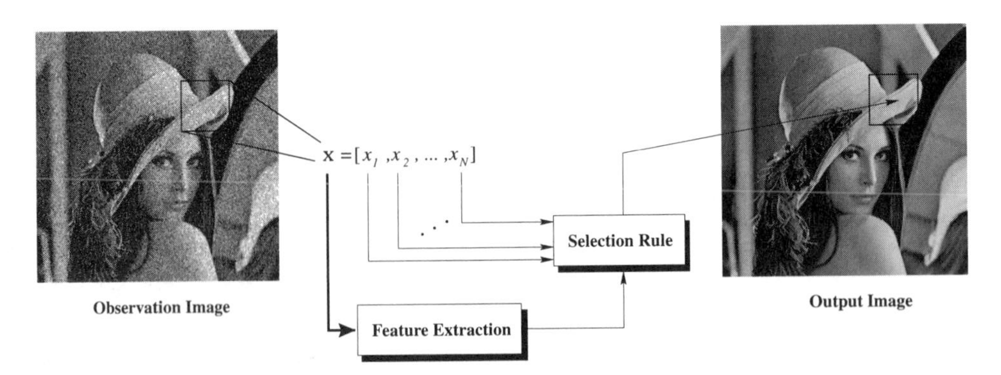

Figure 3.3: In the selection filter operation, a feature is derived from the samples in the current observation window and a selection rule operating on the feature selects one of the observed samples to be the current output.

and the selection function can be written as

$$S : \Omega \mapsto \{1, 2, \ldots, N\}. \tag{3.14}$$

The selection rule S effectively partitions the feature space Ω into N regions, with each region being associated with one of the observation samples. If the current feature lies in the ith partition, the filter output at that window location is simply the observation sample x_i. This general selection filter operation is illustrated in Fig. 3.3.

Selection filters, although a broad filtering class, do limit the output to be one of the N observation samples. A natural question is, therefore, what effect does this restriction have on the filtering process? Is performance significantly limited by the selection constraint? To gauge the effect the selection constraint has on performance, we consider the following image filtering examples.

To investigate an upper bound on performance that the selection constraint imposes, consider the best results that can be achieved by a selection filter in a noise smoothing application. In a simple realization of this case, an observed image is the superposition of a desired underlying image and an additive noise process, $\mathbf{X} = \mathbf{D} + \mathcal{N}$, where $\mathbf{X}$, $\mathbf{D}$, and $\mathcal{N}$ represent the observed image, desired image, and corrupting noise process, respectively. Figures 3.4a and 3.4b show the original 8-bit gray-scale image Lena and a corrupted observation, respectively. The corrupting additive noise process in this case follows a contaminated Gaussian distribution.

As the figure shows, the corrupted image has a rather low signal-to-noise ratio (SNR). Consider now the optimal selection filtering of the corrupted image. Let the optimal selection filter perform the following operation: at each window location, select as the output the observation sample closest in value to the original image pixel. Figures 3.4c and 3.4d show the results of performing this operation utilizing 3×3 and 5×5 observation windows. The mean squared error (MSE) and mean absolute error (MAE) are indicated in the figure caption. Even for a low SNR

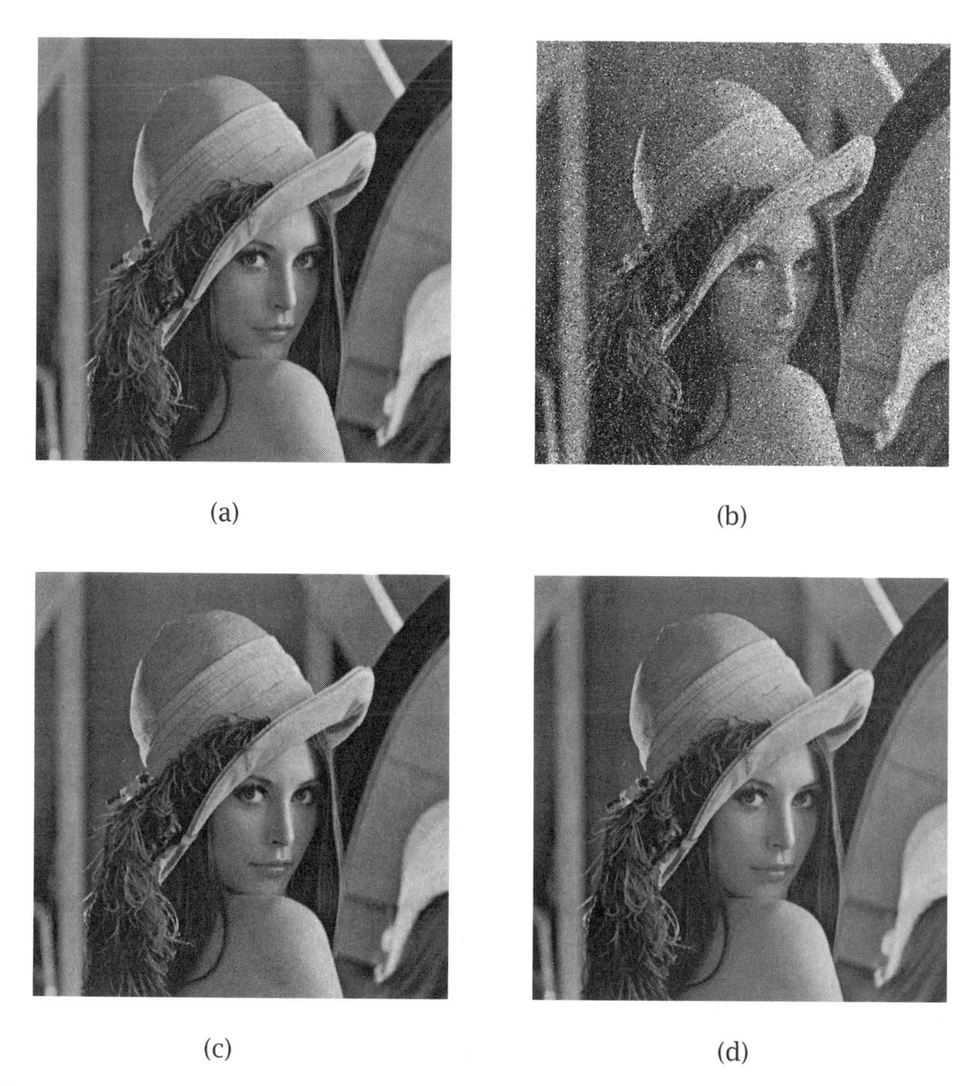

(a) (b)

(c) (d)

Figure 3.4: Optimal selection filter examples: (a) original image Lena, (b) observation image corrupted by contaminated Gaussian noise, and optimal selection filter outputs for (c) 3×3 (MSE = 22.5, MAE = 3.4) and (d) 5×5 (MSE = 4.1, MAE = 1.4) observation windows.

observation and relatively small 3×3 observation window, the resulting output is very close to the original. In the 5×5 observation window case, the output is virtually indistinguishable from the original. Although implementation of the optimal selection filter requires knowledge of the original image, the results indicate that filter performance is not significantly limited by the selection constraint.

The selection constraint, in fact, often improves filter performance. This is particularly true when results are judged subjectively. To illustrate this, we consider the filtering of a color image corrupted by impulses. Here the red, green, and blue tristimulus images are independently corrupted by impulse noise. Figures 3.5a and 3.5b (also see color insert) show the original color image Balloon and an impulse corrupted observation, respectively. To reduce the power of noise, simple averaging is often (mistakenly) applied to a corrupted image. The result of applying a vector averaging operation over a 5×5 moving spatial window is shown in Fig. 3.5c. The averaging operation, while reducing the power of the noise, has the disturbing effect of not only smoothing edges but also introducing new colors into the image. The introduction of colors not present in the original or observation images is often perceptually disturbing and should be avoided. Simply by applying the selection constraint to the filtering operation, much of the objectionable color introduction can be avoided. Figure 3.5d shows the result of applying the selection constraint to the averaging operation. That is, for each observation window, the tristimulus observation vector closest in Euclidean distance to the vector mean is selected as the output. The output realized by applying this constraint is free of color insertions and is therefore subjectively more appealing. As this example illustrates, even for extremely simple filter formulations, the selection constraint can play a valuable role.

3.2.3 Spatial-Rank Ordering

Selecting the appropriate decision feature is the key to defining a selection filter that can be applied in numerous applications and produces desirable results across a broad range of problems. The relationship between the spatial ordering and rank ordering of observed samples defines a large feature space that is particularly useful in numerous image processing applications. Each of these natural orderings contains valuable information. For example, rank ordering is particularly valuable in the design of robust operators, while spatial ordering is helpful when spatial correlations are to be exploited. Operating jointly on the ordering information allows for the design of robust operators that effectively exploit spatial correlations.

To formally relate the spatial ordering and rank ordering of samples in an image processing application, we consider again the typical case in which an observation window passes over an image in a predefined scanning pattern. At each location $\mathbf{n}$ in the image, the observation window covers N samples, which can be indexed according to their spatial location and written in vector form,

$$\mathbf{x}_\ell(\mathbf{n}) = [x_1(\mathbf{n}), x_2(\mathbf{n}), \ldots, x_N(\mathbf{n})]^T. \tag{3.15}$$

The subscript ℓ has now been added to indicate that the samples are indexed according to their natural spatial order within the observation image. A second natural ordering of the observed samples is rank order, in which case the order statistics of the observation samples are obtained,

$$x_{(1)}(\mathbf{n}) \leq x_{(2)}(\mathbf{n}) \leq \cdots \leq x_{(N)}(\mathbf{n}). \tag{3.16}$$

Writing the order statistics in vector form yields the rank-order observation vector,

$$\mathbf{x}_L(\mathbf{n}) = [x_{(1)}(\mathbf{n}), x_{(2)}(\mathbf{n}), \ldots, x_{(N)}(\mathbf{n})]^T. \tag{3.17}$$

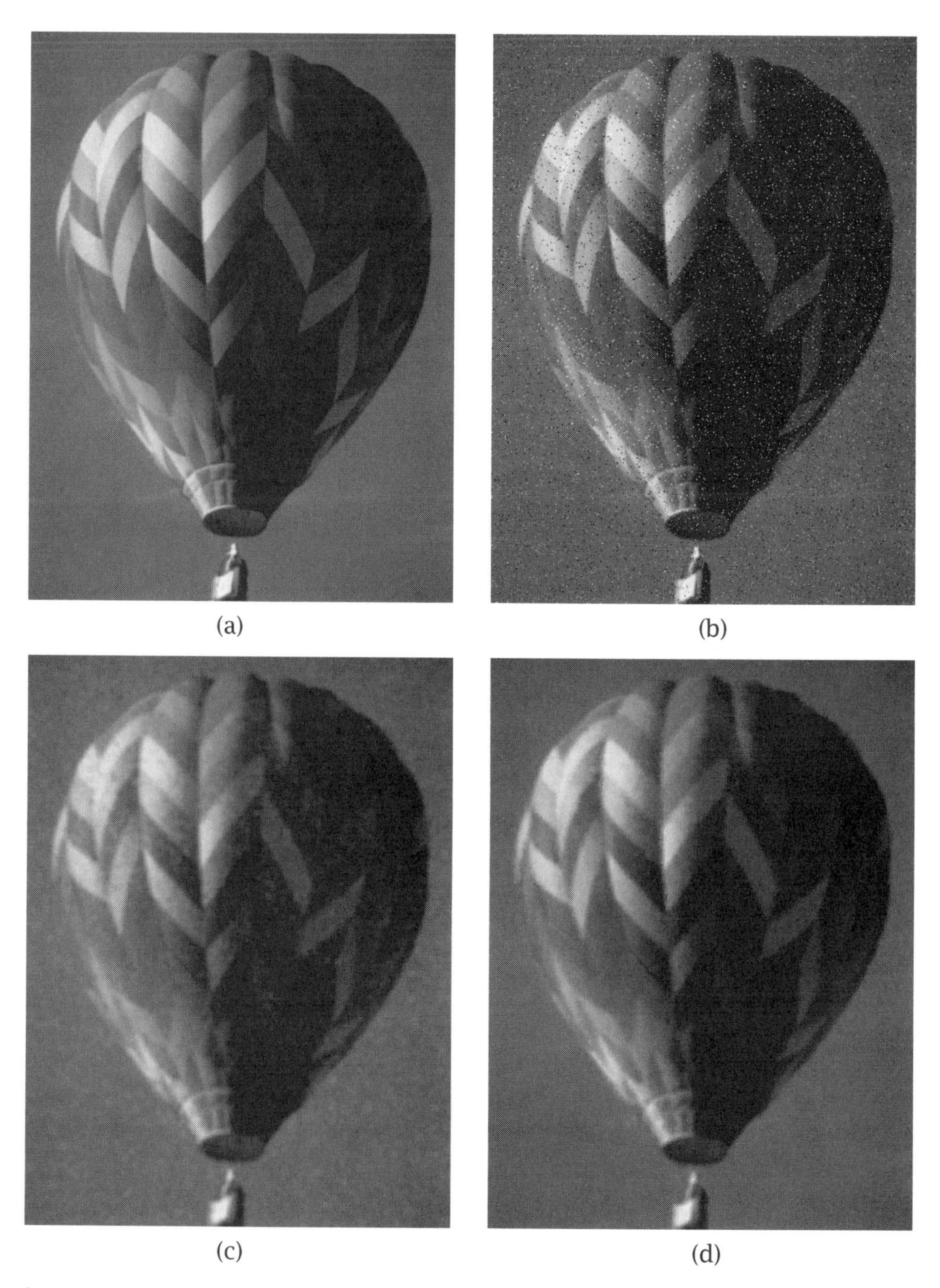

(a) (b)

(c) (d)

Figure 3.5: Simple selection filter constraint examples: (a) original color image Balloon, (b) observation image corrupted by impulses, and simple 5×5 observation window sample mean filtering operation (c) without and (d) with the selection constraint. (See also color insert.)

Again, the spatial location of the observation window is only shown explicitly when needed for clarity. Thus, we write the spatial order and rank order observation vectors as simply $\mathbf{x}_\ell$ and $\mathbf{x}_L$.

Note that both the spatial and rank ordered samples constitute the same set, $X = \{x_1, \ldots, x_N\} = \{x_{(1)}, \ldots, x_{(N)}\}$. The relationship between the original spatial ordering and the rank ordering can be represented by the following binary matrix:

$$\mathfrak{R} = \begin{bmatrix} R_{1,(1)} & \cdots & R_{1,(N)} \\ \vdots & \ddots & \vdots \\ R_{N,(1)} & \cdots & R_{N,(N)} \end{bmatrix}, \tag{3.18}$$

where

$$R_{i,(j)} = \begin{cases} 1 & \text{for } x_i \text{ of rank } j \ (x_i \leftrightarrow x_{(j)}), \\ 0 & \text{otherwise.} \end{cases} \tag{3.19}$$

The matrix $\mathfrak{R}$ contains the full joint SR information of the observation set X. Thus, $\mathfrak{R}$ can be used as a transformation between the two orderings and to extract the marginal vectors $\mathbf{x}_\ell$ and $\mathbf{x}_L$. The transformations yielding the spatial and rank order indexes are given by

$$\mathbf{s} = [1:N]\mathfrak{R}, \qquad \mathbf{r} = \mathfrak{R}[1:N]^T, \tag{3.20}$$

where $[1:N] = [1, \ldots, N]$, and $\mathbf{s} = [s_1, \ldots, s_N]$ and $\mathbf{r} = [r_1, \ldots, r_N]$ are the spatial and rank order index vectors, respectively, that is, $x_{s_j} \leftrightarrow x_{(j)}$ and $x_i \leftrightarrow x_{(r_i)}$ for $i, j = 1, \ldots, N$. Similarly, the spatial and rank ordered samples are related by

$$\mathbf{x}_\ell = \mathfrak{R}\mathbf{x}_L^T, \qquad \mathbf{x}_L = \mathbf{x}_\ell\mathfrak{R}. \tag{3.21}$$

Example 3.1. As an illustrative example, suppose a three-sample observation window is used and a particular spatial order observation vector is given by $\mathbf{x}_\ell = [63, 5, 9]$. This results in the SR matrix

$$\mathfrak{R} = \begin{bmatrix} 0 & 0 & 1 \\ 1 & 0 & 0 \\ 0 & 1 & 0 \end{bmatrix}, \tag{3.22}$$

from which we can obtain the spatial and rank order indexes $\mathbf{s} = [1, 2, 3]\mathfrak{R} = [2, 3, 1]$, $\mathbf{r} = \mathfrak{R}[1, 2, 3]^T = [3, 1, 2]$, and the spatial and rank order samples $\mathbf{x}_\ell = \mathfrak{R}[5, 9, 63]^T = [63, 5, 9]$, $\mathbf{x}_L = [63, 5, 9]\mathfrak{R} = [5, 9, 63]$.

The SR matrix fully relates the spatial and rank orderings of the observed samples. Thus the structure of the SR matrix captures spatial correlations of the data (spatial order information) and indicates which samples are likely to be outliers and which are likely to be reliable (rank order information). To illustrate the structure of the SR matrix for typical signals and to show how this structure changes with the underlying signal characteristics, we examine the statistics of the SR matrix for two types of signals.

Consider first a one-dimensional statistical sequence consisting of a moving average (MA) process. This process is generated by simple FIR filtering of white

Gaussian noise. In the case of white noise samples (no filtering), all spatial-rank order combinations are equally likely and the expected SR matrix is uniform,

$$E[\mathfrak{R}] = \begin{bmatrix} \frac{1}{N} & \cdots & \frac{1}{N} \\ \vdots & \ddots & \vdots \\ \frac{1}{N} & \cdots & \frac{1}{N} \end{bmatrix}, \tag{3.23}$$

where $E[\cdot]$ denotes the expectation operator. Figure 3.6a shows $E[\mathfrak{R}]$ for the white noise case when the window size is $N = 15$. As expected, all spatial order pairs are equally likely. Applying a low-pass FIR filter with a cutoff frequency of $\omega_c = 0.33$ to the noise to generate a MA process yields a time sequence with a greater concentration of low frequency power. This greater concentration of low frequency power gives the time series an increasingly sinusoidal structure, as shown in Fig. 3.6b which is reflected in the resulting SR matrix. As the figure shows, extreme samples are most likely to be located in the first or last observation window location and monotonic observations are more likely than other observations. Decreasing the cutoff of the FIR filter to $\omega_c = 0.25$ (Fig. 3.6c) and $\omega_c = 0.20$ (Fig. 3.6d) increases the sinusoidal nature of the time domain sequence and the structure of the corresponding SR matrices.

Similar results hold for images. Consider the original image and the one contaminated by Gaussian noise shown in Figs. 3.4a and 3.4b. The expected SR matrices for an $N = 15$ one-dimensional observation window passing over these images are shown in Fig. 3.7. As expected, the SR matrix for the original image has a structure that corresponds to the underlying image. The heavily corrupted noisy image, however, has lost much of the underlying structure since all spatial-rank order pairs are nearly equally likely. Thus, just as additive noise tends to decorrelate samples and flatten the power spectral density (PSD) of a signal, it also tends to flatten the expected SR matrix. Appropriate filtering can help restore the SR structure to that of the underlying signal

3.3 Spatial-Rank Order Selection Filters

The relationship between the spatial ordering and the rank ordering of the observation samples contains significant information about the observation set that can be used as the basis for forming the output of a filter. The spatial-rank ordering information is completely represented in the SR matrix. A filter that bases its output decision on SR information can thus be realized by utilizing the SR matrix as the feature in a selection filter formulation.

The remainder of this section develops several spatial-rank order selection filters. We begin by utilizing the full SR information, which leads to the class of permutation filters [Arc95, Bar94]. Subsequent discussion focuses on utilizing partial SR information to reduce the cardinality of the SR feature space. Lastly, extensions to selection SR order filters are discussed. These extensions are based on augmenting the observation set with functions of the observed samples.

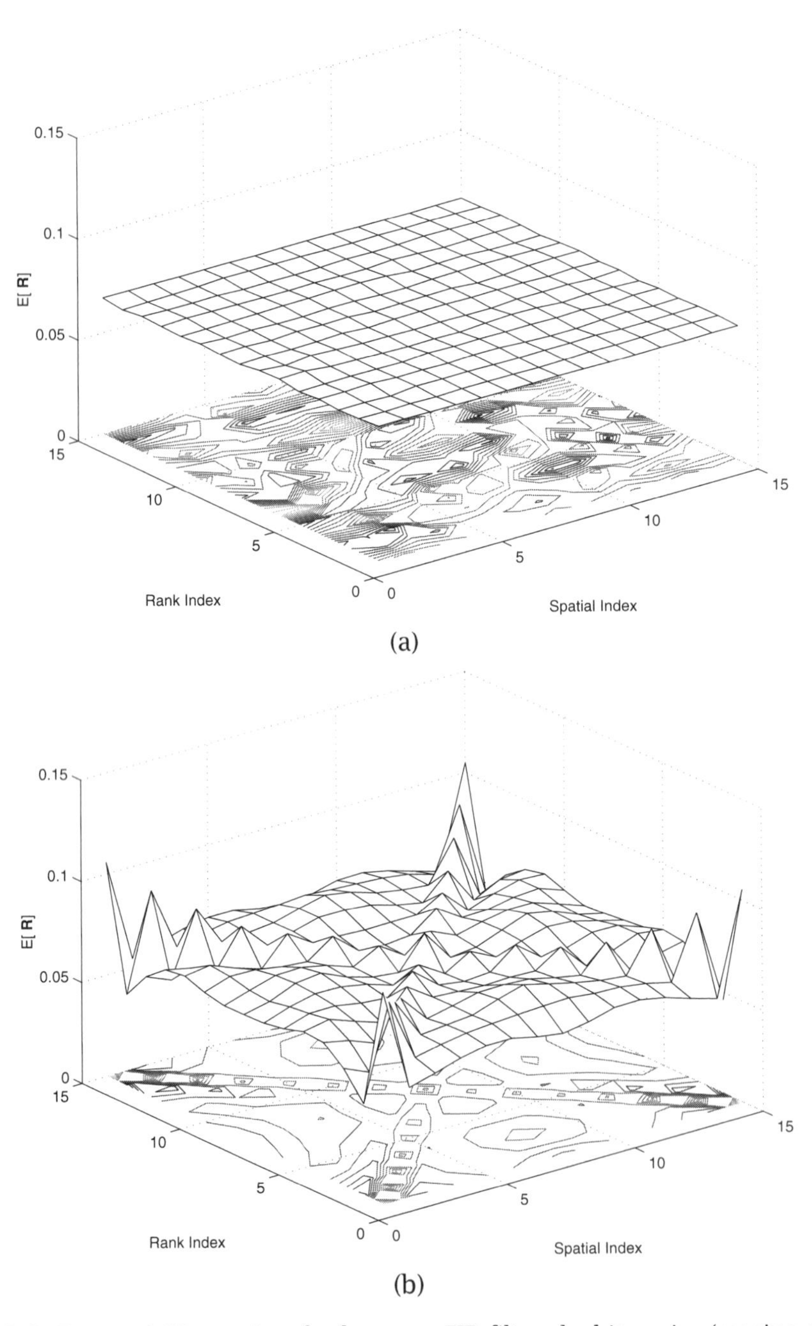

Figure 3.6: Expected SR matrices for low-pass FIR filtered white noise (moving average) processes. The low-pass filter cutoff frequencies are (a) $\omega_c = 1.0$ (no filtering), (b) $\omega_c = 0.33$.

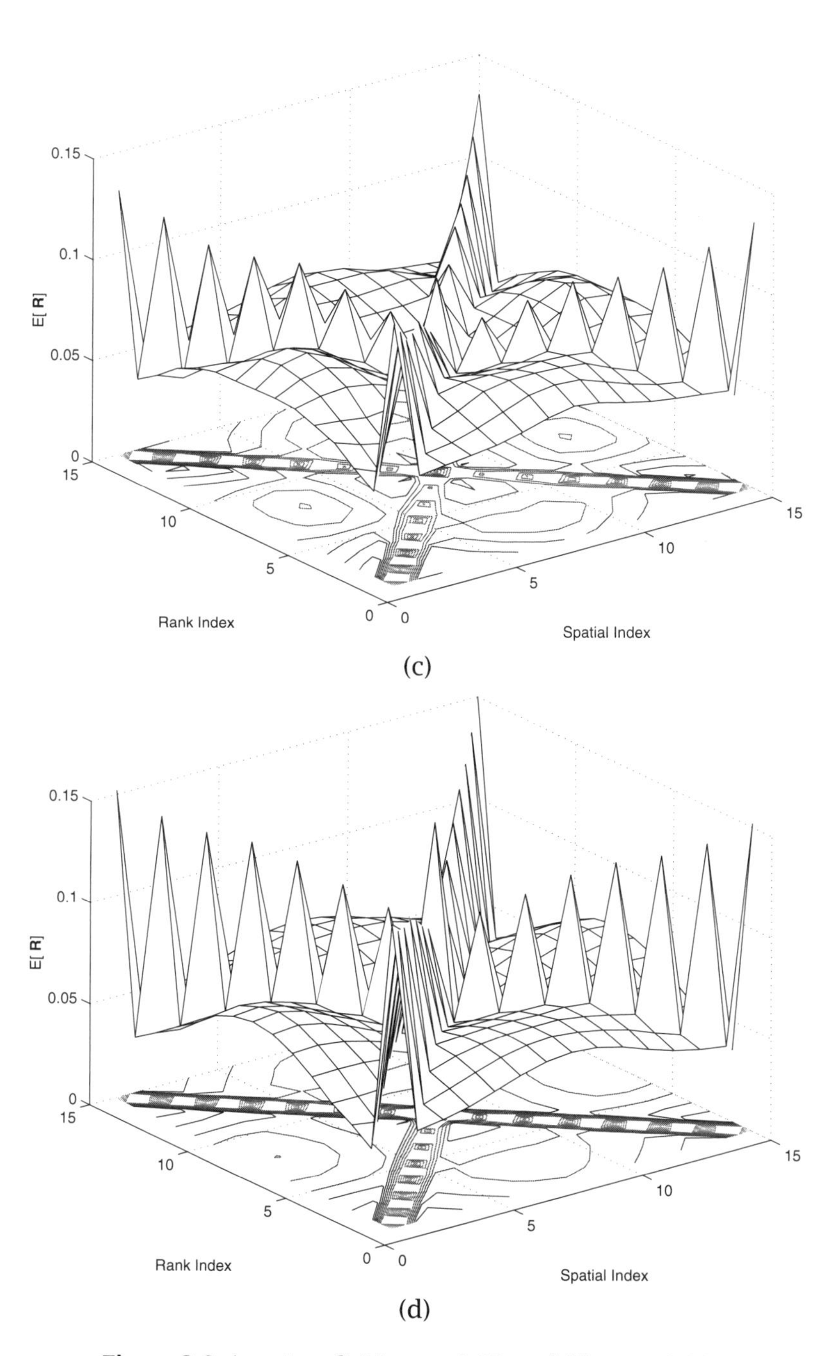

Figure 3.6: (continued) (c) $\omega_c = 0.25$, and (d) $\omega_c = 0.20$.

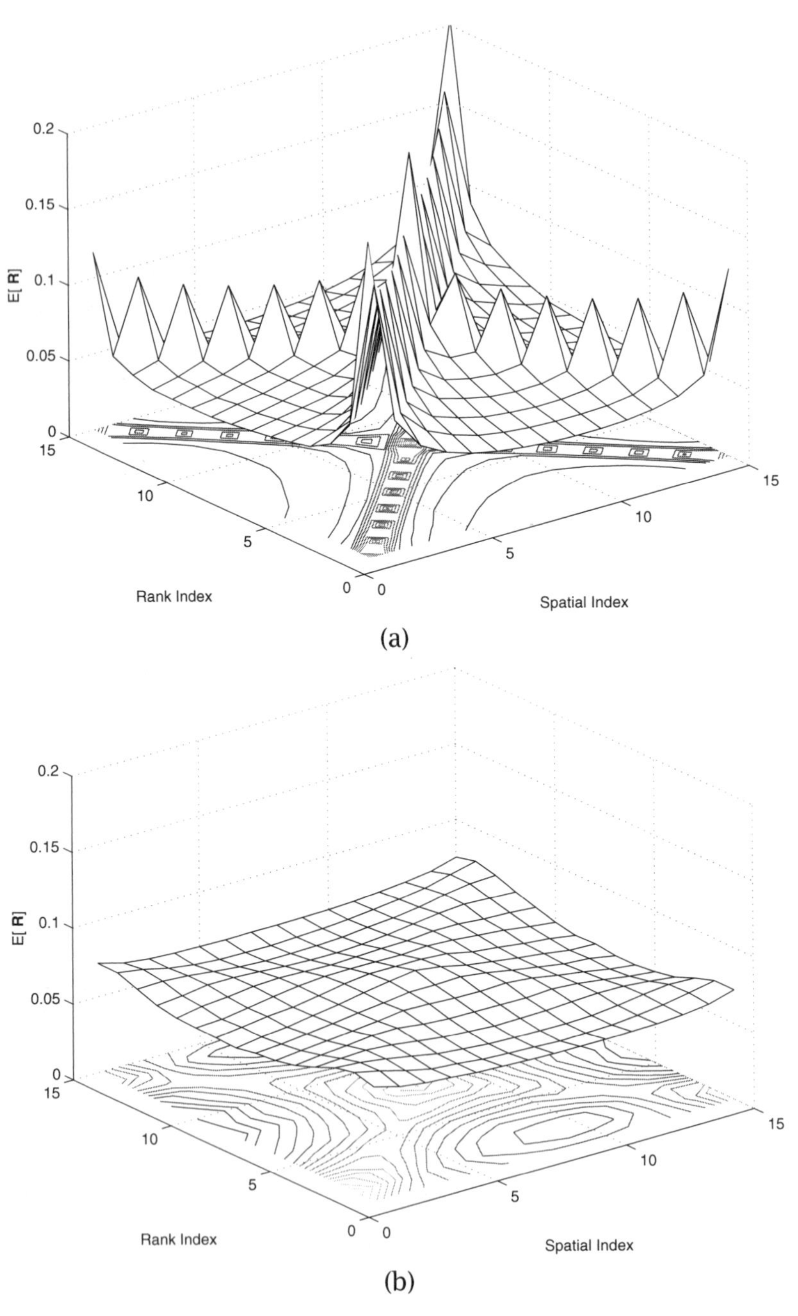

Figure 3.7: $E[\mathbf{R}]$ for (a) the original Lena image of Fig. 3.4a and (b) the one with contaminated Gaussian noise, 3.4b.

Each of the filter cases discussed utilizes full, partial, or augmented SR order information as the feature that output decisions are based upon. Thus, in each case the selection rule $S(\cdot)$ operates on a variation of the SR matrix $\mathfrak{R}$. The focus of this section is on the development of the SR matrix variations and the resulting filters. Optimization of the selection rule is addressed in the following section.

3.3.1 Permutation Filters

The full SR ordering information can be used as the decision feature in a selection filtering framework. In this case, the feature space consists of all possible SR matrices. It is easy to see from the definition of the SR matrix that this feature space is equivalent to the set of $N \times N$ binary matrices that have a single 1 in each row and column,

$$\Omega = \left\{ \mathfrak{R} = \{0,1\}^{N\times N} \mid \text{a single 1 resides in each row and column of } \mathfrak{R} \right\}. \quad (3.24)$$

Since the SR matrix fully relates the spatial and rank orderings of the observation samples, it represents the mapping, or permuting, of the spatial ordering to the rank ordering. Therefore, selection filters that utilize the SR matrix as the observation feature are referred to as permutation filters [Bar94]. From the permutation interpretation and the definition of the SR matrix, it is clear that the cardinality of the feature space in this case is equivalent to the number of possible permutations of N samples, or $\|\Omega\| = N!$.

To define a permutation filter, a selection rule must be chosen that maps each possible SR matrix to an observation sample index,

$$S : \mathfrak{R} \mapsto \{1, 2, \ldots, N\} \qquad \forall \mathfrak{R} \in \Omega. \quad (3.25)$$

Since the SR matrix fully relates the spatial and rank orderings of the observation samples, each partition in Ω can be equivalently associated with either a spatial ordered sample, $x_1, x_2, \ldots, x_N$, or a rank ordered sample, $x_{(1)}, x_{(2)}, \ldots, x_{(N)}$. In subsequent filter definitions, partial SR information will be utilized. In these cases it will be advantageous to associate each feature space partition with a rank ordered sample. For notational consistency, we therefore choose S to map each SR matrix to a rank order index. Utilization of this convention leads directly to the permutation filter definition,

$$F_P(\mathbf{x}_\ell) = x_{(S(\mathfrak{R}))}, \quad (3.26)$$

where $\mathbf{x}_\ell$ is the spatial ordered observation vector. Although not indicated directly in the notation, it should be clear that $\mathfrak{R}$ is dependent on the orderings of the current observation samples.

Permutation filters fully utilize the spatial-rank ordering information of the observed samples and can therefore be successfully applied to applications requiring robust performance and the exploitation of spatial correlations [Arc95, Bar94]. While utilizing the full SR information yields good performance, the $N!$ growth in the cardinality of the feature space limits the size of permutation filter observation window that can be used in practice. Considerable effort has therefore been placed on developing methods that efficiently utilize partial SR information. Accordingly, subsequent subsections investigate filters based on partial use of the SR information as well as extensions to permutation filters.

3.3.2 *M* Permutation Filters

In many filtering applications the use of the full SR information is not required. Acceptable estimates can often be formed with only partial SR information. It may be important, for instance, to know the spatial and rank orderings of certain samples, while the ordering information of other samples may not significantly contribute to forming an appropriate output selection. By utilizing the ordering information on only appropriately selected samples, the cardinality of the features space can be significantly reduced and, consequently, the size of the observation window can be increased. This window size SR information tradeoff has advantages in numerous applications.

To utilize the ordering information of select samples, let the SR ordering matrix be expressed in terms of its rows and columns,

$$\mathfrak{R} = \begin{bmatrix} R_{1,(1)} & \cdots & R_{1,(N)} \\ \vdots & \ddots & \vdots \\ R_{N,(1)} & \cdots & R_{N,(N)} \end{bmatrix} = \begin{bmatrix} \mathfrak{R}_{\ell:1} \\ \mathfrak{R}_{\ell:2} \\ \vdots \\ \mathfrak{R}_{\ell:N} \end{bmatrix} = [\mathfrak{R}_{L:1}, \mathfrak{R}_{L:2}, \ldots, \mathfrak{R}_{L:N}], \tag{3.27}$$

where each $\mathfrak{R}_{\ell:i} = [R_{i,(1)}, R_{i,(2)}, \ldots, R_{i,(N)}]$ is a $1 \times N$ row vector and each $\mathfrak{R}_{L:i} = [R_{1,(i)}, R_{2,(i)}, \ldots, R_{N,(i)}]^T$ is an $N \times 1$ column vector. Note that $\mathfrak{R}_{\ell:i}$ gives the rank of the sample in spatial location i, while $\mathfrak{R}_{L:i}$ gives the spatial index of the ith order statistic; that is, $r_i = \mathfrak{R}_{\ell:i}[1:N]^T$ and $s_i = [1:N]\mathfrak{R}_{L:i}$.

The ordering information of select samples can be obtained by isolating specific rows or columns of $\mathbf{R}$ to form a reduced SR matrix. For instance, we can form a reduced SR matrix by selecting $M \leq N$ rows or columns. In the case of row selection, we have

$$\mathfrak{R}_{\ell:i_1,i_2,\ldots,i_M} = \begin{bmatrix} \mathfrak{R}_{\ell:i_1} \\ \mathfrak{R}_{\ell:i_2} \\ \vdots \\ \mathfrak{R}_{\ell:i_M} \end{bmatrix}, \tag{3.28}$$

where $\mathfrak{R}_{\ell:i_1,i_2,\ldots,i_M}$ is an $M \times N$ matrix that gives the ranks of the M observation samples in spatial locations $i_1, i_2, \ldots, i_M$. To simplify the notation, we express the reduced SR matrix as $\mathfrak{R}_{\boldsymbol{\Theta}}$, where $\boldsymbol{\Theta}$ defines the rows or columns comprising the matrix. Thus Eq. (3.28) can be equivalently expressed as $\mathfrak{R}_{\boldsymbol{\Theta}}$ for $\boldsymbol{\Theta} = [\ell : i_1, i_2, \ldots, i_M]$. The fact that $\mathfrak{R}_{\boldsymbol{\Theta}}$ produced by row selection yields rank information on select spatial samples can be seen by noting that $\mathfrak{R}_{\boldsymbol{\Theta}}[1:N]^T = [r_{i_1}, r_{i_2}, \ldots, r_{i_M}]$.

Example 3.2. To illustrate the formation of the reduced SR matrix, let $N = 5$ and suppose $\mathbf{x}_\ell = [9, 4, 7, 21, 1]$. Then for $M = 3$ and $\Theta = [\ell : 3, 4, 5]$, the resulting full and reduced SR matrices are

$$\mathfrak{R} = \begin{bmatrix} 0 & 0 & 0 & 1 & 0 \\ 0 & 1 & 0 & 0 & 0 \\ 0 & 0 & 1 & 0 & 0 \\ 0 & 0 & 0 & 0 & 1 \\ 1 & 0 & 0 & 0 & 0 \end{bmatrix}$$

and

$$\mathfrak{R}_\Theta = \begin{bmatrix} 0 & 1 & 0 & 0 & 0 \\ 0 & 0 & 1 & 0 & 0 \\ 0 & 0 & 0 & 0 & 1 \end{bmatrix}. \tag{3.29}$$

In this case, the reduced SR matrix gives the ranks of samples in spatial locations 3, 4, and 5, and for the given observation samples we have $[r_3, r_4, r_5] = \mathfrak{R}_\Theta[1, 2, 3, 4, 5]^T = [2, 3, 5]$.

A reduced SR matrix formed by selecting $M \le N$ rows (or columns) thus considers the orderings of M samples from a population of size N. Such orderings are referred to as M permutations. Letting the set of all such orderings constitute the feature space Ω_Θ results in a significant reduction in the feature space cardinality. This is especially true for small M, since $\|\Omega_\Theta\| = N!/(N - M)!$. The use of this feature space leads directly to the definition of an M permutation filter,

$$F_{MP}(\mathbf{x}_\ell) = x_{\left(S(\mathfrak{R}_\Theta)\right)}. \tag{3.30}$$

As noted above, $\mathfrak{R}_\Theta$ contains the rank information of M observation samples. An M permutation filter thus selects a rank sample (order statistic) as the output, where the selection decision is conditioned on the ranks of specific observation samples. Accordingly, this filtering operation was originally referred to as the rank conditioned rank selection (RCRS) filter [Har94], and $M \le N$ is called the filter order.

Noise smoothing, as is illustrated in Sec. 3.5, is one example of an application in which M permutation filters of low order, $M = 1$ or 2, have proven useful. In such applications, the observation sample centrally located in the spatial observation window and its closest neighbors are particularly important. M permutations based on the ranks of these central samples yield simple filters that perform better than many widely used filters and produce results nearly identical to the permutation filter. In applications that require fuller use of the SR information, the order of the M permutation can simply be increased until the desired performance is reached or until the permutation filter ($M = N$) is realized.

A reduced SR matrix can also be constructed, of course, by selecting specific columns from the SR matrix,

$$\mathfrak{R}_\Theta = [\mathfrak{R}_{L:i_1}, \mathfrak{R}_{L:i_2}, \ldots, \mathfrak{R}_{L:i_r}], \tag{3.31}$$

where now $\Theta = [L : i_1, i_2, \ldots, i_r]$. In this case the resulting matrix gives the spatial location of selected order statistics, $[1 : N]\mathfrak{R}_\Theta = [s_{i_i}, s_{i_2}, \ldots, s_{i_M}]$. While column selection can also significantly reduce the cardinality of the feature space, selection filters based on column selected reduced SR matrices have not proven particularly useful. This is especially true for low-order (small M) cases, and we do not consider such filters further. For high-order cases, of course, the fuller set of permutation information is utilized and the performance approaches that of the permutation ($M = N$) filter.

3.3.3 Colored Permutation Filters

A more general method for reducing the cardinality of the SR ordering feature space is to consider selected spatial location or rank ordering equivalents. For instance, in a noise smoothing application with an impulse corrupted image, corrupted samples tend to be in the extremes of the rank ordered observation set; that is, the minimum and maximum samples are generally the outliers. In such cases, the minimum and maximum samples can be considered to be equivalent since both samples are outliers that should be discarded since they yield no information on the underlying signal.

A general framework for establishing equivalences between samples is through coloring [Bar97, Bar00]. Coloring simply designates two or more ordering indexes as equivalent and represents a set of equivalent indexes by a single symbol or index. Since this process represents a number of indices by a single symbol or index, the coloring operation can equivalently be interpreted as a quantization of the indexes. This coloring or quantization can be applied to the rank indexes, spatial indexes, or both sets of indexes simultaneously.

An equivalence can be established in the SR matrix quite simply through the logical OR operation. Thus, if an equivalence is to be established between the maximum and minimum samples, we simply combine the columns corresponding to these samples using the OR operation. This operation is denoted as

$$\mathfrak{R}_{\boldsymbol{\Theta}^c} = [\mathfrak{R}_{L:1} \oplus \mathfrak{R}_{L:N}, \mathfrak{R}_{L:2}, \mathfrak{R}_{L:3}, \ldots, \mathfrak{R}_{L:N-1}], \tag{3.32}$$

where $\boldsymbol{\Theta}^c = [L : (1 \oplus N), 2, 3, \ldots, N - 1]$ indicates the specific coloring performed (or establishes the set of equivalences) and the crossed circle is the logical OR operation that is performed elementwise on vectors.

This operation can be applied simultaneously to rows and columns to establish spatial location and rank order equivalences. Suppose, for instance, that we wish to establish the following equivalences:

1. *the two smallest samples are equivalent*—useful when the two smallest samples are typically outliers that should be discarded

2. *the two largest samples are equivalent*—useful when the two largest samples are typically outliers that should be discarded

3. *the first and last samples in the observation window are equivalent*—useful when a symmetric observation window is used and the signal statistics are symmetric about the center of the observation window.

These equivalences can be represented as $\boldsymbol{\Theta}^c = [\ell : (1 \oplus N), 2, 3 \ldots, N-1; L : (1 \oplus 2), 3, 4, \ldots, N-3, N-2, (N-1 \oplus N)]$. Writing the resulting colored SR matrix elementwise gives

$$\mathfrak{R}_{\boldsymbol{\Theta}^c} = \begin{bmatrix} R_{1,(1)} \oplus R_{1,(2)} \oplus R_{N,(1)} \oplus R_{N,(2)} & R_{1,(3)} \oplus R_{N,(3)} & \cdots & R_{1,(N-2)} \oplus R_{N,(N-2)} & R_{1,(N-1)} \oplus R_{1,(N)} \oplus R_{N,(N-1)} \oplus R_{N,(N)} \\ R_{2,(1)} \oplus R_{2,(2)} & R_{2,(3)} & \cdots & R_{2,(N-2)} & R_{2,(N-1)} \oplus R_{2,(N)} \\ R_{3,(1)} \oplus R_{3,(2)} & R_{3,(3)} & \cdots & R_{3,(N-2)} & R_{3,(N-1)} \oplus R_{3,(N)} \\ \vdots & \vdots & \vdots & \vdots & \vdots \\ R_{N-1,(1)} \oplus R_{N-1,(2)} & R_{N-1,(3)} & \cdots & R_{N-1,(N-2)} & R_{N-1,(N-1)} \oplus R_{N-1,(N)} \end{bmatrix}. \quad (3.33)$$

Example 3.3. To illustrate the construction of a colored SR matrix for a specific set of observation samples, consider again the observation $\mathbf{x}_\ell = [9, 4, 7, 21, 1]$. In this case $N = 5$ and the coloring defined above reduces to $\boldsymbol{\Theta}^c = [\ell : (1 \oplus 5), 2, 3, 4;\ L : (1 \oplus 2), 3, (4 \oplus 5)]$. Utilizing the full SR matrix $\mathfrak{R}$, repeated from Eq. (3.29), and the colored SR matrix expression in Eq. (3.33), we see

$$\mathfrak{R} = \begin{bmatrix} 0 & 0 & 0 & 1 & 0 \\ 0 & 1 & 0 & 0 & 0 \\ 0 & 0 & 1 & 0 & 0 \\ 0 & 0 & 0 & 0 & 1 \\ 1 & 0 & 0 & 0 & 0 \end{bmatrix},$$

$$\mathfrak{R}_{\boldsymbol{\Theta}^c} = \begin{bmatrix} 0 \oplus 0 \oplus 1 \oplus 0 & 0 \oplus 0 & 1 \oplus 0 \oplus 0 \oplus 0 \\ 0 \oplus 1 & 0 & 0 \oplus 0 \\ 0 \oplus 0 & 1 & 0 \oplus 0 \\ 0 \oplus 0 & 0 & 0 \oplus 1 \end{bmatrix}$$

$$= \begin{bmatrix} 1 & 0 & 1 \\ 1 & 0 & 0 \\ 0 & 1 & 0 \\ 0 & 0 & 1 \end{bmatrix}. \quad (3.34)$$

The resulting colored SR matrix indicates the following: (1) there is at least one sample in spatial position 1 or 5 that has rank(s) 1 or (and) 2; (2) there is at least one sample in spatial position 1 or 5 that has rank(s) 4 or (and) 5; (3) the sample in spatial position 2 has rank 1 or 2; (4) the sample in spatial position 3 has rank 3; and (5) the sample in spatial position 4 has rank 4 or 5.

Coloring offers a general framework for establishing equivalences that, if properly chosen, can significantly reduce the cardinality of the feature space while having a minimal effect on filter performance. In addition to selecting an appropriate coloring procedure a priori, new coloring optimization methods can be employed [Bar00]. Additionally, given a set of coloring equivalences, group theoretic methods exist for counting and indexing all possible colored SR matrices

[Bar97, Nij75, Rob84]. Given this indexing of colored SR matrices, a selection rule can be established that partitions the colored SR space and yields a colored permutation filter,

$$F_{\mathrm{CP}}(\mathbf{x}_\ell) = x_{(S(\mathfrak{R}_{\mathbf{\Theta}^{\mathrm{c}}}))}. \tag{3.35}$$

3.3.4 Rank Conditioned Median

The row-column selection and coloring techniques can be combined to significantly reduce the cardinality of the feature space. Moreover, by combining the techniques we can define simple, intuitive filtering operations in which the filter characteristics are controlled by a small number of parameters that can be set experimentally. One such filter is the rank conditioned median (RCM) filter [Har94], which is defined in terms of row selection and coloring in the following.

The central sample in the spatial observation window, as noted above, often yields the most information about the desired underlying signal to be estimated. Thus, let δ denote the spatial index of the central observation sample (typically $\delta = (N+1)/2$). Since this sample generally yields the most information on the desired output, through row selection we can consider the rank of x_δ as the feature. In this case, $\mathbf{\Theta} = [\ell : \delta]$ and we have $\mathfrak{R}_{\mathbf{\Theta}} = \mathfrak{R}_{\ell:\delta}$, which gives the desired rank information, $\mathfrak{R}_{\mathbf{\Theta}}[1:N]^T = r_\delta$.

Suppose further that it is known a priori that the $k-1$ largest and $k-1$ smallest samples can be considered to be outliers that yield no information on the desired output. These $2(k-1)$ noninformation-bearing samples should thus be considered equivalent. This equivalence can be established through coloring, which when combined with the desired row selection yields

$$\begin{aligned}\mathbf{\Theta}^{\mathrm{c}} = {} & [\ell : \delta;\ L : (1 \oplus 2 \oplus \cdots \oplus k-1 \oplus N-k+2 \oplus N-k+3 \oplus \cdots \oplus N), \\ & k, k+1, \ldots, N-k+1],\end{aligned} \tag{3.36}$$

and the resulting reduced SR matrix

$$\begin{aligned}\mathfrak{R}_{\mathbf{\Theta}^{\mathrm{c}}} = {} & \big[\, (R_{\delta,(1)} \oplus R_{\delta,(2)} \oplus \cdots \oplus R_{\delta,(k-1)} \oplus R_{\delta,(N+2-k)} \oplus R_{\delta,(N+3-k)} \oplus \cdots \oplus R_{\delta,(N)}\big), \\ & R_{\delta,(k)}, R_{\delta,(k+1)}, \ldots, R_{\delta,(N+1-k)}\,\big].\end{aligned} \tag{3.37}$$

Although this expression appears to be somewhat cumbersome, it has a very simple interpretation. To see this, let ξ denote a color, or a quantization symbol, representing the set of equivalent rank indexes, $\xi \leftrightarrow \{1, 2, \ldots, k-1, N+2+k, N+3+k, \ldots, N\}$. Then we can write $r_\delta^{\mathrm{c}} = \mathfrak{R}_{\mathbf{\Theta}^{\mathrm{c}}}[\xi, k, k+1, \ldots, N+1-k]$, where r_δ^{c} is the colored rank of sample x_δ. That is, if $k \le r_\delta^{\mathrm{c}} \le N+1-k$, then $r_k^{\mathrm{c}} = r_k$ gives the true rank of x_k. If, however, $r_\delta^{\mathrm{c}} = \xi$, then all that is known is that $r_k \in \{1, 2, \ldots, k-1, N+2+k, N+3+k, \ldots, N\}$.

A selection filter based on the defined colored SR matrix thus selects an output based solely on the colored rank of sample x_δ,

$$F_{\mathrm{CP}}(\mathbf{x}_\ell) = x_{(S(\mathfrak{R}_{\mathbf{\Theta}^{\mathrm{c}}}))} = x_{(S(r_\delta^{\mathrm{c}}))}, \tag{3.38}$$

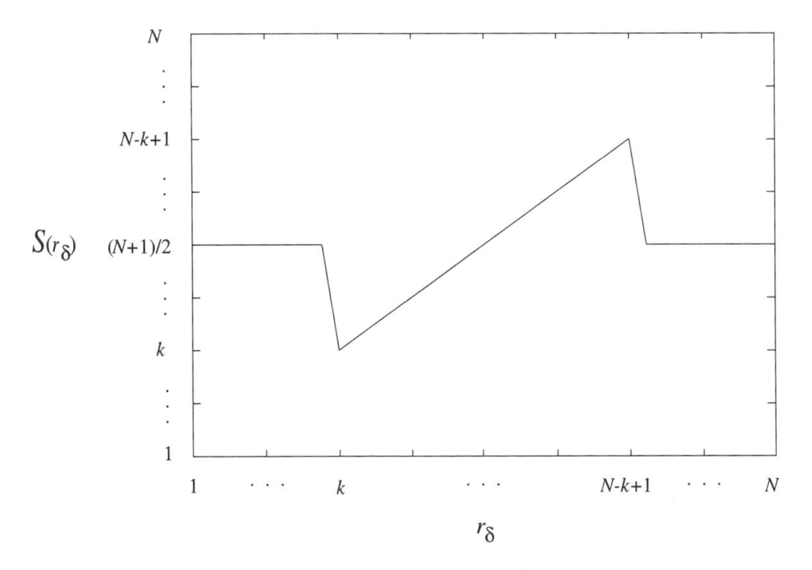

Figure 3.8: The function $S(r_\delta)$ corresponding to the RCM filter with parameter k. (Reproduced with permission from [Har94]. © 1994 IEEE.)

where the selection rule, in this case, can equivalently operate on the colored SR matrix $\Re_{\Theta^c}$ or the colored rank index r_δ^c. By selecting the value of k appropriately, outliers in the observation set can be confined to the set represented by ξ. If the remaining samples are assumed to be free of interference, then a simple and effective colored permutation filter is given by

$$F_{CP}(\mathbf{x}_\ell) = x_{(S(r_\delta^c))} = \begin{cases} x_{(r_\delta^c)} & \text{for } k \le r_\delta^c \le N+1-k, \\ x_{((N+1)/2)} & \text{for } r_\delta^c = \xi. \end{cases} \qquad (3.39)$$

Since δ is the spatial index of the central observation sample, this filter performs the identity operation when x_δ is centrally ranked, $k \le r_\delta^c \le N+1-k$, and outputs the median when this sample is in the tails of the ordered set. That is, the output is the median when $r_\delta^c = \xi$, or equivalently, r_δ is outside the range $[k, N+1-k]$.

This rather simple filter is known as the RCM filter [Har94], which can be represented graphically as in Fig. 3.8. In this figure, the selection rule (rank index) output is plotted as a function of r_δ. Note that the regions in which the output is the median, $S(r_\delta) = (N+1)/2$ for $r_\delta \in [1, k-1] \bigcup [N+2+k, N]$, actually constitute the single point $r_\delta^c = \xi$ in the feature space. An examination of the figure shows that the RCM filter should preserve features well (centrally ranked samples are not altered) while effectively smoothing heavy-tailed noise (samples not centrally ranked are replaced by the median). Additionally, the tradeoff between noise smoothing and detail preservation is controlled by the value of k, or the level of rank coloring introduced.

3.3.5 Extended Permutation Filters

Much of the previous discussion has focused on methods for reducing the cardinality of the SR ordering feature space. This discussion generally assumes that the SR matrix contains more than sufficient information to form an acceptable estimate and that a good estimate can generally be formed from a subset of the SR ordering information. In certain cases, however, the SR ordering information on the original observation samples is, by itself, not sufficient. In these cases, the set of observation samples can be augmented to form a more complete set whose SR ordering information is sufficient to yield an acceptable filter output.

In its most general form, the spatially indexed set of observation samples can be augmented with K functions of the observation samples. This results in an $N + K$ element augmented observation vector,

$$\mathbf{x}'_\ell = [x_1, x_2, \ldots, x_N, x_{N+1}, x_{N+2}, \ldots, x_{N+K}]^T, \tag{3.40}$$

where $x_{N+1} = F_1(\mathbf{x}), x_{N+2} = F_2(\mathbf{x}), \ldots, x_{N+K} = F_K(\mathbf{x})$ are functions of the original N observation samples. These functions, for instance, can be filters that operate on the observation samples and that are not required to be nonlinear or selection type. Thus, the original observation samples could be augmented with a set of linear filter outputs.

A particularly simple augmentation is to append the sample mean $\bar{x}$ to the observation vector, $\mathbf{x}'_\ell = [x_1, x_2, \ldots, x_N, \bar{x}]^T$, where $\bar{x} = 1/N \sum_{i=1}^{N} x_i$. Due to the unique relationship of the mean and the median at the inflection point of an edge, this augmentation is particularly useful in applications where identification of edges is important. Since this property is most easily observed through an example, suppose an $N = 9$ size window is passed over a one-dimensional (temporal) signal. Appending the mean to the spatially (temporally) indexed samples results in a 10 sample augmented observation vector $\mathbf{x}'_\ell = [x_1, x_2, \ldots, x_{10}]^T$, where $x_{10} = \bar{x} = \frac{1}{9} \sum_{i=1}^{N} x_i$. Figure 3.9 shows the value of the sample mean and its rank as a function of window position for the step edge and ramp edge signal cases. Note that for such edges with positive slopes, the rank of the mean is greater than the median on the leading portion of the edge, is less than the median on the trailing portion of the edge, and crosses the median at the edge inflection point [Har96]. Symmetric results hold for edges with negative slopes, and similar results hold for the more general set of concave-convex edges [Har96, Lee87]. The rank of the sample mean can thus be used to implement edge location specific filtering algorithms, such as edge sharpeners.

The ordering of the augmented, or extended, observation set can naturally be incorporated into the SR ordering framework. In this case, the SR information is represented in the augmented SR matrix, $\Re'$, which fully relates the spatial and rank ordering of the extended observation set. The extended permutation filter is then defined as

$$F_{\text{EP}}(\mathbf{x}_\ell) = x_{(S(\Re'))}. \tag{3.41}$$

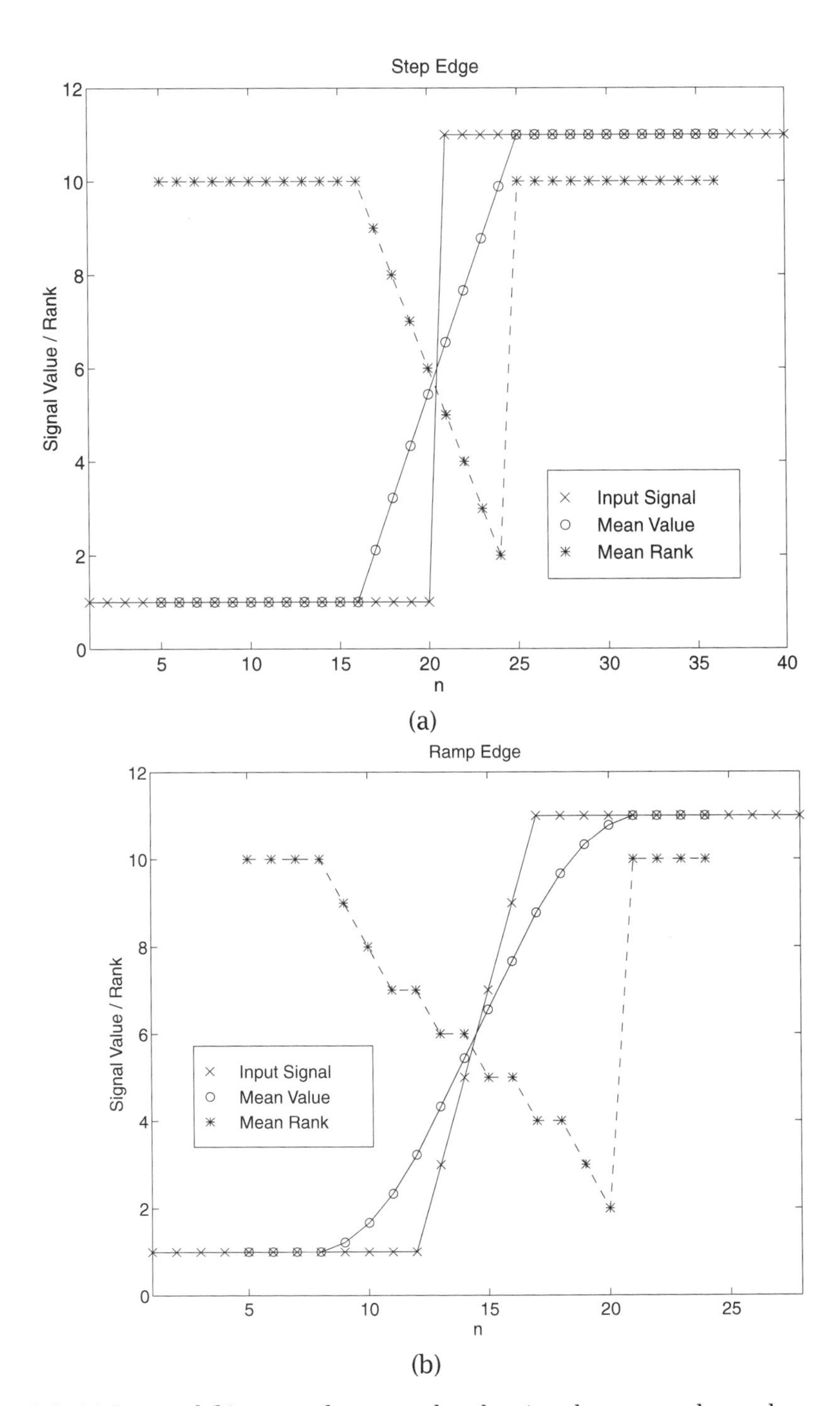

Figure 3.9: (a) Step and (b) ramp edge examples showing the mean value and mean rank, for window size $N = 9$, as a function of window location. Note: The augmented observation vector in this case is $\mathbf{x}'_\ell = [x_1, x_2, \ldots, x_{10}]^T$, where $x_{10} = \bar{x} = 1/9 \sum_{i=1}^{N} x_i$. (Reproduced with permission from [Har96]. © 1996 IEEE.)

It should be noted that an extended permutation filter is not necessarily a strict selection type filter since the selected output can be a function of the input samples, $F_i(\mathbf{x})$, which is not restricted to be selection type. Also, while augmenting the observation set tends to increase the cardinality of the feature space, to $(N+K)!$ in the most general case, reduction of the feature space can be accomplished through row–column selection or coloring. Thus, each of the SR ordering information augmentation and reduction methods can be used in concert to address a specific filtering application.

3.4 Optimization

Each of the SR ordering selection filters discussed in the previous section operate on the same general principle: an observation feature is defined based on full, partial, or extended SR ordering information and a selection rule is set that partitions the SR feature space into N regions; each of the N regions in the SR feature space is associated with a specific order-statistic output. Since the filters all operate on the same general principle, we can define a unified optimization procedure. Numerous optimization methodologies can be adapted, and statistical optimization under the MAE has been investigated [Bar94]. The optimization methodology adopted here is the simpler, and more widely used, least L_η normed error (LNE) strategy. This is a training-based optimization procedure that assumes that a representative data set is available that consists of observed and desired output samples.

Since each of the SR order selection filters operates on the same general principle, we will address the optimization of the generic case in which the SR feature space is represented by Ω. Let the (full-partial-extended) SR matrices comprising Ω be indexed as $\mathfrak{R}_1, \mathfrak{R}_2, \ldots, \mathfrak{R}_{\|\Omega\|}$, so we can write $\Omega = \{\mathfrak{R}_1, \mathfrak{R}_2, \ldots, \mathfrak{R}_{\|\Omega\|}\}$. Also, let the K samples from the training set be indexed in the order that they are observed. In this fashion, the observation vectors can be written as $\mathbf{x}_\ell(\mathbf{n}_1), \mathbf{x}_\ell(\mathbf{n}_2), \ldots, \mathbf{x}_\ell(\mathbf{n}_K)$ and the corresponding desired outputs as $d(\mathbf{n}_1), d(\mathbf{n}_2), \ldots, d(\mathbf{n}_K)$. For the SR selection filter $F(\cdot)$ defined by the selection rule $S(\cdot)$, the LNE over the training sequence is

$$\sum_{i=1}^{K} |d(\mathbf{n}_i) - F(\mathbf{x}_\ell(\mathbf{n}_i))|^\eta = \sum_{i=1}^{K} |d(\mathbf{n}_i) - x_{(S(\mathbf{R}(\mathbf{n}_i)))}|^\eta, \tag{3.42}$$

where $\mathfrak{R}(\mathbf{n}_i) \in \Omega$ is the SR feature at window location $\mathbf{n}_i$. The selection rule that minimizes Eq. (3.42) is referred to as the optimal selection rule and is denoted as $S_{\text{opt}}(\cdot)$.

The LNE can be partitioned according to the SR feature matrices. Let α_i be the index of the feature matrix in Ω corresponding to observation vector $\mathbf{x}_\ell(\mathbf{n}_i)$, i.e., $\mathfrak{R}_{\alpha_i} = \mathfrak{R}(\mathbf{n}_i)$. Additionally, define $\Gamma_{j,K} = \{i \in \{1, 2, \ldots, K\} \mid \alpha_i = j\}$ to be the set of indexes that corresponds to observation samples with SR feature $\mathfrak{R}_j$. The total LNE incurred over the training sequence by estimating the desired signal with the kth order statistic, given that the SR feature vector $\mathfrak{R}_j$ is observed, can be written as

$$\mathcal{E}_j[k] = \sum_{i \in \Gamma_{j,K}} |d(\mathbf{n}_i) - x_{(k)}(\mathbf{n}_i)|^{\eta}. \tag{3.43}$$

The LNE can now be written as a sum of errors, partitioned according to SR feature vector, yielding

$$\sum_{i=1}^{K} |d(\mathbf{n}_i) - F(\mathbf{x}(\mathbf{n}_i))|^{\eta} = \sum_{j=1}^{\|\Omega\|} \mathcal{E}_j[S(\mathfrak{R}_j)]. \tag{3.44}$$

It is easy to show that the LNE in Eq. (3.44) is minimized if and only if each of the $\mathcal{E}_j[S(\mathfrak{R}_j)]$ error terms is minimized. Thus, the optimal RCRS filter classifier is given by

$$S_{\text{opt}}(\mathfrak{R}_j) = k \mid \mathcal{E}_j[k] \leq \mathcal{E}_j[l] \quad \forall l \neq k, \quad \text{for } j = 1, 2, \ldots, \|\Omega\|. \tag{3.45}$$

If there is not a unique minimum error for some j, then a tie-breaking rule must be employed. However, in most practical cases ties are unlikely, given sufficient training.

The optimization can also be performed recursively, updating $S_{\text{opt}}(\cdot)$ as new training samples become available. To do so, define the cumulative partitioned error as

$$C_{j,k}(m) = \sum_{i \in \Gamma_{j,m}} |d(\mathbf{n}_i) - x_{(k)}(\mathbf{n}_i)|^{\eta}. \tag{3.46}$$

The cumulative partitioned error term $C_{j,k}(m)$ contains the total error incurred by outputting the kth order statistic, given the jth SR feature matrix observed, up to index m in the training sequence. These cumulative error terms can be written as a vector yielding

$$\mathbf{C}_j(m) = \left[C_{j,1}(m), C_{j,2}(m), \ldots, C_{j,N}(m)\right]^T. \tag{3.47}$$

The optimal selection rule at index m in the training sequence is determined by the minimum element in $\mathbf{C}_j(m)$ and is given by

$$S^m_{\text{opt}}(\mathfrak{R}_j) = \arg\min_k \{C_{j,k}(m) \mid k = 1, 2, \ldots, N\}, \qquad \text{for } j = 1, 2, \ldots, \|\Omega\|. \tag{3.48}$$

The full iterative optimization procedure can now be summarized as in Table 3.1. The first step in the procedure initializes the decision rule to realize an arbitrary filter, in this case the median, and clears the cumulative error vectors. For each new set of observation values, cumulative error vectors are updated as

$$\mathbf{C}_j(m) = \begin{cases} \lambda \mathbf{C}_j(m-1) + \mathbf{U}(m) & \text{if } j = \alpha_m, \\ \lambda \mathbf{C}_j(m-1) & \text{otherwise}, \end{cases} \tag{3.49}$$

where

$$\begin{aligned} \mathbf{U}(m) = {} & [|d(\mathbf{n}_m) - x_{(1)}(\mathbf{n}_m)|^{\eta}, \\ & |d(\mathbf{n}_m) - x_{(2)}(\mathbf{n}_m)|^{\eta}, \ldots, |d(\mathbf{n}_m) - x_{(N)}(\mathbf{n}_m)|^{\eta}]^T \end{aligned} \tag{3.50}$$

Table 3.1: Recursive Least L_η Normed Error Training Algorithm.

1. Set $m = 1$, $S^0_{\mathrm{opt}}(\mathfrak{R}_j) = (N+1)/2$ and $\mathbf{C}_{j,k}(0) = 0$, for $j = 1, 2, \ldots, \|\Omega\|$ and $k = 1, 2, \ldots, N$.

2. Update each $\mathbf{C}_j(m)$ according to

$$\mathbf{C}_j(m) = \begin{cases} \lambda \mathbf{C}_j(m-1) + \mathbf{U}(m) & \text{for } j = \alpha_m, \\ \lambda \mathbf{C}_j(m-1) & \text{otherwise.} \end{cases}$$

3. Set $S^m_{\mathrm{opt}}(\mathfrak{R}_{\alpha_m}) = \arg\min_k \{C_{\alpha_m,k}(m) | k = 1, 2, \ldots, N\}$.

4. If $m = K$ or filter is sufficiently trained, stop; else increment m and go to 2.

and $\lambda \in (0, 1]$ is an optional forgetting factor. The selection rule for the current SR observation feature is then updated, and the procedure is terminated if the optimization is sufficient or is repeated as new training samples become available.

Advantages of this deterministic training procedure include that : (1) the training process always returns the globally optimal filter for the training set and (2) there is freedom to choose an error norm. In addition, the exponential forgetting factor allows the procedure to accommodate training data with changing statistics [Bar94, Bar92].

3.5 Applications

The performance and advantages of SR selection filters are illustrated through their applications in several common image processing applications. Specifically, we consider the noise smoothing of single frame images and video sequences, image interpolation, and image restoration of smoothed images with additive noise. In the noise smoothing examples the use of reduced SR ordering information through row selection and coloring is investigated. These examples show that good results can often be realized through the use of limited SR ordering information and that increasing the amount of SR ordering information yields minimal improvements in certain applications. The effects of varying the norm utilized in the optimization is also illustrated in these examples. The use of extended SR ordering information is investigated through the interpolation and restoration examples, which show that inclusion of the sample mean in the extended observation set leads to improved performance, especially in applications in which edge enhancement or preservation are critical.

Table 3.2: Results of Filtering Image Lena Corrupted by Impulse Noise (10% and 20%) with SR Selection (M Permutation, L_1 and L_2 Optimized), Median, and Weighted Median Filters.

Filtering Method	10% Impulses		20% Impulses	
	MAE	MSE	MAE	MSE
$M = 1$ permutation (L_1 optimized)	1.67	44.26	2.76	67.69
$M = 1$ permutation (L_2 optimized)	1.93	38.32	3.19	60.15
$M = 2$ permutation (L_1 optimized)	1.59	41.58	2.80	65.93
$M = 2$ permutation (L_2 optimized)	1.90	34.18	3.09	55.18
$M = 3$ permutation (L_1 optimized)	1.43	34.80	2.50	55.55
$M = 3$ permutation (L_2 optimized)	1.64	28.21	2.82	47.29
Median	3.75	44.21	4.34	58.66
Weighted Median (MAE optimized)	2.81	173.79	2.83	60.41
Noisy Image	8.01	992.68	15.66	1937.80

3.5.1 Noise Smoothing

Consider first the smoothing of a single frame image corrupted by impulse noise. Impulse noise arises in numerous practical systems. In the following we utilize a salt-and-pepper model of impulse noise in which corrupted pixels are either saturated (set to white) or depleted (set to black). For such noise processes, filters that utilize rank order information have proven particularly useful. Thus, we compare the results of SR selection filters to two commonly used rank order based filters, the median and weighted median filters.

In a typical impulse-corrupted image, the majority of the pixels are not corrupted (i.e., they retain the desired value). Thus, in practice it is not necessary to utilize the full SR ordering information to identify and appropriately process the image. A pixel can be identified as an outlier, in many cases, by simply examining its rank among the ordered observations. This can be accomplished through row selection of the SR matrix, which results in an M permutation filter.

Table 3.2 gives the MSE and MAE results for the M permutation, median, and weighted median filters operating on the image Lena corrupted with 10% and 20% salt-and-pepper noise. A 5×5 observation window was utilized in each case, with the exception of the median, which utilized a 3×3 window. Since the median is spatially blind, smaller observation windows must be employed to avoid excessive smoothing. The M permutation and weighted median filters were optimized utilizing the left half of the image and observation noise with the same statistics as in the filtered image. Also, the L_1 and L_2 norms were utilized for the M permutation optimization.

An examination of Table 3.2 shows that M permutation filters produce the lowest error measures. This is in agreement with a subjective evaluation of the output images, which are shown in Fig. 3.10 for the 10% corruption case. Note that the median (Fig.3.10e) removes the impulses but excessively smoothes fine detail, while the weighted median (3.10f) retains detail but allows impulses to pass. The M permutation optimized under the L_1 norm (Fig. 3.10c) retains the fine detail while removing most impulses. If full impulse removal is a priority, then the L_2 norm (Fig. 3.10d), which heavily penalizes the passing of an impulse, can be used to yield an M permutation filter that removes virtually all impulses while preserving most fine detail structure. Finally, note that increasing the order of the M permutation filter beyond 1 yields minimal improvement in the output. Thus, the removal of impulse noise can be successfully addressed with the use of partial SR information.

To illustrate how this partial SR information is used in the filter decision process, the selection rule for the $M = 1$ case is shown in Fig. 3.11. Since the $M = 1$ selection rule operates strictly on the rank of the central observation sample, r_δ, we can plot $S_{\text{opt}}(r_\delta)$. The selection rule is plotted for a 9×9 filter operating on Lena corrupted with several impulse probabilities. Note that each of the functions has a linear region in which the input rank equals the output rank, allowing detail to be preserved. However, when r_δ is in the extreme ranks, the output is a rank closer to the median, providing impulse rejection. The break point between detail preservation and impulse rejection moves as a function of impulse probability. For different noise types, such as Gaussian and contaminated Gaussian, the optimum rank selection rule may change significantly [Har94]. For example, with Gaussian noise, the selection function is very linear with a slope that decreases with increasing noise levels [Har94].

Consider next a more general noise corruption process and a signal with changing statistics. A more general noise model is achieved by considering contaminated Gaussian noise, in which a noise process $\Phi(\sigma_1, \sigma_2, p)$ corrupts each pixel with additive Gaussian noise $\mathcal{N}(0, \sigma_1)$ or $\mathcal{N}(0, \sigma_2)$ with probability p and $1 - p$, respectively. For the signal with changing statistics, we considered the 100 frame video sequence Susie. To illustrate the concept of coloring the SR matrix, we filtered the sequence with one, two, three and four color permutation filters and compared the results to that obtained with the stack filter [Coy89, Tab96, Wen86] and L-ℓ filter [Gan91, Pal89, Pal90] processing. Note that both stack and L-ℓ filters utilize partial SR ordering information and that the stack filter contains the median and weighted median filters as subsets. Each filter was designed with a 13 sample spatial window consisting of all of the samples two or fewer city-block steps from the estimate location. This results in the diamond shaped window illustrated in Fig. 3.12, which also shows the spatial equivalences established through coloring. It can be seen that each of these is invariant under spatial rotations of 90°, 180°, and 270°and thus exploits the spatial rotation invariance of most image statistics.

Consider the case in which the video sequence is corrupted by $\Phi(4, 100, 0.085)$ contaminated noise. To illustrate the effect of changing signal statistics on the

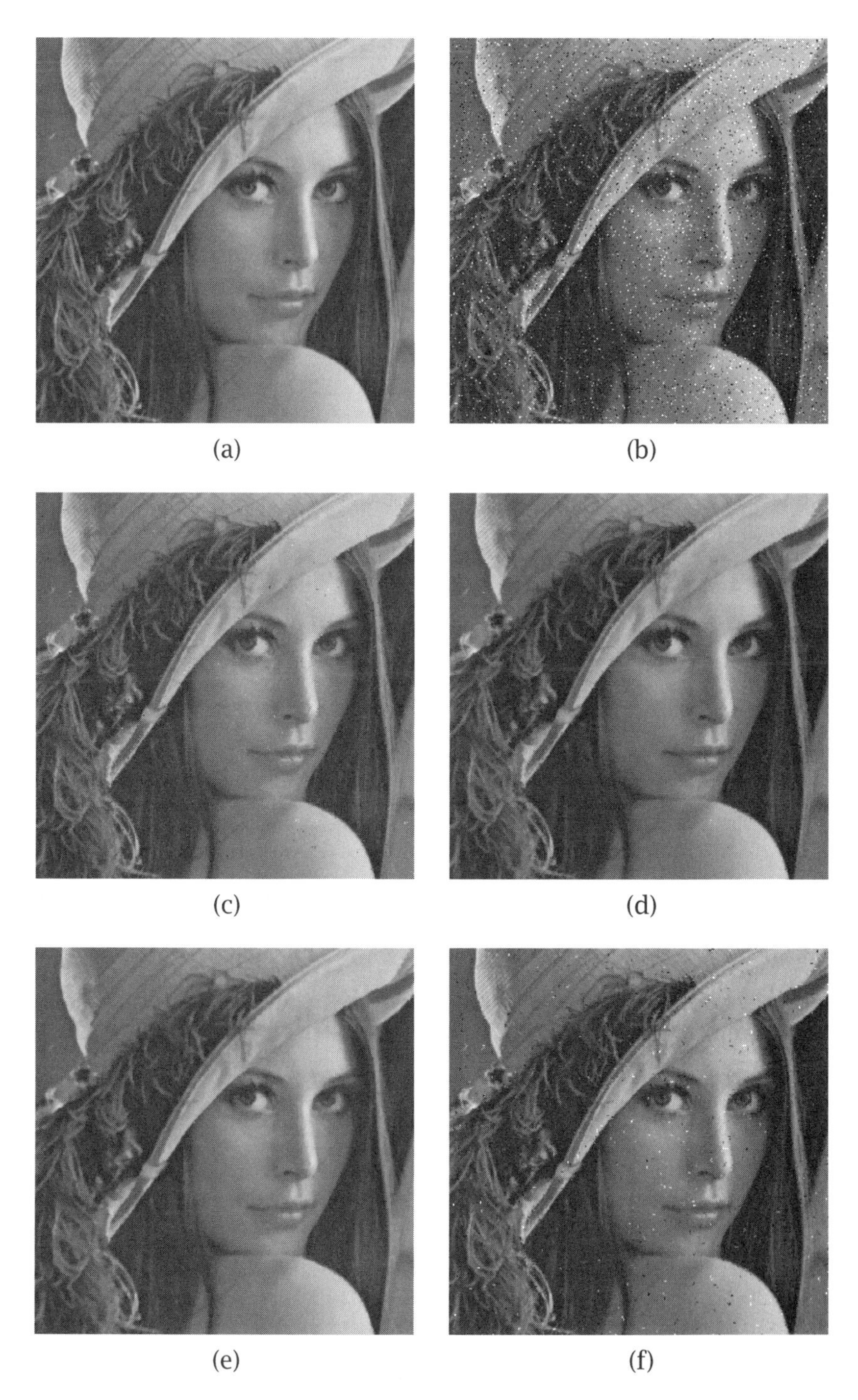

Figure 3.10: Restoration of images with impulse noise: (a) original image and (b) corrupted image with 10% impulses; 5×5 $M = 1$ permutation filter output with (c) L_1 and (d) L_2 optimization, (e) 3×3 median filter output, and (f) 5×5 weighted median filter output.

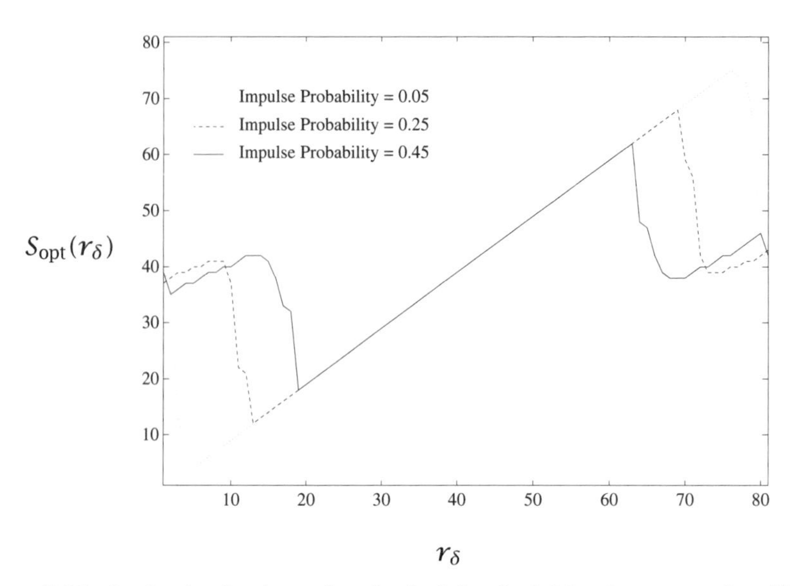

Figure 3.11: Optimal selection rules $S_{\text{opt}}(r_\delta)$ for 9×9 $M = 1$ permutation filter operating on Lena corrupted by impulse noise. (Reproduced with permission from [Har94]. © 1994 IEEE.)

estimates, we optimized each filter on the first five frames of the sequence and then used that to filter all 100 frames. The frame-by-frame estimate MAE values are shown in Fig. 3.12. Each of the filters produces robust sequence estimates in the sense that the estimate errors, in general, do not increase once the filter moves beyond the end of the training set (first five frames). There is a notable exception in the region of frame 50, where the sequence contains rapid motion and each method's performance suffers. The estimate error plot shows that the error is indeed decreased by using a finer quantization (more colors) of the SR matrix, although minimal performance gain is realized by increasing the number of colors beyond two in this case. For cases in which the statistics of the interference are not spatially invariant, such as tone interference, utilizing increasingly fine SR matrix quantizations (more colors) does lead to improved performance [Arc95, Bar97].

To allow for a subjective valuation, we show in Fig. 3.13 the original/noisy, stack, L-ℓ, and 3 color permutation filter estimates of frame 100. The right half of each figure shows the filter estimate, while the left half shows the scaled absolute estimate errors. The figures show that while each filter removed most of the impulses, the stack filter estimate (Fig. 3.13c) contains more residual impulses than the L-ℓ (Fig. 3.13b) or colored permutation (Fig. 3.13d) filter estimates. The estimates and errors also show that the stack and L-ℓ filters smooth details, such as the edges along the phone, hand, and facial features, to a greater extent than the colored permutation filter.

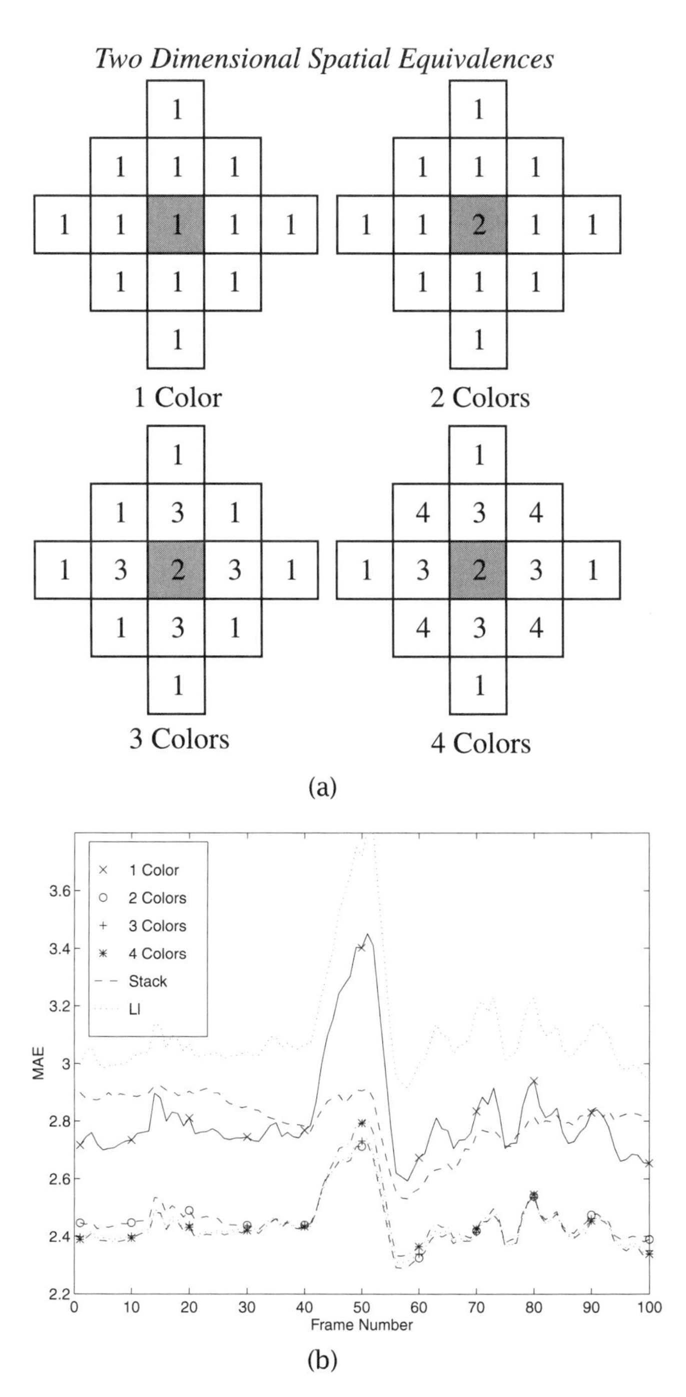

Figure 3.12: (a) Spatial equivalent established through coloring. (b) Individual frame estimate absolute errors for SR selection, stack, and L-ℓ filters operating on the Susie sequence corrupted by $\Phi(4, 100, 0.085)$ contaminated Gaussian noise. (Reproduced with permission from [Bar97]. © 1997 IEEE.)

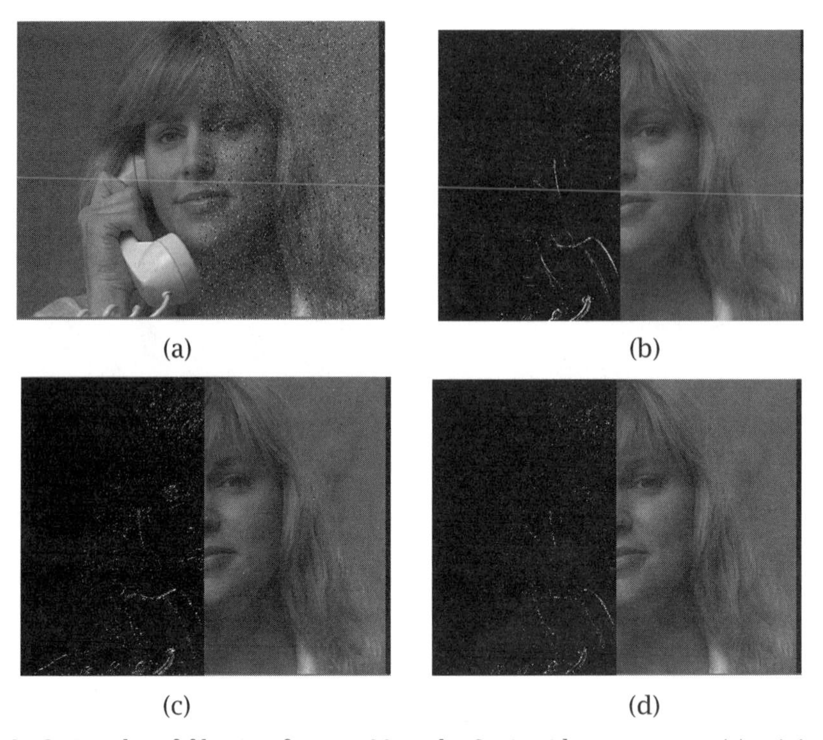

(a) (b)

(c) (d)

Figure 3.13: Results of filtering frame 100 in the Susie video sequence: (a) original (left) and noisy (right), and scaled absolute estimate errors (left) and filter estimates (right) for (b) $L - \ell$, (c) stack, and (d) three color permutation filters. (Reproduced with permission from [Bar97]. © 1997 IEEE.)

3.5.2 Interpolation

Consider next the zooming, or interpolation, of an image. As a simple example, suppose we wish to double the size of an image from $K \times K$ to $2K \times 2K$. A practical method for achieving this result is to insert a blank pixel between each pair of neighboring pixels in the $K \times K$ image, and then estimate the value of the blank pixels based on the surrounding pixels with known value. This results in a polyphase interpolation scheme, in which various filtering methods can be used to estimate the value of the inserted pixels. In the case of image doubling, utilizing the nearest set of known pixel values results in a four sample observation window.

While numerous filtering schemes can be utilized to determine the value of inserted pixels, methods that preserve edges and fine details produce results that are subjectively appealing. Since edge preservation is particularly important, the observation set can be augmented with mean values that, as shown in Sec. 3.3, allow edge inflection points to be identified. To allow identification of directional edges, the four sample observation set can be augmented as $\mathbf{x}'_\ell = [x_1, x_2, x_3, x_4, (x_1 + x_3)/2, (x_2 + x_4)/2]$. The SR ordering information of this extended observation set allows edge inflection points to be identified in the vertical and horizontal directions.

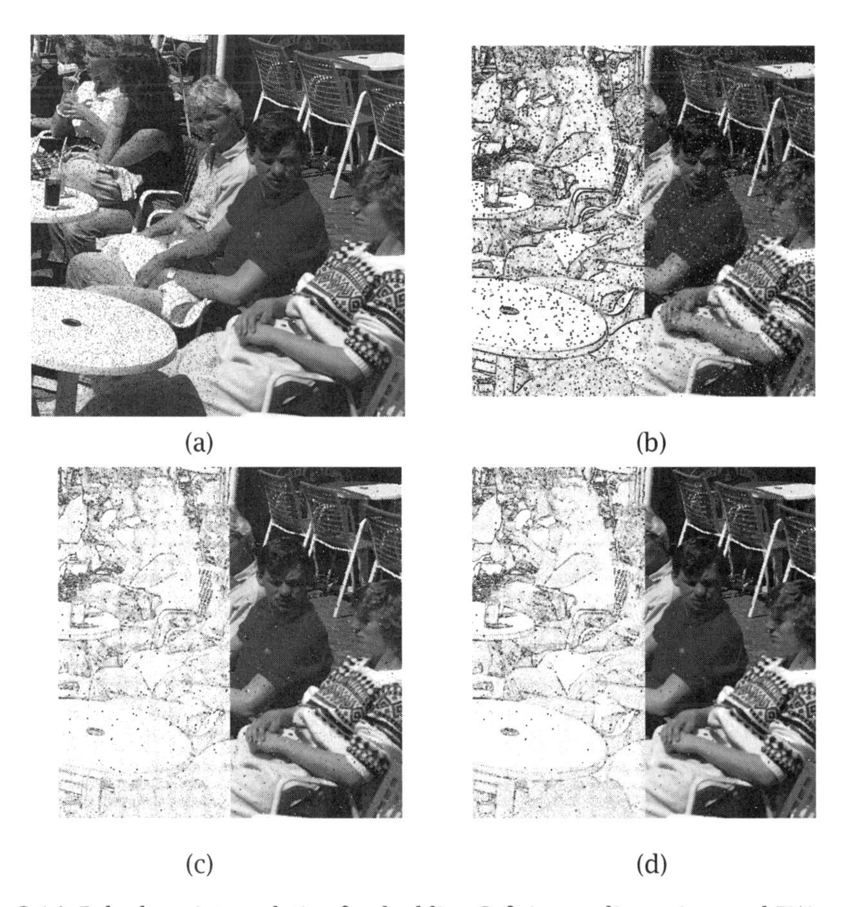

(a) (b)

(c) (d)

Figure 3.14: Polyphase interpolation for doubling Cafe image dimensions and 5% impulses: (a) noisy observation (left) and desired image (right), and interpolation errors (left) and results (right) for (b) linear, (c) median, and (d) extended permutation filters.

Table 3.3 reports the polyphase interpolation error results for linear, median, and extended permutation filter based doubling of the image Cafe corrupted by 5% and 10% impulses. Since the polyphase interpolation results in observation pixels being inserted directly into the interpolated image, 3×3 and 5×5 postfiltering of these samples is examined. Figure 3.14 shows the resulting interpolated images and interpolation errors for the 5% contamination case and 3×3 postfiltering. The error results and images indicate that the linear interpolation (Fig. 3.14b) performs poorly, passing impulses and excessively smoothing edges. While prefiltering can be used to minimize the number of impulses in the linear interpolation, this further exacerbates the smoothing introduced by the linear filter. The nonlinear techniques, in contrast, pass few impulses and retain sharp edge boundaries. This is particularly true for the extended permutation filter (Fig. 3.14d), which produces the fewest interpolation errors and rejects impulses while preserving edge boundaries.

Table 3.3: Polyphase Interpolation Errors for Doubling Cafe Image Dimensions utilizing Four Nearest Neighbors to Estimate Missing Pixel Values and Two Filter Windows to Smooth Noisy Observation Pixels in Expanded Image.

Interpolation Scheme	3×3 Window		5×5 Window	
	MAE	MSE	MAE	MSE
Linear (5% impulses)	19.39	1167.0	20.07	1204.0
Median (5% impulses)	12.15	585.3	13.00	638.2
Extended permutation (5% impulses)	11.45	586.2	12.36	672.3
Linear (10% impulses)	23.80	1614.0	24.20	1617.0
Median (10% impulses)	13.74	748.8	14.44	787.7
Extended permutation (10% impulses)	12.88	744.6	13.49	786.3

3.5.3 Image Restoration

As a final illustration, consider the case of an image that has been both blurred and corrupted by noise, which is common in practical image acquisition systems. Figures 3.15a and 3.15b show desired and acquired versions of the image Aerial captured from an airborne platform. The acquisition process was modeled as a 3×3 blurring mean filter with 2% impulse noise corruption. Thus, restoration of such images requires noise smoothing *and* inversion of the blurring process. Unfortunately, noise smoothing and blur elimination are competing goals, and traditional noise smoothing algorithms tend to worsen, not invert, image blur.

Several specifically designed nonlinear filters have been developed to jointly address noise smoothing and blur inversion, or edge enhancement. Among these techniques are the comparison and selection (CS) filter [Lee87], the lower-upper-middle (LUM) filter [Har93], and the weighted majority of samples with minimum range (WMMR) filter [Lon93]. Utilizing various methods, these filters attempt to identify edge inflection points and apply appropriate processing on each side of an edge while minimizing the effect of outliers. As discussed in Sec. 3.3 and illustrated in the previous example, edge inflection points can be identified through the SR ordering information when the observation set is augmented to include the sample mean. To ensure robust behavior, we can augment the observation set with the alpha trimmed mean, which results in a robust extended permutation filter.

Table 3.4 gives the error measures for the extended permutation, CS, LUM, and WMMR filters operating on the blurred and noisy realization of Aerial. The desired, observed, and filtered images are shown in Fig. 3.15. Note that each method successfully suppresses the impulses in the observed image. In attempting to invert the smoothing, or sharpen the edges, the CS (Fig. 3.15d), LUM (Fig. 3.15e), and WMMR (Fig. 3.15f) filters remove fine detail and produce blocky outputs. The

Table 3.4: Restoration Errors for Image Aerial Blurred by 3×3 Mean Filter and Corrupted by 2% Impulse Noise.

Filtering Method	Error measure	
	MAE	MSE
Extended permutation	7.62	126.36
Comparison and selection	9.42	176.03
Lower-upper-middle	8.72	153.47
WMMR	9.49	188.89
Noisy image	10.14	342.50

extended permutation filter (Fig. 3.15c), in contrast, produces sharp edges and retains fine detail. The extended SR ordering information is thus an appropriate feature for such images, and the selection filter based on this information produces images that are subjectively most appealing and have the lowest quantitative error measures.

3.6 Future Directions

The SR ordering of observation samples contains significant information about the underlying signal or image. This information, as the previous sections show, can be used to define a powerful class of selection filters that can be readily optimized and have proven useful in numerous filtering applications. Moreover, the SR ordering information is scalable, in the sense that partial SR information can be used in less demanding applications while full or extended SR information can be used in specialized or more demanding applications.

Both spatial and rank orderings of samples provide unique information: spatial correlations can be exploited through knowledge of spatial ordering, while rank ordering reveals relative value comparisons between samples. Although rank ordering reveals significant information, strict rank information can often be misleading. For instance, maximum-minimum samples are not necessarily outliers and often may be information bearing samples that should not merely be discarded. Recent research is beginning to address this shortcoming of rank ordering.

The standard SR matrix fully relates the spatial and rank orderings of the observed samples. However, the crisp, or binary, nature of this relation does not allow information regarding sample values or spread to be included in the crisp SR matrix. The relationship between the spatial and rank orderings can be generalized and moved from the binary domain to the real domain through fuzzy set theory [Bar98]. A fuzzy set $\tilde{R}$ is defined as a set of ordered pairs, $\tilde{R} = \{((a,b), \mu_{\tilde{R}}(a,b)) | a \in A, b \in B\}$, where $\mu_{\tilde{R}} : A \times B \mapsto [0,1]$ gives the

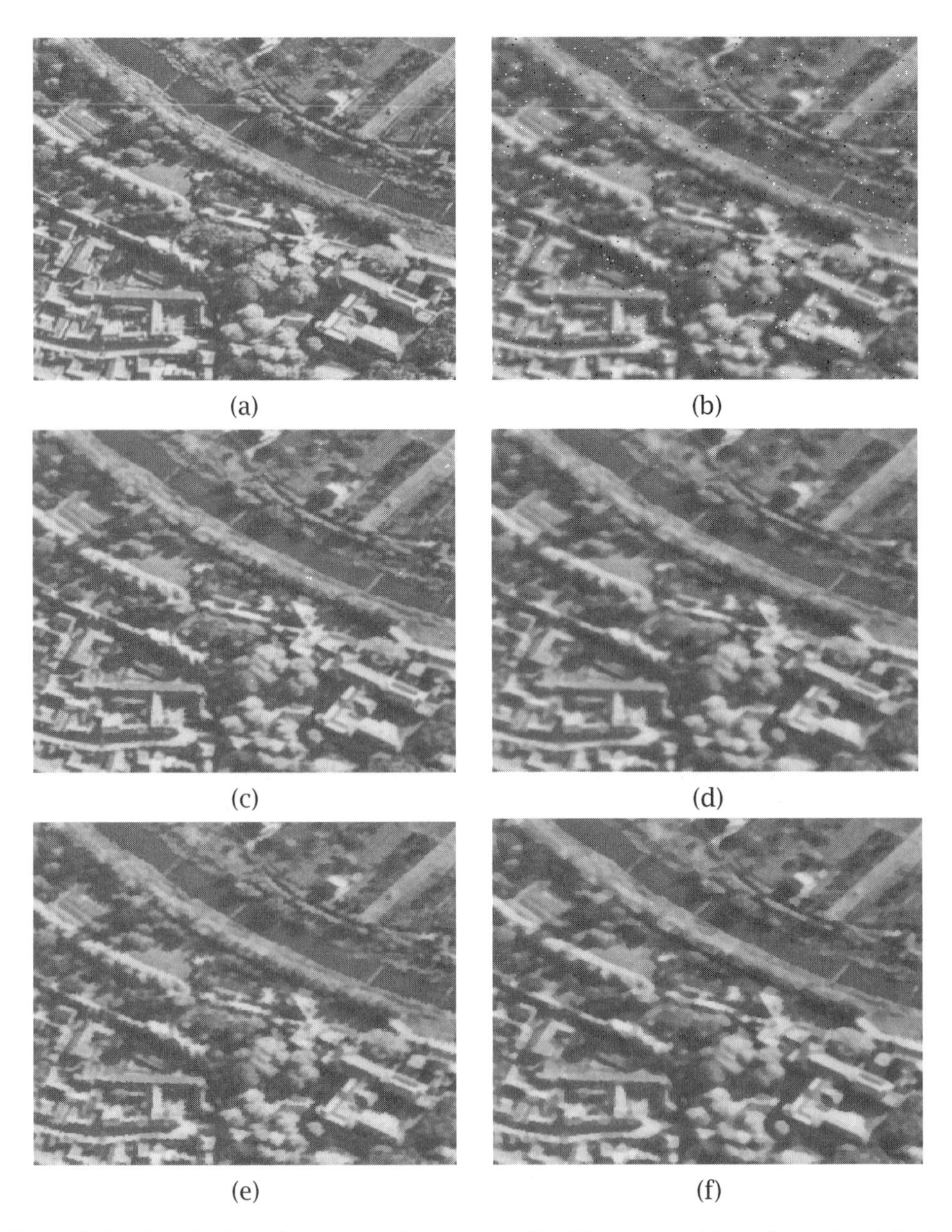

Figure 3.15: Restoration of image Aerial corrupted by blurring and impulse noise: (a) original image and (b) corrupted observation, restored using the filters, (c) extended permutation, (d) comparison and selection (CS), (e) lower-upper-middle (LUM), and (f) weighted majority of samples with minimum range (WMMR). (Reproduced with permission from [Har96]. © 1996 IEEE.)

degree of membership [Ros95]. The spatial and rank ordered samples can now be related through fuzzy sets, $\tilde{R} = \{((x_i, x_{(j)}), \mu_{\tilde{R}}(x_i, x_{(j)}) | x_i \in \mathbf{x}_\ell, x_{(j)} \in \mathbf{x}_L\}$, where $\mu_{\tilde{R}}(x_i, x_{(j)})$ gives the degree of membership between x_i and $x_{(j)}$. This relation can be conveniently represented in a fuzzy SR matrix,

$$\tilde{\mathfrak{R}} = \begin{bmatrix} \tilde{R}_{1,(1)} & \dots & \tilde{R}_{1,(N)} \\ \vdots & \ddots & \vdots \\ \tilde{R}_{N,(1)} & \dots & \tilde{R}_{N,(N)} \end{bmatrix}, \tag{3.51}$$

where $\tilde{R}_{i,(j)} = \mu_{\tilde{R}}(x_i, x_{(j)})$. This fuzzy relation produces a real valued relation matrix, that is, $\tilde{R}_{i,(j)} \in [0,1]$. In this context, $\tilde{R}_{i,(j)}$ denotes the degree to which x_i and $x_{(j)}$ are related by a membership function, such as the commonly used Gaussian or triangular membership functions:

$$\mu_{\mathrm{G}}(a,b) = \mathrm{e}^{-(a-b)^2/2\sigma^2}$$

and

$$\mu_{\mathrm{T}}(a,b) = \begin{cases} 1 - |a-b|/\sigma & \text{for } |a-b| \le \sigma, \\ 0 & \text{otherwise,} \end{cases} \tag{3.52}$$

where $\sigma > 0$ controls the spread of the membership functions.

The fuzzy spatial and rank order indexes, as well as the fuzzy spatial and rank ordered samples, can now be defined in a manner analogous to their crisp counterparts [Bar98],

$$\tilde{\mathbf{s}} = [1:N]\tilde{\mathfrak{R}}, \quad \tilde{\mathbf{r}} = \tilde{\mathfrak{R}}[1:N]^T, \tag{3.53}$$

$$\tilde{\mathbf{x}}_\ell = \tilde{\mathfrak{R}}\mathbf{x}_L^T, \quad \tilde{\mathbf{x}}_L = \mathbf{x}_\ell\tilde{\mathfrak{R}}. \tag{3.54}$$

The process of obtaining the degree of relation between samples is illustrated in Fig. 3.16a for $\mathbf{x}_\ell = [47, 5, 9]$ and a Gaussian membership function ($\sigma = 5$). As illustrated, a Gaussian membership function is centered on each observation sample. The membership between samples x_i and $x_{(j)}$ is determined by the nonlinear Gaussian mapping of the spread $|x_i - x_{(j)}|$ into $[0,1]$. This process results in a fuzzy SR matrix and a set of fuzzy rank indexes, which is illustrated in Fig. 3.16b. The fuzzy SR matrix and fuzzy rank indexes indicate, for instance, not only that $x_2 \le x_3 \le x_1$, but also that x_2 and x_3 are similar in value, while $x_2, x_3 \ll x_1$.

Fuzzy SR matrices thus not only contain the SR information present in the crisp SR matrix, but also indicate the spread between samples. This information can be used to extend the class of SR selection filters to the class of fuzzy SR selection filters, where control over the membership function introduces additional degrees of freedom into the filtering operation. These additional degrees of freedom can be used to improve filter performance. Indeed, extensions to the fuzzy median [Fla00] and affine order statistic filters [Fla98] indicate that considerable improvements in performance can be realized by considering sample spread in conjunction with SR ordering information. The investigation of additional fuzzy extensions to rank order and SR ordering based filters is an active area of research [An00a, An00b] that holds the potential to yield new filtering algorithms with improved performance that can be successfully used in ever more challenging applications.

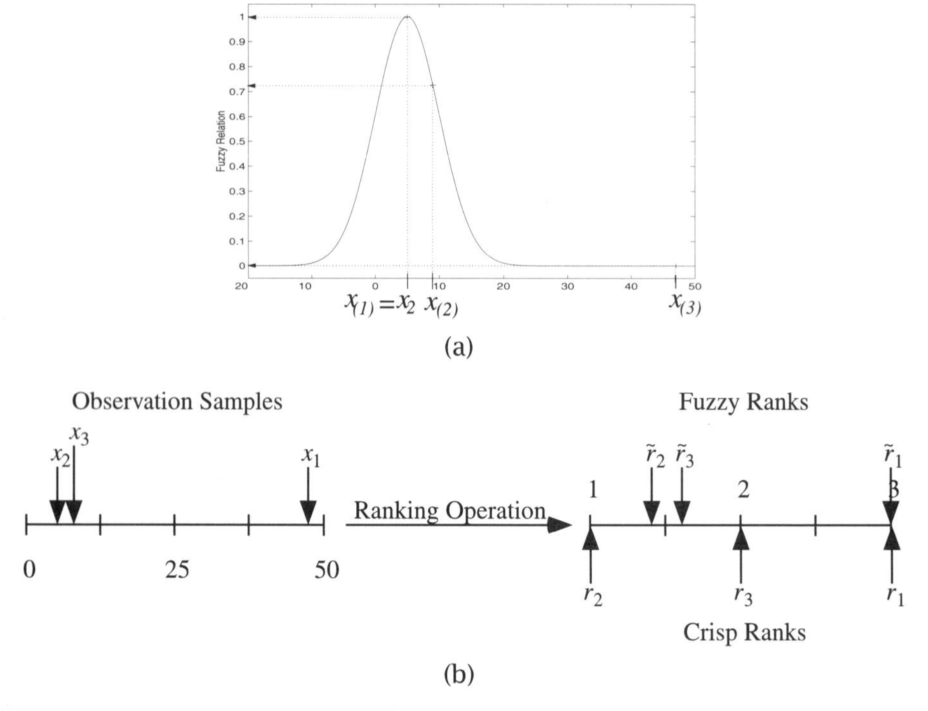

Figure 3.16: Fuzzy membership and fuzzy rank index examples for $\mathbf{x}_\ell = [47, 5, 9]$: (a) obtaining the membership between samples x_i and $x_{(j)}$ and (b) the resulting crisp and fuzzy rank indexes. The crisp ranks indicate only that $x_2 \le x_3 \le x_1$, while the fuzzy ranks indicate that $x_2 < x_3$ are similar in value and that $x_2, x_3 \ll x_1$. (From [Bar98]. © 1998 IEEE.)

References

[An00a] W. An and K. E. Barner. Fuzzy rank condition rank selection filters. *IEEE Trans. Image Process.* Submitted.

[An00b] W. An and K. E. Barner. Statistics of fuzzy order statistics. *IEEE Signal Processing Letters.* Submitted.

[Arc95] G. R. Arce, T. A. Hall, and K. E. Barner. Permutation weighted order statistic filter lattices. *IEEE Trans. Image Process.* **4**(8), 1070–1083 (August 1995).

[Bar92] K. E. Barner. *Permutation Filters: A Group Theoretic Class of Nonlinear Filters.* PhD thesis, University of Delaware, Newark, DE (May 1992).

[Bar94] K. E. Barner and G. R. Arce. Permutation filters: A class of non-linear filters based on set permutations. *IEEE Trans. Signal Process.* **42**(4), 782–798 (April 1994).

[Bar97] K. E. Barner and G. R. Arce. Design of permutation order statistic filters through group colorings. *IEEE Trans. Circ. Syst.* **44**(7), 531–547 (July 1997).

[Bar98] K. E. Barner, A. Flaig, and G. R. Arce. Fuzzy time-rank relations and order statistics. *IEEE Signal Process. Lett.* **5**(10), 252–255 (October 1998).

[Bar00] K. E. Barner. Colored l-ℓ filters with tap bias and their application in speech pitch detection. *IEEE Trans. Signal Process.* **9**(9) (September 2000). To be published.

[Coy89] E. J. Coyle, J.-H. Lin, and M. Gabbouj. Optimal stack filtering and the estimation and structural approaches to image processing. *IEEE Trans. Acoust. Speech Signal Process.* **37**(12), 2037–2066 (December 1989).

[Edg87] F. Y. Edgeworth. A new method of reducing observations relating to several quantities. *Philos. Mag. (Fifth Series)* **24** (1887).

[Fla98] A. Flaig, G. R. Arce, and K. E. Barner. Affine order statistic filters: a data-adaptive filtering framework for nonstationary signals. *IEEE Trans. Signal Process.* **46**(8), 2101–2112 (August 1998).

[Fla00] A. Flaig, K. E. Barner, and G. R. Arce. Fuzzy rank-ordering: theory and applications in filtering and detection. *Signal Process.*, Special Issue on Fuzzy Processing **80**(6), 1017–1036 (June 2000).

[Gan91] P. Gandhi and S. A. Kassam. Design and performance of combination filters. *IEEE Trans. Signal Process.* **39**(7), 1524–1540 (July 1991).

[Har93] R. C. Hardie and C. G. Boncelet, Jr. LUM filters: a class of rank order based filters for smoothing and sharpening. *IEEE Trans. Signal Process.* **41**(3), 1061–1076 (March 1993).

[Har94] R. C. Hardie and K. E. Barner. Rank conditioned rank selection filters for signal restoration. *IEEE Trans. Image Process.* **3**(2), 192–206 (March 1994).

[Har96] R. C. Hardie and K. E. Barner. Extended permutation filters and their application to edge enhancement. *IEEE Trans. Image Process.*, Special Issue on Nonlinear Signal Processing, **5**(6), 855–867 (June 1996).

[Lee87] Y. H. Lee and A. T. Fam. Edge gradient enhancing adaptive order statistic filter. *IEEE Trans. Acoust. Speech Signal Process.* **35**(5), 680–695 (May 1987).

[Lon93] H. G. Longbotham and D. Eberly. The WMMR filters: A class of robust edge enhancers. *IEEE Trans. Signal Process.* **41**(4), 1680–1685 (April 1993).

[Nij75] A. Nijenhuis and H. S. Wilf. *Combinatorial Algorithms.* Academic Press, New York (1975).

[Pal89] F. Palmieri and C. G. Boncelet, Jr. Ll-filters—a new class of order statistic filters. *IEEE Trans. Acoust. Speech Signal Process.* **37**(5), 691–701 (May 1989).

[Pal90] F. Palmieri and C. G. Boncelet, Jr. Frequency analysis and synthesis of a class of nonlinear filters. *IEEE Trans. Acoust. Speech Signal Process.* **38**(8), 1363–1372 (August 1990).

[Rob84] F. S. Roberts. *Applied Combinatorics.* Prentice-Hall, Englewood Cliffs, NJ (1984).

[Ros95] T. J. Ross. *Fuzzy Logic with Engineering Applications.* McGraw-Hill, New York (1995).

[Tab96] I. Tabus, D. Petrescu, and M. Gabbouj. A training framework for stack and Boolean filtering-fast optimal procedures and robustness case study. *IEEE Trans. Signal Process.* **5**(6), 809–826 (June 1996).

[Wen86] P. D. Wendt, E. J. Coyle, and N. C. Gallagher, Jr. Stack filters. *IEEE Trans. Acoust. Speech Signal Process* **34**(8), 898–911 (August 1986).

4

Signal-Dependent Rank-Ordered-Mean (SD-ROM) Filter

Eduardo Abreu

Lucent Technologies
Allentown, Pennsylvania

4.1 Introduction

Images are often corrupted by impulse noise due to a faulty image acquisition device or to channel transmission errors, and much research has been done on removing such noise. The objective is to suppress the noise while preserving the integrity of edges and detail information. To this end, nonlinear methods have been found to provide more satisfactory results than linear techniques. The most frequently used nonlinear method is the median filter [Arc86], which is superior to linear filters in its ability to suppress impulse noise and preserve edges. In median filtering, whether a pixel is corrupted or not, it is replaced with its local median within a window. Although noise suppression is obtained, too much distortion is introduced and the image features and details become blurred, particularly with a large size window. To gain improved performance, many generalizations of the median filter have been proposed [Bov83, Gab92, Pit90]. However, because these filters are still typically implemented uniformly across an image, they tend to modify pixels that are undisturbed by noise. As a result, they still tend to remove details from the image or leave too much impulse noise.

The *signal dependent rank-ordered-mean* (SD-ROM) filter [Abr95, Abr96b] is an efficient nonlinear algorithm to suppress impulse noise from highly corrupted images while preserving details and features.[1] It is applicable to all impulse noise models, including fixed-valued (equal height or salt-and-pepper) and random-valued (unequal height) impulses, covering the whole dynamic range. In the SD-ROM approach, the filtering operation is conditioned on the differences between the input pixel and the remaining rank-ordered pixels in a sliding window. Unlike median filters, the SD-ROM is not implemented uniformly across the image and thus can potentially achieve a better tradeoff between the suppression of noise and the preservation of details and edges without an undue increase in computational complexity. Two versions of the algorithm have been developed. The first version is based on a simple detection–estimation strategy involving thresholds. If a signal sample is detected as a corrupted sample, it is replaced with an estimation of the true value, based on neighborhood information; otherwise, it is kept unchanged. The second version is a more general approach that employs the concept of fuzzy logic and can also effectively restore images corrupted with Gaussian noise and mixed Gaussian and impulse noise.

We also consider the restoration of images corrupted by streaks, where a streak can be any sequence of pixels in the image that has been replaced with random values. We develop first a simple algorithm for the removal of horizontal streaks only and then generalize the technique for the case of general streaks of arbitrary angles, sizes, and shapes.

The outline of this chapter is as follows: First, in Secs. 4.2 and 4.3 we present the impulse noise model that is assumed throughout this chapter and some key definitions, respectively. Next, in Sec. 4.4, the SD-ROM filter structure is introduced and the thresholded version of the algorithm is described. This is followed in Sec. 4.5 by the generalized SD-ROM method, along with several design procedures based on nonrecursive and recursive implementation. Computer simulation results illustrating the performance of the SD-ROM are provided in Sec. 4.6. Finally, in Sec. 4.7 we address the restoration of images corrupted by streaks.

4.2 Impulse Noise Model

In a variety of impulse noise models for images, corrupted pixels are often replaced with values equal to or near the maximum or minimum of the allowable dynamic range [Kun84]. For 8-bit images, this typically corresponds to fixed values near 0 or 255. We consider here a more general noise model in which a noisy pixel can take on arbitrary values in the dynamic range according to some underlying probability distribution. Let $v(\mathbf{n})$ and $x(\mathbf{n})$ denote the luminance values of the original image and the noisy image, respectively, at pixel location $\mathbf{n} = [n_1, n_2]$. Then, for an impulse noise model with error probability p_e, we have

[1]The restoration of images corrupted by impulse noise is also addressed in Chapters 2 and 3.

$$x(\mathbf{n}) = \begin{cases} v(\mathbf{n}) & \text{with probability } 1 - p_e, \\ \eta(\mathbf{n}), & \text{with probability } p_e, \end{cases}$$

where $\eta(\mathbf{n})$ is an identically distributed, independent random process with an arbitrary underlying probability density function. For the computer simulation examples shown in this chapter, we generated corrupted images using both fixed-valued impulse noise (equal heights of 0 or 255 with equal probabilities, also known as salt-and-pepper noise) and impulse noise described by a uniform distribution from 0 to 255. However, the SD-ROM algorithm is not restricted to these cases and applies to other impulse noise models as well, including both additive and multiplicative noise. Moreover, as we show in Sec. 4.6, the method can effectively restore images corrupted with Gaussian noise and mixed Gaussian and impulse noise.

4.3 Definitions

Consider a 3×3 window centered at $x(\mathbf{n})$. We define $\mathbf{w}(\mathbf{n})$ as an 8-element observation vector containing the neighboring pixels of $x(\mathbf{n})$ inside the window [*excluding* $x(\mathbf{n})$, itself]:

$$\begin{aligned} \mathbf{w}(\mathbf{n}) &= [w_1(\mathbf{n}), w_2(\mathbf{n}), \ldots, w_8(\mathbf{n})] \\ &= [x(n_1 - 1, n_2 - 1), x(n_1 - 1, n_2), x(n_1 - 1, n_2 + 1), x(n_1, n_2 - 1), \\ &\quad x(n_1, n_2 + 1), x(n_1 + 1, n_2 - 1), x(n_1 + 1, n_2), x(n_1 + 1, n_2 + 1)], \end{aligned}$$

which corresponds to a left-to-right, top-to-bottom mapping from the 3×3 window to the 1-D vector $\mathbf{w}(\mathbf{n})$.

The observation samples can be also ordered by rank, which defines the vector

$$\mathbf{r}(\mathbf{n}) = [r_1(\mathbf{n}), r_2(\mathbf{n}), \ldots, r_8(\mathbf{n})],$$

where $r_1(\mathbf{n}), r_2(\mathbf{n}), \ldots, r_8(\mathbf{n})$ are the elements of $\mathbf{w}(\mathbf{n})$ arranged in ascending order such that $r_1(\mathbf{n}) \le r_2(\mathbf{n}) \le \ldots \le r_8(\mathbf{n})$.

Next, we define the *rank-ordered mean* (ROM)[2] $m(\mathbf{n})$ as

$$m(\mathbf{n}) = \frac{r_4(\mathbf{n}) + r_5(\mathbf{n})}{2}.$$

Finally, we define the *rank-ordered differences* $\mathbf{d}(\mathbf{n}) \in \mathbf{R}^4$ as

$$\mathbf{d}(\mathbf{n}) = [d_1(\mathbf{n}), d_2(\mathbf{n}), d_3(\mathbf{n}), d_4(\mathbf{n})],$$

[2] Note that the ROM nearly corresponds to the definition of the median filter for an even length window [Pit90], with the important distinction that $\mathbf{w}(\mathbf{n})$ does not include the center pixel of the original 3×3 window.

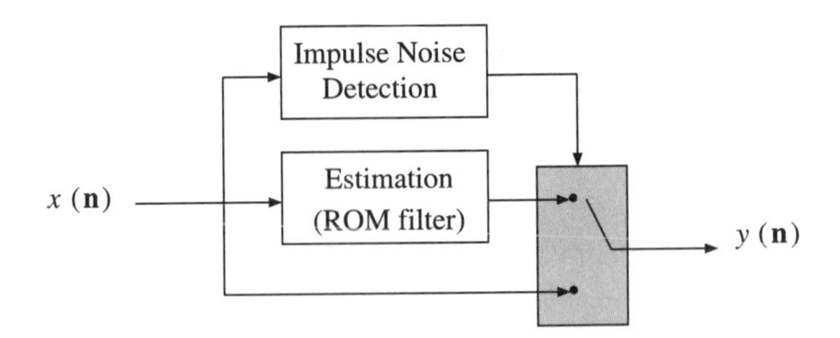

Figure 4.1: The SD-ROM filter structure.

where

$$d_i(\mathbf{n}) = \begin{cases} r_i(\mathbf{n}) - x(\mathbf{n}) & \text{for } x(\mathbf{n}) \le m(\mathbf{n}), \\ x(\mathbf{n}) - r_{9-i}(\mathbf{n}) & \text{for } x(\mathbf{n}) > m(\mathbf{n}), \end{cases}$$

for $i = 1, \ldots, 4$.

The rank-ordered differences provide information about the likelihood of corruption for the current pixel. For example, consider the rank-ordered difference $d_1(\mathbf{n})$: If this value is positive, then the current pixel $x(\mathbf{n})$ is either the smallest or largest value in the current window. If $d_1(\mathbf{n})$ is not only positive but also large, then an impulse is very likely. Together, the differences $d_1(\mathbf{n})$ through $d_4(\mathbf{n})$ reveal even more information about the presence of a corrupted pixel—even for the case when multiple impulses are present in the current window.

For images corrupted with only positive impulse noise, $d_i(\mathbf{n})$ should be redefined simply as

$$d_i(\mathbf{n}) = x(\mathbf{n}) - r_{9-i}(\mathbf{n}), \qquad i = 1, \ldots, 4.$$

Similarly, if only negative impulses are present, $d_i(\mathbf{n})$ is redefined as

$$d_i(\mathbf{n}) = r_i(\mathbf{n}) - x(\mathbf{n}), \qquad i = 1, \ldots, 4.$$

4.4 The SD-ROM Filter

Figure 4.1 shows a block diagram of the SD-ROM filter structure. The purpose of the impulse noise detector is to determine whether it believes the current pixel is corrupted. If a signal sample is detected as a corrupted sample, it is replaced with an estimation of the true value, based on the order statistics of the remaining pixels in the window; otherwise, it is kept unchanged.

The filter operates as follows:

Impulse Noise Detection. The algorithm detects $x(\mathbf{n})$ as a noisy sample if any of the following inequalities are true:

$$d_i(\mathbf{n}) > T_i, \qquad i = 1, \ldots, 4, \tag{4.1}$$

where T_1, T_2, T_3, T_4 are threshold values, with $T_1 < T_2 < T_3 < T_4$.

Estimation of the True Value. If $x(\mathbf{n})$ is detected as a corrupted sample, it is replaced by the ROM filter output $m(\mathbf{n})$; otherwise, it is kept unchanged.

The ROM filter introduced in the method provides improved restoration performance compared to the conventional median filter, particularly for images corrupted with very high percentages of fixed-valued noise such as salt-and-pepper noise. The algorithm, as described above, is conditioned both on the rank-ordered differences *and* the threshold values, so we will also refer to it as the "thresholded SD-ROM." Computer simulations using a large variety of test images have shown that good results are obtained using thresholds selected from the following set of values:

$$T_1 \in \{4, 8, 12\}, \ T_2 \in \{15, 25\}, \ T_3 = 40, \ T_4 = 50.$$

The algorithm performs well even for suboptimally selected thresholds. In fact, the default values

$$T_1 = 8, \ T_2 = 20, \ T_3 = 40, \ T_4 = 50$$

appear to provide satisfactory results with most natural images corrupted with random-valued impulse noise. In general, the experimental results indicate that there is little need to consider threshold values outside the following intervals:

$$T_1 \leq 15, \ 15 \leq T_2 \leq 25, \ 30 \leq T_3 \leq 50, \ 40 \leq T_4 \leq 60.$$

We have found that the selection of the thresholds, although done experimentally, is not a laborious process. Due to the robustness of the algorithm, good results are usually obtained in just one to three trials.

We note other switching schemes for impulse noise removal exist in the literature [Flo94, Kun84, Kim86, Mit94, Sun94]. In these approaches, the output is switched between an identity and a median-based filter, and the decision rules are typically based on a single threshold of a locally computed statistic. These strategies tend to work well for large, fixed-valued impulses but poorly for random-valued impulses, or vice versa. In contrast, the SD-ROM is applicable to any impulse noise type while still employing simple decision rules.

In many cases, improved performance can be obtained if the method is implemented in a recursive fashion. In this case, the sliding window is redefined according to

$$\hat{\mathbf{w}}(\mathbf{n}) = [y_1(\mathbf{n}), \ldots, y_4(\mathbf{n}), w_5(\mathbf{n}), \ldots, w_8(\mathbf{n})],$$

where $y_i(\mathbf{n})$ corresponds to the filter output for each noisy input pixel $w_i(\mathbf{n})$.

Although we have described the algorithm for the case of 2-D signals, the method is general and applies to higher dimensional signals as well as to 1-D signals [Cha98]. Other window sizes and shapes are possible. The procedure in the general case follows similar steps. To detect the impulse noise, we rank order the samples inside a window, excluding the current sample and compare the differences between the current sample and the ordered samples to thresholds. A corrupted sample is replaced with the ROM value.

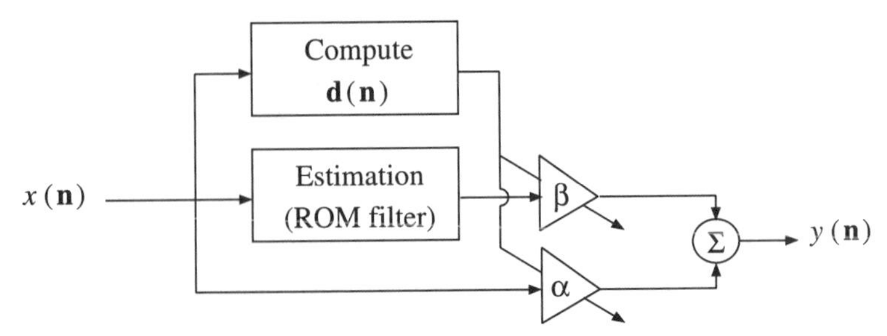

Figure 4.2: The generalized SD-ROM filter structure.

4.5 Generalized SD-ROM Method

As indicated in Fig. 4.1, the filtered output $y(\mathbf{n})$ is switched between the input $x(\mathbf{n})$ and the ROM value $m(\mathbf{n})$. The switching operation is conditioned on the rank-ordered differences $\mathbf{d}(\mathbf{n})$ and fixed threshold values. Using the concept of fuzzy logic, the method can be generalized by redefining $y(\mathbf{n})$ as a linear combination of $x(\mathbf{n})$ and $m(\mathbf{n})$ [Ara95]:

$$y(\mathbf{n}) = \alpha(\mathbf{d}(\mathbf{n}))x(\mathbf{n}) + \beta(\mathbf{d}(\mathbf{n}))m(\mathbf{n}), \tag{4.2}$$

where the weighting coefficients $\alpha(\mathbf{d}(\mathbf{n}))$ and $\beta(\mathbf{d}(\mathbf{n}))$ are constrained so that their sum is normalized to 1, implying that

$$\beta(\mathbf{d}(\mathbf{n})) = 1 - \alpha(\mathbf{d}(\mathbf{n})).$$

The coefficients $\alpha(\mathbf{d}(\mathbf{n}))$ and $\beta(\mathbf{d}(\mathbf{n}))$ are conditioned only on the rank-ordered differences (there are no thresholds). We will also refer to this implementation of the SD-ROM as the "generalized SD-ROM." A block diagram is presented in Fig. 4.2.

Note that the thresholded SD-ROM output is a special case of Eq. (4.2), in which $\alpha(\mathbf{d}(\mathbf{n}))$ can take on only the values 0 and 1. Pictorially, the distinction between the two approaches can be observed by comparing Fig. 4.3 and 4.4 for the simplified case in which $\mathbf{d}(\mathbf{n}) \in \mathbf{R}^2$ where $\mathbf{d}(\mathbf{n}) = [d_1(\mathbf{n}), d_2(\mathbf{n})]$.[3] Note that the region corresponding to $d_i(\mathbf{n}) > d_j(\mathbf{n})$, $i < j$, is outside the domain of $\alpha(\mathbf{d}(\mathbf{n}))$.

The fuzzy nature of the algorithm allows greater flexibility with regards to the removal of highly nonstationary impulse noise. Unlike the thresholded SD-ROM, the generalized method can also effectively restore images corrupted with other noise types such as Gaussian noise and mixed Gaussian and impulse noise. The values of the weighting coefficients $\alpha(\mathbf{d}(\mathbf{n}))$ are obtained by performing optimization using training data. The optimized values are stored in memory and the function $\alpha(\mathbf{d}(\mathbf{n}))$ is implemented as a lookup table. One primary difficulty with this approach is that for $\mathbf{d}(\mathbf{n}) \in \mathbf{R}^4$ and 8-bit images, the number of possible values of $\mathbf{d}(\mathbf{n})$ is very large, so the computational complexity and memory storage requirements associated with optimizing and implementing $\alpha(\mathbf{d}(\mathbf{n}))$ can be very

[3] For the example of Fig. 4.4, the coefficients were trained to restore the Lena image corrupted with 20% random-valued impulse noise using the least-squares design methodology presented in Sec. 4.5.1.

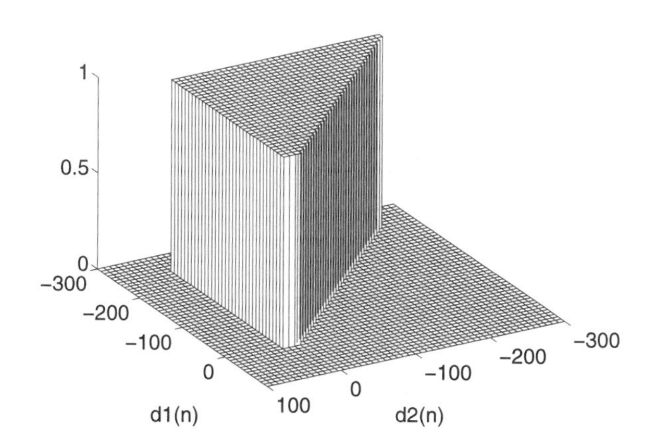

Figure 4.3: The function $\alpha(\mathbf{d}(\mathbf{n}))$ for the thresholded SD-ROM with $\mathbf{d}(\mathbf{n}) \in \mathbf{R}^2$.

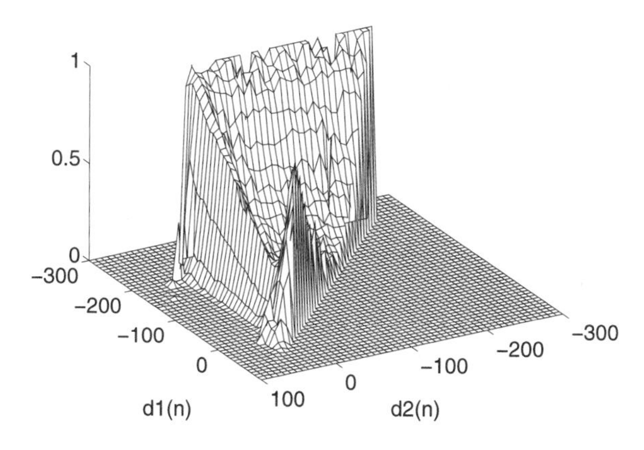

Figure 4.4: The function $\alpha\big(\mathbf{d}(\mathbf{n})\big)$ for the generalized SD-ROM with $\mathbf{d}(\mathbf{n}) \in \mathbf{R}^2$.

high. To overcome this problem, we simplify the method by partitioning the $\mathbf{R}^4$ vector space into M nonoverlapping regions A_i, $i = 1, \ldots, M$, and for each region we assign a constant value to $\alpha(\mathbf{d}(\mathbf{n}))$. Figure 4.5 illustrates this idea for $\mathbf{d}(\mathbf{n}) \in \mathbf{R}^2$ and $M = 16^2$.

We observe that the function $\alpha(\mathbf{d}(\mathbf{n}))$ shown in Fig. 4.5 is a piecewise constant approximation to the smooth function of Fig. 4.4. In the thresholded case, $M = 2$. We restrict all region boundaries to be parallel to the coordinate axes so that each region A_i can be represented as a Cartesian product of four scalar regions. Each scalar region is defined by an interval of the form $(q_{i-1}, q_i]$, where the q_i's are *decision levels* that define a distinct partition on $\mathbf{R}$. We will denote by α_i and β_i, respectively, the values of $\alpha(\mathbf{d}(\mathbf{n}))$ and $\beta(\mathbf{d}(\mathbf{n}))$ associated with the region A_i:

$$\begin{aligned} \alpha_i &= \alpha(\mathbf{d}(\mathbf{n}))|_{\mathbf{d}(\mathbf{n}) \in A_i}, \qquad i = 1, \ldots, M \\ \beta_i &= \beta(\mathbf{d}(\mathbf{n}))|_{\mathbf{d}(\mathbf{n}) \in A_i}, \qquad i = 1, \ldots, M. \end{aligned}$$

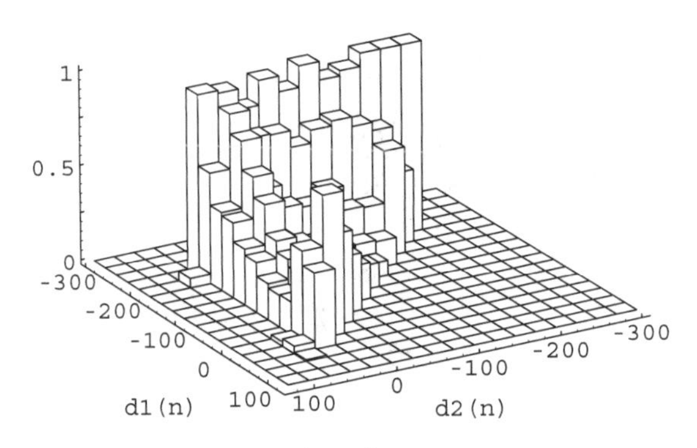

Figure 4.5: The simplified function $\alpha\big(\mathbf{d}(\mathbf{n})\big)$ for the generalized SD-ROM with $\mathbf{d}(\mathbf{n}) \in \mathbf{R}^2$.

Table 4.1: Scalar Partitions Used for Generalized SD-ROM ($M = 6^4 = 1296$). Partitions Shown for Each Scalar Dimension of $\mathbf{d}(\mathbf{n})$, with $T_1 = 8$, $T_2 = 20$, $T_3 = 40$, and $T_4 = 50$.

Scalar	Partition
$d_1(\mathbf{n})$	$[-\infty,\ T_1 - 20,\ T_1 - 5,\ T_1,\ T_1 + 5,\ T_1 + 20,\ \infty]$
$d_2(\mathbf{n})$	$[-\infty,\ T_2 - 20,\ T_2 - 5,\ T_2,\ T_2 + 5,\ T_2 + 20,\ \infty]$
$d_3(\mathbf{n})$	$[-\infty,\ T_3 - 30,\ T_3 - 10,\ T_3,\ T_3 + 10,\ T_3 + 30,\ \infty]$
$d_4(\mathbf{n})$	$[-\infty,\ T_4 - 30,\ T_4 - 10,\ T_4,\ T_4 + 10,\ T_4 + 30,\ \infty]$

Equation (4.2) becomes

$$y(\mathbf{n}) = \alpha_i x(\mathbf{n}) + \beta_i m(\mathbf{n}), \qquad i : \mathbf{d}(\mathbf{n}) \in A_i. \tag{4.3}$$

Experimental results for this partitioning strategy are presented in Sec. 4.6 using the scalar partitions shown in Table 4.1, although other partitions are possible. Interestingly, we have found in practice that the exact locations of the partitions has very little impact on the final restoration results as long as M is sufficiently large. This characteristic contrasts with the thresholded technique, wherein the values of the thresholds $\{T_k\}$ do impact the final results and must be selected more judiciously.

We now address the design of the weighting coefficients $\{\alpha_i\}$ and $\{\beta_i\}$ for both nonrecursive and recursive implementations, using the normalization

$$\beta_i = 1 - \alpha_i, \qquad i = 1, \dots, M. \tag{4.4}$$

For nonrecursive implementation, the algorithm operates on the pixels of the original noisy image only. In contrast, for the recursive case, pixels from the original image are systematically replaced by the output of previous filtering operations, and consequently, the sliding window $\mathbf{w}(\mathbf{n})$ can contain original and "restored" image pixels.

4.5.1 Least-Squares Design

Let $v(\mathbf{n})$, $x(\mathbf{n})$, and $y(\mathbf{n})$ represent the pixel values of the training image, the noise-corrupted training image, and the filtered output image, respectively. Because the determination of $\mathbf{d}(\mathbf{n})$ is independent of the filtering action for nonrecursive implementation, it is possible to derive the weighting coefficients $\{\alpha_i\}$ and $\{\beta_i\}$ that minimize the total squared error, given by

$$E_t = \sum_{\mathbf{n}} [y(\mathbf{n}) - v(\mathbf{n})]^2$$

for a given collection of training data. This equation can be expanded as

$$\begin{aligned} E_t &= \sum_{\mathbf{n}} \{\alpha(\mathbf{d}(\mathbf{n}))\, x(\mathbf{n}) + [1 - \alpha(\mathbf{d}(\mathbf{n}))]\, m(\mathbf{n}) - v(\mathbf{n})\}^2 \\ &= \sum_{i=1}^{M} \sum_{\mathbf{n}:\mathbf{d}(\mathbf{n})\in A_i} [\alpha_i x(\mathbf{n}) + (1 - \alpha_i) m(\mathbf{n}) - v(\mathbf{n})]^2 . \end{aligned} \tag{4.5}$$

Note that the inner sum represents the contribution of each particular α_i to the total squared error, so the entire expression can be minimized by independently minimizing each of these terms, for $i = 1, \ldots, M$. Accordingly, we take the partial derivative of E_t with respect to each α_i and arrive at

$$\frac{\partial E_t}{\partial \alpha_i} = \sum_{\mathbf{n}:\mathbf{d}(\mathbf{n})\in A_i} 2\,[x(\mathbf{n}) - m(\mathbf{n})]\,[\alpha_i x(\mathbf{n}) + (1 - \alpha_i) m(\mathbf{n}) - v(\mathbf{n})] .$$

Setting this term equal to zero and solving for α_i leads to a simple expression for the globally optimal weighting coefficients for each i, given by

$$\alpha_i^* = \frac{- \sum_{\mathbf{n}:\mathbf{d}(\mathbf{n})\in A_i} [x(\mathbf{n}) - m(\mathbf{n})]\,[m(\mathbf{n}) - v(\mathbf{n})]}{\sum_{\mathbf{n}:\mathbf{d}(\mathbf{n})\in A_i} [x(\mathbf{n}) - m(\mathbf{n})]^2} \tag{4.6}$$

for $i = 1, \ldots, M$.

If the training data and noise model are representative, the algorithm is likely to perform well on actual noisy images. Moreover, experimental results indicate that the performance of the algorithm is extremely robust with respect to the types of images and percentage of impulse noise on which the algorithm is trained.

4.5.2 Least-Mean-Square Design

Another possible design strategy is to use a variant of the least-mean-square (LMS) algorithm to compute the weighting coefficients [Ara95, Hay91, Kot92, Pal90, Pit90]. In this case, the objective is to design the weighting coefficients iteratively while continuously imposing the normalization constraint described in Eq. (4.4).

First, let

$$e(\mathbf{n}) = v(\mathbf{n}) - y(\mathbf{n})$$

denote the instantaneous error at pixel location $\mathbf{n}$. If we define

$$h(\mathbf{n}) = x(\mathbf{n}) - m(\mathbf{n}),$$

then $e^2(\mathbf{n})$ is given by

$$\begin{aligned} e^2(\mathbf{n}) &= \{v(\mathbf{n}) - \alpha(\mathbf{d}(\mathbf{n}))x(\mathbf{n}) - [1 - \alpha(\mathbf{d}(\mathbf{n}))]\, m(\mathbf{n})\}^2 \\ &= [(v(\mathbf{n}) - \alpha(\mathbf{d}(\mathbf{n}))h(\mathbf{n}) - m(\mathbf{n})]^2 . \end{aligned}$$

Taking the expected value given that $\mathbf{d}(\mathbf{n}) \in A_i$, we arrive at a conditional mean-square error (MSE),

$$J_i = E\left[e^2(\mathbf{n}) \mid \mathbf{d}(\mathbf{n}) \in A_i\right],$$

from which the total MSE J can be computed as

$$J = \sum_{i=1}^{M} J_i.$$

Similar to the least-squares design in Eq. (4.5), the overall error can be decomposed into M terms when the algorithm is implemented nonrecursively because the determination of $\mathbf{d}(\mathbf{n})$ is disconnected from the filtering operation. Thus, we can easily minimize the total MSE by separately minimizing each J_i using M distinct LMS algorithms. To this end, we take the partial derivative of J with respect to α_i:

$$\nabla J_i = \frac{\partial J}{\partial \alpha_i} = E\left[-2v(\mathbf{n})h(\mathbf{n}) + 2\alpha_i h(\mathbf{n})^2 + 2h(\mathbf{n})m(\mathbf{n}) \mid \mathbf{d}(\mathbf{n}) \in A_i\right].$$

As with traditional LMS, a noisy estimate to the gradient can be computed by simply dropping the expectation operator to obtain

$$\hat{\nabla} J_i = 2\left[\alpha_i h(\mathbf{n})^2 + h(\mathbf{n})m(\mathbf{n}) - v(\mathbf{n})h(\mathbf{n})\right].$$

By using the noisy gradient and substituting $x(\mathbf{n}) - m(\mathbf{n})$ back into $h(\mathbf{n})$, we arrive at the following update procedure for the weighting coefficients

$$\alpha_i^{\mathbf{n}+\mathbf{1}} = \alpha_i^{\mathbf{n}} + \mu_i\, [x(\mathbf{n}) - m(\mathbf{n})]\, e(\mathbf{n}), \tag{4.7}$$

$$\beta_i^{\mathbf{n}+\mathbf{1}} = 1 - \alpha_i^{\mathbf{n}+\mathbf{1}}, \tag{4.8}$$

for $i = 1, \ldots, M$. In both equations, the vector $\mathbf{1}$ defines the movement of the sliding window $w(\mathbf{n})$. Using the traditional assumptions for the derivation of LMS convergence [Hay91], it can easily be shown that the following conditions are sufficient for steady state convergence of the algorithm in the mean and mean square:

$$0 < \mu_i < \frac{1}{E\left[(x(\mathbf{n}) - m(\mathbf{n}))^2 \mid \mathbf{d}(\mathbf{n}) \in A_i\right]}, \qquad i = 1, \ldots, M.$$

Simulations indicate that when applied to noisy image data this approach produces results nearly identical to the least-squares design procedure described in Sec. 4.5.1. For example, when restoring the Miramar image corrupted by 20% impulse noise, the difference in the optimized weighting coefficients is less than 2%.

4.5.3 Recursive Design

We now consider the design of the weighting coefficients for the case when the algorithm is to be implemented in a recursive fashion. The idea of applying the median filter recursively has been examined previously [Nod82] and was shown to produce a more highly correlated signal with increased blurring. However, because the SD-ROM approach selectively controls the filtering operation according to the amount of detected noise, better noise suppression is generally obtained without additional blurring when the algorithm is implemented in a recursive manner. This modification, while providing better restoration overall, precludes a closed form expression for the optimal weighting coefficients. This consequence transpires due to the dependence of the rank-ordered differences on past outputs of the filter. As an alternative, we outline two heuristic strategies based on the least-squares and the LMS design techniques presented in the two previous subsections.

The first strategy is simply to utilize the straightforward nonrecursive least-squares design methodology of Sec. 4.5.1 while simultaneously applying the filter recursively to the training data. Specifically, for $i = 1, \ldots, M$ we generate a current estimate of α_i^* from Eq. (4.6) using a subset of the available training data. This estimate can be computed by dividing two running sums according to the numerator and denominator of Eq. (4.6). When "enough" data have been considered, the training pixels are updated according to Eq. (4.3) using the current value of α_i. For subsequent samples of the training data, $x(\mathbf{n})$ in Eq. (4.6) is replaced by its corresponding filtered output. This process continues until all of the training data have been considered, at which time the weighting coefficients are fixed to their most recently updated value.

Similarly, the LMS algorithm described in Sec. 4.5.2 can be implemented recursively using Eqs. (4.7) and (4.8), with the only difference being that the ROM filter output $m(\mathbf{n})$ now depends on past outputs. As a consequence, the current filter output impacts future values of $\mathbf{d}(\mathbf{n})$, so the MSE expression can no longer be decomposed into M independent terms. Despite this fact, for computer simulations on real image data, the weighting coefficients converge to values nearly identical to those obtained with the recursive updating strategy that we first described. Moreover, in Table 4.2 we show that these approaches, while lacking tractability in analysis, can experimentally converge to solutions that are better than their nonrecursive counterparts. This improvement is typically observed when the images are highly corrupted. For cases when the MSE does increase (usually for impulse noise less than 10%), the increases are small and are typically offset by a rather significant drop in the number of missed impulses.

4.6 Experimental Results

We illustrate in this section the restoration performance of the SD-ROM technique. The following experiments were conducted: (1) restoration of images corrupted by various noise types for quantitative and perceptual evaluation of the noise sup-

Table 4.2: Comparison of PSNRs for Restored Lena Image Using Nonrecursive and Recursive Design. Results Shown for 20%, Random-valued Impulse Noise, and $M = 1296$.

Nonrecursive		Recursive	
Least squares	LMS	Least squares	LMS
31.55 dB	31.58 dB	32.95 dB	32.83 dB

Table 4.3: Restoration of Lena Image Corrupted with 20% Fixed- and Random-valued Impulse Noise.

Algorithm	MAE	MSE	PSNR
	Fixed-valued impulses		
Median filter (3×3)	4.29	90.25	28.57 dB
Median filter (5×5)	5.20	85.95	28.78 dB
SD-ROM ($M = 2$)	1.32	29.23	33.47 dB
SD-ROM ($M = 1296$ outside training set)	1.33	22.30	34.65 dB
SD-ROM ($M = 1296$ inside training set)	1.20	17.51	35.70 dB
	Random-valued impulses		
Median filter (3×3)	4.39	68.63	29.76 dB
Median filter (5×5)	5.50	89.82	28.59 dB
SD-ROM ($M = 2$)	1.62	36.45	32.51 dB
SD-ROM ($M = 1296$ outside training set)	1.91	32.96	32.95 dB
SD-ROM ($M = 1296$ inside training set)	1.78	29.94	33.37 dB

pression and edge preservation capabilities of the SD-ROM method, (2) an assessment of the SD-ROM with respect to the percentage of impulse noise corruption, and (3) a demonstration of the robustness of the method with respect to the type of data used for training. For all the examples shown the SD-ROM was implemented recursively.

Experiment 1. Initially, the 8-bit, 512×512 image Lena was corrupted with 20% impulse noise. We considered both fixed-valued impulse noise (equal heights of 0 or 255 with equal probabilities) and randomly distributed impulse noise (uniformly distributed from 0 to 255). Table 4.3 shows the mean absolute errors (MAEs), mean-square error, and peak signal-to-noise ratios (PSNRs) of the images restored, with the various methods presented, along with the traditional median filter. The PSNR is defined as

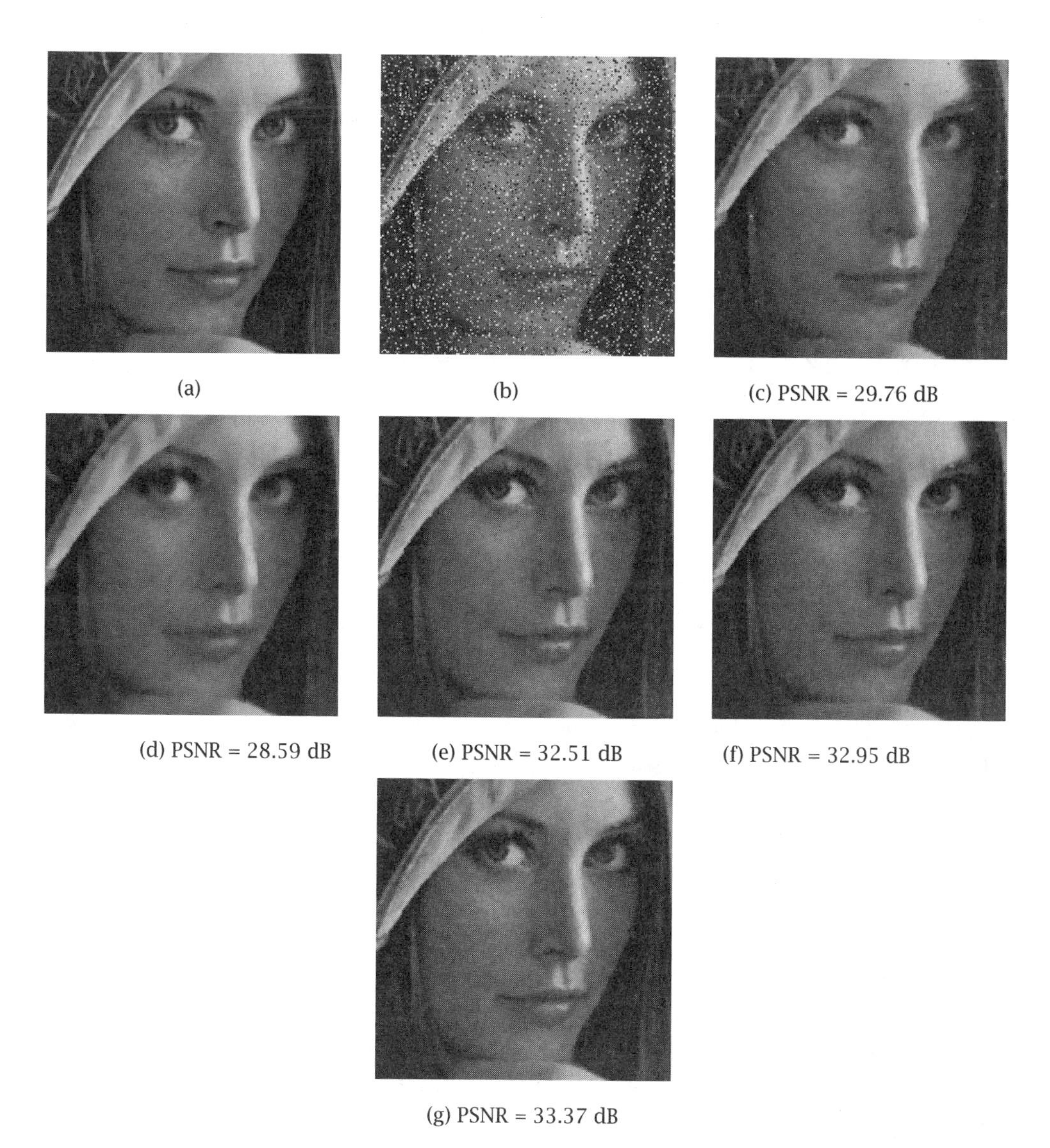

(a) (b) (c) PSNR = 29.76 dB

(d) PSNR = 28.59 dB (e) PSNR = 32.51 dB (f) PSNR = 32.95 dB

(g) PSNR = 33.37 dB

Figure 4.6: Face region from Lena image: (a) original and (b) image corrupted with 20% random-valued impulse noise, which was restored using (c) the 3×3 and (d) the 5×5 median filters and the SD-ROM with (e) $M = 2$, (f) $M = 1296$, outside the training set, and (g) $M = 1296$, inside the training set.

$$\text{PSNR} = 10 \log_{10} \left(\frac{\sum_{\mathbf{n}} 255^2}{\sum_{\mathbf{n}} [y(\mathbf{n}) - v(\mathbf{n})]^2} \right) \text{ dB}.$$

Portions of the images corresponding to the case of randomly distributed impulse noise can be seen in Fig. 4.6.

We notice that even the simple thresholded approach ($M = 2$) achieves better restoration than the median filter. Another important observation is that the performance of the generalized SD-ROM ($M = 1296$) is only moderately diminished

Table 4.4: Restoration of Lena Image Corrupted with 40% Fixed-valued Impulse Noise.

Algorithm	MAE	MSE	PSNR
Median filter (3×3)	9.95	865.91	18.76 dB
Median filter (5×5)	6.16	137.65	26.74 dB
SD-ROM ($M = 1296$ outside training set)	2.87	59.77	30.37 dB

when the optimized coefficients are used to restore an image outside the training set. For instance, in this experiment the coefficients trained on the Miramar image were used to restore Lena.

Next, we show in Table 4.4 and Fig. 4.7 the restoration of the Lena image corrupted with 40% fixed-valued impulse noise. For such a high percentage of impulse noise corruption the generalized SD-ROM version is strongly recommended. We note that the good performance obtained in Fig. 4.7d was due in part to the ROM filter employed in the method. In this particular example, if the ROM is replaced with the median, the PSNR of the restored image is reduced by 2 dB, with a considerable increase in the number of missed impulses. By retaining the noisy pixel as part of the neighborhood window, the median filter also introduces edge jitter into the estimate [Nod84]. This distortion can be a dominant source of error for the median filter when impulse noise is present. In contrast, the ROM filter employed for signal estimation does not include the center pixel in its window $w(\mathbf{n})$ and consequently is less hindered by this particular artifact.

Finally, we assessed the performance of the SD-ROM (generalized version only) with respect to other noise types. Restoration results corresponding to the Miramar image are summarized in Table 4.5 for both Gaussian noise ($\sigma^2 = 100$) and mixed Gaussian noise ($\sigma^2 = 100$) and random-valued impulse noise (20%). For the latter case, zoomed portions of the images obtained are displayed in Fig. 4.8. These results show that while the SD-ROM was originally designed for removal of impulse noise, the generalized version also works well for images corrupted with Gaussian or mixed noise.

Experiment 2. In this experiment, restoration performance was assessed according to the local density of the corrupted pixels in the Lena image. This type of evaluation determines filter effectiveness when a large percentage of a particular window's pixels are corrupted. While every pixel in the test image is equally likely to be an impulse, the actual number of corrupted pixels in a given 3×3 window will vary by location. Therefore, to better assess performance, we computed the PSNRs of the restored image with respect to the number of corrupted pixels in the 3×3 neighborhood of a given spatial location (Fig. 4.9). Notice that, on average, when more than half of the pixels in the current window are corrupted, the SD-ROM is able to improve the PSNR of the corrupted image by over 10 dB. This behavior can be attributed largely to the recursive implementation of the algorithms. When used recursively, even if more than half the pixels in a particular 3×3 window are

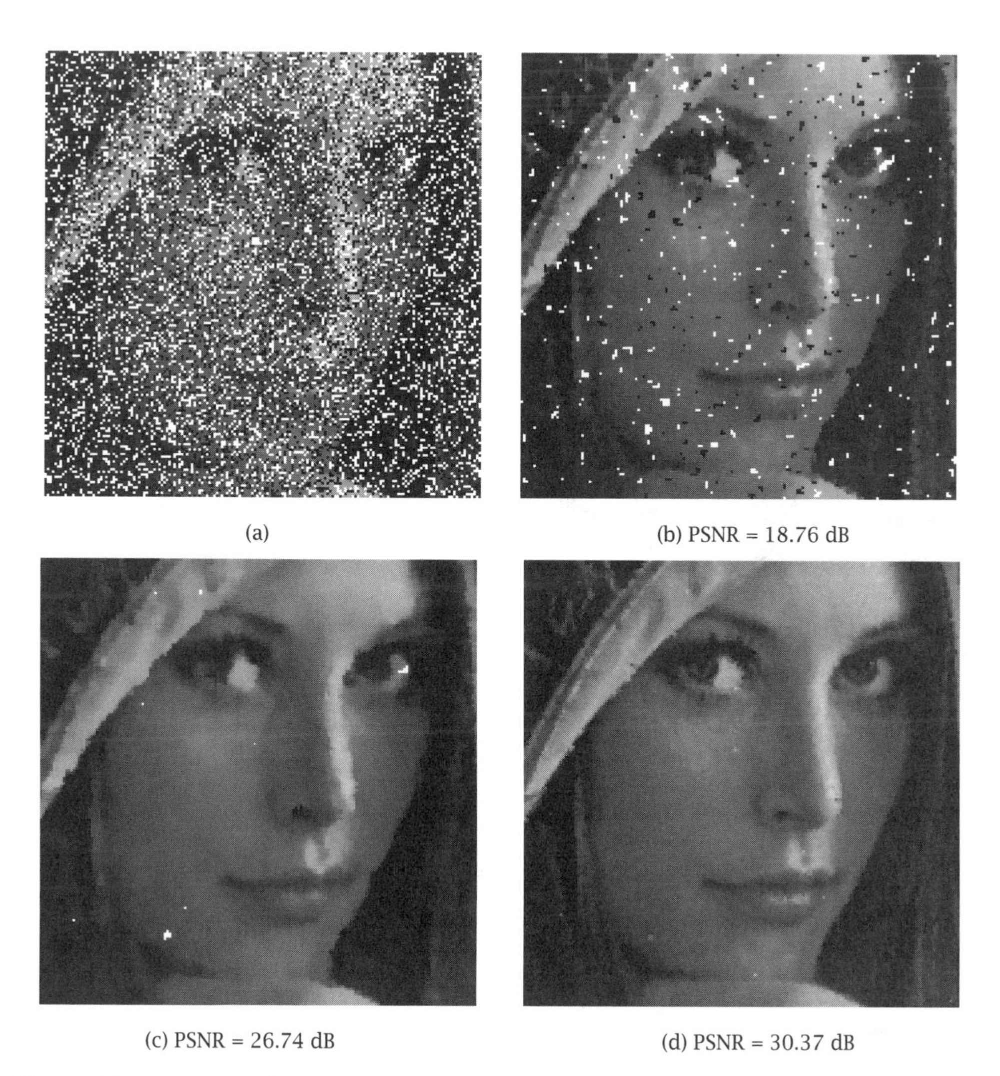

(a) (b) PSNR = 18.76 dB

(c) PSNR = 26.74 dB (d) PSNR = 30.37 dB

Figure 4.7: Face region from Lena image (a) corrupted with 40% fixed-valued impulse noise, which was restored using (b) the 3×3 and (c) the 5×5 median filters and (d) the SD-ROM with $M = 1296$.

corrupted, many of them may be restored by the time the sliding window reaches the center pixel in that region. In contrast, the performance of the 3×3 median filter drops off markedly when the number of corrupted pixels exceeds four. Unlike the SD-ROM, the median filter cannot be used recursively without degrading the overall performance due to excessive blurring.

Experiment 3. Another characteristic of the SD-ROM algorithm that we now illustrate is the robustness of the weighting coefficients with respect to the probability of impulse noise. Table 4.6 shows the PSNRs of the restored Lena image when degraded by random-valued impulse noise with various probabilities ranging from

Table 4.5: Restoration of Miramar Image Corrupted with Gaussian and Mixed Noise.

Algorithm	MAE	MSE	PSNR
	Gaussian noise[a]		
Median filter (3×3)	9.46	168.89	25.85 dB
SD-ROM (M = 1296 outside training set)	6.38	66.00	29.94 dB
	Mixed noise[b]		
Median filter (3×3)	11.14	237.24	24.38 dB
SD-ROM (M = 1296 outside training set)	8.72	155.11	26.22 dB

[a] $\sigma^2 = 100$.

[b] Random-valued impulse noise (20%) plus Gaussian noise ($\sigma^2 = 100$).

Table 4.6: Comparison of PSNRs for Restored Lena Image Using Various Percentages of Random-valued Impulse Noise (Uniform Distribution).

Algorithm	Percentage of impulse noise (in dB)				
	10%	15%	20%	25%	30%
Median filter (3×3)	32.14	31.01	29.76	28.01	26.20
SD-ROM (M = 2)	35.18	33.94	32.51	31.18	29.87
SD-ROM (M = 1296)[a]	36.02	34.44	32.95	31.77	30.49
SD-ROM (M = 1296)[b]	36.64	34.72	32.95	31.52	29.99

[a] Trained for correct percentage of impulse noise (outside training set).

[b] Trained for 20% impulse noise (outside training set).

10 to 30%. Three cases are shown: one in which there was no training (the thresholded case), one in which the training was for the correct probability of corruption, and one in which the training was for 20% impulse noise, irrespective of the actual corruption percentage. We observe that the algorithm performance is only marginally degraded when the training is on impulse noise that is considerably less or more likely than the actual corruption. While in some cases the method actually performs slightly better when trained on a percentage of impulse noise other than the true percentage, this phenomena can be explained by the fact that the restoration is performed outside the training data.

4.7 Restoration of Images Corrupted by Streaks

The removal of streaks in images is a problem that sometimes arises in image restoration [Ban93]. The objective is to suppress the streaks while preserving edges and detail information. Traditionally, the median filter has been used for

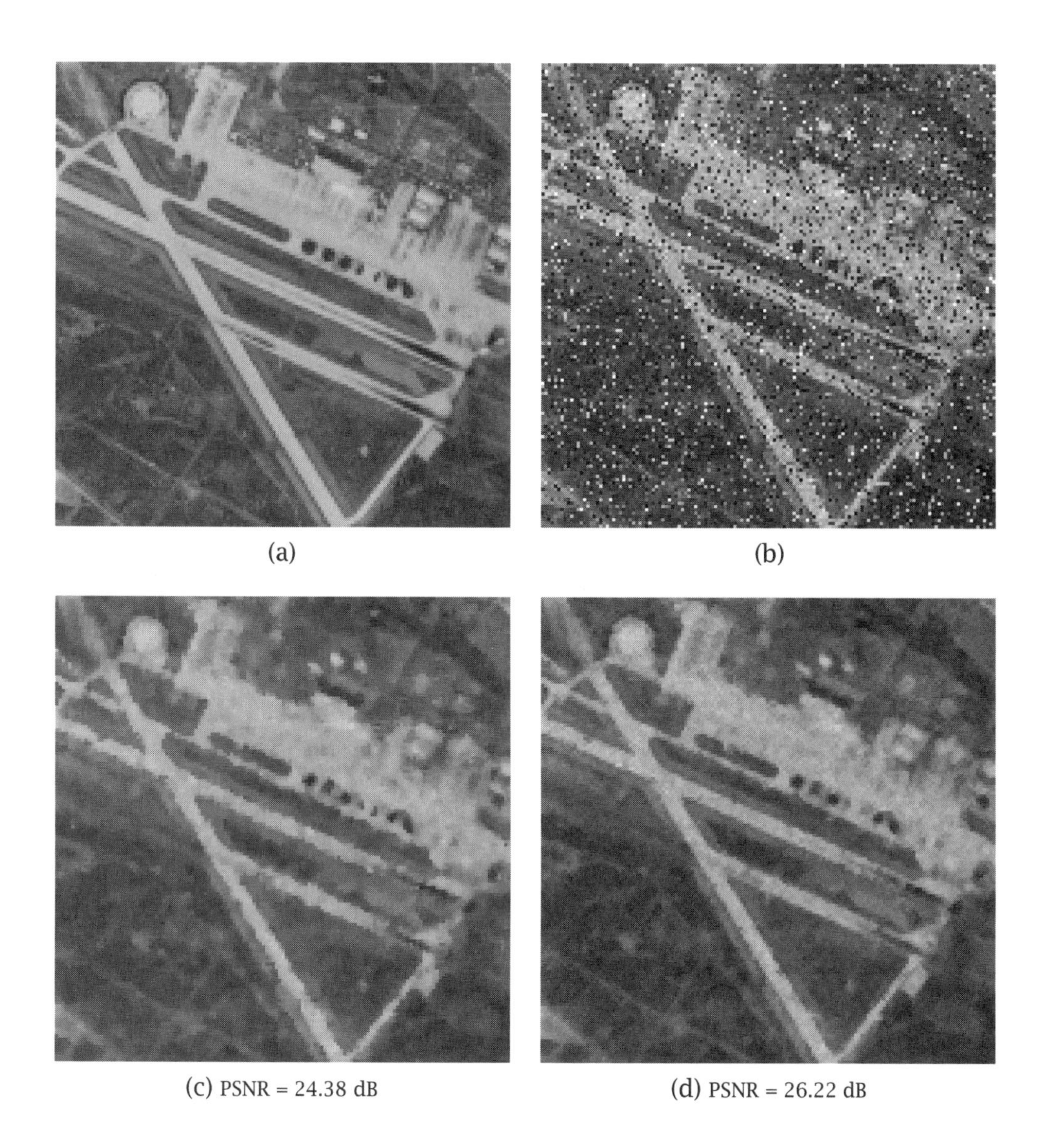

(a) (b)

(c) PSNR = 24.38 dB (d) PSNR = 26.22 dB

Figure 4.8: Airport region from Miramar image: (a) original and (b) corrupted with mixed Gaussian noise ($\sigma^2 = 100$) and 20% random-valued impulse noise, which was restored using (c) the 3×3 median filter and (d) the SD-ROM with $M = 1296$.

this application, but since it is implemented uniformly across an image, important image features such as edges and details may be destroyed, particularly with a large size window. We present here an SD-ROM-based algorithm [Abr96a] to restore such corrupted images. Only those pixels of the image that are detected as belonging to the streaks are replaced with an estimated value using neighborhood information. A streak can be any sequence of pixels in the image that has been replaced with random values. First, a simple method is developed for the case of images corrupted by horizontal streaks affecting entire rows in the image. The

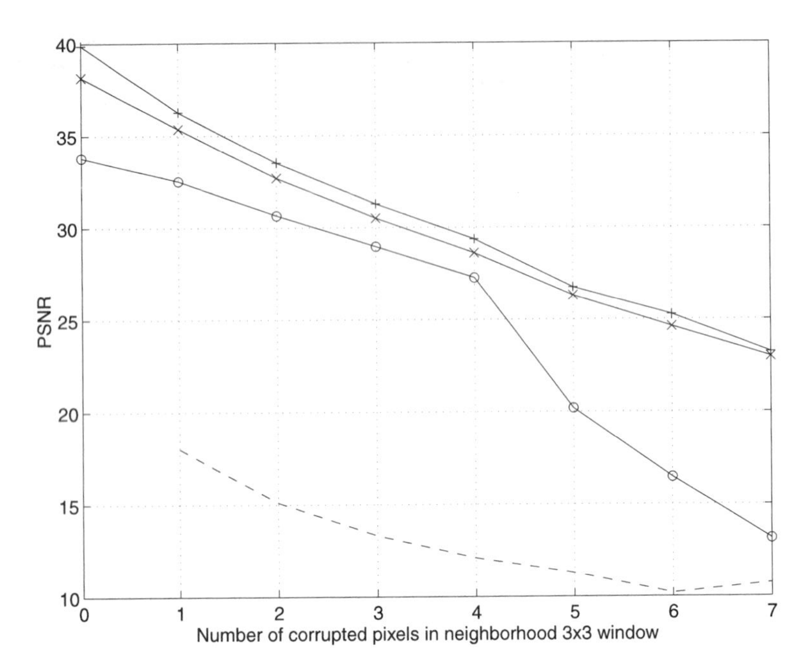

Figure 4.9: Comparison of the PSNRs for the restored Lena image computed according to the number of corrupted pixels in 3×3 neighborhood windows. The curves correspond to the median filter (circles), the SD-ROM for $M = 2$ (multiplication signs) and $M = 1296$ (crosses), and the degraded image itself (dashed line). The original test image was degraded with 20% random-valued impulse noise (uniform distribution).

approach is then generalized for applications involving a wide variety of streaks of arbitrary sizes, angles, curved streaks, etc. Streaks that occur naturally in the image, however, are preserved.

4.7.1 Removal of Horizontal Streaks

We consider first the case of images corrupted by horizontal streaks only (vertical streaks can be treated similarly). The streaks are created by replacing entire rows in the image with random values in the dynamic range. The restoration is implemented in three steps. In the first step the algorithm detects pixels with high probability of corruption and provides a coarse estimation of the location of the streaks. In the second step the assumption of horizontal streaks is introduced to improve the estimation obtained in the first step. Finally, in the third step the pixels in the detected streaks are replaced with an estimation of the true values.

Step 1: Detection of Pixels with a High Probability of Corruption. The detection part of the thresholded SD-ROM algorithm is used for the detection of pixels with a high probability of corruption. When streaks are present in the image, the SD-ROM is capable of effectively detecting pixels with a high probability of belonging to a streak. The algorithm detects $x(\mathbf{n})$ as a noisy sample if any of the inequalities in Eq. (4.1)

(a) (b)

Figure 4.10: The Lena image (a) corrupted by horizontal streaks and (b) restored using the SD-ROM-based approach.

Table 4.7: Restoration of Lena Image Corrupted by Horizontal Streaks.

Algorithm	MAE	MSE	PSNR
Median filter (3×1)	3.59	162.2	26.0 dB
Median filter (5×1)	3.60	63.2	30.1 dB
SD-ROM-based approach	0.47	6.4	40.0 dB

are true. For the case of horizontal streaks the SD-ROM can be implemented either recursively or nonrecursively, and the exact values of the thresholds T_i have little impact on the overall performance. Suggested values are $T_1 = 8$, $T_2 = 20$, $T_3 = 40$, and $T_4 = 50$.

Step 2: Detection of Streaks. A row is considered to be a streak if more than about 30% of its pixels are detected as corrupted in Step 1.

Step 3: Estimation of True Values. Each pixel in a streak is replaced with the mean of its nearest pixels outside the streak located at the same distance from the pixel being replaced.

As an illustrative example we show in Figs. 4.10 and 4.11 and in Table 4.7 the restoration results for the Lena image corrupted by horizontal streaks.

4.7.2 Removal of General Streaks

We address now the restoration of images corrupted by any streak type, including streaks of arbitrary angles and sizes, curved streaks, etc. Similar to the case of

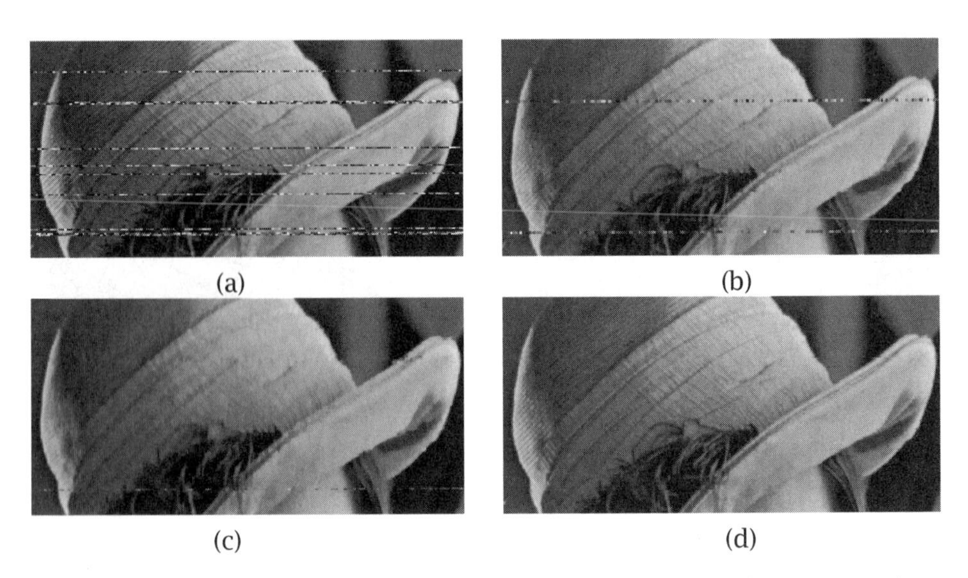

(a) (b) (c) (d)

Figure 4.11: (a) Corrupted image (zoomed portion of Fig. 4.10a), which was restored using (b) the 3×1 and (c) the 5×1 median filters, and (d) the SD-ROM-based approach.

horizontal streaks, the restoration is implemented in three steps. Steps 1 and 3 remain the same. Since no assumptions are made about the nature of the streaks, Step 2 is modified as discussed next.

At the end of Step 1, a binary image is obtained in which the binary values indicate whether a pixel has been detected as corrupted (1) or not (0). This binary image represents a noisy estimation of the location of the streaks. To improve this estimation we filter the binary image by three different nonrecursive nonlinear filters $F1$, $F2$, and $F3$, connected serially. Filters $F1$ and $F3$ operate on 3×3 sliding windows, while filter $F2$ operates on a 5×5 sliding window. Let the input and output pixels inside the 3×3 and 5×5 windows be denoted by x_i and y_i, respectively ($i = 1, \ldots, 9$ for a 3×3 window and $i = 1, \ldots, 25$ for a 5×5 window), with the i's corresponding to a left-to-right, top-to-bottom sequence.

Filter $F1$ (3×3). The output y_5 is 1 if the center pixel x_5 is 1 or if x_i and x_j are 1, $(i, j) \in \{(1,6),(1,9),(1,8),(2,7),(2,8),(2,9),(3,4),(3,7),(3,8),(4,6),(4,9),(6,7)\}$. For all other cases y_5 is equal to zero.

Filter $F2$ (5×5). The output y_{13} is 1 if the center pixel x_{13} is 1. In addition, y_{13} and y_l are both 1 if x_i, x_j, and x_k are 1, $(i, j, k, l) \in \{(1,7,25,19)$, (2,7,24,19), (2,8,24,18), (3,8,23,18), (4,8,22,18), (4,9,22,17), (5,9,21,17), (10,9,16,17), (10,14,16,12), (15,14,11,12), (20,14,6,12), (20,19,6,7), (20,7,6,19), (20,12,6,14), (15,12,11,14), (10,12,16,14), (10,17,16,9), (5,17,21,9), (4,17,22,9), (4,18,22,8), (3,18,23,8), (2,18,24,8), (2,19,24,7), (1,19,25,7)$\}$. For all other cases y_i is equal to x_i.

Filter $F3$ (3×3). The output y_5 is zero if all pixels except the center pixel x_5 are zero; otherwise, y_5 is equal to x_5.

Intuitively, filters $F1$ and $F2$ insert missing 1's along streak locations, and filter $F3$ deletes wrongly detected 1's outside streak locations.

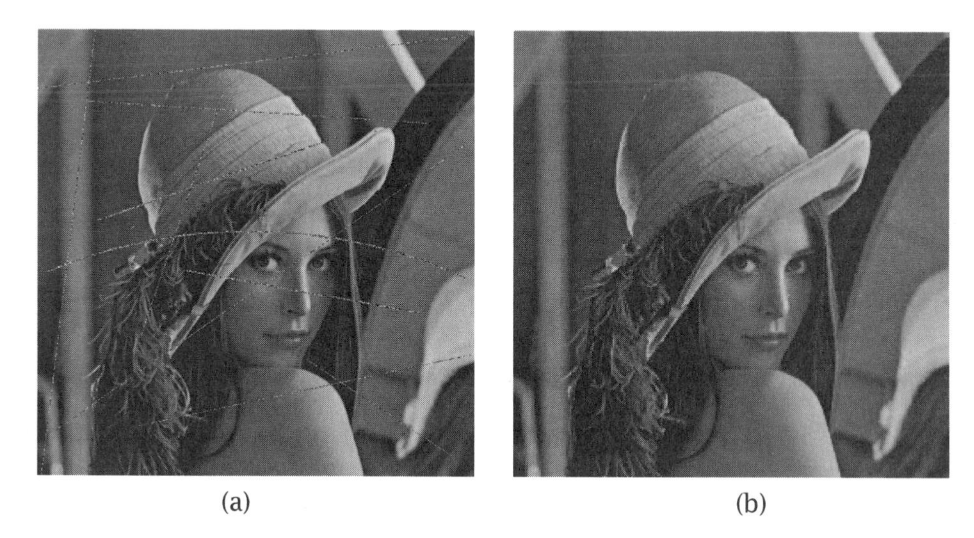

(a) (b)

Figure 4.12: The Lena image (a) corrupted by general streaks and (b) restored using the SD-ROM-based approach.

Table 4.8: Restoration of Lena Image Corrupted by General Streaks.

Algorithm	MAE	MSE	PSNR
Median filter (3×3)	3.28	38.8	32.3 dB
Median filter (5×5)	4.71	70.0	29.7 dB
SD-ROM-based approach	0.32	11.3	37.6 dB

Remarks:

1. In some cases, especially for images corrupted with thick streaks, improved results are obtained if Step 1, Step 2, or both are performed twice.

2. For general streaks the performance of the algorithm becomes more dependent on the threshold values selected in Step 1. Lower values ($T_1 = 4$, $T_2 = 15$, $T_3 = 30$, and $T_4 = 40$) are suitable for smooth or highly corrupted images. If excessive blurring is observed in the restored image, the thresholds can be increased up to about $T_1 = 15$, $T_2 = 30$, $T_3 = 50$, and $T_4 = 80$.

Figure 4.12 and Table 4.8 illustrate the performance of this algorithm as applied to the restoration of the Lena image corrupted by general streaks.

4.8 Concluding Remarks

In this chapter the SD-ROM method for the removal of impulse noise from image data has been presented, in which the filtering operation is conditioned on the rank-ordered differences, defined as the differences between the input pixel and the remaining rank-ordered pixels in a sliding window. Two algorithms have been described, one based on a simple detection–estimation strategy involving thresholds and another incorporating fuzzy rules. In the latter case, strategies for the design of the weighting coefficients have been presented for recursive and nonrecursive implementation, including a least-squares derivation for the nonrecursive case, which leads to a close form expression for the optimal weighting coefficients. Computer simulation examples have been included to illustrate the effectiveness of the SD-ROM method using several distinct noise types, including impulsive, Gaussian, and mixed impulsive and Gaussian. Finally, a simple algorithm for restoration of images corrupted by streaks, based on the SD-ROM approach, has been described.

References

[Arc86] G. R. Arce, N. C. Gallagher, and T. Nodes. Median filters: theory and applications. In *Advances in Computer Vision and Image Processing*, (T. Huang, ed.), JAI Press, Greenwich, CT (1986).

[Abr96b] E. Abreu, M. Lightstone, S. K. Mitra, and K. Arakawa. A new efficient approach for the removal of impulse noise from highly corrupted images. *IEEE Trans. Image Process.*, Special Issue on Nonlinear Image Processing, **5**(6), 1012–1025 (June 1996).

[Abr95] E. Abreu and S. K. Mitra. A signal-dependent rank ordered mean (SD-ROM) filter—a new approach for removal of impulses from highly corrupted images. In *Proc. Intl. Conf. on Acoustics, Speech, and Signal Processing*, Vol. 4, pp. 2371–2374 (Detroit, MI, May 1995).

[Abr96a] E. Abreu and S. K. Mitra. A simple algorithm for restoration of images corrupted by streaks. In *Proc. IEEE Intl. Symp. on Circuits and Systems*, pp. 730–733 (Atlanta, GA, May 1996).

[Ara95] K. Arakawa. A median filter based on fuzzy rules. *IEICE Trans.*, **J78**-A(2), 123–131 (February 1995) In Japanese.

[Ban93] G. J. F. Banon and A. L. B. Candeias. Restoration of NOAA images by mathematical morphology. In *Proc. SIBGRAPI VI*, pp. 139–145 (1993). In Portuguese.

[Bov83] A. C. Bovik, T. Huang, and D. C. Munson. A generalization of median filtering using linear combinations of order statistics. *IEEE Trans. Acoust. Speech Signal Process.* **31**, 1342–1350 (December 1983).

[Cha98] C. Chandra, M. S. Moore, and S. K. Mitra. An efficient method for the removal of impulse noise from speech and audio signals. In *Proc. IEEE Intl. Symp. on Circuits and Systems*, Vol. 4, pp. 206–209 (Monterey, CA, June 1998).

[Flo94] D. A. F. Florêncio and R. W. Schafer. Decision-based median filter using local signal statistics. In *Visual Communication and Image Processing*, Proc. SPIE Vol. 2308, pp. 268–275 (1994).

[Gab92] M. Gabbouj, E. J. Coyle, and N. C. Gallagher. An overview of median and stack filters. *Circ. Syst. Signal Process.* **11** 7–45 (1992).

[Hay91] S. Haykin. *Adaptive Filter Theory*. Prentice Hall, Englewood Cliffs, NJ (1991).

[Kun84] A. Kundu, S. K. Mitra, and P. P. Vaidyanathan. Application of two-dimensional generalized mean filtering for removal of impulsive noises from images. *IEEE Trans. Acoust. Speech Signal Process.* **32**, 600–609 (June 1984).

[Kot92] C. Kotropoulos and I. Pitas. Constrained adaptive LMS L-filters. *Signal Process.* **26**(3), 335–358 (March 1992).

[Kim86] V. Kim and L. Yaroslavskii. Rank algorithms for picture processing. *Comput. Vision Graph. Image Process.* **35**, 234–258 (1986).

[Mit94] S. K. Mitra and T-H. Yu. A new nonlinear algorithm for the removal of impulse noise from highly corrupted images. In *Proc. IEEE Intl. Symp. on Circuits and Systems*, Vol. 3, pp. 17–20 (London, May 1994).

[Nod82] T. A. Nodes and N. C. Gallagher, Jr. Median filters: some modifications and their properties. *IEEE Trans. Acoust. Speech, Signal Process.* **30**(5), 739–746 (October 1982).

[Nod84] T. A. Nodes and N. C. Gallagher, Jr. The output distribution of median type filters. *IEEE Trans. Commun.* **32**(5), 532–541 (May 1984).

[Pal90] F. Palmieri. Adaptive recursive order statistic filters. In *Proc. Intl. Conf. on Acoustics, Speech and Signal Processing*, Vol. 3, pp. 1229–1232 (Albuquerque, NM, April 1990).

[Pit90] I. Pitas and A. N. Venetsanopoulos. *Nonlinear Digital Filters: Principles and Applications*. Kluwer, Boston, MA (1990).

[Sun94] T. Sun and Y. Neuvo. Detail-preserving median based filters in image processing. *Patt. Recog. Lett.* **15**, 341–347 (April 1994).

5

Nonlinear Mean Filters and Their Applications in Image Filtering and Edge Detection

CONSTANTINE KOTROPOULOS, MICHAEL PAPPAS, AND IOANNIS PITAS

Department of Informatics
Aristotle University of Thessaloniki
Thessaloniki, Greece

5.1 Introduction

Noise filtering and edge detection are two of the most important tasks in image processing. Noise removal in images is a particularly difficult task due to the nonstationary nature of images and the different types of noises (e.g., additive and signal-dependent noise) that corrupt images.

In this chapter, we study the class of nonlinear mean filters for noise removal and edge detection. Nonlinear mean filters can be considered as an alternative to the median filter and its extensions and/or generalizations (e.g., max/median, multistage median, median hybrid, trimmed mean, L-filters, etc.) [Ast97, Pit90]. All of these filters offer improvements in certain statistical properties at the expense of a lower ability to remove impulse noise.

The image processing applications of nonlinear mean filters, defined by the $\mathcal{L}_p$ mean filter, were first proposed by Kundu, Mitra, and Vaidyanathan [Kun84]. Further generalizations of the $\mathcal{L}_p$ mean filter were later advanced by Pitas and Venetsanopoulos [Pit86a, Pit86b, Pit86c]. The nonlinear mean filters remove the impulse noise very effectively, especially when the impulses occur with a high probability. They have a very simple structure and are suitable for real-time processing applications. The class of nonlinear mean filters can be treated as a generalization of linear filters (e.g., the arithmetic mean filter). They are closely related to homomorphic filters, perhaps one of the oldest classes of nonlinear filters. From a statistical point of view, nonlinear mean filters rely on the nonlinear means that are well-known location estimators [Ken73]. The latter approach is found to be more fruitful because it produces a general filter structure that encompasses the filters that are based on order statistics, the homomorphic filters, and the morphological filters in addition to the nonlinear mean filters.

The nonlinear mean filters are first defined and their statistical as well as edge preservation properties reviewed in Sec. 5.2. Their performance in the presence of impulse noise is outlined as well. In Sec. 5.3, $\mathcal{L}_p$ mean filters are proven to be a general filter structure able to suppress signal-dependent noise. A class of edge detectors obtained from the difference of nonlinear mean filters is studied in Sec. 5.4. Next, soft grayscale morphological filters based on $\mathcal{L}_p$ mean filters are described in Sec. 5.5. $\mathcal{L}_2$ mean filters are shown to be optimal for both multiplicative Rayleigh speckle and signal-dependent Gaussian speckle noise. These properties supported the application of the $\mathcal{L}_2$ mean filter, the signal-adaptive $\mathcal{L}_2$ mean filter, and the $\mathcal{L}_2$ learning vector quantizer in ultrasonic image processing and analysis, which are reviewed in Sec. 5.6. The use of $\mathcal{L}_p$ mean filters as approximators of max and min operators for positive and negative p, respectively, is exploited in Sec. 5.7 to design analog implementations of sorting networks. Edge preserving filtering by combining nonlinear means and order statistics is studied in Sec. 5.8.

5.2 Nonlinear Mean Filters

The nonlinear mean of the N numbers x_i, $i = 1, 2, \dots, N$, is defined by

$$y = f(x_1, x_2, \dots, x_N) = g^{-1}\left(\frac{\sum_{i=1}^{N} a_i\, g(x_i)}{\sum_{i=1}^{N} a_i}\right), \tag{5.1}$$

where $g(x)$ is, in general, a single-valued analytic nonlinear function and a_i are weights. The properties of the nonlinear mean depend on the function $g(x)$ and the weights a_i. Equation (5.2) indicates several choices of $g(x)$ that result in filters playing a special role in image processing:

$$g(x) = \begin{cases} x & \text{arithmetic mean } \overline{x}, \\ 1/x & \text{harmonic mean } y_{\mathrm{H}}, \\ \ln x & \text{geometric mean } y_{\mathrm{G}}, \\ x^p,\ p \in \mathbf{R} - \{-1, 0, 1\} & \mathcal{L}_p \text{ mean } y_{\mathcal{L}_p}, \end{cases} \tag{5.2}$$

where **R** is the set of real numbers. Other functions $g(x)$ can also be considered [Ken73]. Accordingly, the one-dimensional nonlinear mean filter of length $N = 2v + 1$ is defined by

$$y_l = f(x_{l-v}, \ldots, x_l, \ldots, x_{l+v}), \quad l \in \mathbf{Z}, \tag{5.3}$$

where **Z** is the set of all integer numbers. Similar definitions can be given for N even and for two-dimensional nonlinear mean filters. If the weights are constant, the nonlinear mean filters reduce to the well-known homomorphic filters [Ars83, Opp63]. If they are not constant, other classes of nonlinear filters can be obtained. If $a_i = x_i^p$ and $g(x) = x$, the contraharmonic mean results, that is,

$$y_{CH_p} = \frac{\sum_{i=1}^{N} x_i^{p+1}}{\sum_{i=1}^{N} x_i^{p}}, \tag{5.4}$$

which can be interpreted as an arithmetic mean having data-dependent weights $a_i = x_i^p$ [Pit86a]. The nonlinear means previously described have the following well-known property [Ken73], which can be used in their analysis:

$$\min\{x_i\} \le y_{CH_{-p}} \le y_{\mathcal{L}_{-p}} \le y_{\mathrm{H}} \le y_{\mathrm{G}} \le \overline{x} \le y_{\mathcal{L}_p} \le y_{CH_p} \le \max\{x_i\}. \tag{5.5}$$

The statistical properties of nonlinear mean filters depend on the output mean $m_y = \mathrm{E}\{y\}$ and the output variance $\sigma_y^2 = \mathrm{E}\{(y - m_y)^2\}$, where $\mathrm{E}\{.\}$ denotes the expected value operator [Pap84]. Let $m_i = \mathrm{E}\{x_i\}$, $i = 1, \ldots, N$. The function $f(x_1, \ldots, x_N)$ can be approximated around $(m_1, \ldots, m_N)$ by a truncated Taylor series that includes up to second-order terms, that is,

$$\begin{aligned} y = {} & f(m_1, \ldots, m_N) + \sum_{i=1}^{N} f_i'(x_i - m_i) + \frac{1}{2} \sum_{i=1}^{N} f_i''(x_i - m_i)^2 \\ & + \sum_{i=1}^{N} \sum_{j=i+1}^{N} f_{ij}''(x_i - m_i)(x_j - m_j), \end{aligned} \tag{5.6}$$

where

$$f_i' = \frac{\partial f}{\partial x_i} |_{m_1, \ldots, m_N}, \tag{5.7}$$

$$f_i'' = \frac{\partial^2 f}{\partial x_i^2} |_{m_1, \ldots, m_N}, \tag{5.8}$$

$$f_{ij}'' = \frac{\partial^2 f}{\partial x_i \partial x_j} |_{m_1, \ldots, m_N}. \tag{5.9}$$

Such an expansion is valid if the x_i, $i = 1, \ldots, N$, form random vectors that are concentrated near $(m_1, \ldots, m_N)$, that is, when the variances $\sigma_{x_i}^2$, $i = 1, \ldots, N$, are small and the marginal probability density functions (pdfs) $p(x_i)$ are either short tailed (e.g., uniform) or medium tailed (e.g., Gaussian). In the latter case, the approximation is valid in a probabilistic sense, since for Gaussian random

Table 5.1: Performance of Nonlinear Filters Expressed as σ_y^2/σ_x^2. Filter Weights: $a_i = 1$, $i = 1, \ldots, N$.

pdf	$\overline{x}$	Geometric mean	$\mathcal{L}_p$ mean
Uniform	$\frac{1}{N}$	$\frac{1}{N} - \frac{(N-1)^2\sigma_x^2}{4N^2m^2}$	$\frac{1}{N} - \frac{(p-1)^2(N-1)^2\sigma_x^2}{4N^2m^2}$
Gaussian	$\frac{1}{N}$	$\frac{1}{N} - \frac{(N-1)^2\sigma_x^2}{4N^2m^2}$	$\frac{1}{N} - \frac{(p-1)^2(N-1)^2\sigma_x^2}{4N^2m^2}$
pdf	$\overline{x}$	CH_p mean	Median
Uniform	$\frac{1}{N}$	$\frac{1}{N} - \frac{p^2(N-1)^2\sigma_x^2}{N^2m^2}$	$\frac{3}{N+2}$
Gaussian	$\frac{1}{N}$	$\frac{1}{N} - \frac{p^2(N-1)^2\sigma_x^2}{N^2m^2}$	$\frac{\pi}{2N+\pi-2}$

variables (RVs), the probability of occurrence of large differences $(x_i - m_i)$ is low. Following an approach similar to that described for functions of a single RV [Pap84], we obtain

$$m_y \approx f(m_1, \ldots, m_N) + \frac{1}{2}\sum_{i=1}^{N} f_i'' \sigma_{x_i}^2 + \sum_{i=1}^{N}\sum_{j=i+1}^{N} f_{ij}'' C_{ij}, \tag{5.10}$$

where $C_{ij} = \mathrm{E}\{(x_i - m_i)(x_j - m_j)\}$ is the covariance of the RVs x_i and x_j, $i = 1, \ldots, N$ and $j = i+1, \ldots, N$. By neglecting the moments of RVs of order higher than 2, the following approximation for the variance of the nonlinear mean filter output is found:

$$\sigma_y^2 \approx \sum_{i=1}^{N} f_i'^2 \sigma_{x_i}^2 + 2\sum_{i=1}^{N}\sum_{j=i+1}^{N} f_i' f_j' C_{ij} - \left(\frac{1}{2}\sum_{i=1}^{N} f_i''\sigma_{x_i}^2 + \sum_{i=1}^{N}\sum_{j=i+1}^{N} f_{ij}'' C_{ij}\right)^2. \tag{5.11}$$

The partial derivatives f_i', f_i'', and f_{ij}'' for the geometric, $\mathcal{L}_p$, and contraharmonic means can easily be computed [Pit86a]. Simpler expressions can be obtained if the RVs x_i, $i = 1, \ldots, N$, are independent identically distributed (i.i.d.) with mean m and variance σ_x^2 because $C_{ij} = 0$. Table 5.1 summarizes the performance of the arithmetic (i.e., the moving average), geometric , contraharmonic , and $\mathcal{L}_p$ means, expressed as the ratio σ_y^2/σ_x^2. The corresponding figure of merit for the median filter is also included for comparison purposes. From an inspection of Table 5.1, it can be seen that the nonlinear mean filters have smaller output variances than the arithmetic mean over homogeneous image regions. Therefore, they suppress the noise better than the arithmetic mean filter. It is well known that the median filter has larger variance than the arithmetic mean $\overline{x}$ for additive white uniform or Gaussian noise [Jus81, Pit90] (see also Table 5.1). Thus, the nonlinear mean filters outperform the median filter for the aforementioned noise pdfs, since they have a smaller output variance than the median filter.

For white noise having a long-tailed pdf, the approximations of Eqs. (5.10) and (5.11) are not valid. The white noise having a long-tailed pdf most frequently encountered in image processing is impulse noise. In the following, the performance of nonlinear mean filters in the presence of impulse noise is studied.

Impulse noise consists of very large positive or negative impulses of short duration; both are easily detected by the eye and degrade the image quality. Filtering by the nonlinear means y_G, y_H, $y_{\mathcal{L}_{-p}}$, and $y_{CH_{-p}}$ tends to reduce the average signal level, as can be seen by inequalities (5.5). Thus, the aforementioned nonlinear means can be used for positive impulse suppression. On the other hand, the nonlinear means $y_{\mathcal{L}_p}$ and y_{CH_p} tend to increase the signal level, so they can be used for negative impulse suppression. Let the background signal and the impulses have means m and $M = \beta m$ ($\beta \gg 1$), respectively. Let us also consider the following impulse noise model for positive impulses:

$$x_i = \begin{cases} \beta m & \text{with probability } q, \\ m & \text{with probability } 1 - q. \end{cases} \tag{5.12}$$

The positive impulses are considered to be suppressed when the following inequality for the expected value of the output of the nonlinear mean filter holds:

$$\mathrm{E}\{|y|\} \leq \alpha m, \qquad \alpha = 1 + \varepsilon, \qquad \varepsilon \geq 0. \tag{5.13}$$

If the $\mathcal{L}_{-p}$ mean filter has sufficiently high p so that

$$q \leq \frac{\alpha^{-p} - 1}{\beta^{-p} - 1}, \tag{5.14}$$

it can be proven that it satisfies the inequality (5.13). Thus, it removes the positive impulses. Similarly, a CH_{-p} filter whose coefficient p satisfies

$$q \leq \frac{\alpha - 1}{\alpha - 1 + \beta^{-p}(\beta - \alpha)} \tag{5.15}$$

removes the positive impulses having probability of occurrence q [Pit86b]. $\mathcal{L}_p$ and CH_p filters can be used for the removal of negative impulses having mean $M = \beta m$ ($0 < \beta \ll 1$) if appropriately high p is chosen to satisfy the following inequalities:

$$q \leq \frac{1 - \alpha^p}{1 - \beta^p} \tag{5.16}$$

$$q \leq \frac{1 - \alpha}{1 - \alpha + \beta^p(\alpha - \beta)}, \tag{5.17}$$

respectively [Pit86a]. The major drawback in the use of nonlinear mean filters for impulse noise removal is that they cannot remove positive and negative impulses simultaneously. In contrast to median filters, whose performance deteriorates rapidly by increasing the probability q of impulse occurrence or by decreasing the filter window extent, nonlinear mean filters effectively suppress one kind of impulse noise even with high probability of occurrence ($q \gg 0.4$). The capability of the 3×3 CH_{-2} mean filter to remove positive impulses has been demonstrated elsewhere [Pit86a]. When the probability of noise corruption is not very high (e.g., $q < 0.1$), two separate detection rules for positive and negative impulses can be employed [Kun84]. Let $\{(i-k, j-l) \in \mathcal{A}\}$ denote the image pixels within the filter window centered on pixel (i, j). We declare that the central pixel is corrupted by negative impulse noise if

$$\overline{x} - x_{ij} > \frac{\max_{\mathcal{A}}\{x_{i-k,j-l}\}}{3}. \tag{5.18}$$

Similarly, the central pixel is corrupted by positive impulse noise if

$$\overline{x} - x_{ij} < \frac{\min_{\mathcal{A}}\{x_{i-k,j-l}\} - M}{3}, \tag{5.19}$$

where M is the height of positive impulses. If none of the above tests passes, the central pixel is noise-free.

For white noise, Eq. (5.10) shows that the nonlinear mean response in the presence of a noisy edge follows approximately the filter response for the same edge without noise when the noise levels are small enough. Thus, nonlinear means $y_{\mathcal{L}_{-p}}$, $y_{CH_{-p}}$, $y_{\mathcal{L}_p}$, and y_{CH_p} tend to preserve edges even if they are noisy.

5.3 Signal-Dependent Noise Filtering by Nonlinear Means

Signal-dependent noise is a special kind of noise, modeled by

$$x = s + h(s)n, \tag{5.20}$$

where n is white noise, s is the noise-free signal, and $h(s)$ is the point spread function of the imaging system. In Eq. (5.20), x is the (output) signal that is corrupted by the noise. Two usual kinds of signal-dependent noise are multiplicative noise, which results in a noisy signal of the form $x = s + \kappa s n$, where κ is a constant [Ars83, Opp75, Pra78], and film-grain noise which yields a noisy signal $x = s + \kappa s^{\gamma} n$, where κ and γ are parameters describing the noise [Ars83, Pra78]. Signal-dependent noise cannot be removed by conventional linear techniques, as is done with additive noise. Homomorphic filtering has been used for the removal of multiplicative [Opp63] and film-grain noise [Ars83]. Such techniques attempt to decouple the noise from the signal and transform the signal-dependent noise to additive noise so that conventional linear techniques, for example, Wiener filtering [Pra78], can be used to remove the additive noise. As we mentioned in Sec. 5.2, homomorphic filters can be considered as a special case of the nonlinear means defined by Eq. (5.1). It can be shown [Pit86a] that the multiplicative noise is the only kind of signal-dependent noise modeled by Eq. (5.20) that is completely reduced to additive noise by homomorphic filtering by an analytic function g, that is, $g(s + h(s)n) = g(s) + \mathcal{N}(n)$. In this case, the analytic function is the natural logarithm, that is, $g(x) = \ln x$. For film-grain noise, the homomorphic filtering by g results in output signal $g(x)$ that can be expanded to a Taylor series around s, that is,

$$g(x) = g(s) + g'(s)\,\kappa\, s^{\gamma}\, n + \frac{1}{2} g''(s)\kappa^2 s^{2\gamma} n^2 + \ldots, \tag{5.21}$$

where $g'(s)$ and $g''(s)$ are the first- and second-order derivatives of g at s, respectively, and the higher-order terms are not explicitly described for notational simplicity. If g is chosen so that $g'(s) = s^{-\gamma}$, or

$$g(s) = \frac{1}{1-\gamma} s^{1-\gamma}, \tag{5.22}$$

that is, if we filter x by an $\mathcal{L}_p$ mean filter with $p = 1 - \gamma$, then

$$g(x) = g(s) + \underbrace{\kappa\, n + \frac{1}{2}\gamma\, \kappa^2 \left(\frac{s^{\gamma} n}{s}\right)^2 s^{-\gamma+1} + \ldots}_{\mathcal{N}(s,n)}. \tag{5.23}$$

Equation (5.23) reveals that $\mathcal{N}(s,n)$ is still signal dependent, although it includes the term n that is the additive noise. However, its dependence on s is significantly reduced since the higher-order terms of $\mathcal{N}(s,n)$ progressively vanish when the noise levels are below the signal level, that is, $|s^{\gamma} n| \leq |s|$. The term n can be reduced by the use of linear techniques. Accordingly, the use of nonlinear means in the presence of signal-dependent noise modeled by Eq. (5.20) leads at least to the partial decoupling of signal and noise.

5.4 Edge Detectors Based on Nonlinear Means

An edge can be roughly defined as the border between two homogeneous regions of different luminance values. Such a definition implies that an edge corresponds to a local variation of image luminance (but not necessarily vice versa). Several edge detectors employing different criteria have been proposed in the literature [Pit92, Pra78]. Linear filtering techniques have already been used in edge detection [Fri79, Gue83]. A linear edge detector compares the difference of two low-pass filters with a threshold [Gue83]. In Sec. 5.2, we showed that nonlinear mean filtering is more successful than linear filtering in image processing because it removes certain kinds of noise better (e.g., impulse noise) and preserves edge information. Accordingly, the application of nonlinear means in edge detection produces a novel class of edge detectors that have good noise characteristics. Their structure is depicted in Fig. 5.1.

The nonlinear mean filters measure the mean of the luminance; their difference is a measure of the dispersion of the luminance within the filter extent. If it is greater than a certain threshold, the center of the filter extent is declared an edge point. Accordingly, the class of edge detectors under study are essentially

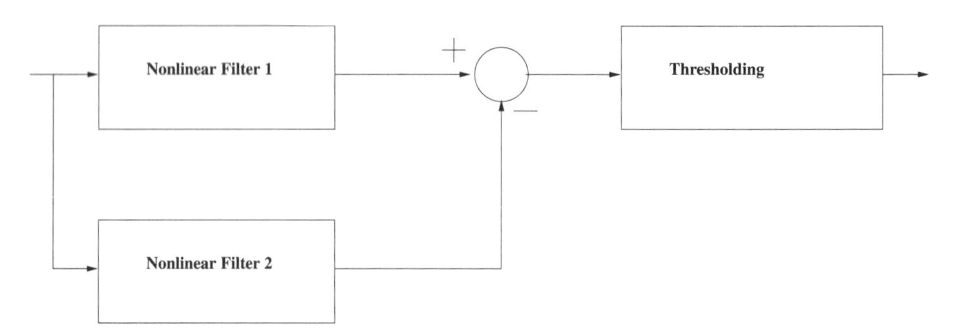

Figure 5.1: Structure of nonlinear edge detector based on nonlinear means.

measures of the local standard deviation of the image, that is, they are local estimators of scale followed by a thresholder. The simplest local estimator of scale is the so-called range, that is, the difference between maximum and minimum nonlinear filters. Filters that approximate the maximum filter are the $\mathcal{L}_p$ and CH_p means; the minimum filter can be approximated by either the $\mathcal{L}_{-p}$ or the CH_{-p} mean. The difference of the $\mathcal{L}_p$ and the $\mathcal{L}_{-p}$ mean filters is the $\mathcal{L}_p$ edge detector. Similarly, the difference of the CH_p and the CH_{-p} filters defines the CH_p edge detector. The performance of the $\mathcal{L}_p$ and CH_p edge detectors on 1-D noisy step edge of height h has been analyzed elsewhere [Pit86b]. The difference of nonlinear means can be approximated by a Taylor series around the point $(m_1, m_2, \ldots, m_N)$. For i.i.d. RVs x_i, $i = 1, \ldots, N$, having mean m and variance σ^2, the expected value of the difference of $\mathcal{L}_p$ and $\mathcal{L}_{-p}$ filters can easily be approximated by

$$m_y = \mathrm{E}\{y_{\mathcal{L}_p} - y_{\mathcal{L}_{-p}}\} \approx \frac{p(N-1)}{Nm}\sigma^2. \tag{5.24}$$

Similarly, it can be shown that the expected value of the difference of CH_p and CH_{-p} filters can be approximated by

$$m_y = \mathrm{E}\{y_{CH_p} - y_{CH_{-p}}\} \approx \frac{2p(N-1)}{Nm}\sigma^2. \tag{5.25}$$

Ideally, m_y should be zero. Equations (5.24) and (5.25) show that the bias of the nonlinear edge detector with contraharmonic mean is twice larger than the bias of the $\mathcal{L}_p$ edge detector. Thus, higher threshold levels are generally required for the contraharmonic edge detector. For small noise levels, the variance of the edge detectors under study is approximately equal to σ^2/N; this crude approximation indicates that these edge detectors have good noise characteristics.

5.5 Grayscale Morphology Using $\mathcal{L}_p$ Mean Filters

The $\mathcal{L}_p$ filters possess some important properties, notably the ability to apply "soft" nonlinearities to a signal that can harden progressively by choosing appropriately the parameter p. Thus, $\mathcal{L}_p$ filters can be a "soft" substitute of classical max/min operators. We now investigate some general properties of these filters, when used to create filters similar to the morphological ones.

Let $\{x_i\}$ be a 1-D sequence. Using a slight modification of the notation presented by Salembier [Sal92], we can define the $\mathcal{L}_p$ dilation as follows:

$$D_p(x_i) = \left(\frac{\sum_{j=1}^{N} a_j x_{i-j}^{p}}{\sum_{j=1}^{N} a_j} \right)^{1/p}, \tag{5.26}$$

where $p \neq 0$. For $p > 0$, $\mathcal{L}_p$ dilation is essentially an $\mathcal{L}_p$ mean filter [Pit86a]. Similarly, we can define the $\mathcal{L}_p$ erosion as

$$E_p(x_i) = \left(\frac{\sum_{j=1}^{N} a_j x_{i-j}^{-p}}{\sum_{j=1}^{N} a_j} \right)^{-1/p}. \tag{5.27}$$

For $p > 0$, $\mathcal{L}_p$ erosion is essentially an $\mathcal{L}_{-p}$ mean filter [Pit86a]. The block diagram of an $\mathcal{L}_p$ dilation or erosion filter is shown in Fig. 5.2a. Filter block $\mathcal{A}$ represents the linear part of the filter, which has coefficients $a'_j = a_j / \sum_{j=1}^{N} a_j$. Due to inequalities (5.5), we conclude that the output of the grayscale morphological erosion with a flat structuring element is always smaller than that of the $\mathcal{L}_p$ erosion, when $a_j = 1$, $j = 1, \ldots, N$. Similarly, the output of the morphological dilation with a flat structuring element is always greater than that of the $\mathcal{L}_p$ dilation. For very large positive values of p, the outputs of the $\mathcal{L}_p$ erosion and $\mathcal{L}_p$ dilation converge to the outputs of the minimum and maximum operators, respectively. Thus, $\mathcal{L}_p$ erosion and $\mathcal{L}_p$ dilation are "soft" morphological operators [Kuo93]. The counterparts of the grayscale opening and closing can be defined using the $\mathcal{L}_p$ dilation and the $\mathcal{L}_p$ erosion in the following ways:

$$y_i = D_p\left(E_p(x_i)\right) = \left(\sum_{j=1}^{N} \frac{a_j}{\sum_{k=1}^{N} a_k x_{i-j-k}^{-p}} \right)^{1/p}, \tag{5.28}$$

$$y_i = E_p\left(D_p(x_i)\right) = \left(\sum_{j=1}^{N} \frac{a_j}{\sum_{k=1}^{N} a_k x_{i-j-k}^{p}} \right)^{-1/p}. \tag{5.29}$$

It should be noted that the effective filter window length used in the opening/closing definitions is not N but $N' = 2N - 1$. The $\mathcal{L}_p$ operators possess some interesting properties, which are listed below.

Property 1. Duality:

$$D_{-p}(x_i) = E_p(x_i), \tag{5.30}$$

$$D_p(x_i^{-1}) = E_p^{-1}(x_i), \tag{5.31}$$

$$E_p(x_i^{-1}) = D_p^{-1}(x_i), \tag{5.32}$$

$$D_p\left(E_p(x_i)\right) = \left[E_p\left(D_p(x_i^{-1})\right)\right]^{-1}, \tag{5.33}$$

$$E_p\left(D_p(x_i)\right) = \left[D_p\left(E_p(x_i^{-1})\right)\right]^{-1}, \tag{5.34}$$

where all superscripts denote powers.

Property 2. If $\{x_i\}$ and $\{y_i\}$ are two sequences with positive terms satisfying $x_i \leq y_i$, $\forall i$, and the filter weights are nonnegative ($a_j \geq 0$), then the following inequalities hold:

$$D_p(x_i) \leq D_p(y_i), \tag{5.35}$$

$$E_p(x_i) \leq E_p(y_i), \tag{5.36}$$

$$E_p\left(D_p(x_i)\right) \leq E_p\left(D_p(y_i)\right), \tag{5.37}$$

$$D_p\left(E_p(x_i)\right) \leq D_p\left(E_p(y_i)\right). \tag{5.38}$$

Property 3. The output of an $\mathcal{L}_p$ dilation filter $D_{p,\mathcal{A}}(x_i)$ having window $\mathcal{A} = \{-v, -v+1, \ldots, v-1, v\}$ applied r times to a signal $\{x_i\}$ will be the same as that of a $D_{p,\mathcal{B}}(x_i)$ filter applied once to the signal $\{x_i\}$, where $\mathcal{B} = \{-rv, -rv+1, \ldots, rv-1, rv\}$ is the new filter window. The new filter weights are given by

$$\frac{b_j}{\sum_{j' \in \mathcal{B}} b_j} = \frac{\overbrace{a_l * a_l * a_l * \cdots * a_l}^{r \text{ terms}}}{\left(\sum_{l' \in \mathcal{A}} a_l\right)^r}, \qquad \forall j \in \mathcal{B} \text{ and } \forall l \in \mathcal{A}, \tag{5.39}$$

where the asterisks denote the linear convolution operators, and a_l and b_j are the filter weights defined on the windows $\mathcal{A}$ and $\mathcal{B}$, respectively. We also assume that $a_l = 0$, $l \notin \mathcal{A}$. For an $\mathcal{L}_p$ erosion, Eq. (5.39) applies exactly as well.

For example, let us assume that the filter window is $\mathcal{A} = \{-2, -1, 0, 1, 2\}$, with $a_i = 1$ for $i \in \mathcal{A}$ but zero otherwise. If $D_{p,\mathcal{B}}(\cdot)$, $\mathcal{B} = \{-4, \ldots, 4\}$, is the filter equivalent to the cascade of two $D_{p,\mathcal{A}}(\cdot)$ filters, then the normalized coefficients b_i' can be calculated from Eq. (5.39) for $m = 2$. Similarly, we could create a filter $D_{p,C}(\cdot)$, $C = \{-6, \ldots, 6\}$, equivalent to the cascade of three $D_{p,\mathcal{A}}(\cdot)$ filters. The normalized coefficients a_i', b_i', and c_i' are shown in Fig. 5.3. Conversely, it would be possible to "break" an $\mathcal{L}_p$ dilation filter with a large filter window into a cascade of $\mathcal{L}_p$ filters having smaller window sizes. A graphical explanation of Property 3 is shown in Fig. 5.2b through 5.2d. In a cascade of $\mathcal{L}_p$ filters having the same parameter p (Fig. 5.2b), convolution takes place between the linear blocks, resulting in a new linear block $\mathcal{B}$ (Fig. 5.2d), since the blocks x^p and $x^{1/p}$ cancel each other (Fig. 5.2c).

In the following, we approximate the mean and the variance of the output of the $\mathcal{L}_p$ opening filter by extending the analysis of Sec. 5.2. The starting point is to approximate the filter output given by Eq. (5.28) by a Taylor series. We observe that the filter output is a function of the following $2N-1$ input signal variables:

$$y_i = D_p(E_p(x_i)) = f(x_{i-2}, \ldots, x_{i-N-1}, \ldots, x_{i-2N}). \tag{5.40}$$

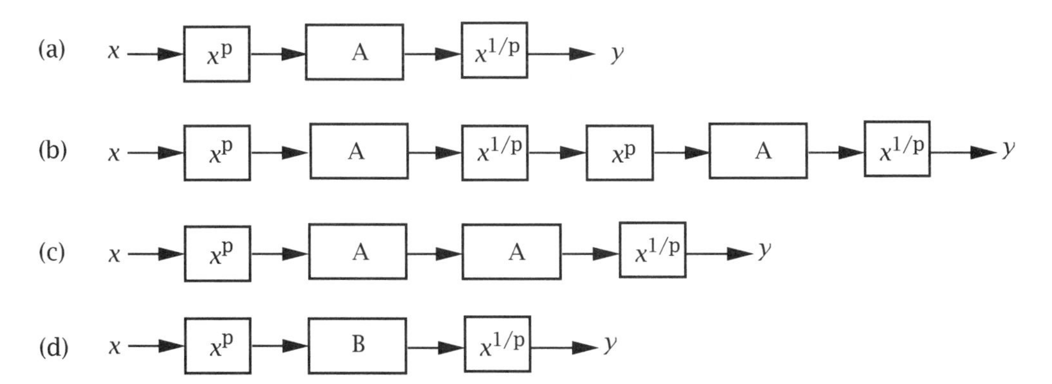

Figure 5.2: (a) $D_{p,\mathcal{A}}(\cdot)$ filter, (b) cascade type $D_{p,\mathcal{A}}\left(D_{p,\mathcal{A}}(\cdot)\right)$ filter, (c) cancellation of the x^p and $x^{1/p}$ blocks, and (d) equivalent $D_{p,\mathcal{B}}(\cdot)$ filter.

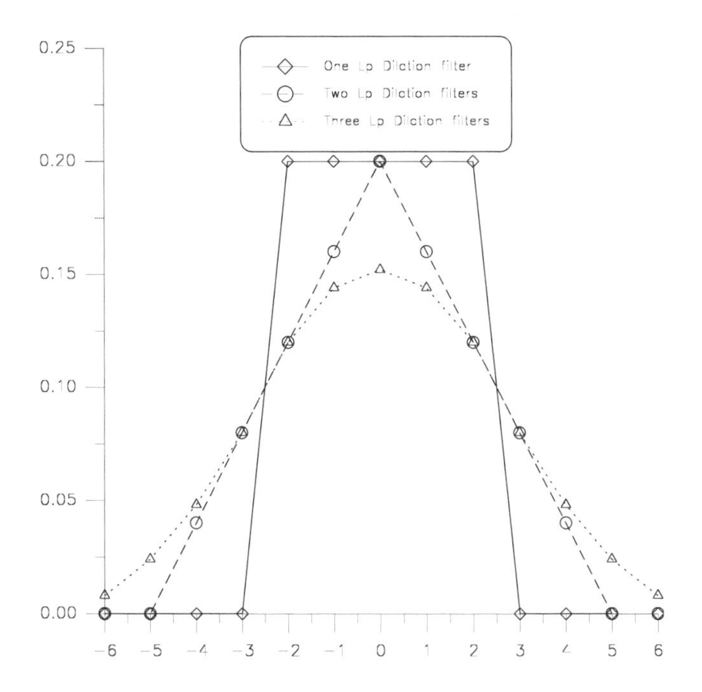

Figure 5.3: Normalized coefficients of a $D_{p,\mathcal{A}}(\cdot)$ dilation filter (diamonds) and of the $D_{p,\mathcal{B}}(\cdot)$ filter (open circles), and the $D_{p,\mathcal{C}}(\cdot)$ filter (triangles), which are equivalent to the cascade of either two or three $D_{p,\mathcal{A}}(\cdot)$ dilation filters, respectively.

Accordingly, Eqs. (5.10) and (5.11) employ the aforementioned variables. If we assume that an input signal of constant value m_x is corrupted by additive white zero-mean short- or medium-tailed noise of variance σ_x^2 and that the filter coefficients are $a_i = 1, i = 1, \ldots, N$, the mean m_y and the variance σ_y^2 of the $\mathcal{L}_p$ opening filtered sequence can be obtained by the following closed forms [Pap95, Pap96a]:

$$m_y \simeq m_x - \frac{(3N^3 - 4N^2 + 1)p + 3N^3 - 2N^2 - 1}{6N^3 m_x}\sigma_x^2, \tag{5.41}$$

$$\frac{\sigma_y^2}{\sigma_x^2} \simeq \frac{2N^2+1}{3N^3} - \left[\frac{(-3N^3+4N^2-1)p - 3N^3+2N^2+1}{6N^3 m_x}\right]^2 \sigma_x^2. \tag{5.42}$$

From Eq. (5.41), we can see that the output mean m_y of the $\mathcal{L}_p$ opening filter is linearly related to the power p. Furthermore, the sign of p does not necessarily imply that an $\mathcal{L}_p$ opened noisy sequence will have negative bias, when applied to constant regions of the original sequence. The bias depends on the size of the filter window; that is, if $p \geq (3N^3 - 2N^2 - 1)/(-3N^3 + 4N^2 - 1)$, the filtered output will be negatively biased, but it will be positively biased otherwise. For $N = 3$, the bias changes sign at $p_c \simeq -1.3478261$.

As discussed in Sec. 5.2, the previous results do not hold for long-tailed noise distributions (e.g., for impulse noise). In the following, some interesting deterministic properties are derived when a signal of constant value m is corrupted by mixed impulse noise of probability q. We assume that positive and negative impulses appear with equal probability $q/2$. Let the positive impulses have height γm with $\gamma \gg 1$ and the negative impulses have height δm with $0 < \delta \ll 1$. The filtering of the input sequence $\{x_i\}$ is considered to be successful when the filtered sequence $\{y_i\}$ satisfies

$$\alpha m \leq y_i \leq \beta m, \qquad \alpha = 1 - \varepsilon,\ \beta = 1 + \varepsilon,\ \varepsilon \geq 0. \tag{5.43}$$

We can make the assumption that $q \simeq k/N$, where k is the total number of impulses in N input samples. In the case of $\mathcal{L}_p$ opening, the substitution of Eq. (5.28) into Eq. (5.43) yields

$$\frac{\beta^{-p} - 1}{(\gamma^{-p} + \delta^{-p})/2 - 1} \leq q \leq \frac{\alpha^{-p} - 1}{(\gamma^{-p} + \delta^{-p})/2 - 1}. \tag{5.44}$$

Similarly, for $\mathcal{L}_p$ closing the substitution of Eq. (5.29) in Eq. (5.43) yields

$$\frac{\alpha^{p} - 1}{(\gamma^{p} + \delta^{p})/2 - 1} \leq q \leq \frac{\beta^{p} - 1}{(\gamma^{p} + \delta^{p})/2 - 1}. \tag{5.45}$$

Equations (5.44) and (5.45), for given α, β, γ, δ, and p, essentially define upper and lower bounds for the probability of impulse occurrence q. By comparing them, we conclude that similar performance between opening and closing is expected when $\gamma = \delta^{-1}$ and $\alpha = \beta^{-1}$.

We demonstrate pictorially the ability of the $\mathcal{L}_p$ openings and closings to remove both impulse and additive white zero-mean Gaussian noise from corrupted images. The reference image used was Lena of size 256×256. The corrupted image was filtered by $\mathcal{L}_p$ openings and closings, where coefficient p took values in the set $\{2, 5, 8\}$. Each operation (e.g., erosion or dilation) had a filter window $\mathcal{A} = \{(a_i, a_j) : i, j \in \{-1, 0, 1\}\}$. The filter coefficients were $a_{ij} = 1$, $\forall i, j \in \mathcal{A}$. The performance of the grayscale opening and closing with the flat structuring set $\mathcal{A}$ was also evaluated. It has been found that in the case of mixed impulse noise and/or Gaussian noise, $\mathcal{L}_p$ operators outperform their morphological counterparts in noise reduction by at least 3 dB. Figures 5.4 and 5.5 show the reference

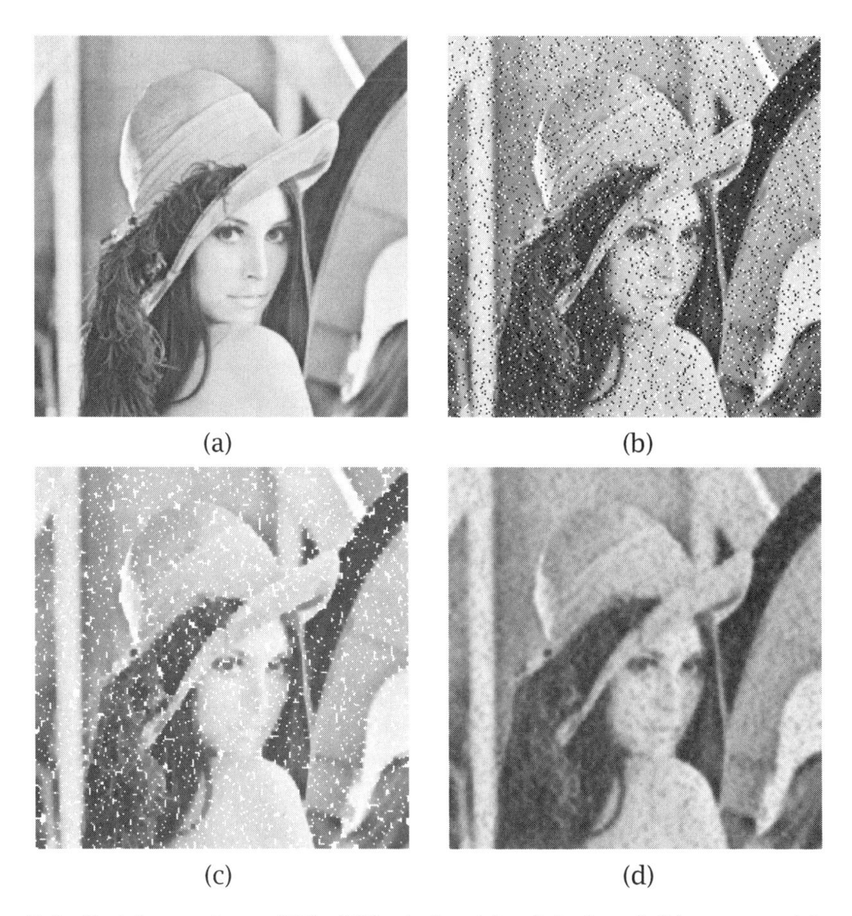

Figure 5.4: Test image Lena, 256×256 pixels: (a) original and (b) corrupted by mixed impulse noise of heights 0.001 and 255 with equal probabilities of appearance (0.05), which was filtered using (c) a grayscale closing filter with a flat 3×3 structuring element and (d) an $E_2(D_2(\cdot))$ filter with a 3×3 window symmetrical around the origin.

and noisy images and those filtered by the grayscale and $\mathcal{L}_p$ morphological filters (for each type of operator, e.g., opening or closing). As can be seen, in many cases $\mathcal{L}_p$ filtering does not exhibit the same blocking effect that characterizes the grayscale opening or closing with a flat structuring set (e.g., in Figs. 5.5c and 5.5d).

5.6 Ultrasonic Image Processing Using $\mathcal{L}_2$ Mean Filters

Speckle noise is a special kind of noise encountered in images formed by laser beams and in radar images, as well as in envelope-detected ultrasound B-mode images. It is an interference effect caused by the scattering of the ultrasound beam from microscopic tissue inhomogeneities [Abb79, Bur78, Goo76]. There is a rich literature on ultrasound image filtering and analysis [Kot92, Kot94a]. Homomorphic filtering and signal adaptive median filtering have also been applied for

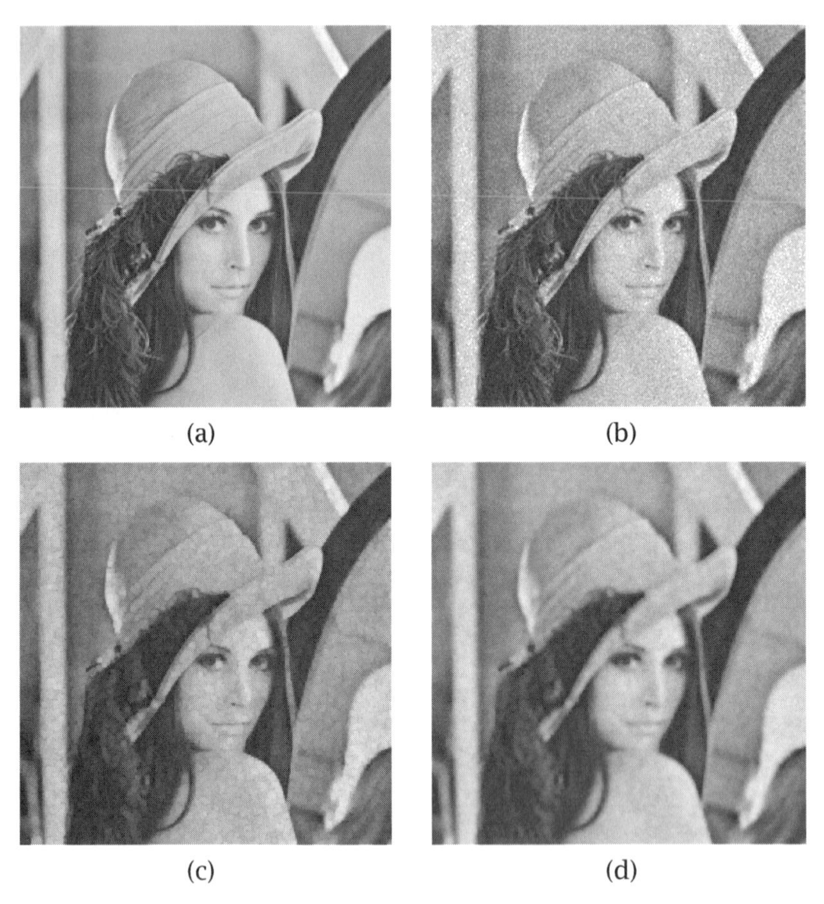

Figure 5.5: Test image Lena 256×256 pixels: (a) original and (b) corrupted by additive zero-mean Gaussian noise with variance $\sigma^2 = 100$, which was filtered using (c) a grayscale opening filter with a flat 3×3 structuring element and (d) a $D_8(E_8(\cdot))$ filter with a 3×3 window symmetrical around the origin.

speckle suppression [Pit90]. The problem of optimum nonlinear signal detection and estimation has been studied for the raw ultrasound signal just after envelope detection in [Kot92]. Such an approach has two distinctive advantages: (a) accurate modeling and (b) computational efficiency. However, a serious cost is paid since the mean gray level and the lateral speckle size depend greatly on the distance to the transducer [Oos85]. The starting point is to model ultrasonic speckle as multiplicative noise [Lou88]. Let x be the envelope-detected observed signal, m be the true lesion signal, and n be a noise term statistically independent of m. It is assumed that the signal m is related to the observation x by

$$x = mn. \tag{5.46}$$

The pdf of the observed RV x is considered to be Rayleigh, that is,

$$p_x(X) = \frac{X}{\sigma^2} \exp\left(-\frac{X^2}{2\sigma^2}\right), \qquad X > 0. \tag{5.47}$$

The expected value of the RV x and its variance are [Pap84]

$$\mathrm{E}\{x\} = \sigma\sqrt{\pi/2} \qquad \mathrm{var}\{x\} = \sigma^2\left(\frac{4-\pi}{2}\right). \tag{5.48}$$

It can easily be proven that, if the signal m is constant and equal to $\sigma\sqrt{\pi/2}$, and the noise term n is a Rayleigh RV having unity expected value and variance $(4-\pi)/\pi$, then the pdf of the RV x is given by Eq. (5.47).

Let us assume that we have a set of N observations $x_1, x_2, \ldots, x_N$, denoted by a vector $\mathbf{x} = [x_1, x_2, \ldots, x_N]^t$ in the observation space $\mathbf{R}^N$. Let $\mathbf{n} = [n_1, n_2, \ldots, n_N]^t$ be a noise vector of N i.i.d. Rayleigh RVs. Let us assume the following hypotheses:

$$\begin{aligned} H_0 &: \mathbf{x} = m_0\mathbf{n}, \\ H_1 &: \mathbf{x} = m_1\mathbf{n}, \end{aligned} \tag{5.49}$$

created by the probabilistic transition mechanisms:

$$p_{x_i|H_k}(X_i|H_k) = \frac{X_i}{\sigma_k^2}\exp\left(-\frac{X_i^2}{2\sigma_k^2}\right), \qquad X_i > 0, \quad i = 1,\ldots,N, \quad k = 0, 1. \tag{5.50}$$

Under the null hypothesis H_0 the constant signal $m_0 = \sigma_0\sqrt{\pi/2}$, while under the alternative hypothesis H_1 we have $m_1 = \sigma_1\sqrt{\pi/2}$. In the general case, the pixel values in the ultrasound B-mode image are correlated. In the following analysis, we treat the pixel values as independent RVs. The Bayes criterion [Van68] leads to the log-likelihood ratio test (LRT) [Kot92]:

$$\sum_{i=1}^{N} X_i^2 \underset{H_0}{\overset{H_1}{\gtrless}} \frac{2\sigma_0^2\sigma_1^2}{\sigma_1^2-\sigma_0^2}\left(\ln\theta - 2N\ln\frac{\sigma_0}{\sigma_1}\right) = \gamma \qquad \text{for } \sigma_1^2 > \sigma_0^2 \tag{5.51}$$

$$\sum_{i=1}^{N} X_i^2 \underset{H_0}{\overset{H_1}{\gtrless}} \frac{2\sigma_0^2\sigma_1^2}{\sigma_0^2-\sigma_1^2}\left(2N\ln\frac{\sigma_0}{\sigma_1} - \ln\theta\right) = \gamma' \qquad \text{for } \sigma_1^2 < \sigma_0^2 \tag{5.52}$$

where θ, γ, and γ' are thresholds. Therefore, for the binary hypothesis testing problem described above, the sufficient statistic is the sum of the squares of the observations, that is,

$$l(\mathbf{X}) = \sum_{i=1}^{N} X_i^2. \tag{5.53}$$

It has been argued [Kot92] that if m_1 can take (1) values only greater that m_0, a uniformly most powerful (UMP) test exists, that is, Eq. (5.51); (2) values only smaller than m_0, a UMP test exists, that is, Eq. (5.52); and (3) values both greater than and smaller than m_0, a UMP test does not exist.

Let $\mathcal{R}_0$ be the decision region under the hypothesis H_0 and $\mathcal{R}_1$ the corresponding decision region under the alternative hypothesis. The probabilities of false alarm and detection are given by

$$P_{\mathrm{F}} = \int_{\mathcal{R}_0} p_{\mathbf{x}|H_0}(\mathbf{X}|H_0)\,d\mathbf{X}, \tag{5.54}$$

$$P_{\mathrm{D}} = \int_{\mathcal{R}_1} p_{\mathbf{x}|H_1}(\mathbf{X}|H_1)\,d\mathbf{X}. \tag{5.55}$$

The plot of $P_{\rm D}$ versus $P_{\rm F}$ for varying parameters γ is defined as the receiver operating characteristic (ROC). $P_{\rm F}$ and $P_{\rm D}$ have been derived theoretically for both Eqs. (5.51) and (5.52) [Kot92]. Let $d = \sigma_1/\sigma_0$ and

$$\gamma = \frac{2d^2\sigma_0^2}{d^2 - 1}(\ln\theta + 2N\ln d). \tag{5.56}$$

For the decision rule given by Eq. (5.51) we have

$$P_{\rm F} = 1 - \mathcal{I}_\Gamma\left(\frac{\gamma}{2\sigma_0^2\sqrt{N}}, N-1\right), \qquad d \geq 1, \tag{5.57}$$

$$P_{\rm D} = 1 - \mathcal{I}_\Gamma\left(\frac{\gamma}{d^2\sigma_0^2\sqrt{N}}, N-1\right), \qquad d \geq 1, \tag{5.58}$$

where $\mathcal{I}_\Gamma(u, M)$ is the incomplete gamma function defined by

$$\mathcal{I}_\Gamma(u, M) \triangleq \int_0^{u\sqrt{M+1}} \frac{x^M}{M!}\exp(-x)\,dx. \tag{5.59}$$

Let us now focus our attention on the estimation of parameter m in Eq. (5.46) when n is multiplicative noise independent of m, which is distributed as follows:

$$p_n(\mathcal{N}) = \frac{\pi\mathcal{N}}{2}\exp\left(-\frac{\pi\mathcal{N}^2}{4}\right), \qquad \mathcal{N} > 0. \tag{5.60}$$

The conditional density function of the observations assuming $m = M$ is given by [Pap84]

$$p_{x|m}(X|M) = \frac{1}{M}p_n\left(\frac{X}{M}\right), \qquad M > 0. \tag{5.61}$$

Let us suppose that we have a set of N observations. The maximum likelihood (ML) estimate of M maximizes the log-likelihood function [Kot92], that is,

$$\hat{m}_{\rm ML} = \frac{\sqrt{\pi}}{2}\sqrt{\frac{1}{N}\sum_{i=1}^{N} X_i^2}. \tag{5.62}$$

It is seen that the ML estimator of the constant signal is the $\mathcal{L}_2$ mean scaled by the factor $\sqrt{\pi}/2$. Let $\pi(N)$ be the following polynomial in N:

$$\pi(N) \triangleq \frac{\Gamma\left(N + \frac{1}{2}\right)}{\sqrt{N}(N-1)!}, \tag{5.63}$$

where $\Gamma(\cdot)$ denotes the gamma function. The expected value and variance of the ML estimator as well as the mean squared estimation error are given by [Kot92]

$$\mathrm{E}\{\hat{m}\} = \pi(N)M, \tag{5.64}$$

$$\mathrm{var}\{\hat{m}\} = [1 - \pi^2(N)]M^2, \tag{5.65}$$

$$\mathrm{E}\{(\hat{m} - M)^2\} = 2[1 - \pi(N)]M^2. \tag{5.66}$$

Moreover, the asymptotic properties of the ML estimate hold [Van68].

In most practical cases, it is unrealistic to consider a constant signal hypothesis. Without any loss of generality, the following binary hypothesis problem is assumed:

$$\begin{aligned} H_1 : x &= mn, \\ H_0 : x &= n, \end{aligned} \tag{5.67}$$

where m and n are RVs. Our aim is to perform detection and estimation based on this model. Since m is an RV, the conditional density of the observations under H_1 is given by

$$p_{x|H_1}(X|H_1) = \int_{\chi_m} p_{x|m,H_1}(X|M,H_1)p_{m|H_1}(M|H_1)\,\mathrm{d}M, \tag{5.68}$$

where χ_m is the domain of the RV m. We make the following assumptions:

1. n is a Rayleigh RV having unity expected value and variance $(4-\pi)/\pi$; that is, its pdf is given by Eq. (5.60).
2. The conditional density of the observations under H_1 and with m known is given by:

$$p_{x|m,H_1}(X|M,H_1) = \frac{1}{M}p_n\left(\frac{X}{M}\right) = \frac{\pi X}{2M^2}\exp\left(-\frac{\pi X^2}{4M^2}\right), \quad M > 0. \tag{5.69}$$

3. The conditional density of m under H_1 is chosen in such a way that it represents a realistic model and offers mathematical tractability. A Maxwell density with parameter λ fulfills both requirements; that is,

$$p_{m|H_1}(M|H_1) = \frac{M^2\exp(-\lambda M^2)}{K}, \tag{5.70}$$

where K ensures that Eq. (5.70) is a pdf,

$$K = \frac{\sqrt{\pi}}{4\lambda^{3/2}}. \tag{5.71}$$

By substitution of Eqs. (5.70) and (5.71) into Eq. (5.68) we obtain a gamma density:

$$p_{x|H_1}(X|H_1) = \pi\lambda X\exp\left(-X\sqrt{\lambda\pi}\right). \tag{5.72}$$

Such a result is very reasonable, since it is known that speckle can be modeled by a gamma density function [Pit90]. Based on N observations the LRT leads to

$$\sum_{i=1}^{N} X_i^2 - 4\sqrt{\frac{\lambda}{\pi}}\sum_{i=1}^{N} X_i \underset{H_0}{\overset{H_1}{\gtrless}} \frac{4}{\pi}(\theta - N\ln 2\Lambda) = \gamma''. \tag{5.73}$$

The maximum a posteriori (MAP) estimate of the signal m is given by [Kot92]

$$\hat{m}_{\mathrm{MAP}}(\mathbf{x}) = \left[\frac{\frac{\pi}{2}\sum_{i=1}^{N} X_i^2}{(N-1) + \left((N-1)^2 + \pi\lambda\sum_{i=1}^{N} X_i^2\right)^{1/2}}\right]^{1/2}. \tag{5.74}$$

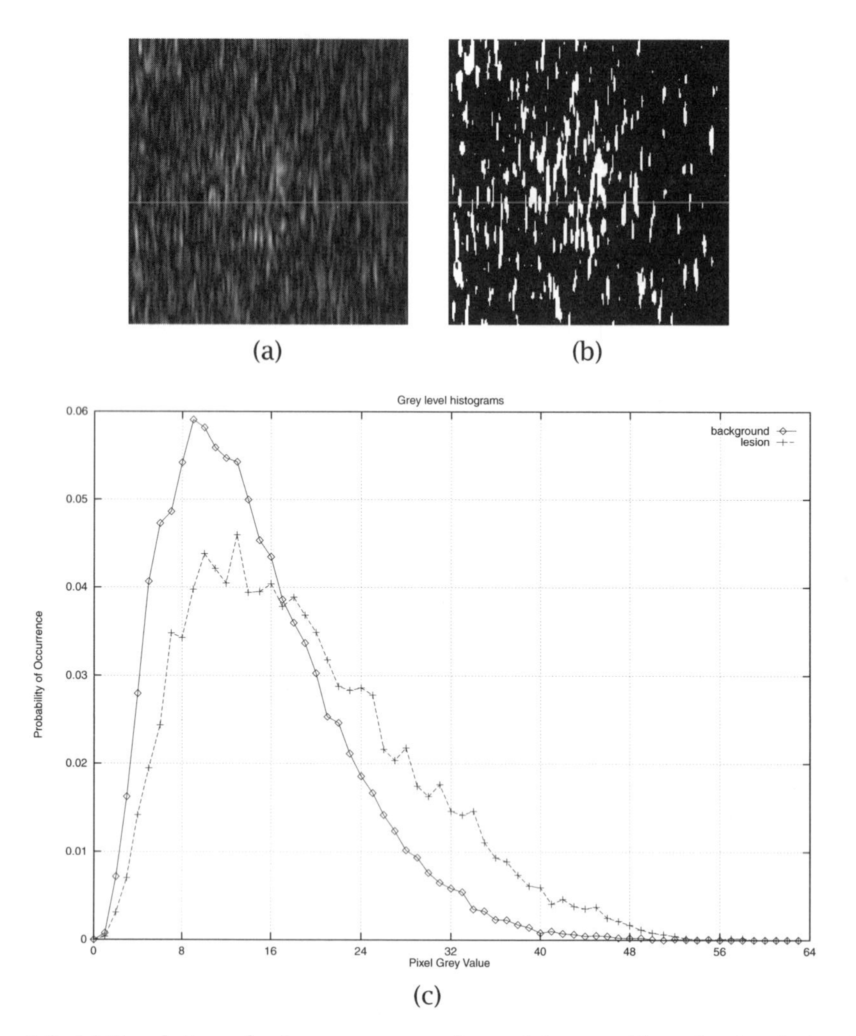

Figure 5.6: (a) Simulation of a homogeneous piece of tissue with a circular lesion in the middle; the lesion/background amplitude is +3 dB and the number density of scatterers in the background and the lesion is $5000/\text{cm}^3$. (b) Thresholded original image. (c) Gray level histograms of the pixels belonging to the lesion and to the background areas.

It can be seen that for $\lambda = 0$, the MAP estimate of m reduces to the form of the ML estimate of the constant signal, that is, to an $\mathcal{L}_2$ mean filter. Indeed,

$$\hat{m}_{\text{MAP}}(\mathbf{x}; \lambda = 0) = \frac{\sqrt{\pi}}{2} \sqrt{\frac{1}{N-1} \sum_{i=1}^{N} X_i^2}. \tag{5.75}$$

The $\mathcal{L}_2$ mean filter has been applied to both simulated ultrasound B-mode images and real ultrasonic images for speckle suppression. Simulated ultrasound B-mode images are used to evaluate the performance of various filters in speckle suppression and to select parameters (such as filter length and thresholds) in the image processing task. Figures 5.6a and 5.7a are simulations of a homogeneous piece of tissue (4×4 cm) with a circular lesion in the middle; the lesion has a diameter of 2 cm. These ultrasound B-mode images were produced and described by Verhoeven et al. [Ver91]. The lesion differs from the background in reflection strength (+3 dB). The background has a number density of scatterers of $5000/\text{cm}^3$. The lesion has a number density of scatterers of either $5000/\text{cm}^3$ (Fig. 5.6a) or $500/\text{cm}^3$ (Fig. 5.7a). In the former case, there is no change in second-order statistics between lesion and background; in the latter case, the lesion is characterized by a sub-Rayleigh distribution. Both simulated images have dimensions 241×241 and resolution 6 bits/pixel. The gray level histograms of the pixels belonging to the lesion area and to the background are plotted in Figs. 5.6c and 5.7c. It can be seen that they are very similar to the Rayleigh pdf.

Two types of rectangular filter windows were employed, one having dimensions proportional to the lateral and axial correlation sizes (15×3) and the other having dimensions inversely proportional to them (3×15). We briefly assess the performance of filtering the original image by the $\mathcal{L}_2$ mean filter and thresholding the filtered image, using as figures of merit the area under the ROC in each case and the probability of detection P_D for a threshold chosen so that the probability of false alarm $P_F \simeq 10\%$.

Figure 5.8a depicts the output of the 15×3 $\mathcal{L}_2$ mean filter applied to the image shown in Fig. 5.6a; Fig. 5.8b shows the thresholded image. The probability of lesion detection in Fig. 5.8b is 3.26% higher than that measured in Fig. 5.6b, and the area under the ROC is 6.45% larger than that measured in Fig. 5.6b as well [Kot92]. Figure 5.9a depicts the output of the 3×15 $\mathcal{L}_2$ mean filter applied to the image shown in Fig. 5.6a; Fig. 5.9b shows the thresholded image. The probability of lesion detection in Fig. 5.9b is 28.532% higher than that measured in Fig. 5.7b; moreover, the area under the ROC is 17.89% larger than that measured in Fig. 5.7b [Kot92].

A representative real ultrasonic image of a liver recorded using a 3 MHz probe is shown in Fig. 5.10a and the output of the $\mathcal{L}_2$ mean filter of dimensions 5×5 is shown in Fig. 5.10b. It is seen that the proposed nonlinear filter suppresses the speckle noise effectively. However, any spatial filtering without adjusting its smoothing performance at each point of the image according to the local image content results in edge blurring. Better edge preservation is attained by the so-called signal-adaptive filters [Pit90]. In the following, we address the design of signal-adaptive $\mathcal{L}_2$ mean filters.

Let us first revise the image formation model. For ultrasonic images in which the displayed image data have undergone excessive manipulation (e.g., logarithmic compression, low- and high-pass filtering, postprocessing, etc.), a realistic image formation model is the signal-dependent one [Lou88]:

$$x = m + m^{1/2} n, \tag{5.76}$$

where n is a zero-mean Gaussian RV. It has been proven that the ML estimate of

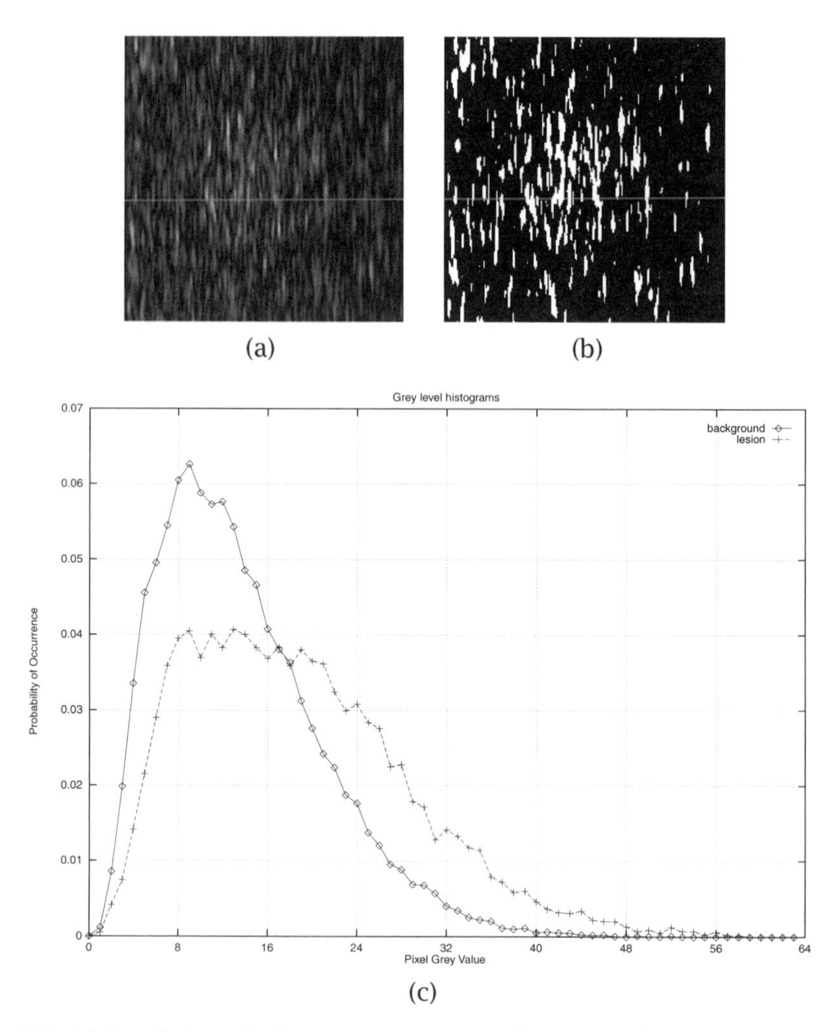

Figure 5.7: (a) Simulation of a homogeneous piece of tissue with a circular lesion in the middle; the lesion/background amplitude is +3 dB and the number density of scatterers in the background is $5000/\text{cm}^3$ and in the lesion is $500/\text{cm}^3$. (b) Thresholded original image. (c) Gray level histograms of the pixels belonging to the lesion and to the background areas.

$m = M$ is given by [Kot94a]

$$\hat{m}_{\text{ML}} = -\frac{\sigma^2}{2} + \sqrt{\frac{\sigma^4}{4} + \frac{1}{N}\sum_{i=1}^{N} X_i^2}, \tag{5.77}$$

which closely resembles the $\mathcal{L}_2$ mean filter given by Eq. (5.62).

The results mentioned above led us to the design of signal-adaptive maximum likelihood filters (i.e., $\mathcal{L}_2$ mean filters) both for multiplicative Rayleigh speckle and for signal-dependent speckle [Kot94a]. The output of the signal-adaptive maximum likelihood filter, that is, the estimate of the original image at (i, j), is

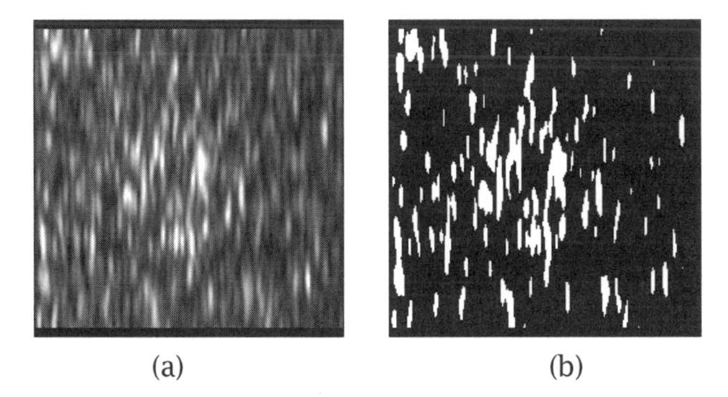

Figure 5.8: (a) Output of the 15×3 $\mathcal{L}_2$ mean filter applied to Fig. 5.6a and (b) result of thresholding.

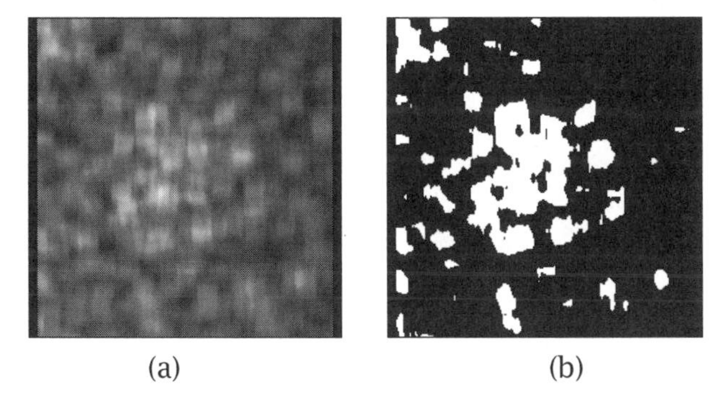

Figure 5.9: (a) Output of the 3×15 $\mathcal{L}_2$ mean filter applied to Fig. 5.7a and (b) result of thresholding.

$$\hat{m}(i,j) = \hat{m}_{\mathrm{ML}}(i,j) + \beta(i,j)[x(i,j) - \hat{m}_{\mathrm{ML}}(i,j)], \tag{5.78}$$

where

$x(i,j)$	is the noisy observation at pixel (i,j)
$\hat{m}_{\mathrm{ML}}(i,j)$	is the maximum likelihood estimate of $m(i,j)$ based on the observations inside the filter window $\mathcal{A}$, $x(i-k, j-l) \in \mathcal{A}$
$\beta(i,j)$	is a weighting factor, approximating the local SNR over the window $\mathcal{A}$
$\hat{m}(i,j)$	is the signal-adaptive filter output at pixel (i,j).

When $\beta(i,j)$ approaches 1, the actual observation is preserved by the suppression of the low-pass component $\hat{m}_{\mathrm{ML}}(i,j)$; when it is close to 0, maximum noise reduction is performed since the high-frequency component is suppressed. Analytic expressions for $\beta(i,j)$ for the multiplicative and the signal-dependent models can be found elsewhere [Kot94a]. Figure 5.10c depicts the output of the signal-adaptive $\mathcal{L}_2$ mean filter applied to the ultrasonic image of a liver of Fig. 5.10a. It is seen that not only edges but additional diagnostically significant image details are preserved.

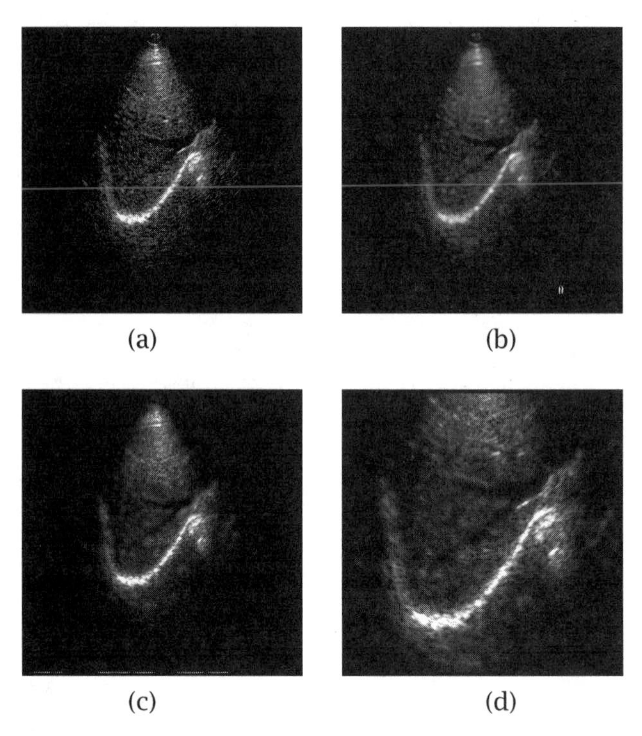

Figure 5.10: (a) Real ultrasonic image of a liver recorded using a 3 MHz probe. (b) Output of the 5×5 $\mathcal{L}_2$ mean filter. (c) Output of the signal-adaptive $\mathcal{L}_2$ mean filter. (d) Zoom-in of the results obtained by using segmentation in conjunction with filtering.

A modification of the signal-adaptive maximum likelihood filter that utilizes segmentation information obtained prior to the filtering process has been proposed as well [Kot94a]. Moreover, the segmentation of ultrasonic images by using a variant of the learning vector quantizer (LVQ) based on the $\mathcal{L}_2$ mean, the so-called $\mathcal{L}_2$ LVQ, was developed, and its convergence properties in the mean and in the mean square were studied [Kot94a]. Such a learning vector quantizer based on the $\mathcal{L}_2$ mean has been created using 49 neurons at the first layer, corresponding to input patterns taken from a block of 7×7 pixels. The second layer consists of 2 to 8 neurons, corresponding to the output classes. A 7×7 window scans the image in a random manner to feed the network with input training patterns. During the recall phase, the 7×7 window scans the entire image to classify each pixel into one of p-many ($p = 2, \ldots, 8$) classes. A parametric image is created containing the class membership of each pixel. The ability of the $\mathcal{L}_2$ LVQ to perform segmentation is shown in Fig. 5.11. Figure 5.11a illustrates the classification performed by the $\mathcal{L}_2$ LVQ on the simulated image; two output classes were used, representing background and lesion. Figure 5.11b illustrates the segmentation of a real ultrasonic image of a liver into six classes by using the $\mathcal{L}_2$ LVQ; each class is shown as a distinct region having a gray value ranging from black to white. The more white a region is, the more important is; in general, important regions are blood

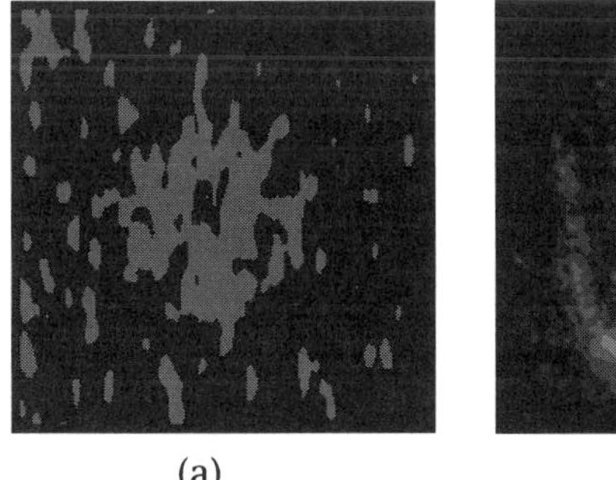

(a)

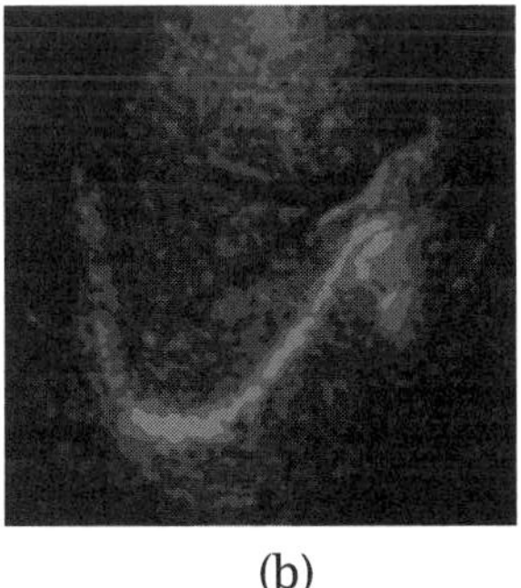

(b)

Figure 5.11: Segmentation of (a) a simulated ultrasound B-mode image and (b) Real ultrasonic image of a liver using the $\mathcal{L}_2$ LVQ neural network.

vessel boundaries, strong reflectors, etc. that should be preserved for diagnostic purposes. Regions having rich texture are shown as light gray; in these regions, a tradeoff between speckle suppression and texture preservation should occur by limiting the maximal filter window size. Image regions in which speckle dominates are shown dark; here, speckle should be efficiently suppressed by allowing the filter window to reach its maximum size.

The result of the overall filtering process using the modified signal-adaptive $\mathcal{L}_2$ mean filter that utilizes the segmentation information provided by the $\mathcal{L}_2$ LVQ is demonstrated in Fig. 5.10d. It is seen that combining segmentation with filtering better preserves the edge information as well as acknowledging image areas containing valuable information that should not be filtered.

5.7 Sorting Networks Using $\mathcal{L}_p$ Mean Comparators

In many areas of signal processing there is a strong need for fast sorting algorithms and structures [Pit90]. A solution exists in the form of *sorting networks*, a special case of sorting algorithms, where the sequence of comparisons performed has homogeneous structure [Knu73]. The basic functional unit of a sorting network is the comparator; the network performance depends mostly on the performance of the type of comparator utilized. In this section a new type of comparator, the $\mathcal{L}_p$ comparator, is proposed that can be implemented using analog circuitry, that is, adders, multipliers, and nonlinear amplifiers that raise the input signal to a power. Thus, it can be used for high-speed analog or hybrid signal processing. The $\mathcal{L}_p$ comparator is based on the $\mathcal{L}_p$ mean. Since sorting networks provide a topology that is independent of the comparator being used, we will "replace" conventional comparators with $\mathcal{L}_p$ comparators and we will determine what additional modifications are needed so that the errors are within acceptable levels [Pap96b].

Figure 5.12c shows an $\mathcal{L}_p$ comparator. When p is large, $\mathcal{L}_{-p}$ converges to the min operator and $\mathcal{L}_p$ to the max operator. If the network of Fig. 5.12a utilizes $\mathcal{L}_p$ comparators, then estimates $\hat{x}_{(i)}$ of the ordered input samples $x_{(i)}$ will be pro-

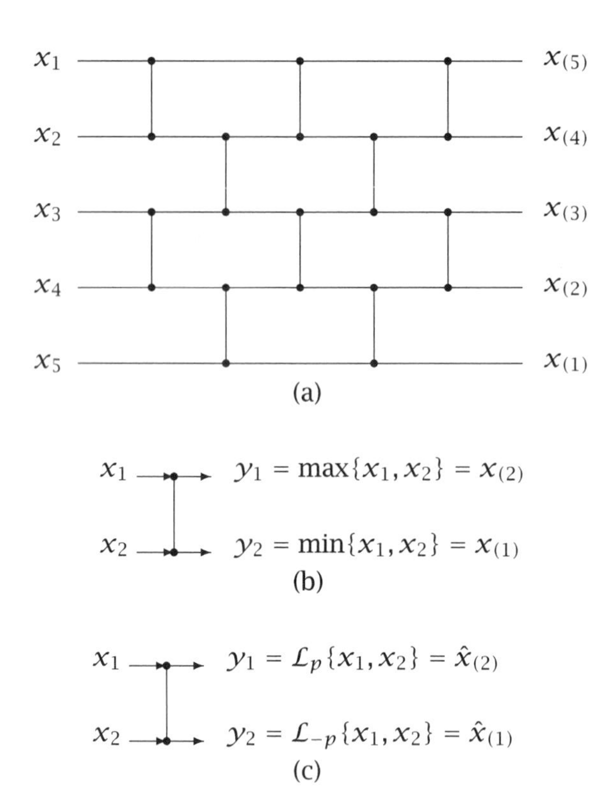

Figure 5.12: (a) An odd-even transposition sorting network of $N = 5$ inputs. (b) A max/min comparator. (c) An $\mathcal{L}_p$ comparator.

duced at the network outputs, where $i = 1, \ldots, N$. However, for small or medium values of p, $\mathcal{L}_p$ comparators introduce errors.

Let us first examine the approximation error introduced by a single $\mathcal{L}_p$ comparator. We assume that the inputs x_1 and x_2 are i.i.d. RVs distributed in the interval $(0, 255)$ obeying some pdf (e.g., uniform). It can be shown that Eq. (5.79) holds for the error $e_{\max}(x_1, x_2)$ between $x_{(i)}$ and its $\mathcal{L}_p$ approximation $\hat{x}_{(i)}$:

$$\begin{aligned} e_{\max}(x_1, x_2) &\approx \max(x_1, x_2) \qquad \text{for } p \gg 1, \\ \frac{1}{2}|x_1 - x_2| \le e_{\max}(x_1, x_2) &\le |x_1 - x_2| \quad \text{for } 0 < p < 1. \end{aligned} \tag{5.79}$$

Accordingly, we argue that

$$e_{\max}(x_1, x_2) = x_{(2)} - \hat{x}_{(2)} \simeq c|x_1 - x_2|, \qquad c \ge 0. \tag{5.80}$$

Similarly, it can be shown that

$$e_{\min}(x_1, x_2) = x_{(1)} - \hat{x}_{(1)} \simeq d|x_1 - x_2|, \qquad d \le 0. \tag{5.81}$$

Equations (5.80) and (5.81) state that the approximation errors can be expressed in terms of the absolute difference of the input samples; this is on par with the

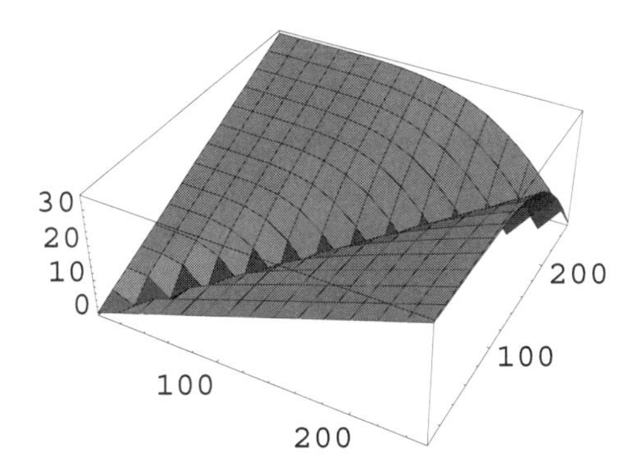

Figure 5.13: Approximation error $e_{\max}(x_1, x_2)$ for $p = 5$.

analysis that follows. The $\mathcal{L}_p$ mean output can be rewritten as

$$\mathcal{L}_p\{x_1, x_2\} = \left(\frac{x_1^p + x_2^p}{2}\right)^{1/p} = \left(\frac{\max^p\{x_1, x_2\} + \min^p\{x_1, x_2\}}{2}\right)^{1/p}. \tag{5.82}$$

Depending on the values of x_1 and x_2, there are two cases: If $\max\{x_1, x_2\} \gg \min\{x_1, x_2\}$, then

$$e_{\max}(x_1, x_2) \simeq \left(1 - \frac{1}{2^{1/p}}\right) \max\{x_1, x_2\}. \tag{5.83}$$

On the other hand, if $x_1 \simeq x_2$, then $e_{\max}(x_1, x_2) \simeq 0$. The approximation error $e_{\max}$ is plotted as a function of the comparator inputs x_1 and x_2 for $p = 5$ in Fig. 5.13 (also see color insert). It is seen that the error is zero across the line $x_1 = x_2$ and increases almost linearly with the distance from this line. Similar conclusions hold for the approximation error $e_{\min}(x_1, x_2)$. We must stress that approximation errors depend largely on the value of the parameter p. Indeed, from Eq. (5.83) it is evident that as p increases, $e_{\max}$ converges to 0 regardless of the values of x_1 and x_2.

From Eqs. (5.80) and (5.81) it can be seen that errors introduced in the inputs of an $\mathcal{L}_p$ comparator accumulate with the errors introduced by the $\mathcal{L}_p$ element itself and propagate throughout the sorting network. As a consequence, the estimates of the largest (smallest) elements are produced at the network output with a negative (positive) bias. For example, if the largest input sample is $x_i = x_{(N)}$, then each time it passes an $\mathcal{L}_p$ comparator, a negative bias will be added to it. For large values of N and small values of p, it may be possible that the ordering of approximated output samples no longer holds. In this scenario, the network is rendered useless.

Error compensation can be introduced in the output of the $\mathcal{L}_p$ comparator:

$$\hat{x}_{(2)} = \mathcal{L}_p\{x_1, x_2\} + c|x_1 - x_2|, \tag{5.84}$$

$$\hat{x}_{(1)} = \mathcal{L}_p\{x_1, x_2\} + d|x_1 - x_2|. \tag{5.85}$$

The constants c and d can be found by minimizing the following functions:

$$E[(e_{\max}\{x_1, x_2\} - c|x_1 - x_2|)^2], \tag{5.86}$$

$$E[(e_{\min}\{x_1, x_2\} - d|x_1 - x_2|)^2]. \tag{5.87}$$

If an input sample x_j has a value close to the median, the number of comparisons with samples that have greater values is about the same as the number of comparisons with samples that have smaller values. If Eqs. (5.80) and (5.81) are taken into account, then it can be stated that the number of times the sample x_j will be negatively biased will be the same as the number of times it will be positively biased. As a consequence, input samples with values close to the median will appear at the network output with much smaller errors. Of course, the bias is a function of both N and p.

$\mathcal{L}_p$ mean comparators can also be used to produce only the data median. The operation of this median selection network is based on the following observation: If a set has N elements, then the total number of its subsets, consisting of ν elements each, equals $C(N,\nu) = \binom{N}{\nu}$, where $\nu = (N+1)/2$. In the ordered set representation, it is easy to verify that $x_{(\nu)}$ represents the median. We can show that there is at least one ν-element subset C_i of the N-element set that satisfies

$$\max\{C_i\} = x_{(\nu)}, \quad i = 1, \ldots, C(N,\nu), \tag{5.88}$$

while the remaining ν-element subsets satisfy

$$\max\{C_j\} \geq \max\{C_i\}, \quad j \neq i, \quad i,j \in \{1, \ldots, C(N,\nu)\}. \tag{5.89}$$

If the $\binom{N}{\nu}$ maxima $y_j = \max\{C_j\}$ are computed, then by Eqs. (5.88) and (5.89),

$$x_{(\nu)} = \min\{y_j : 1 \leq j \leq C(N,\nu)\}. \tag{5.90}$$

The $\mathcal{L}_p$ implementation requires $C(N,\nu)$ $\mathcal{L}_p$ comparators of ν inputs each and one $\mathcal{L}_{-p'}$ comparator of $C(N,\nu)$ inputs. The raise-to-a-power elements required in the evaluation of maxima can be reduced to 1 by moving them to the network input; Fig. 5.14 depicts a median sorting network of this type. It is evident that the number of nonlinear amplifiers has decreased. It is also possible to choose $p \ll p'$ to reduce the implementation complexity of the $\mathcal{L}_p$ elements. It has been found that median approximation networks based on Eqs. (5.88) and (5.90) reduce the error significantly compared to odd-even transposition sorting networks.

5.8 Edge Preserving Filtering by Combining Nonlinear Means and Order Statistics

In this section, we define a general class of nonlinear order statistic (NOS) filters by extending the definition given in Eq. (5.1) [Pit86c]. We define as an NOS filter the following operation:

$$y = f\left(\frac{\sum_{i=1}^{N} a_i g(x_{(i)})}{\sum_{i=1}^{N} a_i}\right), \tag{5.91}$$

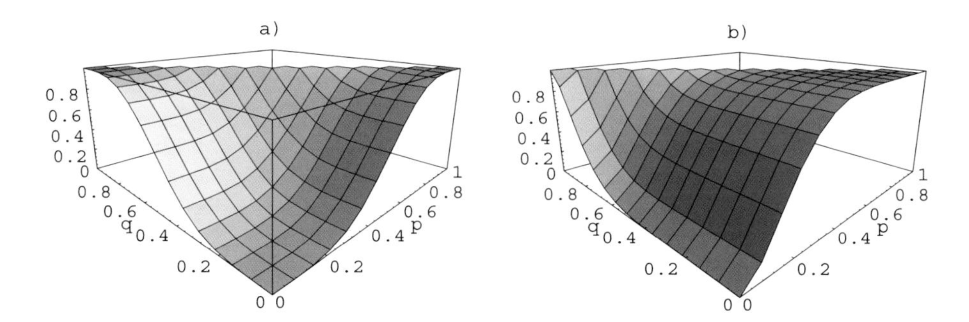

Figure 1.2: Breakdown probability functions of (a) 9-point median and (b) second order statistic filters.

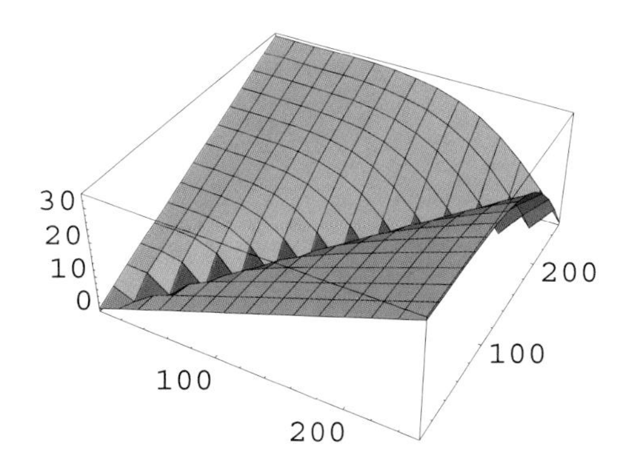

Figure 5.13: Approximation error $e_{\max}(x_1, x_2)$ for $p = 5$.

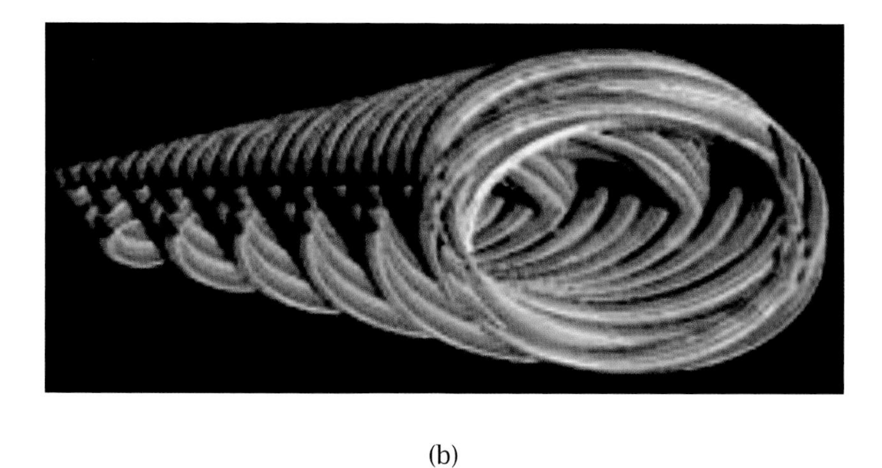

(b)

Figure 11.4: Construction of (b) the fractal form "shell," based on Algorithm 11.2.

(a) (b) (c) (d)

Figure 3.5: Simple selection filter constraint examples: (a) original color image Balloon, (b) observation image corrupted by impulses, and simple 5×5 observation window sample mean filtering operation (c) without and (d) with the selection constraint.

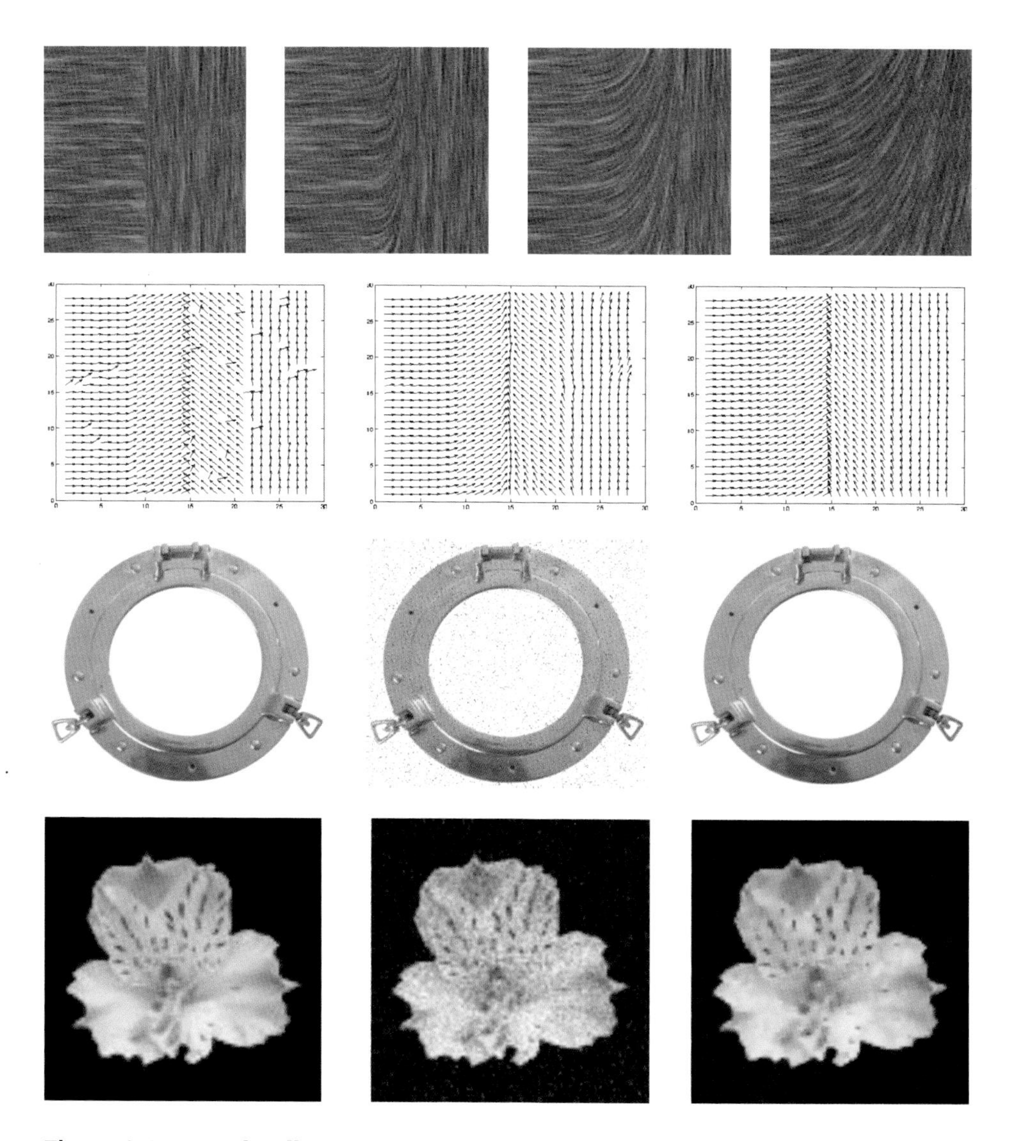

Figure 8.4: Examples illustrating the ideas of direction diffusion (see text for details).

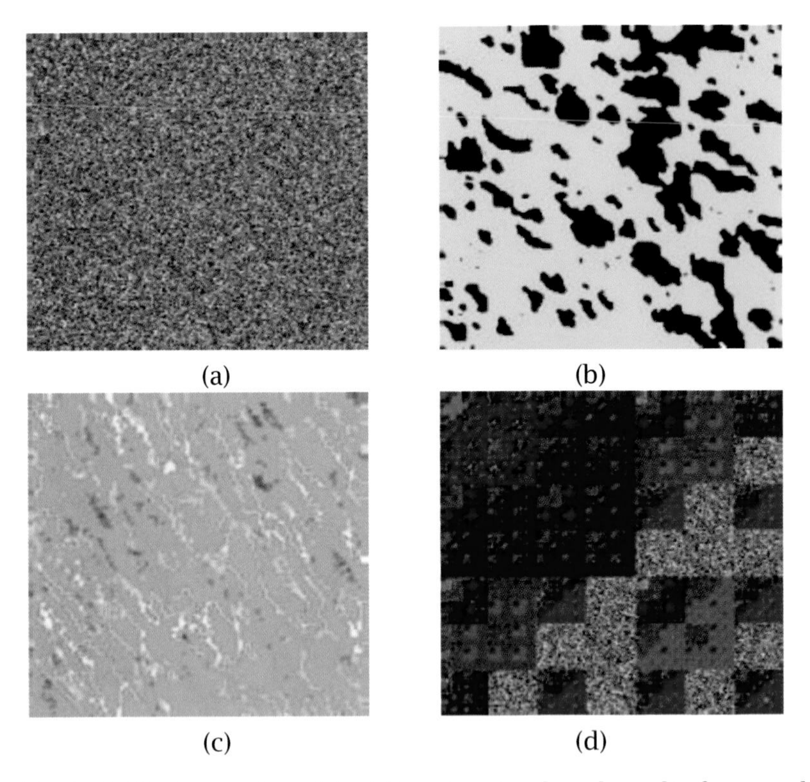

Figure 11.5: (a) Initial image G_4, a 2D white noise signal, and results from applying CL Rules (b) 1, (c) 2, (d) 3 for $m = 4$ iterations.

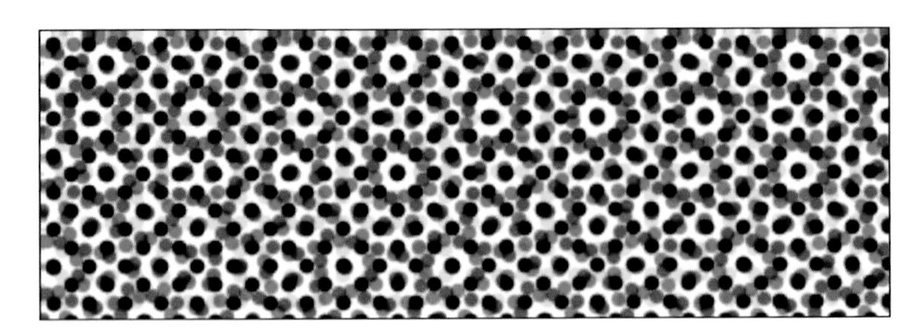

Figure 13.5: Rosette pattern created by setting the CMYK channels to screen angles 15°, 75°, 0°, and 45°, respectively.

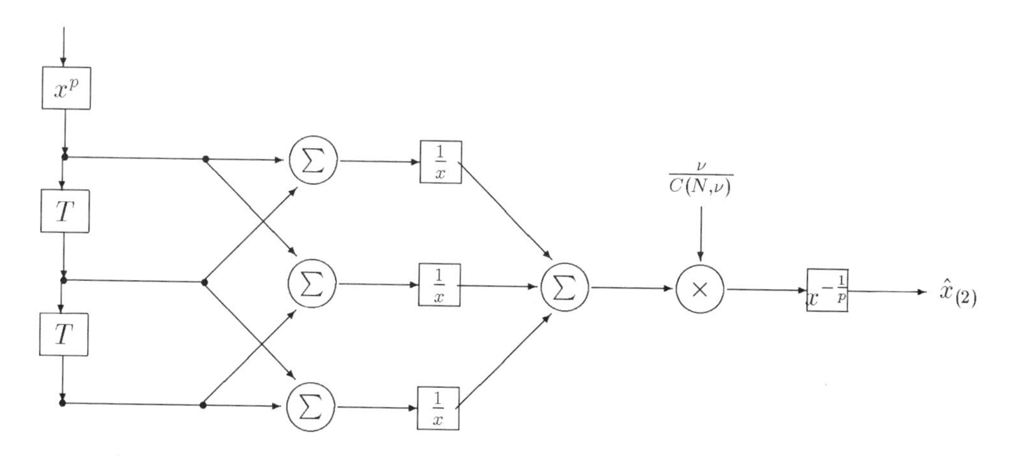

Figure 5.14: Median approximation sorting network utilizing delay elements.

where $g(\cdot)$ and $f(\cdot)$ are memoryless single-valued nonlinear functions and a_i are weights that may or may not be independent of x_i, $i = 1,\ldots,N$. The numbers $g_{(i)}(x_i)$, $i = 1,\ldots,N$ are the numbers $g_i = g(x_i)$, $i = 1,\ldots,N$, ordered according to their magnitude (i.e., order statistics). The smallest of them is $g_{(1)}$ and the largest is $g_{(N)}$. Many known nonlinear filters are special cases of Eq. (5.91). If the nonlinear functions $g(\cdot)$ and $f(\cdot)$ are removed, then the class of order statistic filters [Bov83, Dav81, Jus81] results. Let $\nu = (N+1)/2$. The median filter is a special case of Eq. (5.91) for $f(x) = g(x) = x$, $a_\nu = 1$, and $a_i = 0$ for $i = 1,\ldots,N$, $i \neq \nu$. Similarly, it can be shown that the ith order statistic, the alpha-trimmed mean [Bed84], and the morphological filters [Ser86] are special cases of the NOS filter. The class of nonlinear means results if $f(x) = g^{-1}(x)$. We realize that NOS filters offer a large freedom in the choice of the functions $f(\cdot)$ and $g(\cdot)$ and the weights a_i. This freedom combined with the possibility of ordering the data creates a great variety of filters structures. This fact explains the versatility of the NOS filters and their ability to adapt to many different tasks ranging from filtering of signal-dependent noise to edge detection.

Let us first examine the ability of NOS filters to suppress signal-dependent noise. A useful representation of the signal-dependent noise is given by

$$x = t(s) + r(s)n, \tag{5.92}$$

which extends Eq. (5.20) used in Sec. 5.3. In Eq. (5.92) $t(s)$ is a nonlinear detector response function. Multiplicative noise $x = s + \kappa s n$ is a special case of Eq. (5.92). Photoelectron noise is another type of noise that can be modeled by Eq. (5.92); that is,

$$x = \kappa_1 s^{\gamma_1} + \kappa_2 s^{\gamma_2} n. \tag{5.93}$$

By applying the analysis of Sec. 5.3 to the image formation model described by Eq. (5.93), it can be shown [Pit86c] that $f(\cdot)$ and $g(\cdot)$ should be chosen as follows:

$$g(x) = \frac{\gamma_1}{\kappa_2 (\gamma_1 - \gamma_2)\kappa_1^{-\gamma_2/\gamma_1}} x^{(\gamma_1-\gamma_2)/\gamma_2}, \tag{5.94}$$

$$f(x) = \left(\frac{\kappa_2(\gamma_1 - \gamma_2)}{\kappa_1 \gamma_1} x \right)^{1/(\gamma_1 - \gamma_2)} . \tag{5.95}$$

For example, the image formation model that describes sufficiently well the behavior of commercial television orthicons is [Pit86c]

$$x = s^{0.7} + s^{0.35} n. \tag{5.96}$$

For the model of Eq. (5.96), Eqs. (5.94) and (5.95) are simplified as

$$g(x) = 2x^{0.5}, \qquad f(x) = (0.5x)^{2.857}. \tag{5.97}$$

By using Eq. (5.97), the image formation model reduces to the form of Eq. (5.23), which is an additive noise model, if the higher-order noise terms are neglected. The removal of the noise n can be done by an $\mathcal{L}$-filter whose weights can be found elsewhere [Bov83] or by applying adaptive signal processing algorithms [Kot94b].

Another combination of nonlinear means and order statistics is exploited in the so-called $\mathcal{L}_p$ mean median filter ($\mathcal{L}_p$ MMF) [Nie91]. The $\mathcal{L}_p$ MMF filters belong to a more general filter class, the so-called mapping order statistic filters (MOSF). Let us define the signal subsequence within the filter window of length $N = 2v + 1$ centered at the point l as $\mathcal{A}(l) = \{x_i \mid l - v \le i \le l + v\}$. Let $\mathcal{A}_1(l), \mathcal{A}_2(l), \ldots, \mathcal{A}_r(l)$ be r subsets of $\mathcal{A}(l)$ such that $\mathcal{A}(l) = \mathcal{A}_1(l) \cup \mathcal{A}_2(l) \cup \ldots \cup \mathcal{A}_r(l)$, where $\cup$ denotes the union of two sets. The subsets $\mathcal{A}_1(l), \mathcal{A}_2(l), \ldots, \mathcal{A}_r(l)$ need not be disjoint sets. Let $\Phi_j(\cdot)$ for $j = 1, 2, \ldots, r$ be some functions that map every set $\mathcal{A}_j(l)$ to a real number; that is,

$$\phi_j(l) = \Phi_j(\mathcal{A}_j(l)). \tag{5.98}$$

The output of the MOSF filter at point l is defined as a linear combination of the order statistics of $\phi_1(l), \phi_2(l), \ldots, \phi_r(l)$. Let us choose $r = 3$ and

$$\begin{aligned} \mathcal{A}_1(l) &= \{x_{l-v}, x_{l-v+1}, \ldots, x_{l-1}\}, \\ \mathcal{A}_2(l) &= \{x_l\}, \\ \mathcal{A}_3(l) &= \{x_{l+1}, x_{l+2}, \ldots, x_{l+v}\}. \end{aligned} \tag{5.99}$$

If the mapping functions are chosen to be $\mathcal{L}_p$ means, that is,

$$\begin{aligned} \phi_1 &= \left(\frac{1}{\nu} \sum_{i=1}^{v} x_{l-i}^p \right)^{1/p}, \\ \phi_2 &= x_l, \\ \phi_3 &= \left(\frac{1}{\nu} \sum_{i=1}^{v} x_{l+i}^p \right)^{1/p}, \end{aligned} \tag{5.100}$$

the $\mathcal{L}_p$ MMF output is defined by

$$y(l) = \sum_{j=1}^{3} \phi_{(j)}(l). \tag{5.101}$$

The main idea in developing the $\mathcal{L}_p$ MMF is to use a median filter appended to the outputs of the $\mathcal{L}_p$ mean subfilters to emphasize only the mean value at one side of the filtering point which is closer to the actual signal value to that of the mean value at the other side.

Let us consider the following signal model:

$$x_i = \begin{cases} m & \text{uncorrupted with probability } 1 - q_n - q_p, \\ S_{\min} & \text{corrupted by positive impulse noise with probability } q_p, \\ S_{\max} & \text{corrupted by negative impulse noise with probability } q_n, \end{cases} \tag{5.102}$$

where $m > 0$, $S_{\max} \gg m$, and $S_{\min}$ is assumed to be zero without loss of generality. It has been proven [Nie91] that the probability that the reconstruction using an $\mathcal{L}_p$ MMF filter will fail is given by

$$P_{\text{MMF}}(q_n, p_m) = P_p(q_p) + P_n(q_n), \tag{5.103}$$

where

$$P_n(q_n) = q_n^{v+1}(q_n^{N-1} + 2 - 2q_n^v), \tag{5.104}$$

$$P_p(q_p) = \left[1 - (1 - q_p)^v\right]\left[(2q_p - 1)(1 - q_p)^v + 1\right]. \tag{5.105}$$

For positive impulses, the probability of failure of an $\mathcal{L}_p$ mean filter with window length $N = 2v + 1$ is given by

$$P_{\mathcal{L}_p} = 1 - (1 - q)^{2v+1}. \tag{5.106}$$

It has been found that the $\mathcal{L}_p$ MMF is more efficient that the median filter in suppressing impulse noise, especially at higher q_n. In the case of positive impulse noise, the probability of failure for the $\mathcal{L}_p$ MMF is always less than that for the $\mathcal{L}_p$ mean having the same window size. Being able to suppress negative impulses and offering a smaller probability of failure for positive impulses, the $\mathcal{L}_p$ MMF is superior to the $\mathcal{L}_p$ mean filter.

5.9 Summary

In this chapter, the class of nonlinear mean filters has been reviewed. Their statistical properties and their performance in the presence of impulse and signal-dependent noise have been derived. The edge preserving properties of the nonlinear filters have been studied as well. Combinations of nonlinear mean filters have been used for edge detection. The optimality of $\mathcal{L}_2$ mean filters in Rayleigh multiplicative speckle and signal-dependent Gaussian speckle supports important applications in ultrasonic image processing and analysis that have thoroughly been described. Motivated by the fact that $\mathcal{L}_p$ mean filters are approximators of the max/min filters for positive and negative p, respectively, soft grayscale morphological filters have been derived. Moreover, the substitution of traditional max/min operators with $\mathcal{L}_p$ means has resulted in sorting networks suitable for analog implementations. A general filter structure encompassing the filters that are based on order statistics, the homomorphic filters, the nonlinear mean filters, and the morphological filters has been outlined. Finally, $\mathcal{L}_p$ mean median filters have been reviewed for edge preserving filtering.

References

[Abb79] J. G. Abbott and F. L. Thurstone. Acoustic speckle: Theory and experimental analysis. *Ultrason. Imag.* **1**, 303–324 (1979).

[Ars83] H. H. Arsenault and M. Denis. Image processing in signal-dependent noise. *Can. J. Phys.* **61**, 309–317 (1983).

[Ast97] J. Astola and P. Kuosmanen. *Fundamentals of Nonlinear Digital Filters.* CRC Press, Boca Raton, FL (1997).

[Bed84] J. B. Bednar and T. L. Watt. Alpha-trimmed means and their relationship to median filters. *IEEE Trans. Acoust. Speech Signal Process.* ASSP-**32**(1), 145–153 (February 1984).

[Bov83] A. C. Bovik, T. S. Huang, and D. C. Munson, Jr. A generalization of median filtering using linear combinations of order statistics. *IEEE Trans. Acoust. Speech Signal Process.* ASSP-**31**(6), 1342-1349 (December 1983).

[Bur78] C. B. Burckhardt. Speckle in ultrasound B-mode scans. *IEEE Trans. Son. Ultrason.* SU-**25**(1), 1–6 (January 1978).

[Dav81] H. A. David. *Order Statistics.* Wiley, New York (1981).

[Fri79] R. W. Fries and J. W. Modestino. Image enhancement by stochastic homomorphic filtering. *IEEE Trans. Acoust. Speech Signal Process.* ASSP-**27**(6), 625–637 (December 1979).

[Goo76] J. W. Goodman. Some fundamental properties of speckle. *J. Opt. Soc. Amer. A* **66**, 1145–1150 (November 1976).

[Gue83] W. Geuen. An advanced edge detection method based on two stage procedure. In *Signal Processing II: Principles and Applications* (W. Schuessler, ed.). Elsevier, Amsterdam (1983).

[Jus81] B. J. Justusson. Median filtering: Statistical properties. In *Two-Dimensional Digital Signal Processing*, Vol. II (T. S. Huang, ed.). Springer, New York (1981).

[Ken73] B. G. Kendall. *The Advanced Theory of Statistics*, Vol. 1. Griffin, London (1973).

[Knu73] D. E. Knuth. *The Art of Computer Programming*, Vol. 3. Addison-Wesley, Reading, MA (1973).

[Kot92] C. Kotropoulos and I. Pitas. Optimum nonlinear signal detection and estimation in the presence of ultrasonic speckle. *Ultrason. Imag.* **14**(3), 249–275 (July 1992).

[Kot94a] C. Kotropoulos, X. Magnisalis, I. Pitas, and M. G. Strintzis. Nonlinear ultrasonic image processing based on signal-adaptive filters and self-organizing neural networks. *IEEE Trans. Image Process* **3**(1), 65–77 (January 1994).

[Kot94b] C. Kotropoulos and I. Pitas. Adaptive nonlinear filters for digital signal/image processing. In *Control and Dynamic Systems*, Vol. 67 (C. Leondes, ed.), pp. 263–317. Academic Press, San Diego (1994).

[Kuo93] P. Kuosmanen. Soft morphological filtering. Report A270, Department of Mathematical Sciences, Univ. of Tampere, Finland (April 1993).

[Kun84] A. Kundu, S. K. Mitra, and P. P. Vaidyanathan. Application of two-dimensional generalized mean filtering for removal of impulse noises from images. *IEEE Trans. Acoust. Speech Signal Process.* ASSP-**32**(3), 600–609 (June 1984).

[Lou88] A. Loupas. Digital image processing for noise reduction in medical ultrasonics. Ph.D dissertation, University of Edinburgh, Scotland (July 1988).

[Nie91] X. Nie and R. Unbehauen. Edge preserving filtering by combining nonlinear mean and median filters. *IEEE Trans. Signal Process.* **39**(11), 2552–2554 (November 1991).

[Oos85] B. J. Oosterveld, J. M. Thijssen, and W. A. Verhoef. Texture of B-mode echograms: 3-D simulations and experiments of the effects of diffraction. *Ultrason. Imag.* **7**, 142–160 (1985).

[Opp63] A. V. Oppenheim, R. W. Schafer, and T. G. Stockham. Nonlinear filtering of multiplied and convolved signals. *Proc. IEEE* **56**, 1264–1294 (August 1963).

[Opp75] A. V. Oppenheim, and R. W. Schafer. *Digital Signal Processing.* Prentice-Hall, Englewood Cliffs, NJ (1975).

[Pap84] A. Papoulis. *Probability, Random Variables and Stochastic Processes.* McGraw-Hill, New York (1984).

[Pap95] M. Pappas and I. Pitas. Grayscale morphology using nonlinear L_p mean filters. In *Proc. IEEE Workshop on Nonlinear Signal and Image Processing*, Vol. I, pp. 34–37 (Neos Marmaras, Chalkidiki, Greece, June 1995).

[Pap96a] M. Pappas and I. Pitas, Soft morphological operators based on nonlinear L_p mean operators. In *Proc. Int. Symposium on Mathematical Morphology*, pp. 187–193 (Atlanta, GA, May 1996).

[Pap96b] M. Pappas and I. Pitas. Sorting networks using nonlinear L_p mean comparators. In *Proc. Int. Symposium on Circuits and Systems*, Vol. II, pp. 1–4 (Atlanta, GA, May 1996).

[Pit86a] I. Pitas and A. N. Venetsanopoulos. Nonlinear mean filters in image processing. *IEEE Trans. Acoust. Speech Signal Process.* ASSP-**34**(34), 573–584 (June 1986).

[Pit86b] I. Pitas and A. N. Venetsanopoulos. Edge detectors based on nonlinear means. *IEEE Trans. Patt. Anal. Machine Intell.* PAMI-**8**(4), 538–550 (July 1986).

[Pit86c] I. Pitas and A. N. Venetsanopoulos. Nonlinear order statistic filters for image filtering and edge detection. *Signal Process.* **10**, 395–411 (1986).

[Pit90] I. Pitas and A.N. Venetsanopoulos. *Nonlinear Digital Filters: Principles and Applications.* Kluwer, Higham, MA (1990).

[Pit92] I. Pitas. *Digital Image Processing Algorithms.* Prentice-Hall, London (1992).

[Pra78] W. K. Pratt. *Digital Image Processing.* Wiley, New York (1978).

[Sal92] P. Salembier. Adaptive rank order based filters. *Signal Process.* **27**(1), 1–25 (April 1992).

[Ser86] J. Serra. Introduction to mathematical morphology. *Comput. Vision Graph. Image Process.* **35**, 283–305 (1986).

[Van68] H. L. Van Trees. *Detection, Estimation and Modulation Theory.* Wiley, New York (1968).

[Ver91] J. T. M. Verhoeven, J. M. Thijssen, and A. G. M. Theeuwes. Lesion detection by echographic image processing: Signal-to-noise ratio imaging. *Ultrason. Imag.* **13**, 238–251 (1991).

6

Two-Dimensional Teager Filters

Stefan Thurnhofer

Lucent Technologies
Huntington Beach, California

6.1 Introduction

The Teager filter [Kai90] is a homogeneous quadratic Volterra filter. It has the property that sinusoidal inputs generate constant outputs that are approximately proportional to the square of the input frequency. This filter and modifications of it have been used successfully in image enhancement applications [Mit91]. In this chapter, we present a framework that describes a more general category of homogeneous quadratic Volterra filters to which the Teager filter belongs. We analyze their properties and derive a two-dimensional version of the Teager filter. We conclude the chapter with application examples of image enhancement, image interpolation, and image halftoning.

6.2 Discrete Volterra Series and Properties

Discrete Volterra filters are defined by the input-output relationship [Sic92]

$$y(n) = \mathcal{V}\{x(n)\} \tag{6.1}$$

$$= \sum_{p=0}^{\infty} y_p(n) \tag{6.2}$$

$$= h_0 + \sum_{p=1}^{\infty} \overline{h}_p(x(n)) \tag{6.3}$$

$$= h_0 + \sum_{p=1}^{\infty} \sum_{k_1=-\infty}^{\infty} \cdots \sum_{k_p=-\infty}^{\infty} h_p(k_1, \ldots, k_p) \prod_{q=1}^{p} x(n-k_q), \tag{6.4}$$

which we can also view as a discrete-time nonlinear system with an input sequence $x(n)$ and an output sequence $y(n)$. These relationships will also be used in Chapters 7 and 14. In general, the pth-order Volterra kernels $h_p(k_1, \ldots, k_p)$ can assume arbitrary complex values and have infinite extent. In this chapter, however, we consider only real-valued kernels; we also assume that $y_0 = h_0 = 0$ and that input and output signal are real. To illustrate the meaning of Eq. (6.4), we examine the first two terms of the sum over p:

$$\begin{aligned} y(n) &= \sum_{k_1=-\infty}^{\infty} h_1(k_1)x(n-k_1) \\ &\quad + \sum_{k_1=-\infty}^{\infty} \sum_{k_2=-\infty}^{\infty} h_2(k_1, k_2)x(n-k_1)x(n-k_2) \qquad (6.5) \\ &= y_1(n) + y_2(n). \qquad (6.6) \end{aligned}$$

The second term in the above equation, involving the double summation

$$y_2(n) = \sum_{k_1=-\infty}^{\infty} \sum_{k_2=-\infty}^{\infty} h_2(k_1, k_2)x(n-k_1)x(n-k_2), \tag{6.7}$$

defines a homogeneous quadratic Volterra filter [Pic82].

Clearly, the expression for the first term $y_1(n)$ in Eq. (6.5) is identical to the well-known time-domain description of a linear time-invariant discrete-time system given by the convolution of $x(n)$ with $h_1(n)$. The second-order term $y_2(n)$ can be interpreted as a two-dimensional convolution of $h_2(n_1, n_2)$ with products of samples of the input sequence. This becomes even clearer if we define

$$\bar{x}(n_1, n_2) = x(n_1)x(n_2) \tag{6.8}$$

and write

$$y(n) = h_1(n) * x(n) + h_2(n_1, n_2) ** \bar{x}(n_1, n_2)\Big|_{n=n_1=n_2}, \tag{6.9}$$

where the single and double asterisks denote one- and two-dimensional convolutions, respectively. For example, consider the system given by $y(n) = 2x(n) + x(n-1) + 2x(n-1)x(n) - 3x(n)^2 + x(n+1)x(n-2)$. Using the notation for unit-sample sequences,

$$\delta(n) = \begin{cases} 1 & \text{for } n = 0, \\ 0 & \text{for } n \neq 0, \end{cases} \qquad \delta(n_1, n_2) = \begin{cases} 1 & \text{for } n_1 = n_2 = 0, \\ 0 & \text{otherwise,} \end{cases} \tag{6.10}$$

the Volterra kernels for these systems are then $h_1(n) = 2\delta(n) + \delta(n-1)$ and $h_2(n_1, n_2) = 2\delta(n_1 - 1, n_2) - 3\delta(n_1, n_2) + \delta(n_1 + 1, n_2 - 2)$.

We write the general expression for the pth-order term using the convolution notation

$$y_p(n) = h_p(n_1, \ldots, n_p) \underset{(p)}{*} [x(n_1) \ldots x(n_p)]\Big|_{n=n_1=\cdots=n_p}, \tag{6.11}$$

where the asterisk and its subscript denote p-dimensional convolution. Even though Volterra systems are nonlinear in the input sequence, they are linear in the kernels; it is this observation that allows us to use frequency domain techniques for the analysis and design of these systems. We define the spectrum of the Volterra kernels as the multidimensional Fourier transform of $h_p(n_1, \ldots, n_p)$ [Bil89a, Pit90],

$$H_p(e^{j\omega_1}, \ldots, e^{j\omega_p}) = \mathcal{F}\{h_p(n_1, \ldots, n_p)\} \tag{6.12}$$

$$= \sum_{k_1=-\infty}^{\infty} \cdots \sum_{k_p=-\infty}^{\infty} h_p(n_1, \ldots, n_p) e^{-j(n_1\omega_1 + \ldots + n_p\omega_p)}, \tag{6.13}$$

with the inverse relationship

$$h_p(n_1, \ldots, n_p) = \mathcal{F}^{-1}\{H_p(e^{j\omega_1}, \ldots, e^{j\omega_p})\} \tag{6.14}$$

$$= \frac{1}{(2\pi)^p} \int_0^{2\pi} \cdots \int_0^{2\pi} H_p(e^{j\omega_1}, \ldots, e^{j\omega_p}) e^{j(n_1\omega_1 + \ldots + n_p\omega_p)} \, d\omega_1 \ldots d\omega_p. \tag{6.15}$$

The symbol $\mathcal{F}$ denotes Fourier transform in one or more dimensions, depending on the context. For the case $p = 1$, these relations define the standard frequency response $H_1(e^{j\omega_1})$. For all other values of p, we follow the same idea and call $H_p(e^{j\omega_1}, \ldots, e^{j\omega_p})$ the pth-order generalized frequency responses [Bil89a] or, for simplicity, the pth-order frequency response. Similarly, the $h_p(n_1, \ldots, n_p)$ are called the pth-order impulse responses. The interpretation of these functions, however, is quite different for $p = 1$ and $p > 1$; we study them in detail in Sec. 6.3. We rewrite Eq. (6.9) using the frequency responses of input and system as

$$\begin{aligned} y(n) &= \mathcal{F}^{-1}\{H_1(e^{j\omega})X(e^{j\omega})\} \\ &\quad + \mathcal{F}^{-1}\{H_2(e^{j\omega_1}, e^{j\omega_2})\bar{X}(e^{j\omega_1}, e^{j\omega_2})\}\Big|_{n=n_1=n_2}, \end{aligned} \tag{6.16}$$

where $X(e^{j\omega}) = \mathcal{F}\{x(n)\}$ and $\bar{X}(e^{j\omega_1}, e^{j\omega_2}) = \mathcal{F}\{\bar{x}(n_1, n_2)\} = X(e^{j\omega_1})X(e^{j\omega_2})$. Unless noted otherwise, we use lower case letters for sequences in time or space and upper case letters for their Fourier transforms.

6.2.1 Two-Dimensional Discrete Filters

The two-dimensional Volterra series is a straightforward extension of Eq. (6.4) [Ram88] with the input and output signals being two-dimensional sequences. We are interested only in the case of finite sequences. Also, to limit implementation complexity, we consider only the first- and second-order terms of the filter equation [Mit91, Ram86, Ram88, Ram93, Sic92]. Thus,

$$\begin{aligned} y(n_1,n_2) &= y_1(n_1,n_2) + y_2(n_1,n_2) && (6.17)\\ &= \sum_{k_1}\sum_{k_2} h_1(k_1,k_2)x(n_1-k_1,n_2-k_2)\\ &\quad + \sum_{k_1}\sum_{k_2}\sum_{k_3}\sum_{k_4} h_2(k_1,k_2,k_3,k_4)\\ &\quad x(n_1-k_1,n_2-k_2)x(n_1-k_3,n_2-k_4) && (6.18)\\ &= h_1(n_1,n_2) \underset{(2)}{*} x(n_1,n_2)\\ &\quad + h_2(n_1,n_2,n_3,n_4) \underset{(4)}{*} \bar{x}(n_1,n_2,n_3,n_4)\Big|_{\substack{n_1=n_3,\\ n_2=n_4}} && (6.19) \end{aligned}$$

with $\bar{x}(n_1,n_2,n_3,n_4) = x(n_1,n_2)x(n_3,n_4)$. We assume again that the zero-order term $y_0(n_1,n_2)$ equals zero and that symbols like x or y can represent both one- and two-dimensional sequences, depending on the context. In Eq. (6.18), we have omitted the limits of the sums; typically, $-N_1 \le k_1, k_3 \le N_1$ and $-N_2 \le k_2, k_4 \le N_2$. In the following we omit the limits for simplicity, unless their values cannot be inferred easily.

Again using the Fourier transform notation, we can rewrite Eq. (6.19) as

$$\begin{aligned} y(n_1,n_2) &= \mathcal{F}^{-1}\{H_1(e^{j\omega_1},e^{j\omega_2})X(e^{j\omega_1},e^{j\omega_2})\}\\ &\quad + \mathcal{F}^{-1}\{H_2(e^{j\omega_1},e^{j\omega_2},e^{j\omega_3},e^{j\omega_4})\\ &\quad \bar{X}(e^{j\omega_1},e^{j\omega_2},e^{j\omega_3},e^{j\omega_4})\}\Big|_{\substack{n_1=n_3,\\ n_2=n_4}} && (6.20) \end{aligned}$$

with $\bar{X}(e^{j\omega_1},e^{j\omega_2},e^{j\omega_3},e^{j\omega_4}) = X(e^{j\omega_1},e^{j\omega_2})X(e^{j\omega_3},e^{j\omega_4})$. It is also worth noting that the convergence behavior of Volterra filters with infinite impulse response is not well understood, and only sufficient but not necessary conditions for stability are known [Rug81, Sch80].

6.2.2 Properties of the Impulse Response

The impulse responses can be written in various forms. In fact, due to an inherent redundancy of Volterra filters, we have some freedom to modify the coefficients without affecting the overall system behavior. For example, the systems $\hat{y}(n) = x(n-1)x(n+1)$ and $\tilde{y}(n) = x(n+1)x(n-1)$ are identical, but their second-order impulse responses are not: $\hat{h}_2(n_1,n_2) = \delta(n_1-1,n_2+1)$ and $\tilde{h}_2(n_1,n_2) = \delta(n_1+1,n_2-1)$. The ambiguity is resolved by the symmetrized impulse response, defined by [Chu79, Zha93]

$$h_p^{\mathrm{s}}(n_1,\ldots,n_p) = \frac{1}{p!}\sum_{\substack{\text{all permutations}\\ \text{of } n_1,\ldots,n_p}} h_p(n_1,\ldots,n_p). \qquad (6.21)$$

Any kernel can be replaced by its symmetrized version without changing the input-output relationship of the system. The impulse responses $\tilde{h}_2(n_1, n_2)$ and $\hat{h}_2(n_1, n_2)$ both lead to the same symmetrized kernel $\tilde{h}_2^s(n_1, n_2) = \hat{h}_2^s(n_1, n_2) = 1/2[\delta(n_1 - 1, n_2 + 1) + \delta(n_1 + 1, n_2 - 1)]$, which belongs to the system $\hat{y}(n) = \tilde{y}(n) = 1/2[x(n-1)x(n+1) + x(n+1)x(n-1)]$. For a second-order system, this means that [Sch80]

$$h_2^s(n_1, n_2) = h_2^s(n_2, n_1), \tag{6.22}$$

which we will use frequently in the following sections.

6.3 Interpretation of Frequency Responses

For linear systems, there exists a simple dependence of the output spectrum on the input spectrum and the frequency response of the system:

$$Y(e^{j\omega}) = H(e^{j\omega})X(e^{j\omega}). \tag{6.23}$$

This equation suggests an intuitive explanation of the system behavior. Frequency components that are present in the input signal are either suppressed or amplified by the corresponding values of the system frequency response. In the case of nonlinear Volterra systems the interpretation of the frequency response it is not as obvious and can lead to novel interpretations of filter operations, as will be seen in Chapter 14. In Eq. (6.13), we defined the Fourier transform of the pth-order kernel. If we consider the second-order frequency response $H_2(e^{j\omega_1}, e^{j\omega_2}) = \mathcal{F}\{h_2(n_1, n_2)\}$, the problem is how to interpret the shape and form of $H_2(e^{j\omega_1}, e^{j\omega_2})$ and how, for instance, to specify a quadratic lowpass filter in the (ω_1, ω_2) domain. Some approaches have been published [Bil89b, Pic82, Vin87, Zha93], but most of them are limited to more general statements about certain example systems. Furthermore, they do not explain the process that leads to the output spectrum, that is, the transition from the (ω_1, ω_2) domain of $H_2(e^{j\omega_1}, e^{j\omega_2})\bar{X}_2(e^{j\omega_1}, e^{j\omega_2})$ to the ω domain of $Y(e^{j\omega})$.

First, we investigate how the input spectrum determines the output signal for a homogeneous quadratic filter. From Eq. (6.16), we write

$$\begin{aligned} y_2(n) &= \mathcal{F}^{-1}\{Y_2(e^{j\omega})\} & (6.24)\\ &= \frac{1}{4\pi^2}\iint H_2(e^{j\omega_1}, e^{j\omega_2})X(e^{j\omega_1})X(e^{j\omega_2})e^{jn(\omega_1+\omega_2)}\,d\omega_1\,d\omega_2 & (6.25)\\ &= \frac{1}{4\pi^2}\iint \bar{Y}_2(e^{j\omega_1}, e^{j\omega_2})e^{jn(\omega_1+\omega_2)}\,d\omega_1\,d\omega_2, & (6.26) \end{aligned}$$

where we have used $\bar{Y}_2(e^{j\omega_1}, e^{j\omega_2}) = H_2(e^{j\omega_1}, e^{j\omega_2})X(e^{j\omega_1})X(e^{j\omega_2})$. If we view the integrals approximately as a linear combination of the integrand's values, then we can say that the output sequence is a linear combination of weighted one-dimensional exponentials, with each contributing to one particular output frequency. Unlike in the standard formula for the inverse Fourier transform, the power of the exponentials is $n(\omega_1 + \omega_2)$. Roughly speaking, this means that all of the (ω_1,ω_2) combinations for which $\omega_1+\omega_2 = \omega_0$ = constant will contribute to the output frequency ω_0. We derive this result in a mathematically more rigorous way in Sec. 6.5. Frequency domain considerations of Volterra filters will also play an important role in Chapters 7 and 14.

We want to illustrate this mapping by studying a very simple example:

Example 6.1. Consider the simplest homogeneous quadratic Volterra filter, defined by $y(n) = x^2(n)$. Let the input be a single sinusoid $x(n) = \sin(\omega_0 n)$. We find the output analytically as $y(n) = \sin^2(\omega_0 n) = [1 - \cos(2\omega_0 n)]/2$ and the magnitudes of the transforms as

$$|X(e^{j\omega})| = \pi\,[\delta(\omega-\omega_0) + \delta(\omega - 2\pi + \omega_0)]$$

and

$$|Y(e^{j\omega})| = \pi\left[\delta(\omega) + \frac{1}{2}\delta(\omega - 2\omega_0) + \frac{1}{2}\delta(\omega - 2\pi + 2\omega_0)\right]$$

for $0 \le \omega \le 2\pi$.

Note that the frequency components at ω_0 and $2\pi - \omega_0$ in the input have been suppressed at the output, whereas new ones have been generated at dc, $2\omega_0$, and $2\pi - 2\omega_0$. To explain this, we examine the impulse response of the system and its frequency response:

$$\begin{aligned} h_2(n_1,n_2) &= \delta(n_1,n_2), \\ H_2(e^{j\omega_1},e^{j\omega_2}) &= 1. \end{aligned}$$

The contour plot of $\bar{Y}_2(e^{j\omega_1},e^{j\omega_2}) = X(e^{j\omega_1})X(e^{j\omega_2})$ in Fig. 6.1 shows the four δ functions between 0 and 2π. For each of these components, we find a line $\omega_1+\omega_2$ = const, such that the δ-function lies on it. For instance, the component $\delta(\omega_1-\omega_0,\omega_2-\omega_0)$ is on the line $\omega_1+\omega_2 = 2\omega_0$. Thus, it will contribute to the frequency $2\omega_0$ in the output signal. In Fig. 6.1, this line is shown with a small arrow, indicating a projection from the two-dimensional plane onto a frequency axis. Similarly, the components $\delta(\omega_1-\omega_0,\omega_2-2\pi+\omega_0)$ and $\delta(\omega_1-2\pi+\omega_0,\omega_2-\omega_0)$ are mapped, along $\omega_1+\omega_2 = 2\pi$, which accounts for the dc component with twice the height of the other ones. The last δ function $\delta(\omega_1 - 2\pi + \omega_0, \omega_2 - 2\pi + \omega_0)$ is mapped to $4\pi - 2\omega_0$, which is equivalent to $2\pi - 2\omega_0$.

6.4 The Teager Algorithm and One-Dimensional Extensions

6.4.1 Teager Algorithm

In 1990, Kaiser [Kai90] introduced an operator defined by the input-output relation

$$y(n) = \mathcal{T}\{x(n)\} = x^2(n) - x(n-1)x(n+1), \tag{6.27}$$

which he called "Teager's algorithm" [Tea80, Tea90]. For sinusoidal inputs, this operator yields a constant output that is an estimate of the physical energy of a pendulum oscillating with the same frequency and amplitude as the input sinusoid [Kai90]. This property justifies the use of the Teager algorithm as an energy operator. Mathematically, if the input is $x(n) = A\cos(\omega_0 n + \phi)$, where ϕ is an arbitrary initial phase, then the operator $\mathcal{T}$ gives

$$y(n) = A^2\sin^2(\omega_0) \approx A^2\omega_0^2. \tag{6.28}$$

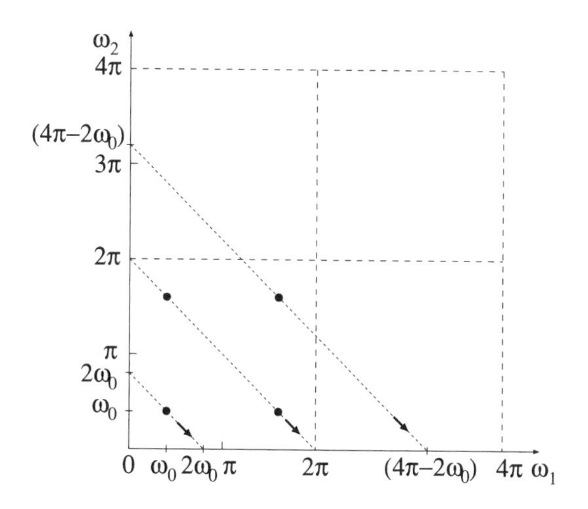

Figure 6.1: Contour plot indicating the locations of the δ functions of $\tilde{Y}(e^{j\omega_1}, e^{j\omega_2})$, which are mapped along the dashed lines onto the one-dimensional frequency axis, thus accounting for the frequency components of the output signal.

The approximation can be made for small frequencies. If $\omega_0 < \pi/8$ then the relative error is always less than 5%. This result is constant in time and very suitable for tracking the energy of an oscillation, for instance, chirp and mixed frequency signals. The Teager algorithm has also been applied to the problem of detecting transient signals in a noisy background [Dun93]. Various properties of the continuous time counterpart of Eq. (6.27) have been derived [Kai93].

The Teager filter is obviously a second-order Volterra filter, and we can study it by using the tools derived in the previous sections. We start by considering the impulse response and its Fourier transform:

$$h_2(n_1, n_2) = \delta(n_1, n_2) - \delta(n_1 + 1, n_2 - 1) \tag{6.29}$$

$$H_2(e^{j\omega_1}, e^{j\omega_2}) = 1 - e^{j\omega_1} e^{-j\omega_2} \tag{6.30}$$

$$\begin{aligned} |H_2(e^{j\omega_1}, e^{j\omega_2})| &= |1 - e^{j\omega_1} e^{-j\omega_2}| \\ &= \sqrt{2 - 2\cos(\omega_1 - \omega_2)}. \end{aligned} \tag{6.31}$$

Figure 6.2 shows the responses. The dots in Fig. 6.2a indicate the location of the δ functions and the numbers in parentheses are the weighting factors. The magnitude response for $0 \le \omega_1, \omega_2 \le 2\pi$ is shown in Fig. 6.2b; by comparing with Fig. 6.1, it is clear how the four δ functions of $X(e^{j\omega_1})X(e^{j\omega_2})$ are weighted by the filtering operation in Eq. (6.25). Since $H_2(e^{j\omega_1}, e^{j\omega_2}) = 0$ for $\omega_1 = \omega_2$, the two δ functions at (ω_0, ω_0) and $(2\pi - \omega_0, 2\pi - \omega_0)$ are completely suppressed by the filter. Only the ones that are mapped to the dc component remain. Thus, the output signal is a constant. From this simple analysis, we can draw the conclusion that any system for which $H_2(e^{j\omega}, e^{j\omega}) = 0$ will map a single sinusoid onto a constant output.

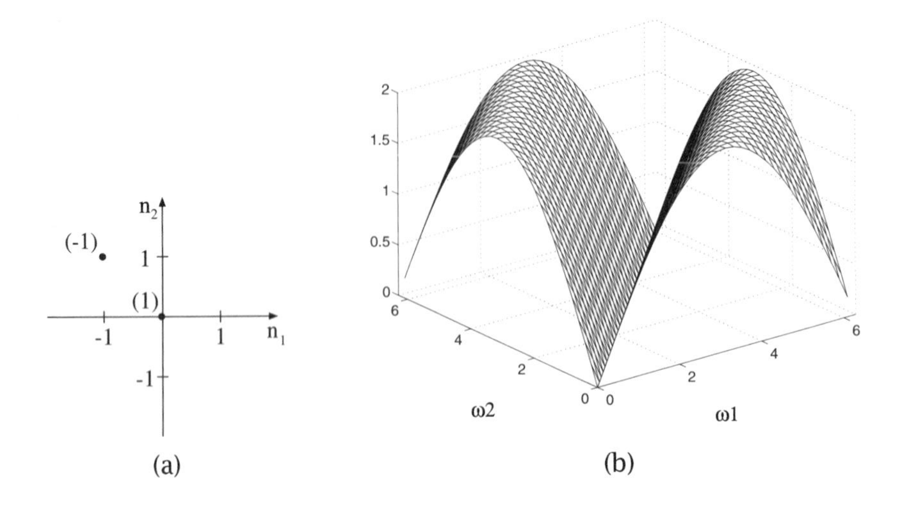

Figure 6.2: (a) Impulse and (b) magnitude responses of the Teager filter; the latter also shows $H_2(e^{j\omega_1}, e^{j\omega_2})$ for the extension of the Teager algorithm for $l = 2$.

The Teager filter does not have a symmetric impulse response since $h_2(n_1, n_2) \neq h_2(n_2, n_1)$. The equivalent system after symmetrizing the kernel is given by

$$y(n) = x^2(n) - \frac{1}{2}x(n+1)x(n-1) - \frac{1}{2}x(n-1)x(n+1), \tag{6.32}$$

and the responses are

$$\begin{aligned} h_2(n_1, n_2) &= \delta(n_1, n_2) - \frac{1}{2}\delta(n_1 + 1, n_1 - 1) \\ &\quad - \frac{1}{2}\delta(n_1 - 1, n_2 + 1), \end{aligned} \tag{6.33}$$

$$H_2(e^{j\omega_1}, e^{j\omega_2}) = 1 - \cos(\omega_1 - \omega_2). \tag{6.34}$$

Their plots are shown in Fig. 6.3. This frequency response is now even and real, since $h_2(n_1, n_2)$ is now symmetric (even).

6.4.2 Extension of the Teager Algorithm

The Teager algorithm maps a sinusoidal input onto a constant output that is approximately proportional to the square of the input frequency. In some cases, however, it might be desirable to generalize this dependence and make the exponent variable. We investigate this problem by formalizing the intuitive explanations of Sec. 6.4.1. Let $x(n) = \sin(\omega_0 n)$ be the input signal. We consider unit magnitude signals here since we are interested in the dependence of the filter output on the input frequency only. The input spectrum is then

$$X(e^{j\omega}) = -j\pi\delta(\omega - \omega_0) + j\pi\delta(\omega - 2\pi + \omega_0)$$

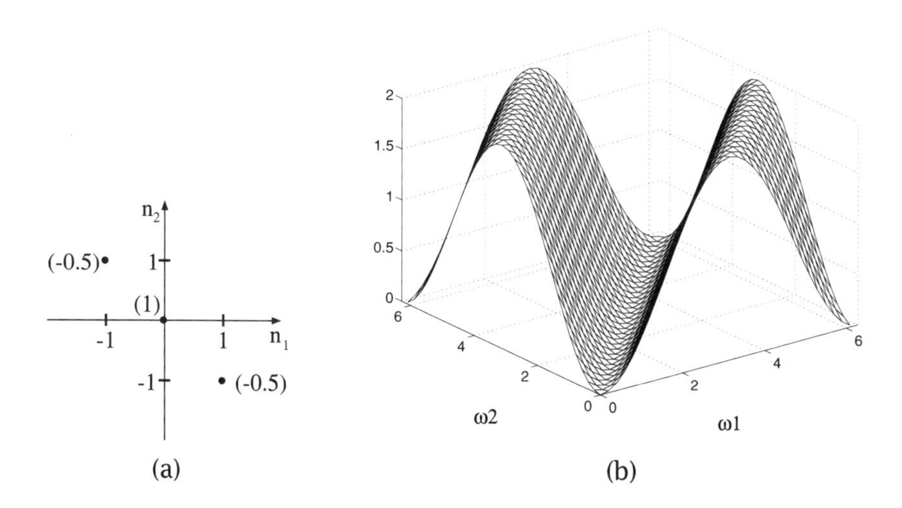

Figure 6.3: (a) Impulse and (b) magnitude responses of the symmetrized Teager filter.

for $0 \leq \omega \leq 2\pi$. Here and in the following, we do not consider the response outside the interval from 0 to 2π; this simplifies notation considerably without losing information. Furthermore, we have

$$\begin{aligned} X(e^{j\omega_1})X(e^{j\omega_2}) &= -\pi^2\delta(\omega_1 - \omega_0, \omega_2 - \omega_0) \\ &\quad -\pi^2\delta(\omega_1 - 2\pi + \omega_0, \omega_2 - 2\pi + \omega_0) \\ &\quad +\pi^2\delta(\omega_1 - \omega_0, \omega_2 - 2\pi + \omega_0) \\ &\quad +\pi^2\delta(\omega_1 - 2\pi + \omega_0, \omega_2 - \omega_0). \end{aligned} \tag{6.35}$$

The first two terms in Eq. (6.35) will be cancelled by $H_2(e^{j\omega_1}, e^{j\omega_2})$, since we assume that

$$H_2(e^{j\omega_1}, e^{j\omega_2}) = 0 \qquad \text{for } \omega_1 = \omega_2, \tag{6.36}$$

similar to the standard Teager filter. We use the abbreviations

$$H_2(e^{j\omega_0}, e^{j(2\pi-\omega_0)}) = \alpha, \tag{6.37}$$

$$H_2(e^{j(2\pi-\omega_0)}, e^{j\omega_0}) = \beta, \tag{6.38}$$

and thus

$$\begin{aligned} \bar{Y}_2(e^{j\omega_1}, e^{j\omega_2}) &= \pi^2\alpha\delta(\omega_1 - \omega_0, \omega_2 - 2\pi + \omega_0) \\ &\quad +\pi^2\beta\delta(\omega_1 - 2\pi + \omega_0, \omega_2 - \omega_0). \end{aligned} \tag{6.39}$$

With Eq. (6.25), we compute the output sequence:

$$y(n) = y_2(n) = \frac{1}{4}(\alpha + \beta). \tag{6.40}$$

Table 6.1: Impulse Response Coefficients on Diagonal $n_1 = -n_2$ for $l = 1, \ldots, 7$; Case $l = 2$ Corresponds to the Teager Filter.

n_1	$l = 1$	$l = 2$	$l = 3$	$l = 4$	$l = 5$[a]	$l = 6$	$l = 7$[a]
−10	−0.00420	0	0.00004	0	−0.00000	0	0.00000
−9	−0.00525	0	0.00007	0	−0.00000	0	0.00000
−8	−0.00679	0	0.00012	0	−0.00001	0	0.00000
−7	−0.00916	0	0.00020	0	−0.00002	0	0.00000
−6	−0.01311	0	0.00039	0	−0.00006	0	0.00002
−5	−0.02046	0	0.00084	0	−0.00022	0	0.00018
−4	−0.03663	0	0.00220	0	−0.00113	0	0.00316
−3	−0.08513	0	0.00808	0	−0.01469	−0.03125	−0.04748
−2	−0.42466	0	0.07275	0.12500	0.16168	0.18750	0.20577
−1	−0.42466	−0.50000	−0.50929	−0.50000	−0.48504	−0.46875	−0.45270
0	1.27298	1.00000	0.84882	0.75000	0.67906	0.62500	0.58205

[a]The number 0.00000 denotes a very small but nonzero value.

Suppose we would like to map a sinusoidal input onto a constant signal approximately proportional to the input frequency. We can equate $y(n)$ from Eq. (6.40) with $\sin(\omega_0) \approx \omega_0$. Additionally, we make the assumption that $H_2(e^{j\omega_1}, e^{j\omega_2})$ is symmetric about $\omega_1 = \omega_2$. This yields $\alpha = \beta = 2\sin(\omega_0)$. To obtain $H_2(e^{j\omega_1}, e^{j\omega_2})$ from this one-dimensional slice, we substitute

$$\omega_0 = \frac{\omega_1 + \omega_2}{2} + \pi. \tag{6.41}$$

Thus,

$$H_2(e^{j\omega_1}, e^{j\omega_2}) = \left| 2\sin\left(\frac{\omega_1 + \omega_2}{2} + \pi\right) \right| \tag{6.42}$$

$$= \sqrt{2 - 2\cos(\omega_1 - \omega_2)}, \tag{6.43}$$

which, incidentally, is equal to the magnitude of $H_2(e^{j\omega_1}, e^{j\omega_2})$ for the original Teager filter in Eq. (6.31) and is therefore also shown in Fig. 6.2b. Without computing the absolute value in Eq. (6.42), $H_2(e^{j\omega_1}, e^{j\omega_2})$ would have an odd symmetry about the main diagonal, that is, $\alpha = -\beta$.

Our next step is to compute the quadratic kernel of this filter. Because $H_2(e^{j\omega_1}, e^{j\omega_2})$ is constant along lines parallel to the diagonal $\omega_1 = \omega_2$, $h_2(n_1, n_2)$ will be nonzero only for $n_1 = -n_2$. We tabulate the resulting coefficients in Table 6.1.

We can now derive the general case in which the desired output is given by

$$y(n) = y_2(n) = \sin^l(\omega_0).$$

Again using Eqs. (6.40) and (6.41) and the symmetry assumption, we obtain

$$H_2(e^{j\omega_1}, e^{j\omega_2}) = \left| 2\sin^l\left(\frac{\omega_1 - \omega_2}{2}\right) \right|.$$

Table 6.1 lists the first 11 coefficients of the impulse responses for $1 \leq l \leq 7$. All responses are symmetric. For even values of l, we obtain finite length impulse responses since the corresponding $H_2(e^{j\omega_1}, e^{j\omega_2})$ is smooth and does not have any sharp transitions.

6.5 Spectrum of the Output Signal

6.5.1 First Method

We give two equations for the output spectrum of homogeneous quadratic Volterra filters. The first one is mathematically simpler and quite useful for many applications [Zha93]. The second is in the next subsection, and even though it is more complicated, it facilitates the understanding of the filtering mechanism in the frequency domain.

Starting with Eq. (6.26),

$$y(n) = \frac{1}{4\pi^2} \int_0^{2\pi} \int_0^{2\pi} \bar{Y}_2(e^{j\omega_1}, e^{j\omega_2}) e^{jn(\omega_1+\omega_2)} \, d\omega_1 \, d\omega_2,$$

and after some modifications the output spectrum is [Zha93]

$$\begin{aligned} Y(e^{j\omega}) &= Y_2(e^{j\omega}) \\ &= \frac{1}{2\pi} \int_0^{2\pi} \bar{Y}_2(e^{j\omega_1}, e^{j(\omega-\omega_1)}) \, d\omega_1 & (6.44) \\ &= \frac{1}{2\pi} \int_0^{2\pi} H_2(e^{j\omega_1}, e^{j(\omega-\omega_1)}) X(e^{j\omega_1}) X(e^{j(\omega-\omega_1)}) \, d\omega_1. & (6.45) \end{aligned}$$

The corresponding equation for two-dimensional filters can be derived analogously:

$$\begin{aligned} Y(e^{j\omega_1}, e^{j\omega_2}) &= Y_2(e^{j\omega_1}, e^{j\omega_2}) \\ &= \frac{1}{2\pi^2} \int_0^{2\pi} \int_0^{2\pi} \bar{Y}_2(e^{ju}, e^{jv}, e^{j(\omega_1-u)}, e^{j(\omega_2-v)}) \, du \, dv & (6.46) \\ &= \frac{1}{2\pi^2} \int_0^{2\pi} \int_0^{2\pi} H_2(e^{ju}, e^{jv}, e^{j(\omega_1-u)}, e^{j(\omega_2-v)}) \\ &\quad X(e^{ju}, e^{jv}) X(e^{j(\omega_1-u)}, e^{j(\omega_2-v)}) \, du \, dv. & (6.47) \end{aligned}$$

6.5.2 Second Method

The integration in the previous section was carried out parallel to the ω_1 and ω_2 axes. According to the intuitive explanation given in Sec. 6.3, however, it appears to be more natural to integrate parallel to the diagonal lines $\omega_2 = \omega_0 - \omega_1$, where ω_0 is a free parameter. All of the values on such a line contribute to the same output frequency and we can directly compute the output spectrum.

We state the result of the derivation in the following theorem.

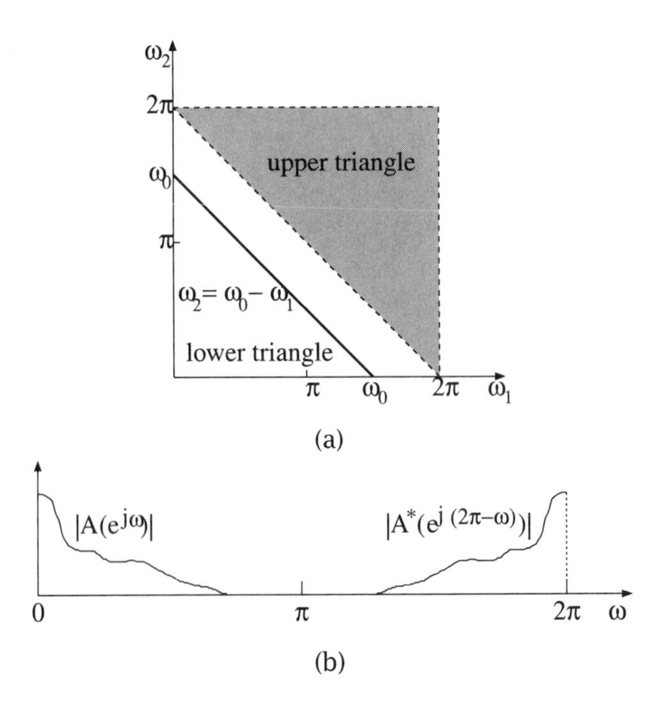

Figure 6.4: Pictorial explanation of the output spectrum.

Theorem 6.1. *The spectrum of the real output signal of a one-dimensional homogeneous quadratic Volterra filter is given by*

$$Y(\mathrm{e}^{j\omega}) = Y_2(\mathrm{e}^{j\omega}) = \frac{1}{2\pi}[A(\mathrm{e}^{j\omega}) + A^*(\mathrm{e}^{j(2\pi-\omega)})] \tag{6.48}$$

for $0 \le \omega \le 2\pi$, *where*

$$\begin{aligned} A(\mathrm{e}^{j\omega}) &= \int_0^{\omega} \bar{Y}_2(\mathrm{e}^{ju}, \mathrm{e}^{j(\omega-u)})\,\mathrm{d}u \\ &= \int_0^{\omega} H_2(\mathrm{e}^{ju}, \mathrm{e}^{j\omega-u})X(\mathrm{e}^{ju})X(\mathrm{e}^{j(\omega-u)})\,\mathrm{d}u. \end{aligned}$$

$X(e^{j\omega})$ *is the input spectrum and* $H_2(\mathrm{e}^{j\omega_1}, \mathrm{e}^{j\omega_2})$ *is the second-order frequency response of the system.*

The proof of Theorem 6.1 is given in the appendix of this chapter (Sec. 6.9.1).

Theorem 6.1 shows the dependence of $Y(\mathrm{e}^{j\omega})$ on $A(\mathrm{e}^{j\omega})$, which is essentially obtained by integrating over the lower triangle in Fig. 6.4a. By flipping, shifting and conjugating it, we obtain $A^*(\mathrm{e}^{j(2\pi-\omega)})$, as shown in Fig. 6.4b. Even though Theorem 6.1 defines $Y(\mathrm{e}^{j\omega})$ and $A(\mathrm{e}^{j\omega})$ only for $0 \le \omega \le 2\pi$, we find the spectrum for all other frequencies by periodic extension. The entire spectrum is therefore completely defined by the lower triangle in Fig. 6.4a. The upper triangle does not contain any new information.

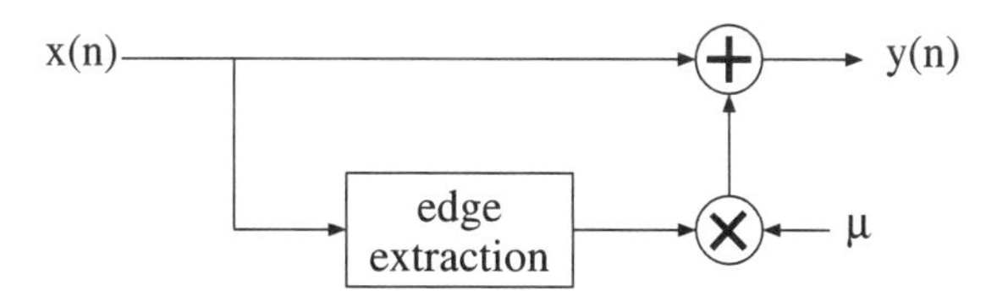

Figure 6.5: Block diagram of the unsharp masking technique. (Reproduced with permission from [Thu96c]. © 1996 IEEE.)

6.6 Mean-Weighted Highpass Filters

In the previous sections, we developed the basic tools that are necessary for working with homogeneous quadratic Volterra filters in one and two dimensions. We are now interested in the applications of these filters to image processing problems, specifically the enhancement of edges. We make use of a property of the human visual system (HVS) whereby it is able to discern details and brightness variations in images depending on masking by the surrounding area [Bof86]. According to Weber's law, the just noticeable difference (JND) in brightness is proportional to the average intensity of the surrounding pixels [Rub69, Wan93]. In other words, the HVS sees details more easily in dark regions, whereas bright portions tend to mask details. Even though this rule does not hold for very small average intensity values, we can still use the proportionality as a good overall estimate.

As a direct consequence of Weber's law, image processing systems (image enhancement, image coding, image restoration, etc.) should treat the darker regions of an image very carefully because a human observer would perceive any imperfections more easily than in the brighter areas. This is especially problematic when images are noisy and corrupted, such as, after transmission through a wireless channel. For example, consider the structure of Fig. 6.5. The input image is processed by a block that extracts edges and features. The output is then scaled by an appropriate factor μ and added back to the original image. This method is generally referred to as unsharp masking [Jai89, Lim90] and is quite effective for enhancing low contrast images. The edge extraction block in Fig. 6.5 is often implemented as a linear highpass filter such as a discrete Laplacian operator [Jai89]:

$$\begin{aligned} y(n_1,n_2) &= 4x(n_1,n_2) - x(n_1-1,n_2) - x(n_1+1,n_2) \\ &\quad - x(n_1,n_2-1) - x(n_1,n_2+1). \end{aligned} \tag{6.49}$$

The main advantage of using this particular filter is its computational simplicity. The highpass filter enhances those portions of the image that contain mostly high frequency information, that is, edges and textured regions. The perceptual impression is improved because the image appears sharper and better defined.

An apparent problem of this technique is that it does not discriminate between actual image information and noise. Thus, noise is enhanced as well. Unfortunately, visible noise tends to be mostly in the medium to high frequency range.

The contrast sensitivity function (CSF) of the human visual system shows that the eye (and the higher level processing systems in the visual cortex) is less sensitive to low frequencies [Rog83]. The peak sensitivity is reached at about 3 to 4 cycles per degree and falls off at both lower and higher frequencies. Therefore, noise around this peak is quite visible and the unsharp masking technique will make it even worse. Since the CSF is measured in cycles per degree, it also depends on the viewing distance. On the other hand, the digital highpass filter employed in a scheme like the one in Fig. 6.5 is designed in cycles per pixel. Thus, no matter which specifications we use for the filter, that is, which cutoff frequency, there will always be a viewing distance where the passband region includes the peak of the CSF. The amplified high frequency noise will then be very visible and deteriorate the perceptual image quality.

To eliminate this problem while still preserving the simplicity of the algorithm, we make use of Weber's law and modify the unsharp masking method such that the image enhancement is dependent on the local average pixel intensity. In bright regions we can enhance the image more because noise and other gray level fluctuations are much less visible. On the other hand, in darker portions we want to suppress the enhancement process since it might deteriorate image quality. This simple idea indicates the need for a highpass filter that depends on the local mean, similar to

$$H(\mathrm{e}^{j\omega}) \propto H_{\mathrm{h}}(\mathrm{e}^{j\omega}) \cdot (\text{local mean}), \tag{6.50}$$

where $H_{\mathrm{h}}(\mathrm{e}^{j\omega})$ is a linear highpass filter. The most obvious solution to the problem is to multiply $H_{\mathrm{h}}(\mathrm{e}^{j\omega})$ with a linear local mean estimator (i.e., an averaging filter) [Yu94]. However, the computational expense for such a structure would be quite high, and a much more efficient way is to use quadratic Volterra filters. In the following sections, we derive the properties of a subclass of homogeneous quadratic Volterra filters, that approximate mean-weighted highpass filters.

6.6.1 One-Dimensional Mean-Weighted Highpass Filters

It has been shown [Mit91] that the Teager filter in Eq. (6.27) is approximately equal to a product of local mean and highpass:

$$y(n) = \mathcal{T}\{x(n)\} \approx k_n[2x(n) - x(n-1) - x(n+1)], \tag{6.51}$$

where $k_n = [x(n-1) + x(n) + x(n+1)]/3$ estimates the local mean. Additionally, two-dimensional extensions of the Teager filter were introduced, that were found intuitively and could be decomposed in a way analogous to Eq. (6.51). These filters were used successfully for image enhancement. The one-dimensional Teager filter belongs to a class of Volterra filters that has the following property, where μ_x represents the mean of the input sequence $x(n)$ [Thu95b, Thu96c]:

Theorem 6.2. *A one-dimensional homogeneous second-order Volterra filter can be approximated by a mean-weighted highpass filter, that is,*

$$Y(\mathrm{e}^{j\omega}) \approx 2\mu_x X(\mathrm{e}^{j\omega}) H_2(\mathrm{e}^{j\omega}, \mathrm{e}^{j0}) \tag{6.52}$$

if $H_2(e^{j\omega}, e^{j0})$ *has highpass characteristics and*

$$H_2(e^{j0}, e^{j0}) = \sum_{k_1}\sum_{k_2} h_2(k_1, k_2) = 0, \tag{6.53}$$

$$H_2(e^{j\omega}, e^{j0}) = H_2(e^{j0}, e^{j\omega}), \tag{6.54}$$

$$S^h = \sum_{\substack{k_1 \; k_2 \\ k_1 \neq k_2}}\sum h_2(k_1, k_2)[2h_2(k_1, k_1) + h_2(k_2, k_2)] + \sum_{\substack{k_1 \; k_2 \; k_3 \\ k_1 \neq k_2 \\ k_2 \neq k_3}}\sum\sum h_2(k_1, k_2)[h_2(k_1, k_3) + h_2(k_2, k_3)] \geq 0. \tag{6.55}$$

The approximation error is then given by

$$Y_e(e^{j\omega}) = \frac{1}{2\pi}\int_0^{2\pi} H_2(e^{j\omega_1}, e^{j(\omega-\omega_1)})\hat{X}(e^{j\omega_1})\hat{X}(e^{j(\omega-\omega_1)})\, d\omega_1, \tag{6.56}$$

where $\hat{X}(e^{j\omega}) = X(e^{j\omega}) - 2\pi\mu_x\delta(\omega)$ *is the mean-removed version of* $X(e^{j\omega})$.

The proof of Theorem 6.2 is given in the appendix of this chapter (Sec. 6.9.2). The basic idea is to decompose the general expression of the homogeneous quadratic Volterra filters given in Eq. (6.7) into the two terms

$$y_1(n) = \mu_x \sum_{k_1}\sum_{k_2} h_2(k_1, k_2)[\hat{x}(n - k_1) + \hat{x}(n - k_2)],$$

$$y_2(n) = \mathcal{V}(\hat{x}) = \sum_{k_1}\sum_{k_2} h_2(k_1, k_2)\hat{x}(n - k_1)\hat{x}(n - k_2),$$

where $\hat{x}(n) = x(n) - \mu_x$ is the mean-removed version of $x(n)$. We then show under which conditions $y_2(n)$ can be neglected compared to $y_1(n)$.

We need to make some remarks about Theorem 6.2. First, we implicitly assume here and in the following that the input signal's (positive) mean is larger than its standard deviation. Should this not be true, then the signal has to be scaled and shifted accordingly before processing. Second, we can approximate the mean μ_x in Eq. (6.52) by a local averaging filter if the kernel of the Volterra filter is finite. In this case, only pixels in a local neighborhood influence the result, and thus only the local characteristics of the signal play a role for the computation of the output. Third, the assumption that $H_2(e^{j\omega}, e^{j0})$ has highpass properties is not necessary in general. Even though this is important for the kind of applications that we have in mind in this chapter, any filter for which Eqs. (6.53) through (6.55) are true can be approximated as a mean-weighted filter. For the most general case, we can relax the restrictions even further and drop the symmetry condition in Eq. (6.54), but then we must rewrite Eq. (6.52) as

$$Y(e^{j\omega}) \approx \mu_x X(e^{j\omega})[H_2(e^{j\omega}, e^{j0}) + H_2(e^{j0}, e^{j\omega})].$$

We will see, however, that the symmetry condition is automatically fulfilled for important subclasses of mean-weighted highpass filters. Therefore, the only truly

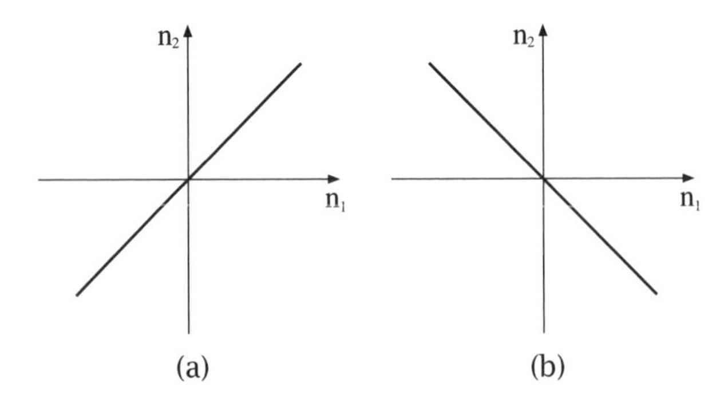

Figure 6.6: Locations of nonzero impulse response coefficients of special cases of mean-weighted highpass filters: (a) class I and (b) class II filters. (Reproduced with permission from [Thu96c]. © 1996 IEEE.)

necessary conditions are given in Eqs. 6.53 and 6.55, which are both written as conditions on the kernel. Even though Eq. (6.55) appears to be fairly complicated, we can greatly simplify it for certain special cases. In the first class, filters consist of squares of pixels.

Definition 6.1. If $h_2(n_1, n_2) = 0$ for all $n_1 \neq n_2$ and $\sum_{n_1} \sum_{n_2} h_2(n_1, n_2) = 0$ and if $H_2(e^{j\omega}, e^{j0})$ has highpass characteristics, then we call the filter a class I discrete-time homogeneous quadratic Volterra filter, or class I filter for short.

The following theorem states the approximate behavior of class I filters.

Theorem 6.3. *Class I filters can be approximated as mean-weighted highpass filters.*

Proof. Equation (6.53) holds by assumption and Eq. (6.54) is true since we have $h_2(n_1, n_2) = \sum_{k=-\infty}^{\infty} h_2(k, k)\delta(n_1 - k, n_2 - k)$, which leads to

$$H_2(e^{j\omega_1}, e^{j\omega_2}) = \sum_{k=-\infty}^{\infty} h_2(k, k) e^{-jk\omega_1 - jk\omega_2},$$

and thus $H_2(e^{j\omega}, e^{j0}) = \sum_k h_2(k, k) e^{-jk\omega} = H_2(e^{j0}, e^{j\omega})$.

It remains to show that $S^h \geq 0$. In Eq. (6.55) all the sums yield zero because $h_2(k_1, k_2)$ is nonzero only for $k_1 = k_2$. Therefore, we obtain $S^h = 0$ here.

Note that these systems are not necessarily symmetric. Examples are $y(n) = x^2(n) - 0.5x^2(n-1) - 0.5x^2(n-3)$ and $y(n) = x^2(n-1) - x^2(n-2)$. The first filter can approximately be written as $\mu_x[2x(n) - x(n-1) - x(n-3)]$ and the second as $2\mu_x[x(n-1) - x(n-2)]$. Figure 6.6a shows the possible locations for nonzero coefficients of the impulse response in the n_1-n_2 plane.

The second family of filters is more important for our applications.

Definition 6.2. If $h_2(n_1, n_2) = 0$ for all $n_1 \neq -n_2$ and $\sum_{n_1} \sum_{n_2} h_2(n_1, n_2) = 0$ and if $H_2(e^{j\omega}, e^{j0})$ has highpass characteristics, then we call the filter a class II discrete-time homogeneous quadratic Volterra filter, or class II filter for short.

The following theorem states the approximate behavior of class II filters:

Theorem 6.4. *Class II filters can be approximated as mean-weighted highpass filters.*

Proof. Here, we can again simplify S^h by substituting $k_2 = -k_1$:

$$\begin{aligned} S^h &= \sum_{\substack{k_1 \\ k_1 \neq 0}} h_2(k_1, -k_1) \left[2 \underbrace{h_2(k_1, k_1)}_{=0} + \underbrace{h_2(-k_1, -k_1)}_{=0} \right] \\ &+ \sum_{\substack{k_1 \\ k_3 \neq -k_1 \\ k_1 \neq 0}} \sum_{k_3} h_2(k_1, -k_1) \left[h_2(k_1, k_3) + h_2(-k_1, k_3) \right]. \end{aligned}$$

We rewrite this equation further since $h_2(k_1, k_3)$ is nonzero only if $k_3 = -k_1$, which is not part of the summation, and thus $h_2(k_1, k_3)$ is zero for all cases. Taking into account that $h_2(-k_1, k_3) \neq 0$ only if $k_3 = k_1$ finally yields

$$\begin{aligned} S^h &= \sum_{\substack{k_1 \\ k_1 \neq 0}} h_2(k_1, -k_1) h_2(-k_1, k_1) \\ &= \sum_{\substack{k \\ k \neq 0}} h_2^2(k, -k) \geq 0. \end{aligned}$$

For the last step, we used the symmetry condition of $h_2(k_1, k_2)$.

We do not have to assume that $H_2(\mathrm{e}^{j\omega}, \mathrm{e}^{j0}) = H_2(\mathrm{e}^{j0}, \mathrm{e}^{j\omega})$ since it is true because of the symmetry of $h_2(n_1, n_2)$. Writing $h_2(n_1, n_2) = \sum_{k=-\infty}^{\infty} h_2(k, -k)\delta(n_1 - k, n_2 + k)$ and with $h_2(k, -k) = h_2(-k, k)$, we obtain

$$\begin{aligned} H_2(\mathrm{e}^{j\omega_1}, \mathrm{e}^{j\omega_2}) &= \sum_{k=-\infty}^{\infty} h_2(k, -k) \mathrm{e}^{-jk\omega_1 + jk\omega_2} \\ &= h_2(0,0) + 2 \sum_{k=1}^{\infty} h_2(k, -k) \cos(k\omega_1 - k\omega_2). \end{aligned} \tag{6.57}$$

Thus, $H_2(\mathrm{e}^{j\omega}, \mathrm{e}^{j0}) = \sum h_2(k, -k)\mathrm{e}^{-jk\omega}$ and $H_2(\mathrm{e}^{j0}, \mathrm{e}^{j\omega}) = \sum h_2(k, -k)\mathrm{e}^{jk\omega}$. We substitute $l = -k$ and use the symmetry condition for $h_2(n_1, n_2)$:

$$H_2(\mathrm{e}^{j0}, \mathrm{e}^{j\omega}) = \sum h_2(l, -l)\mathrm{e}^{-jl\omega} = H_2(\mathrm{e}^{j\omega}, \mathrm{e}^{j0}).$$

Again, the locations of the nonzero coefficients of the impulse response are indicated in Fig. 6.6. The Teager filter in Eq. (6.32) is an example for a simple system from this class. Since the original definition of the Teager algorithm in Eq. (6.27) does not have a symmetric impulse response, we do not use it here.

The Teager filter maps sinusoidal inputs to constant outputs, and it is straightforward to show that in fact every class II system has this property. Using Eq. (6.57), we can see that $H_2(e^{j\omega}, e^{j\omega}) = \sum h_2(k, -k) = H_2(e^{j0}, e^{j0})$, which is assumed to be zero. Thus, in $X(e^{j\omega_1})X(e^{j\omega_2})$ only the component that is mapped to dc will be preserved.

6.6.2 Two-Dimensional Mean-Weighted Highpass Filters

We consider the extension of the previous results to two-dimensional filters for applications in image processing. Since the class II filters of Theorem 6.4 inherently lead to symmetric systems, they preserve the localization of edges (similar to zero-phase linear filters). Because of their symmetric system equation, class II filters always have real and even frequency responses. The symmetry of the impulse response itself is, as always, implied.

The fundamental building block of class II systems is $x(n-k)x(n+k)$. The overall system consists of a linear combination of several of these terms for different k. Thus, each term is a product of samples that are equally far away to the left and right of the center location n. We extend this idea to the two-dimensional case simply by defining the basic building block of the 2D filter as $x(n_1-k_1, n_2-k_2)x(n_1+k_1, n_2+k_2)$; that is, it consists of pixels centered around the current pixel (n_1, n_2). For the impulse response, we can easily derive the simple property

$$h_2(n_1, n_2, n_3, n_4) \neq 0 \quad \text{only if} \quad n_1 = -n_3 \quad \text{and} \quad n_2 = -n_4. \tag{6.58}$$

Analogous to Theorem 6.4, we also require that

$$\sum_{k_1}\sum_{k_2}\sum_{k_3}\sum_{k_4} h_2(k_1, k_2, k_3, k_4) = H_2(e^{j0}, e^{j0}, e^{j0}, e^{j0}) = 0. \tag{6.59}$$

Additionally, we can show that

$$H_2(e^{j\omega_1}, e^{j\omega_2}, e^{j0}, e^{j0}) = H_2(e^{j0}, e^{j0}, e^{j\omega_1}, e^{j\omega_2}). \tag{6.60}$$

To prove this, we write the kernel as

$$h_2(n_1, n_2, n_3, n_4) = \sum_{k_1}\sum_{k_2} h_2(k_1, k_2, -k_1, -k_2)\delta(n_1 - k_1, n_2 - k_2, n_3 + k_1, n_4 + k_4).$$

Then we find that

$$H_2(e^{j\omega_1}, e^{j\omega_2}, e^{j\omega_3}, e^{j\omega_4}] = \sum_{k_1}\sum_{k_2} h_2(k_1, k_2, -k_1, -k_2)e^{-jk_1\omega_1 - jk_2\omega_2 + jk_1\omega_3 + jk_2\omega_4}, \tag{6.61}$$

and thus

$$\begin{aligned} H_2(e^{j\omega_1}, e^{j\omega_2}, e^{j0}, e^{j0}) &= \sum_{k_1}\sum_{k_2} h_2(k_1, k_2, -k_1, -k_2)e^{-jk_1\omega_1 - jk_2\omega_2} \\ &= H_2(e^{j0}, e^{j0}, e^{j\omega_1}, e^{j\omega_2}), \end{aligned}$$

where we have used the symmetry of the kernel, that is,

$$h_2(n_1, n_2, n_3, n_4) = h_2(n_3, n_4, n_1, n_2). \tag{6.62}$$

Using Eqs. (6.58), (6.59), and (6.60), we arrive at the corresponding extension to Eq. (6.52) as

$$Y(e^{j\omega_1}, e^{j\omega_2}) \approx 2\mu_x X(e^{j\omega_1}, e^{j\omega_2}) H_2(e^{j\omega_1}, e^{j\omega_2}, e^{j0}, e^{j0}). \tag{6.63}$$

Assuming that $H_2(e^{j\omega_1}, e^{j\omega_2}, e^{j0}, e^{j0})$ has highpass characteristics in either the ω_1 or ω_2 direction or in both, we conclude that the class II two-dimensional systems can be approximated by mean-weighted highpass filters.

The proof of Eq. (6.63) follows the same idea as the proof of Theorem 6.2 in the appendix of this chapter. Basically, we use the general reconstruction formula for two-dimensional quadratic systems in Eq. (6.47) and substitute $X(e^{j\omega_1}, e^{j\omega_2}) = \hat{X}(e^{j\omega_1}, e^{j\omega_2}) + 4\pi^2 \mu_x \delta(\omega_1, \omega_2)$, where μ_x represents the mean of $x(n_1, n_2)$. Using $H_2(e^{j0}, e^{j0}, e^{j0}, e^{j0}) = 0$, we obtain

$$\begin{aligned} Y(e^{j\omega_1}, e^{j\omega_2}) = {} & \mathcal{V}\{\hat{X}(e^{j\omega_1}, e^{j\omega_2})\} + \mu_x \hat{X}(e^{j\omega_1}, e^{j\omega_2}) \\ & [H_2(e^{j\omega_1}, e^{j\omega_2}, e^{j0}, e^{j0}) + H_2(e^{j0}, e^{j0}, e^{j\omega_1}, e^{j\omega_2})]. \end{aligned}$$

Then, we show that $\mathcal{V}\{\hat{X}(e^{j\omega_1}, e^{j\omega_2})\}$ contributes much less to the overall result and thus can be neglected. Resubstituting $X(e^{j\omega_1}, e^{j\omega_2})$ finally yields Eq. (6.63).

6.7 Least-Squares Design of Edge Extracting Filters

6.7.1 Characterization of Edge Extracting Filters

We have developed the theory behind the class of filters that we want to use for edge extraction. We know their structure and properties in both space and frequency domains, and we also have justified why, at least in principle, this family would be advantageous for image enhancement. In the next step, we need to develop a performance measure that somehow allows us to find the best filter from this class [Thu95a]. Even though essentially all mean-weighted highpass filters would extract edges of an image in some way, there are still important differences that we have to take into account. For instance, a problem that would not occur in one-dimensional signal processing but is of importance in image processing is isotropy. The filter should find edges independent of their orientation. Horizontal or vertical boundaries must lead to the same response, and therefore the degree to which a filter is isotropic is critical for the resulting image quality.

In this section, we introduce a method of characterizing the dependence of the filter output on the input frequency and the orientation of the input. We use rotated sinusoids with rotation and frequency as free parameters:

$$x(n_1, n_2) = \sin[\omega_0(an_1 + bn_2)], \tag{6.64}$$

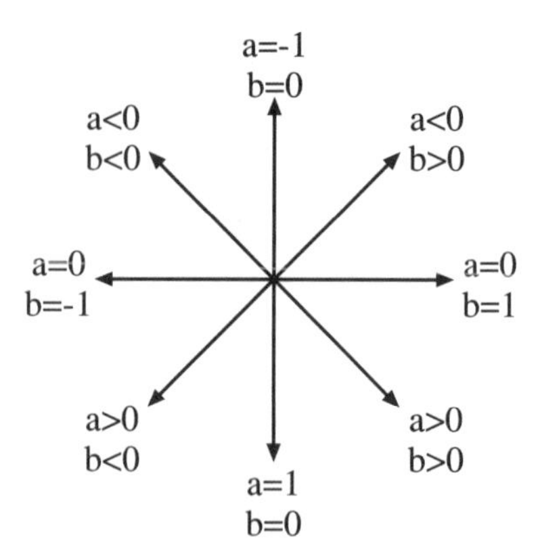

Figure 6.7: Parameters a and b determine the orientation of the input sinusoid; they are interlinked by $a^2 + b^2 = 1$. (Reproduced with permission from [Thu96c]. © 1996 IEEE.)

where a and b determine the orientation and $a^2 + b^2 = 1$. In Fig. 6.7, the possible orientations are indicated with the corresponding values for these parameters. Both of them must be in the interval $[-1, 1]$, and because they are interlinked, only one of them is sufficient to define the orientation. Also, for our purposes half of the possible directions in Fig. 6.7 are redundant, so we will only use

$$-1 \leq a \leq 1,$$
$$b = \sqrt{1 - a^2}.$$

With Eqs. (6.64) and (6.47), we compute the output spectrum. We find

$$\begin{aligned}
&\bar{Y}_2(e^{ju}, e^{jv}, e^{j(\omega_1 - u)}, e^{j(\omega_2 - v)}) \\
&= H_2(e^{ju}, e^{jv}, e^{j(\omega_1 - u)}, e^{j(\omega_2 - v)}) X(e^{ju}, e^{jv}) X(e^{j(\omega_1 - u)}, e^{j(\omega_2 - v)}) \\
&= -4\pi^4 H_2(e^{ju}, e^{jv}, e^{j(\omega_1 - u)}, e^{j(\omega_2 - v)}) \\
&\quad [-\delta(u - \omega_0 a, v - \omega_0 b) + \delta(u + \omega_0 a, v + \omega_0 b)] \\
&\quad [-\delta(\omega_1 - u - \omega_0 a, \omega_2 - v - \omega_0 b) + \delta(\omega_1 - u + \omega_0 a, \omega_2 - v + \omega_0 b)]
\end{aligned}$$

and

$$\begin{aligned}
Y_2(e^{j\omega_1}, e^{j\omega_2}) = \pi^2 [& -H_2(e^{j\omega_0 a}, e^{j\omega_0 b}, e^{j\omega_0 a}, e^{j\omega_0 b}) \\
& \quad \delta(\omega_1 - 2\omega_0 a, \omega_2 - 2\omega_0 b) \\
& + H_2(e^{j\omega_0 a}, e^{j\omega_0 b}, e^{-j\omega_0 a}, e^{-j\omega_0 b}) \delta(\omega_1, \omega_2) \\
& + H_2(e^{-j\omega_0 a}, e^{-j\omega_0 b}, e^{j\omega_0 a}, e^{j\omega_0 b}) \delta(\omega_1, \omega_2) \\
& - H_2(e^{-j\omega_0 a}, e^{-j\omega_0 b}, e^{-j\omega_0 a}, e^{j\omega_0 b}) \\
& \quad \delta(\omega_1 + 2\omega_0 a, \omega_2 + 2\omega_0 b)].
\end{aligned} \tag{6.65}$$

As expected, the output consists of components with zero (dc) and twice the input frequency. We can easily show, however, that only the dc component will be nonzero. We use Eqs. (6.61) and (6.58) and obtain

$$H_2(e^{j\omega_1}, e^{j\omega_2}, e^{j\omega_1}, e^{j\omega_2}) = \sum_{k_1}\sum_{k_2} h_2(k_1, k_2, -k_1, -k_2) = 0. \tag{6.66}$$

Thus, only the terms with $\delta(\omega_1, \omega_2)$ remain in Eq. (6.65). This is not too surprising since the class II two-dimensional filters were derived as an extension of their one-dimensional counterparts in Sec. 6.6.1, and for those we have already shown this property. Thus, we write

$$Y_2(e^{j\omega_1}, e^{j\omega_2}) = Y_2^{(\mathrm{dc})}(e^{j\omega_0}, a)\delta(\omega_1, \omega_2), \tag{6.67}$$

with

$$\begin{aligned} Y_2^{(\mathrm{dc})}(e^{j\omega_0}, a) = \pi^2[&H_2(e^{j\omega_0 a}, e^{j\omega_0 b}, e^{-j\omega_0 a}, e^{-j\omega_0 b}) \\ &-H_2(e^{-j\omega_0 a}, e^{-j\omega_0 b}, e^{j\omega_0 a}, e^{j\omega_0 b})]. \end{aligned} \tag{6.68}$$

6.7.2 Basis Filters

By combining the second half of Eq. (6.18) with Eq. (6.58), we obtain the input-output relation for class II filters:

$$\begin{aligned} y(n_1, n_2) = &\sum_{k_1}\sum_{k_2} h_2(k_1, k_2, -k_1, -k_2) \\ &\times x(n_1 - k_1, n_2 - k_2)x(n_1 + k_1, n_2 + k_2), \end{aligned} \tag{6.69}$$

with $\sum_{k_1}\sum_{k_2} h_2(k_1, k_2, -k_1, -k_2) = 0$. Obviously, any filter from this class is a linear combination of terms like $x(n_1 - k_1, n_2 - k_2)x(n_1 + k_1, n_2 + k_2)$. This is based on the property of Volterra filters that they are linear in the kernels. We make use of this observation and introduce the concept of basis filters for the design of Volterra filters.

We define an expression

$$y_{k_1,k_2}(n_1, n_2) = x^2(n_1, n_2) - x(n_1 + k_1, n_2 + k_2)x(n_1 - k_1, n_2 - k_2) \tag{6.70}$$

as a basis filter for class II filters, where $0 \le k_1 < \infty$, $-\infty < k_2 < \infty$, and $k_1 + |k_2| \ne 0$ (i.e., k_1 and k_2 cannot be equal to zero at the same time). We consider only unique pairs of k_1 and k_2. If we would also permit $k_1 < 0$, then some of the filters would be identical, since $y_{k_1,k_2}(n_1, n_2) = y_{-k_1,-k_2}(n_1, n_2)$. Using Eq. (6.70), we find the expression equivalent to Eq. (6.69) as

$$y(n_1, n_2) = \sum_{k_1}\sum_{k_2} \theta_{k_1,k_2} y_{k_1,k_2}(n_1, n_2), \tag{6.71}$$

where the θ_{k_1,k_2} represent the design coefficients. We proof this equivalence by rewriting Eq. (6.69) as

$$
\begin{aligned}
y(n_1,n_2) &= -\sum_{k_1}\sum_{k_2} h_2(k_1,k_2,-k_1,-k_2)x^2(n_1,n_2) \\
&\quad + \sum_{k_1}\sum_{k_2} h_2(k_1,k_2,-k_1,-k_2)x(n_1-k_1,n_2-k_2)x(n_1+k_1,n_2+k_2) \\
&= \sum_{k_1}\sum_{k_2} \theta_{k_1,k_2}x^2(n_1,n_2) \\
&\quad - \sum_{k_1}\sum_{k_2} \theta_{k_1,k_2}x(n_1-k_1,n_2-k_2)x(n_1+k_1,n_2+k_2) \\
&= \sum_{k_1}\sum_{k_2} \theta_{k_1,k_2}y_{k_1,k_2}(n_1,n_2),
\end{aligned}
$$

with the restrictions for k_1 and k_2 as before and where we have substituted $\theta_{k_1,k_2} = -h_2(k_1,k_2,-k_1,-k_2)$. Therefore, the set of filters in Eq. (6.70) completely describes all possible filters from class II, and it is not redundant since none of the filters in Eq. (6.70) can be written in terms of any other from this set. Furthermore, the basis filters have the advantage that for each of them the sum of their coefficients equals zero, and thus Eq. (6.53) holds by design and the coefficients θ_{k_1,k_2} can have arbitrary values. The relation between the kernel coefficients and the θ_{k_1,k_2} is

$$
h_2(0,0,0,0) = \sum_{k_1}\sum_{k_2} \theta_{k_1,k_2} \tag{6.72}
$$

$$
h_2(n_1,n_2,-n_1,-n_2) = \begin{cases} -\theta_{n_1,n_2} & \text{for } n_1+|n_2| \neq 0, \\ & 0 \leq n_1 < \infty, \\ & -\infty < n_2 < \infty \\ 0 & \text{for } n_1 < 0. \end{cases} \tag{6.73}
$$

This definition generates a nonsymmetric kernel that directly represents the system we need to implement.

In the frequency domain, the linear dependence on the basis filters is preserved by the Fourier transform. Since we know the frequency characteristics of each of the basis filters, we must find the coefficients in such a way that the overall system has a certain desired frequency response.

We are interested in designing class II systems that operate on a relatively small neighborhood. This keeps the computational complexity low for both design and implementation. We choose a 5×5 region of support, since it is a reasonable compromise between the computational complexity of the filter and the degree of freedom for the design. We obtain a total of 12 basis filters for $-2 \leq k_2 \leq 2$ and $0 \leq k_1 \leq 2$ in Eq. (6.70). We find their frequency responses $H_2^{(k_1,k_2)}(e^{j\omega_1}, e^{j\omega_2}, e^{j\omega_3}, e^{j\omega_4})$ as

$$
\begin{aligned}
H_2^{(0,1)}(e^{j\omega_1},e^{j\omega_2},e^{j\omega_3},e^{j\omega_4}) &= 1-\cos(\omega_2-\omega_4), \\
H_2^{(0,2)}(e^{j\omega_1},e^{j\omega_2},e^{j\omega_3},e^{j\omega_4}) &= 1-\cos(2\omega_2-2\omega_4), \\
H_2^{(1,-2)}(e^{j\omega_1},e^{j\omega_2},e^{j\omega_3},e^{j\omega_4}) &= 1-\cos(\omega_1-2\omega_2-\omega_3+2\omega_4), \\
H_2^{(1,-1)}(e^{j\omega_1},e^{j\omega_2},e^{j\omega_3},e^{j\omega_4}) &= 1-\cos(\omega_1-\omega_2-\omega_3+\omega_4),
\end{aligned}
$$

$$
\begin{aligned}
H_2^{(1,0)}(e^{j\omega_1}, e^{j\omega_2}, e^{j\omega_3}, e^{j\omega_4}) &= 1 - \cos(\omega_1 - \omega_3), \\
H_2^{(1,1)}(e^{j\omega_1}, e^{j\omega_2}, e^{j\omega_3}, e^{j\omega_4}) &= 1 - \cos(\omega_1 + \omega_2 - \omega_3 - \omega_4) \\
H_2^{(1,2)}(e^{j\omega_1}, e^{j\omega_2}, e^{j\omega_3}, e^{j\omega_4}) &= 1 - \cos(\omega_1 + 2\omega_2 - \omega_3 - 2\omega_4), \\
H_2^{(2,-2)}(e^{j\omega_1}, e^{j\omega_2}, e^{j\omega_3}, e^{j\omega_4}) &= 1 - \cos(2\omega_1 - 2\omega_2 - 2\omega_3 + 2\omega_4), \\
H_2^{(2,-1)}(e^{j\omega_1}, e^{j\omega_2}, e^{j\omega_3}, e^{j\omega_4}) &= 1 - \cos(2\omega_1 - \omega_2 - 2\omega_3 + \omega_4), \\
H_2^{(2,0)}(e^{j\omega_1}, e^{j\omega_2}, e^{j\omega_3}, e^{j\omega_4}) &= 1 - \cos(2\omega_1 - 2\omega_3), \\
H_2^{(2,1)}(e^{j\omega_1}, e^{j\omega_2}, e^{j\omega_3}, e^{j\omega_4}) &= 1 - \cos(2\omega_1 + \omega_2 - 2\omega_3 - \omega_4), \\
H_2^{(2,2)}(e^{j\omega_1}, e^{j\omega_2}, e^{j\omega_3}, e^{j\omega_4}) &= 1 - \cos(2\omega_1 + 2\omega_2 - 2\omega_3 - 2\omega_4).
\end{aligned}
$$

6.7.3 Least-Squares Design

For each of the basis filters, we obtain the response to a sinusoidal input according to Eq. (6.68). For simplicity, we denote them $B_{k_1,k_2}(\omega_0, a)$ instead of $Y_2^{\mathrm{dc},(k_1,k_2)}(e^{j\omega_0}, a)$.

$$
\begin{aligned}
B_{0,1}(\omega_0, a) &= 2\pi^2[1 - \cos(2\omega_0 b)], \\
B_{0,2}(\omega_0, a) &= 2\pi^2[1 - \cos(4\omega_0 b)], \\
B_{1,-2}(\omega_0, a) &= 2\pi^2[1 - \cos(2\omega_0 a - 4\omega_0 b)], \\
B_{1,-1}(\omega_0, a) &= 2\pi^2[1 - \cos(2\omega_0 a - 2\omega_0 b)], \\
B_{1,0}(\omega_0, a) &= 2\pi^2[1 - \cos(2\omega_0 a)], \\
B_{1,1}(\omega_0, a) &= 2\pi^2[1 - \cos(2\omega_0 a + 2\omega_0 b)], \\
B_{1,2}(\omega_0, a) &= 2\pi^2[1 - \cos(2\omega_0 a + 4\omega_0 b)], \\
B_{2,-2}(\omega_0, a) &= 2\pi^2[1 - \cos(4\omega_0 a - 4\omega_0 b)], \\
B_{2,-1}(\omega_0, a) &= 2\pi^2[1 - \cos(4\omega_0 a - 2\omega_0 b)], \\
B_{2,0}(\omega_0, a) &= 2\pi^2[1 - \cos(4\omega_0 a)], \\
B_{2,1}(\omega_0, a) &= 2\pi^2[1 - \cos(4\omega_0 a + 2\omega_0 b)], \\
B_{2,2}(\omega_0, a) &= 2\pi^2[1 - \cos(4\omega_0 a + 4\omega_0 b)].
\end{aligned}
$$

where $b = \sqrt{1 - a^2}$. Then we sample these continuous functions in $-1 \le a \le 1$ and $0 \le \omega_0 \le \pi/2$. The reason that we do not use frequencies beyond $\pi/2$ is that they are of negligible importance in real-world images since images tend to be oversampled. We choose a sample spacing of $\Delta a = 0.05$ and $\Delta\omega_0 = \pi/50$, which yields 201×26 samples. We order the samples into column vectors in the following manner. The first 201 elements are for $\omega_0 = 0$ and from $a = -1$ up to $a = 1$. After that, we take the 201 samples for $\omega_0 = \pi/50$ and so on until we reach $\omega_0 = \pi/2$ and $a = 1$. We number the functions $B_{k_1,k_2}(\omega_0, a)$ in the order given before from 1 to 12 and obtain vectors $\mathbf{b}_1$ through $\mathbf{b}_{12}$, which we combine into the matrix $\mathbf{B}$:

$$
\mathbf{B} = (\mathbf{b}_1\, \mathbf{b}_2\, \ldots\, \mathbf{b}_{12}). \tag{6.74}
$$

Table 6.2: Parameter Vectors for Design of Two-Dimensional Version of the Teager Filter; $\hat{\boldsymbol{\theta}}_T$ is the exact filter, $\hat{\boldsymbol{\theta}}_3$, the scaled filter, and $\hat{\boldsymbol{\theta}}_a$ the approximate solution.

k	$\hat{\boldsymbol{\theta}}_T$	$\hat{\boldsymbol{\theta}}_3$	$\hat{\boldsymbol{\theta}}_a$
0	0.01323	1.01641	1.0
1	0.00061	0.04733	0.0
2	−0.00011	−0.00897	0.0
3	0.00616	0.47368	0.5
4	0.01325	1.01759	1.0
5	0.00616	0.47368	0.5
6	−0.00011	−0.00897	0.0
7	−0.00021	−0.01674	0.0
8	−0.00012	−0.00961	0.0
9	0.00054	0.04197	0.0
10	−0.00012	−0.00961	0.0
11	−0.00012	−0.01674	0.0

We write the linear combination of the basis filters with coefficients θ_{k_1,k_2} as a matrix product $\mathbf{B}\boldsymbol{\theta}$, where

$$\boldsymbol{\theta} = (\theta_{0,1}\ \theta_{0,2}\ \dots\ \theta_{2,1}\ \theta_{2,2})^T. \tag{6.75}$$

Thus, we can express the design problem as the linear matrix equation

$$\mathbf{d} = \mathbf{B}\hat{\boldsymbol{\theta}} + \mathbf{e}. \tag{6.76}$$

The desired behavior of the ideal filter is contained in $\mathbf{d}$, which we sample in the same fashion as described above for $B_{k_1,k_2}(\omega_0, a)$, and $\mathbf{e}$ represents the error vector. We find the optimal solution $\hat{\boldsymbol{\theta}}$ that minimizes the error in the least-squares sense [Men87]:

$$\hat{\boldsymbol{\theta}} = \left[\mathbf{B}^T\mathbf{B}\right]^{-1}\mathbf{B}^T\mathbf{d}. \tag{6.77}$$

6.7.4 Two-Dimensional Teager filter

We want to design a two-dimensional version of the Teager filter defined in Eq. (6.27). In Eq. (6.28), we stated that the one-dimensional filter responds with an output $y(n) = \sin^2(\omega_0)$ if the input is a sinusoid $x(n) = \sin(\omega_0 n)$. Now we need to find a two-dimensional system that outputs a constant signal $y(n_1, n_2) = \sin^2(\omega_0)$ for a sinusoidal input and that is, of course, independent of the orientation of the excitation.

We set $d(\omega_0, a) = \sin^2(\omega_0)$ and convert $d(\omega_0, a)$ into the column vector $\mathbf{d}$. Using Eq. (6.77), yields the parameter vector $\hat{\boldsymbol{\theta}}_T$ given in Table 6.2. If we relax the

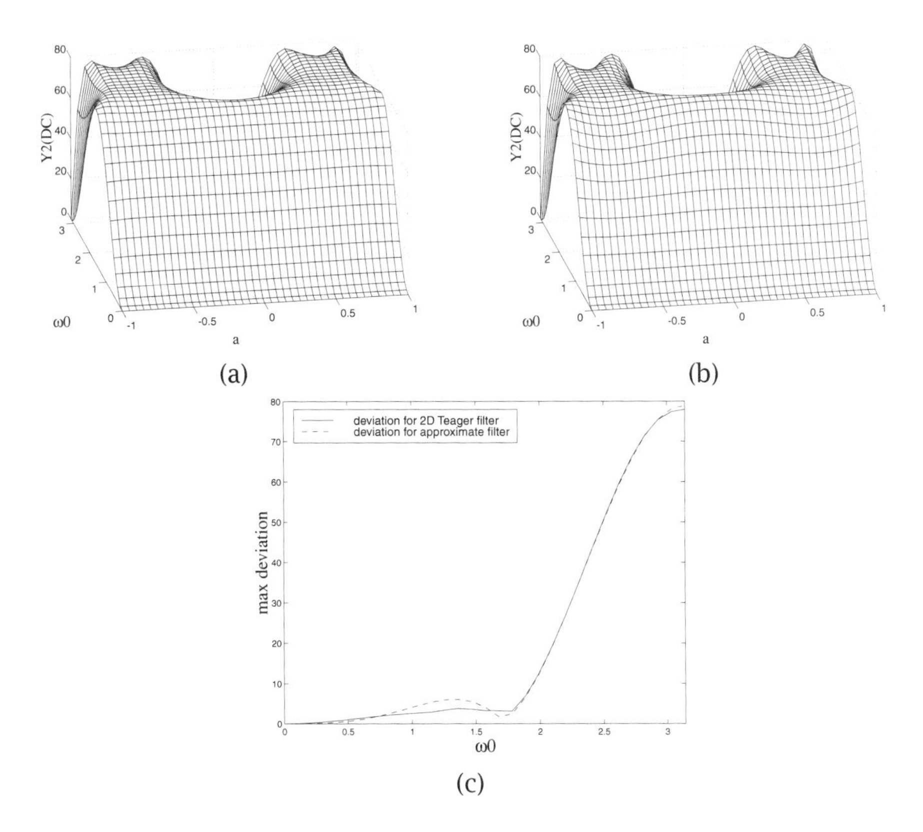

Figure 6.8: Characteristic functions as defined in Eq. (6.68) for (a) the two-dimensional Teager filter and (b) its approximation; (c) maximum deviation of the functions from the desired curve for all values of a. (Parts a and b reproduced with permission from [Thu96c]. © 1996 IEEE.)

restrictions somewhat and allow the filter output to be scaled, that is, $d(\omega_0, a) = \alpha \sin^2(\omega_0)$, we can normalize the parameters such that $\sum_i \hat{\boldsymbol{\theta}}(i) = 3$; this yields $\hat{\boldsymbol{\theta}}_3$ in Table 6.2. Figure 6.8a shows the characteristic function $Y_2^{(\mathrm{dc})}(\mathrm{e}^{j\omega_0}, a)$ for this filter. The design goal of virtually perfectly isotropical behavior has been achieved up to $\omega_0 = \pi/2$.

The advantage of normalizing the parameter vector is that the elements of $\hat{\boldsymbol{\theta}}_3$ can easily be approximated by simple numbers. The rightmost column in Table 6.2 shows the approximation $\hat{\boldsymbol{\theta}}_a$ using only the coefficients 1 and 0.5. Thus, we obtain the input-output relationship for the approximate two-dimensional Teager filter:

$$\begin{aligned} y(n_1, n_2) = {} & 3x^2(n_1, n_2) - \frac{1}{2}x(n_1+1, n_2+1)x(n_1-1, n_2-1) \\ & -\frac{1}{2}x(n_1+1, n_2-1)x(n_1-1, n_2+1) \\ & -x(n_1+1, n_2)x(n_1-1, n_2) \\ & -x(n_1, n_2+1)x(n_1, n_2-1). \end{aligned} \tag{6.78}$$

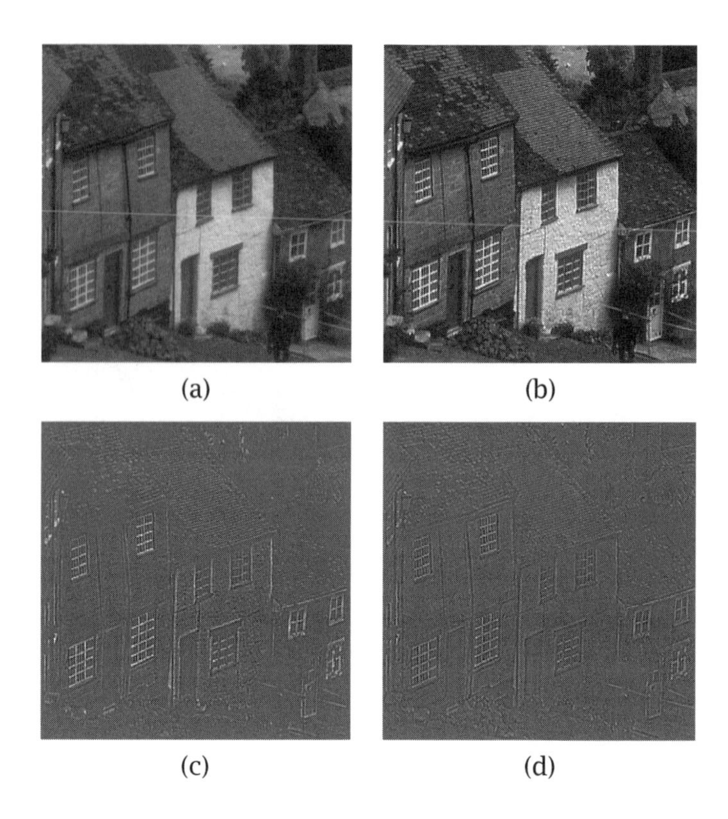

Figure 6.9: (a) Original image, (b) enhanced image after unsharp masking, (c) edge image obtained by applying the Volterra filter in Eq. (6.78) to the image in (a), and (d) output of the Laplacian filter.

This equation is considerably simpler to implement than the full description in $\hat{\boldsymbol{\theta}}_T$. At the same time, the behavior of this system, as shown in Figs. 6.8b and 6.8c, does not deviate too much from the ideal one. It is still extremely isotropic for almost the entire range of interest from 0 to $\pi/2$.

6.7.5 Image Enhancement Example

Figure 6.9 shows an example of edge enhancement. The original image in Fig. 6.9a is somewhat blurry. It was enhanced using the unsharp masking technique and the approximate 2D Teager filter in Eq. (6.78); the enhanced image, Fig. 6.9b, is noticeably sharper. The outputs of the Teager and Laplacian filters are shown in Figs. 6.9c and 6.9d, respectively. The Laplacian filter shows a uniform response to edges independent of background intensity, whereas the Teager filter output is weaker in darker regions (e.g., the darker areas of the roof) and stronger in brighter areas (e.g., the bright wall), as expected.

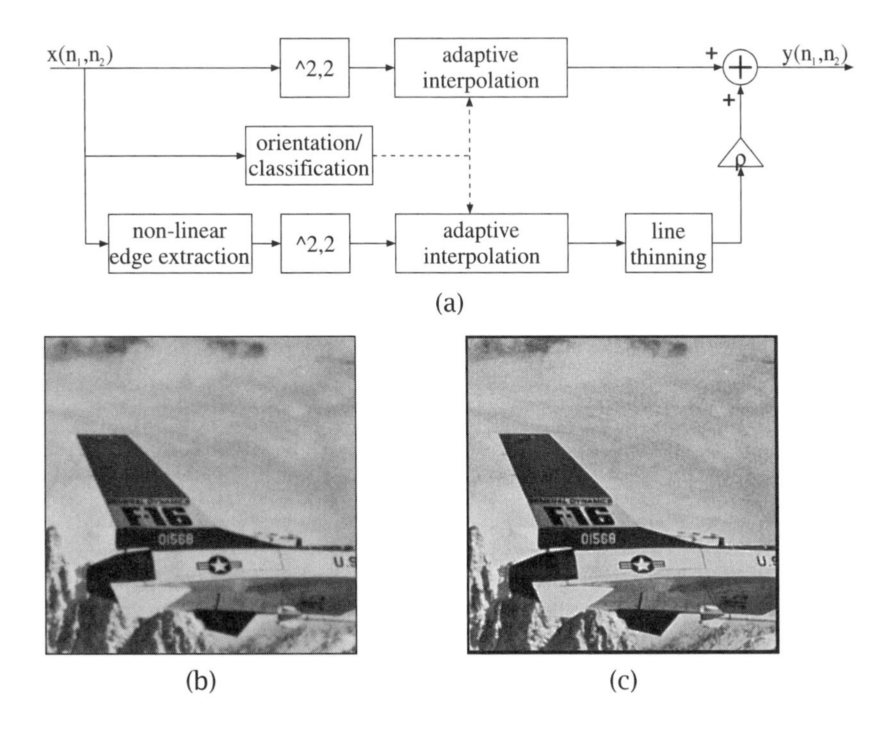

Figure 6.10: (a) Block diagram of edge enhanced image interpolation, and images zoomed by (b) bilinear interpolation and (c) edge enhanced technique. (Reproduced with permission from [Thu96a]. ©1996 SPIE.)

6.7.6 Application in Image Interpolation

The edge enhancement technique described in the previous section can be extended to image interpolation [Thu96a]. As shown in the block diagram in Fig. 6.10a, both the original image and the edge image created by the Volterra filter in Eq. (6.78) are upsampled and interpolated. The enlarged edge image is then processed by the "line thinning" block, which reduces the width of the extracted edges. After a final scaling operation, the edge information is added back to the zoomed image. The interpolation block adapts to the local image characteristics and processes different image regions in different ways. In particular, edges are interpolated only along their direction and not across, which preserves sharpness. The results of standard bilinear interpolation and this technique are compared in Figs. 6.10b and 6.10c.

6.7.7 Application in Image Halftoning

Another application of the quadratic Volterra in Eq. (6.78) can be found in [Thu96b]. The technique of error diffusion is used to halftone an image, to requantize an image into fewer (usually two) levels of intensity or color while minimizing the impact on perceptual quality. In the standard method, the quantization error is filtered

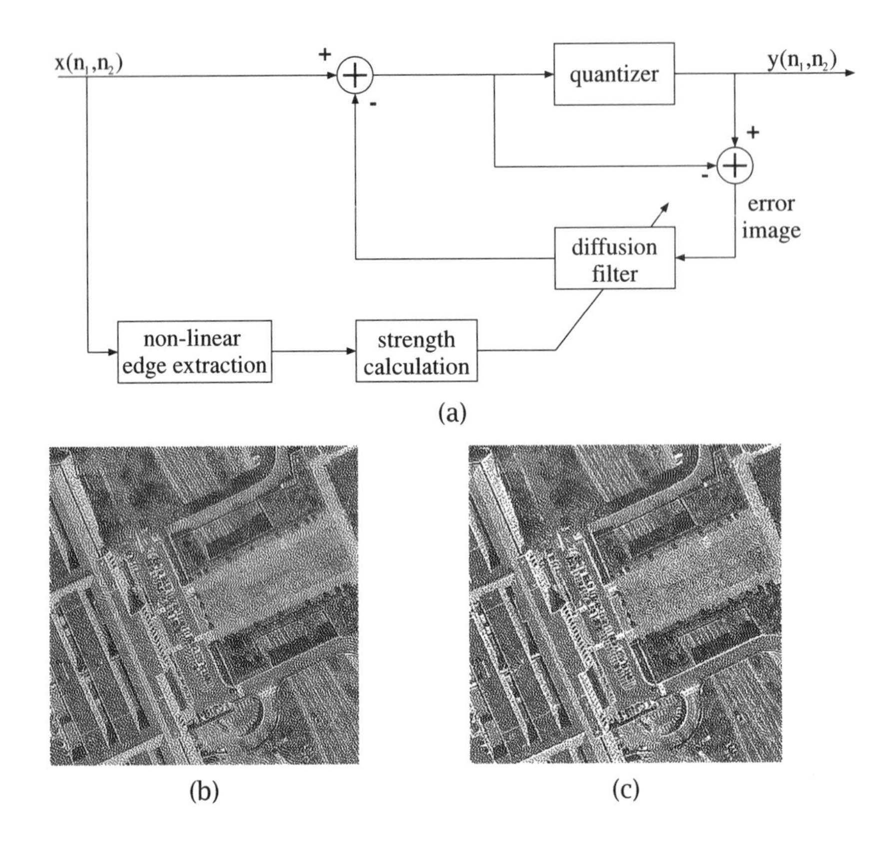

Figure 6.11: (a) Block diagram of modified error diffusion technique, (b) example of halftoning using standard error diffusion, and (c) result of modified error diffusion. (Reproduced with permission from [Thu96b]. © 1996 SPIE.)

by a diffusion filter in a feedback loop. The modification is shown in Fig. 6.11a. The "strength calculation" block essentially converts the output of the Volterra filter to its absolute value. The result is then used to adapt the diffusion filter in such a way that the final image appears sharper. Examples allowing comparison of standard and modified error diffusion are shown in Figs. 6.11b and 6.11c.

6.8 Summary

In this chapter, we have studied homogeneous quadratic Volterra filters and, in particular, the Teager filter and its extensions in one and two dimensions. We have presented an intuitive interpretation of the frequency response of these filters, which facilitates a better understanding of their properties. Certain subclasses of quadratic Volterra filters are natural extensions of the Teager filter. Two-dimensional Teager filters have properties that are desirable for image enhancement since they can be approximated as mean-weighted highpass filters. We have demonstrated their application in image sharpening, interpolation, and halftoning. The framework presented here also provides a foundation for different filter designs (e.g., lowpass filters) and applications that require a dependence on the local image intensity.

6.9 Appendix

6.9.1 Proof of Theorem 6.1

We start with the relationship

$$y_2(n) = \frac{1}{4\pi^2} \iint \mathcal{D}\bar{Y}_2(e^{j\omega_1}, e^{j\omega_2}) e^{jn(\omega_1+\omega_2)} \, d\omega_1 \, d\omega_2,$$

where $\bar{Y}_2(e^{j\omega_1}, e^{j\omega_2}) = H(e^{j\omega_1}, e^{j\omega_2})X(e^{j\omega_1})X(e^{j\omega_2})$ and $\mathcal{D} = (0, 2\pi) \times (0, 2\pi)$. Figure 6.12 shows the integration region graphically. Instead of integrating parallel to the axes, we split the region into two parts, the lower and upper triangles. For each of these portions, we compute the integral separately, and at the end of the derivation we combine the results. As shown in Fig. 6.12, we integrate over the lower triangle along the line $\omega_2 = \omega_0 - \omega_1$ first, where ω_0 is a free parameter between 0 and 2π, and then along the orthogonal direction indicated by the arrow:

$$\begin{aligned} y_{\mathrm{lt}}(n) &= \frac{1}{4\pi^2} \iint\limits_{\text{lower}} \bar{Y}_2(e^{j\omega_1}, e^{j\omega_2}) e^{jn(\omega_1+\omega_2)} \, d\omega_1 \, d\omega_2 \\ &= \frac{1}{4\pi^2} \int_0^{2\pi} \int_0^{\omega_0} \bar{Y}_2(e^{j\omega_1}, e^{j(\omega_0-\omega_1)}) e^{jn\omega_0} \, d\omega_1 \, d\omega_0 \\ &= \frac{1}{2\pi} \int_0^{2\pi} \left[\frac{1}{2\pi} A(e^{j\omega_0}) \right] e^{jn\omega_0} \, d\omega_0, \end{aligned}$$

where we have used the abbreviation $A(e^{j\omega_0}) = \int_0^{\omega_0} \bar{Y}_2(e^{j\omega_1}, e^{j(\omega_0-\omega_1)}) \, d\omega_1$.

For the integral over the upper triangle, we substitute

$$\begin{aligned} \omega_1 &= 2\pi - u, & u &= 2\pi - \omega_1, \\ \omega_2 &= 2\pi - v + u, & v &= 4\pi - \omega_2 - \omega_1. \end{aligned}$$

The basic idea behind this is to integrate along the line $\omega_2 = 2\pi + \omega_0 - \omega_1$ and then in the orthogonal direction. This yields

$$\begin{aligned} y_{\mathrm{ut}}(n) &= \frac{1}{4\pi^2} \iint\limits_{\text{upper}} \bar{Y}_2(e^{j\omega_1}, e^{j\omega_2}) e^{jn(\omega_1+\omega_2)} \, d\omega_1 \, d\omega_2 \\ &= \frac{1}{4\pi^2} \int_{2\pi}^{0} \int_{v}^{0} \bar{Y}_2(e^{j(2\pi-u)}, e^{j(2\pi-v+u)}) e^{jn(4\pi-v)} \, du \, dv \\ &= \frac{1}{4\pi^2} \int_0^{2\pi} \int_0^{v} H(e^{j(2\pi-u)}, e^{j[2\pi-(v-u)]}) X(e^{j(2\pi-u)}) \\ &\quad X(e^{j[2\pi-(v-u)]}) \, du \, e^{-jnv} \, dv. \end{aligned}$$

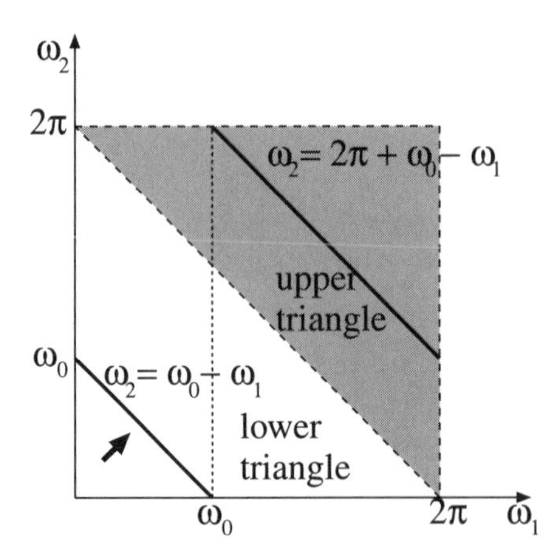

Figure 6.12: Integration process for the derivation of the output spectrum.

Since $X(\mathrm{e}^{j\omega})$ and $H(\mathrm{e}^{j\omega_1}, \mathrm{e}^{j\omega_2})$ are Fourier transforms of real signals, we can use the symmetry relationships $X(\mathrm{e}^{j(2\pi-\omega)}) = X^*(\mathrm{e}^{j\omega})$ and $H(\mathrm{e}^{j(2\pi-\omega_1)}, \mathrm{e}^{j(2\pi-\omega_2)}) = H^*(\mathrm{e}^{j\omega_1}, \mathrm{e}^{j\omega_2})$. We obtain

$$\begin{aligned} y_{\mathrm{ut}}(n) &= \frac{1}{4\pi^2}\int_0^{2\pi}\int_0^{v} H^*(\mathrm{e}^{ju}, \mathrm{e}^{j(v-u)})X^*(\mathrm{e}^{ju})X^*(\mathrm{e}^{j(v-u)})\,\mathrm{d}u\,\mathrm{e}^{-jnv}\,\mathrm{d}v \\ &= \frac{1}{4\pi^2}\int_0^{2\pi} A^*(\mathrm{e}^{jv})\mathrm{e}^{-jnv}\,\mathrm{d}v \\ &= \frac{1}{4\pi^2}(-1)\int_{2\pi}^{0} A^*(\mathrm{e}^{j(2\pi-\omega_0)})\mathrm{e}^{jn\omega_0}\,\mathrm{d}\omega_0. \end{aligned}$$

For this last step, we have substituted $\omega_0 = 2\pi - v$. Finally, we combine the results for the upper and lower parts:

$$y(n) = \frac{1}{2\pi}\int_0^{2\pi}\left\{\frac{1}{2\pi}[A(\mathrm{e}^{j\omega_0}) + A^*(\mathrm{e}^{j(2\pi-\omega_0)})]\right\}\mathrm{e}^{jn\omega_0}\,\mathrm{d}\omega_0.$$

Note that this is an inverse Fourier transform, which means that the expression inside the braces must be the desired expression for the output spectrum.

6.9.2 Proof of Theorem 6.2

With the general definition of homogeneous quadratic Volterra filters,

$$y(n) = \mathcal{V}_2(x) = \sum_{k_1}\sum_{k_2} h_2(k_1, k_2)x(n-k_1)x(n-k_2),$$

and $\mu_x = E(x)$ and $\hat{x}(n) = x(n) - \mu_x$, we write

$$y(n) = \sum_{k_1}\sum_{k_2} h_2(k_1, k_2)\,[\hat{x}(n-k_1) + \mu_x]\,[\hat{x}(n-k_2) + \mu_x]$$

$$= \sum_{k_1}\sum_{k_2} h_2(k_1,k_2)\hat{x}(n-k_1)\hat{x}(n-k_2)$$
$$+\mu_x \sum_{k_1}\sum_{k_2} h_2(k_1,k_2)\left[\hat{x}(n-k_1)+\hat{x}(n-k_2)\right] + \mu_x^2 \sum_{k_1}\sum_{k_2} h_2(k_1,k_2).$$

Assuming that $H_2(e^{j0}, e^{j0}) = 0$, then $\sum_{k_1}\sum_{k_2} h_2(k_1,k_2) = 0$. We decompose $y(n)$ into a sum of $y_1(n)$ and $y_2(n)$, where

$$y_1(n) = \mu_x \sum_{k_1}\sum_{k_2} h_2(k_1,k_2)\left[\hat{x}(n-k_1)+\hat{x}(n-k_2)\right],$$
$$y_2(n) = \mathcal{V}_2(\hat{x}) = \sum_{k_1}\sum_{k_2} h_2(k_1,k_2)\hat{x}(n-k_1)\hat{x}(n-k_2).$$

To analyze the effects of both of these terms on the overall result, we assume that $\hat{x}(n)$ can be modeled as independent identically distributed (i.i.d.) noise with uniform distribution between $-\Delta$ and Δ. Thus, $E[\hat{x}] = 0$ and $\sigma_{\hat{x}}^2 = E[\hat{x}^2] = 1/3\Delta^2$. We now show under which conditions $\sigma_{y_1}^2 \geq \sigma_{y_2}^2$.

Because $y_1(n)$ is a linearly filtered version of a zero mean signal, $y_1(n)$ is also zero mean. Therefore,

$$\begin{aligned}\sigma_{y_1}^2 &= E[y_1^2]\\ &= \mu_x^2 \sum_{k_1}\sum_{k_2}\sum_{k_3}\sum_{k_4} h_2(k_1,k_2)h_2(k_3,k_4)\Big\{E[x(n-k_1)x(n-k_3)]\\ &\quad +E[x(n-k_1)x(n-k_4)]\\ &\quad +E[x(n-k_2)x(n-k_3)] + E[x(n-k_2)x(n-k_4)]\Big\}.\end{aligned}$$

Using the i.i.d. assumption for $\hat{x}(n)$, the autocorrelation is given by

$$E[x(n-\alpha)x(n-\beta)] = R_x(\alpha-\beta) = \begin{cases} 0 & \text{for } \alpha \neq \beta,\\ \frac{1}{3}\Delta^2 & \text{for } \alpha = \beta,\end{cases}$$

so the variance becomes

$$\begin{aligned}\sigma_{y_1}^2 &= \mu_x^2\frac{\Delta^2}{3}\Big\{\sum_{k_1}\sum_{k_2}\sum_{k_4} h_2(k_1,k_2)h_2(k_1,k_4) + \sum_{k_1}\sum_{k_2}\sum_{k_3} h_2(k_1,k_2)h_2(k_3,k_1)\\ &\quad +\sum_{k_1}\sum_{k_2}\sum_{k_4} h_2(k_1,k_2)h_2(k_2,k_4) + \sum_{k_1}\sum_{k_2}\sum_{k_3} h_2(k_1,k_2)h_2(k_3,k_2)\Big\}\\ &= \mu_x^2\frac{\Delta^2}{3}\Big\{\sum_{k_1}\sum_{k_2}\sum_{k_3} h_2(k_1,k_2)[h_2(k_1,k_3)\\ &\quad +h_2(k_3,k_1) + h_2(k_2,k_3) + h_2(k_3,k_2)]\Big\}.\end{aligned}$$

With the symmetry condition of $h_2(n_1,n_2)$, that is, $h_2(n_1,n_2) = h_2(n_2,n_1)$,

$$\sigma_{y_1}^2 = \frac{2}{3}\mu_x^2\Delta^2 \sum_{k_1}\sum_{k_2}\sum_{k_3} h_2(k_1,k_2)[h_2(k_1,k_3) + h_2(k_2,k_3)]$$

$$= \frac{2}{3}\mu_x^2\Delta^2\Big\{2\sum_k h_2^2(k,k) + \sum_{\substack{k_1\\k_1\neq k_2}}\sum_{k_2} h_2^2(k_1,k_2)$$

$$+ \sum_{\substack{k_1\\k_1\neq k_2}}\sum_{k_2} h_2(k_1,k_2)[2h_2(k_1,k_1) + h_2(k_2,k_2)]$$

$$+ \sum_{\substack{k_1\\k_1\neq k_2\\k_2\neq k_3}}\sum_{k_2}\sum_{k_3} h_2(k_1,k_2)[h_2(k_1,k_3) + h_2(k_2,k_3)]\Big\}.$$

Denoting the last two terms by S^h and neglecting the factor 2 of the first sum, we approximate $\sigma_{\tilde{y}_1}^2$ by

$$\sigma_{\tilde{y}_1}^2 \geq \frac{2}{3}\mu_x^2\Delta^2\Big\{\sum_k h_2^2(k,k) + \sum_{\substack{k_1\\k_1\neq k_2}}\sum_{k_2} h_2^2(k_1,k_2) + S^h\Big\}. \tag{6.79}$$

For $y_2(n)$ we find the expected value as

$$\begin{aligned} E[y_2(n)] &= \sum_{k_1}\sum_{k_2} h_2(k_1,k_2)E[\hat{x}(n-k_1)\hat{x}(n-k_2)] \\ &= \sum_k h_2(k,k)E[\hat{x}^2(n-k)] \\ &= \frac{\Delta^2}{3}\sum_k h_2(k,k). \end{aligned}$$

Thus, the variance is

$$\begin{aligned} \sigma_{\tilde{y}_2}^2 &= E\left[y_2^2\right] - E[y_2(n)]^2 \\ &= E\left[\sum_{k_1}\sum_{k_2}\sum_{k_3}\sum_{k_4} h_2(k_1,k_2)h_2(k_3,k_4)x(n-k_1)x(n-k_2)x(n-k_3)x(n-k_4)\right] \\ &\quad -\frac{\Delta^4}{9}\sum_{k_1}\sum_{k_2} h_2(k_1,k_1)h_2(k_2,k_2). \end{aligned} \tag{6.80}$$

Using $E[\hat{x}^4] = \int_{-\Delta}^{\Delta} \hat{x}^4/(2\Delta)\,\mathrm{d}x = \Delta^4/5$, we rewrite the expression for $\sigma_{\tilde{y}_2}^2$ as

$$\begin{aligned} \sigma_{\tilde{y}_2}^2 &= \frac{\Delta^4}{5}\sum_k h_2^2(k,k) + \frac{\Delta^4}{9}\sum_{\substack{k_1\\k_1\neq k_3}}\sum_{k_3} h_2(k_1,k_1)h_2(k_3,k_3) + \frac{\Delta^4}{9}\sum_{\substack{k_1\\k_1\neq k_2}}\sum_{k_2} h_2^2(k_1,k_2) \\ &\quad + \frac{\Delta^4}{9}\sum_{\substack{k_1\\k_1\neq k_2}}\sum_{k_2} \underbrace{h_2(k_1,k_2)h_2(k_2,k_1)}_{=h_2^2(k_1,k_2)} - \frac{\Delta^4}{9}\sum_{k_1}\sum_{k_2} h_2(k_1,k_1)h_2(k_2,k_2). \end{aligned} \tag{6.81}$$

Note that the expression inside the first sum of Eq. (6.80) is zero whenever one of the variables $k_1, \ldots, k_4$ is different from all the others. If, for example, k_1 is different from $k_2, \ldots, k_4$, then

$$\begin{aligned} &E[x(n-k_1)x(n-k_2)x(n-k_3)x(n-k_4)] \\ &\quad = E[x(n-k_1)]E[x(n-k_2)x(n-k_3)x(n-k_4)], \end{aligned}$$

which is zero. Therefore, we only have to consider the cases in which $k_1 = k_2 = k_3 = k_4$ or when pairs of these variables are equal. This leads to the first four sums in Eq. (6.81). For the second of them, we write

$$\sum_{k_1}\sum_{\substack{k_3 \\ k_1 \neq k_3}} h_2(k_1,k_1)h_2(k_3,k_3) = \sum_{k_1}\sum_{k_3} h_2(k_1,k_1)h_2(k_3,k_3) - \sum_k h_2^2(k,k),$$

and therefore

$$\begin{aligned} \sigma_{y_2}^2 &= \frac{4}{45}\Delta^4 \sum_k h_2^2(k,k) + \frac{2}{9}\Delta^4 \sum_{k_1}\sum_{\substack{k_2 \\ k_1 \neq k_2}} h_2^2(k_1,k_2) \\ &\leq \frac{2}{9}\Delta^4 \left(\sum_k h_2^2(k,k) + \sum_{k_1}\sum_{\substack{k_2 \\ k_1 \neq k_2}} h_2^2(k_1,k_2) \right). \end{aligned} \tag{6.82}$$

We combine Eqs. (6.79) and (6.82) and compute the ratio of the variances:

$$\begin{aligned} \frac{\sigma_{y_1}^2}{\sigma_{y_2}^2} &\geq 3\left(\frac{\mu_x}{\Delta}\right)^2 \left(1 + \frac{S^h}{\sum_k h_2^2(k,k) + \sum_{k_1}\sum_{\substack{k_2 \\ k_1 \neq k_2}} h_2^2(k_1,k_2)} \right) \\ &\geq 3\left(\frac{\mu_x}{\Delta}\right)^2. \end{aligned}$$

The last step is based on the assumption that $S^h \geq 0$. This means that whenever $\mu_x \geq \Delta$, the contribution of $y_1(n)$ to the overall filter output is greater than that of $y_2(n)$, that is, $\sigma_{y_1}^2 \geq \sigma_{y_2}^2$. Even if μ_x is only slightly larger than Δ, the ratio will be significant. For example, assume that $\mu_x = 2\Delta$; then $\sigma_{y_1}^2 \geq 12\sigma_{y_2}^2$.

Thus, the filter output can be approximated by the result of $y_1(n)$:

$$y(n) \approx \mu_x \sum_{k_1}\sum_{k_2} h_2(k_1,k_2)[\hat{x}(n-k_1) + \hat{x}(n-k_2)],$$

and again with $x(n) = \hat{x}(n) + \mu_x$,

$$\begin{aligned} y(n) &\approx \mu_x \sum_{k_1}\sum_{k_2} h_2(k_1,k_2)[x(n-k_1) + x(n-k_2)] - 2\mu_x^2 \sum_{k_1}\sum_{k_2} h_2(k_1,k_2) \\ &= \mu_x \sum_{k_1}\sum_{k_2} h_2(k_1,k_2)[x(n-k_1) + x(n-k_2)]. \end{aligned}$$

The corresponding expression in the frequency domain can easily be derived by starting from the general spectral input–output relationship,

$$Y(e^{j\omega}) = \frac{1}{2\pi}\int_0^{2\pi} H_2(e^{j\omega_1}, e^{j(\omega-\omega_1)})X(e^{j\omega_1})X(e^{j(\omega-\omega_1)})\, d\omega_1,$$

and replacing $X(e^{j\omega_1})$ with $\hat{X}(e^{j\omega_1}) + 2\pi\mu_x\delta(\omega_1)$, which yields

$$Y(e^{j\omega}) \approx \mu_x[H_2(e^{j\omega}, e^{j0})\hat{X}(e^{j\omega}) + H_2(e^{j0}, e^{j\omega})\hat{X}(e^{j\omega})].$$

With the symmetry assumption $H_2(e^{j\omega}, e^{j0}) = H_2(e^{j0}, e^{j\omega})$ and $H_2(e^{j0}, e^{j0}) = 0$, we resubstitute $X(e^{j\omega})$, which yields the desired result:

$$Y(e^{j\omega}) \approx 2\mu_x X(e^{j\omega})H_2(e^{j\omega}, e^{j0}).$$

References

[Bil89a] S. A. Billings and K. M. Tsang. Spectral analysis for non-linear systems, Part I: Parametric non-linear spectral analysis. *Mechan. Syst. Signal Process.* **3**(4), 319–339 (1989).

[Bil89b] S. A. Billings and K. M. Tsang. Spectral analysis for non-linear systems, Part II: Interpretation of non-linear frequency response functions. *Mechan. Syst. Signal Process.* **3**(4), 341–359 (1989).

[Bof86] K. R. Boff, L. Kaufman, and J. P. Thomas. *Handbook of Perception and Human Performance, Vol. I: Sensory Processes and Perception.* Wiley, New York (1986).

[Chu79] L. O. Chua and C. Y. Ng. Frequency domain analysis of nonlinear systems: General theory. *Electron. Circ. Syst.* **3**, 165–185 (July 1979).

[Dun93] R. B. Dunn, T. F. Quatieri, and J. F. Kaiser. Detection of transient signals using the energy operator. In *Proc. IEEE Intl. Conf. on Acoustics, Speech and Signal Processing*, pp. III-145–III-148 (Minneapolis, MN, April 1993).

[Jai89] A. K. Jain. *Fundamentals of Digital Image Processing.* Prentice-Hall, Englewood Cliffs, NJ (1989).

[Kai90] J. F. Kaiser. On a simple algorithm to calculate the "energy" of a signal. In *Proc. IEEE Intl. Conf. on Acoustics, Speech and Signal Processing*, pp. 381–384 (Albuquerque, NM, April 1990).

[Kai93] J. F. Kaiser. Some useful properties of the Teager's energy operator. In *Proc. IEEE Intl. Conf. on Acoustics, Speech and Signal Processing*, pp. III-149–III-152 (Minneapolis, MN, April 1993).

[Lim90] J. S. Lim. *Two-Dimensional Signal and Image Processing.* Prentice-Hall, Englewood Cliffs, NJ (1990).

[Men87] J. M. Mendel. *Lessons in Digital Estimation Theory.* Prentice-Hall, Englewood Cliffs, NJ (1987).

[Mit91] S. K. Mitra, H. Li, I. S. Lin, and T.-H. Yu. A new class of nonlinear filters for image enhancement. In *Proc. IEEE Intl. Conf. on Acoustics, Speech and Signal Processing*, pp. 2525–2528 (Toronto, 1991).

[Pic82] B. Picinbono. Quadratic filters. In *Proc. IEEE Intl. Conf. on Acoustics, Speech and Signal Processing*, pp. 298–301 (1982).

[Pit90] I. Pitas and A. N. Venetsanopoulos. *Nonlinear Digital Filters.* Kluwer, Boston, MA (1990).

[Ram86] G. F. Ramponi. Edge extraction by a class of second-order nonlinear filters. *Electron. Lett.* **22**, 482–484 (April 1986).

[Ram88] G. F. Ramponi, G. L. Sicuranza, and W. Ukovich. A computational method for the design of 2-D Volterra filters. *IEEE Trans. Circ. Syst.* **35**, 1095–1102 (September 1988).

[Ram93] G. Ramponi and P. Fontanot. Enhancing document images with a quadratic filter. *Signal Process.* **33**, 23–34 (July 1993).

[Rog83] B. E. Rogowitz. The human visual system: A guide for the display technologist. In Proc. SID **24**, 235–252 (1983).

[Rub69] M. L. Rubin and G. L. Walls. *Fundamentals of Visual Science.* Charles C. Thomas, Springfield, IL (1969).

[Rug81] W. J. Rugh. *Nonlinear System Theory: The Volterra/Wiener Approach.* Johns Hopkins University Press, Baltimore, MD (1981).

[Sic92] G. L. Sicuranza. Quadratic filters for signal processing. *Proc. IEEE* **80**, 1263–1285 (August 1992).

[Sch80] M. Schetzen. *The Volterra and Wiener Theories of Nonlinear Systems.* Wiley, New York (1980).

[Tea80] H. M. Teager. Some observations on oral air flow during phonation. *IEEE Trans. Acoust. Speech and Signal Process.* ASSP-**28**, 599–601 (October 1980).

[Tea90] H. M. Teager and S. M. Teager. Evidence for nonlinear sound production mechanisms in the vocal tract. In *Speech Production and Speech Modelling* (W. J. Hardcastle and A. Marchal, eds.), pp. 241–261. Kluwer (1990).

[Thu95a] S. Thurnhofer and S. K. Mitra. Designing quadratic Volterra filters for nonlinear edge enhancement. In *Proc. Intl. Conf. on Digital Signal Processing*, pp. 320–325 (Limassol, Cyprus, June 1995).

[Thu95b] S. Thurnhofer and S. K. Mitra. Quadratic Volterra filters with mean-weighted highpass characteristics. In *Proc. IEEE Workshop on Nonlinear Signal and Image Processing*, pp. 368–371 (Halkidiki, Greece, June 1995).

[Thu96a] S. Thurnhofer and S. K. Mitra. Edge-enhanced image zooming. *Opt. Eng.* **35**, 1862–1869 (July 1996).

[Thu96b] S. Thurnhofer and S. K. Mitra. Detail-enhanced error diffusion. *Opt. Eng.* **35**, 2592–2598 (September 1996).

[Thu96c] S. Thurnhofer and S. K. Mitra. A general framework for quadratic Volterra filters for edge enhancement. *IEEE Trans. Image Process.* Special issue on nonlinear image processing. **5**, 950–963 (June 1996).

[Wan93] Y. Wang and S. I. Mitra. Image representation using block pattern models and its image processing applications. *IEEE Trans. Pattern Anal. Machine Intell.* **15**, 321–336 (April 1993).

[Vin87] T. Vinh, T. Chouychai, H. Liu, and M. Djouder. Second order transfer function: Computation and physical interpretation. In *Proc. SPIE International Modal Analysis Conf.*, pp. 587–592 (London, 1987).

[Yu94] T.-H. Yu and S. K. Mitra. Unsharp masking with nonlinear filters. In *Proc. 7th European Signal Processing Conference—EUSIPCO '94*, pp. 1485–1488 (Edinburgh, Scotland, September 1994).

[Zha93] H. Zhang. *Frequency Domain Estimation and Analysis for Nonlinear Systems.* Ph.D. thesis, University of Sheffield, Sheffield, England (1993).

7

Polynomial and Rational Operators for Image Processing and Analysis

GIOVANNI RAMPONI

Dipartimento di Elettrotecnica Elettronica Informatica
Università degli Studi di Trieste
Trieste, Italy

7.1 Introduction

A wide class of nonlinear operators can be devised for image processing applications, based on polynomial and rational functions of the pixels (luminance or color) of an image. This chapter shows that this approach can be exploited successfully for image enhancement (contrast sharpening, edge-preserving noise smoothing), image analysis (texture segmentation, edge extraction), and image format conversion (interpolation).

The attractive side of polynomial and rational expressions under a mathematical viewpoint is that they come as the most obvious extension to linear operators, which in turn are the first expertise required of a person working in the digital signal processing area. The properties of polynomial-based functions are well established, and they also enjoy the advantage of sharing some of these properties with linear operators (adaptive techniques can be devised in a similar way).

On the other hand, the most important obstacle to the diffusion of such operators is their computational complexity, which increases very rapidly with the degree of the nonlinearity and with the size of the data support involved. However, in most applications extremely simple operators, of degree at most three and operating on a very small support, can provide satisfactory results.

This chapter is organized as follows: The formalisms of polynomial and rational filters are dealt with first, and then Sec. 7.2 provides an overview of the theoretical properties of operators that are based on the Volterra series and of their extensions to two dimensions. Section 7.3 indicates a set of possible applications of polynomial filters for contrast enhancement, texture segmentation, and edge extraction. Rational operators occupy the remainder of the chapter, with a number of applications such as noise smoothing, image interpolation and contrast enhancement discussed in Sec. 7.4. Finally, some conclusions are drawn in Sec. 7.5, which also indicates important issues remaining in the field.

7.2 Theoretical Survey of Polynomial and Rational Filters

A thorough study of the theory of polynomial operators is far beyond the scope of this chapter, so only a few fundamental concepts are given to familiarize the reader with this approach. For a deeper understanding, the most important references are the book by Schetzen [Sch80] and the very recent book by Mathews and Sicuranza [Mat00].

7.2.1 Volterra Series Expansions for Discrete Nonlinear Time-Invariant Systems

A continuous nonlinear time-invariant (NLTI) system without memory can be represented by a Taylor series. Similarly, a continuous NLTI system with memory is described by the Volterra series [Vol87]. Through orthogonalization of the complete set of Volterra operators the Wiener theory is obtained [Sch80, Sch81, Wie58]. Exploiting such mathematical tools in the digital signal processing arena requires a discrete truncated version of the Volterra series [Mat91, Mat00, Sic92].

In the discrete case, the input-output relation of a Volterra system was given in Chapter 6 and is reported here for convenience:

$$y(n) = h_0 + \sum_{p=1}^{\infty} \bar{h}_p(x(n)) \,. \tag{7.1}$$

The so-called Volterra kernels $\bar{h}_p = h_p(i_1, \dots, i_p)$ are symmetric functions of their arguments. The term h_0 represents an offset component, $h_1(i_1)$ is the impulse response of a digital noncausal IIR filter, and $h_p(i_1, \dots, i_p)$ is a generalized pth-order impulse response.

7.2.2 Classes of Polynomial Filters

In the most general case, each term in the summation of Eq. (7.1) has the expression

$$\bar{h}_p[x(n)] = \sum_{i_1=-\infty}^{\infty} \cdots \sum_{i_p=-\infty}^{\infty} h_p(i_1,\ldots,i_p)x(n-i_1)\cdots x(n-i_p). \qquad (7.2)$$

In practice however, it is necessary to resort to more restricted classes of polynomial filters. By changing to 0 the lower limits of the summations in Eq. (7.2) we get a causal system, and if the upper limits are changed to a finite value N the system has a finite memory. In this case, $h_1(i_1)$ is an FIR filter; for the higher-order kernels, the effect of the nonlinearity on the output depends only on the present and a finite set of past input values. The truncated Volterra series of order P is obtained by replacing the infinity in the summation of Eq. (7.1) with the finite integer P; the class of truncated (or finite-support, finite-memory) Volterra filters results.

If, however, an infinite memory is required while still avoiding the obvious problem of infinite summations, the class of finite-order recursive Volterra filters may be employed. The latter is constituted of operators whose output is expressed as a function of input values and of previous output values as well [Mat00]:

$$y(n) = f[x(n), x(n-1), \ldots, x(n-N_1), y(n-1), y(n-2), \ldots, y(n-N_2)]. \qquad (7.3)$$

The simplest recursive filter is the bilinear operator:

$$y(n) = \sum_{i=0}^{N_1} a_i x(n-i) + \sum_{i=1}^{N_2} b_i y(n-i) + \sum_{i=0}^{N_1}\sum_{j=1}^{N_2} c_{i,j} x(n-i)y(n-j)\,. \qquad (7.4)$$

A bilinear operator has the advantage that it shares with the nonrecursive Volterra filter the property of being linear in its coefficients. On the other hand, like any recursive operator it can show unstable behavior [Lee94].

7.2.3 Properties of Polynomial Filters

Among the various properties of Volterra operators, a few are worth citing even in this simplified overview. First, we observe that the output of the system is linear with respect to the kernel coefficients; hence, an extension of the optimum filter theory is allowed, and adaptation algorithms can be devised. For fixed-coefficient filters, the identification of each kernel requires a set of p products of unit impulses conveniently placed on the filter support as input excitation. An example is the bi-impulse response of a quadratic filter, described elsewhere [Ram90].

Second, due to their nonlinearity, Volterra operators can be devised that have a peculiar property: the response to a change in the input signal is an increasing function of the local average of the signal. In terms of image processing applications, this means that we can obtain a larger response to luminance steps located in brighter zones of an image. This feature is in agreement with Weber's law and will be exploited in the following sections. Moreover, according to the amplitude dependence, a threshold can be defined so that steps with amplitude above it are considered to be relevant details of the image and thus are amplified while steps below it are considered to be noise and thus are reduced in amplitude. A suitable threshold can be determined by trial and error procedures or by knowledge of the statistics of the noise. The resulting edge-preserving effect is exploited in the rational operator for noise smoothing described in Sec. 7.4.1.

Finally, it was already shown in Chapter 6, Sec. 6.2, that the output of a Volterra system can be expressed by sums of multidimensional convolutions. This property permits the derivation of a frequency domain interpretation of a polynomial operator, which was introduced for a quadratic operator in Sec. 6.3. This interpretation will be used in Chapter 14 and has important implications here. Let us apply a p-dimensional sampled complex sinusoid

$$v(n_1, \dots, n_p) = \mathrm{e}^{j\omega_1 n_1} \cdots \mathrm{e}^{j\omega_p n_p}$$

as the input sequence to a causal linear filter described by the p-dimensional impulse response $h_p(i_1, \dots, i_p)$. The output sequence is given by

$$w(n_1, \dots, n_p) = H_p(\mathrm{e}^{j\omega_1}, \dots, \mathrm{e}^{j\omega_p})\mathrm{e}^{j\omega_1 n_1} \dots \mathrm{e}^{j\omega_p n_p}, \tag{7.5}$$

where

$$H_p(\mathrm{e}^{j\omega_1}, \dots, \mathrm{e}^{j\omega_p}) = \sum_{i_1=0}^{\infty} \cdots \sum_{i_p=0}^{\infty} h_p(i_1, \dots, i_p)\mathrm{e}^{-j\omega_1 i_1} \dots e^{-j\omega_p i_p} \tag{7.6}$$

is the continuous frequency response of the p-dimensional causal linear filter.

The output of the Volterra operator of the pth order is then derived, by assuming $v(n_1, \dots, n_p) = u(n_1) \cdots u(n_p)$ and $n_1 = \cdots = n_p = n$,

$$y(n) = w(n, \dots, n) = H_p(\mathrm{e}^{j\omega_1}, \dots, \mathrm{e}^{j\omega_p})\mathrm{e}^{j(\omega_1 + \cdots + \omega_p)n}, \tag{7.7}$$

so that the frequency contributions at the output are recovered at the angular frequency $\omega_1 + \cdots + \omega_p$. Hence, it is immediately recognized that new frequencies appear at the output that are not present at the input; this property is the basis for the operation of the rational interpolators presented in Sec. 7.4.2. These interpolators exploit such new high-frequency components to yield sharper details than those obtainable by a linear operator.

7.2.4 Rational Filters

It should be noticed that the kernel complexity in each component of Eq. (7.1) is N^p coefficients. Due to the symmetry property, the number of independent coefficients is expressed by the binomial factor

$$N_p = \begin{pmatrix} N+p-1 \\ p \end{pmatrix} = \frac{(N+p-1)!}{(N+p-1-p)!p!}.$$

For example, if $p = 2$ and $N = 9$, $N^p = 81$ and $N_p = 45$; if $p = 2$ and $N = 25$, $N^p = 625$ and $N_p = 325$. This clearly shows one of the most important problems encountered when trying to exploit the potentialities of the Volterra series: the number of parameters grows at a very fast rate with the degree of the nonlinearity and with the memory of the system. There are two different approaches to circumventing this problem, the recursive operators already mentioned and a new class of polynomial-based filters, the rational function operators.

Rational functions (the ratio of two polynomials) were recently proposed to represent the input–output relation in a nonlinear signal processing system [Leu94]. One of the motivations for their introduction was to overcome another limitation typical of the more conventional polynomial approach, its poor ability to extrapolate beyond its domain of validity. Similar to a polynomial function, a rational function is a universal approximator—with enough parameters and enough data to optimize them, it can approximate any function arbitrarily well. Moreover, it can achieve the desired level of accuracy with a lower complexity, and it possesses better extrapolation capabilities. It also has been demonstrated that a linear adaptive algorithm can be devised for determining the parameters of this structure; with this approach, the two problems of estimating the direction of arrival of plane waves on an array of sensors and of detecting radar targets in clutter were tackled [Leu94].

A rational function, used as a filter, can be expressed as

$$y(n) = \frac{a_0 + \sum_{i=0}^{N-1} a_{1i}x(n-i) + \sum_{i=0}^{N-1}\sum_{j=0}^{N-1} a_{2ij}x(n-i)x(n-j) + \ldots}{b_0 + \sum_{i=0}^{N-1} b_{1i}x(n-i) + \sum_{i=0}^{N-1}\sum_{j=0}^{N-1} b_{2ij}x(n-i)x(n-j) + \ldots}. \tag{7.8}$$

A rational function with a linear numerator and a linear denominator was used by Leung and Haykin [Leu94]. A major obstacle for these functions can again be their complexity: if a high order is used, many parameters will be required, and this will cause slow convergence.

7.2.5 Extension to 2-D and Multidimensional Discrete Systems

The complexity of the extension of polynomial and rational operators to the multidimensional case is in general extremely large. However, for applications in image and image sequence processing, often filters defined on very small supports yield remarkable results; moreover, a low-order filter is often sufficient to obtain significant improvements over conventional linear filters.

The arrangement of the independent variables into suitable vectors is the basis of the representation of a multidimensional Volterra filter; to derive a compact notation, we must conveniently order the input data and the filter coefficients. For example, an expression for a quadratic 2-D Volterra filter was given in Sec. 6.2.1.

Analogous expressions can be derived for rational operators by separately manipulating the numerator and the denominator of their input–output equation.

7.3 Applications of Polynomial Filters

In the 1-D case, polynomial filters have been successfully applied for modeling nonlinear systems, quadratic detectors (Teager's operator, see Chapter 6), echo cancellation, cancellation of nonlinear intersymbol interference, channel equalization in communications, nonlinear prediction, etc.

Applications to image processing are in enhancement (image sharpening, edge-preserving smoothing, processing of document images), analysis (edge extraction, texture discrimination), and communications (nonlinear prediction, nonlinear interpolation of image sequences). Overviews of some of these applications are described next.

7.3.1 Contrast Enhancement

In the linear unsharp masking method, a fraction of the highpass-filtered version $v(m,n)$ of the input image $x(m,n)$ is used as a correction signal and added to the original image, resulting in the enhanced image $y(m,n)$:

$$y(m,n) = x(m,n) + \lambda v(m,n),$$

where

$$\begin{aligned} v(m,n) &= 4x(m,n) - x(m-1,n) - x(m+1,n) \\ &\quad -x(m,n-1) - x(m,n+1). \end{aligned}$$

This method is very sensitive to noise due to the presence of the highpass filter. Polynomial unsharp masking techniques, in which a nonlinear filter is substituted for the highpass linear operator in the signal sharpening path, can solve this problem. Different polynomial functions can be used. In the Teager-based operator, details are amplified in bright regions, where the human visual system is less sensitive to luminance changes (Weber's law), and reduced noise sensitivity is achieved in dark areas. The correction signal in this case is [Mit91]

$$\begin{aligned} v(m,n) &= 2x^2(m,n) - x(m-1,n+1)x(m+1,n-1) \\ &\quad -x(m-1,n-1)x(m+1,n+1). \end{aligned} \tag{7.9}$$

In the cubic unsharp masking approach, the sharpening action is performed only if opposite sides of the filtering mask are each deemed to correspond to a different object [Ram96a], thus avoiding noise amplification:

$$\begin{aligned} v(m,n) &= [x(m-1,n) - x(m+1,n)]^2 \\ &\quad \times [2x(m,n) - x(m-1,n) - x(m+1,n)] \\ &\quad + [x(m,n-1) - x(m,n+1)]^2 \\ &\quad \times [2x(m,n) - x(m,n-1) - x(m,n+1)). \end{aligned}$$

Figure 7.1 shows the results of the unsharp masking approaches to image contrast enhancement. Figure 7.1a is a portion of the original Lena test image; 7.1b was

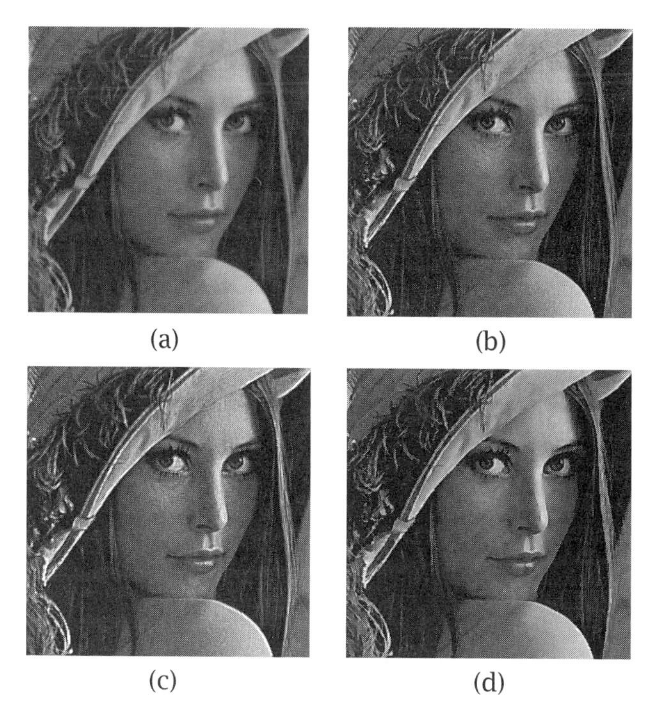

Figure 7.1: (a) Original test image, and contrast enhanced versions obtained using unsharp masking: (b) linear, (c) Teager-based, and (d) cubic methods. (Reproduced with permission from [Ram96a]. © 1996 SPIE.)

obtained using the linear method, 7.1c the Teager-based method, and 7.1d the cubic method.

Expressions of a similar type can be used for $v(m, n)$. For example, when the data are noisy a more powerful edge sensor is needed and the Sobel operator can be used [Ram96a].

7.3.2 Texture Segmentation

The segmentation of different types of textures present in an image can be performed based on local estimates of second- and higher-order statistics [Mak94]. In particular, third-order moments are the best features to use in noisy texture discrimination. A pth order statistical moment estimator is a special polynomial operator; for example, for $p = 3$

$$\begin{aligned} m_{x,3}(\mathbf{n}; \mathbf{i}, \mathbf{j}) &= E\{x(\mathbf{n})x(\mathbf{n}+\mathbf{i})x(\mathbf{n}+\mathbf{j})\} \\ &\simeq \frac{1}{M^2} \sum_{\mathbf{n} \in X} x(\mathbf{n})x(\mathbf{n}+\mathbf{i})x(\mathbf{n}+\mathbf{j}), \end{aligned} \tag{7.10}$$

where $\mathbf{n} = (n_1, n_2)$, $\mathbf{i} = (i_1, i_2)$, and $\mathbf{j} = (j_1, j_2)$. The moments are evaluated within an $M \times M$ image block X.

A class of third-order moments exists that are insensitive to white additive noise. Consider for simplicity of notation the 1-D case. If the available data result from an ideal signal corrupted by additive noise, $x(n) = s(n) + d(n)$, then

$$\begin{aligned} m_{x,3}(i,j) &= m_{s,3}(i,j) + m_{d,3}(0,0)\delta(i)\delta(j) \\ &\quad + m_{s,1}\sigma_d^2[\delta(i) + \delta(j) + \delta(i-j)]. \end{aligned} \tag{7.11}$$

To render the estimate independent of the noise, we just need to satisfy the simple constraints $i \neq 0$, $j \neq 0$, $i \neq j$. The selected local moments, after averaging, training, and clustering, permit us to segment composite textures in noise. Examples of the results achievable are presented in Fig. 7.2. The original image in Fig. 7.2a was formed by three different textures from Brodatz's album. Figure 7.2b shows the same image but with added zero-mean Gaussian noise; the SNR is 0 dB.

Results of segmentation using quadratic filters that estimate second-order moments are presented in Figs. 7.2c and 7.2d for the uncorrupted and the noisy image, respectively; it is evident that second-order moments are well suited for the uncorrupted image but do not work well in the presence of noise. Figures 7.2e and 7.2f show segmentation results obtained from the same images as above but now using cubic filters also. The selected third-order moments make the operator able to discern the noisy textures with good precision (Fig. 7.2f), but it is also apparent that the result on the uncorrupted data (Fig. 7.2e) is slightly worse than the one achieved with second-order moments only. This is due to the higher instability and variance of the output of high-order operators and indicates that the latter should be used only for the noisy case.

7.3.3 Edge Extraction

An important step in image analysis is to determine the position of the edges of the objects in the image. However, edge extraction is much more difficult in the presence of noise. Almost all edge extractor algorithms that have been proposed tend to perform roughly in the same manner when noise is absent but yield very different outputs in the presence of noise. Zero crossings of the second derivative (Laplacian) of the signal are often used to detect edges. The most popular algorithm in this category is the Marr-Hildreth operator [Mar80], also known as the Laplacian-of-Gaussian (LoG), which executes a lowpass (Gaussian) filtering before evaluating the Laplacian, in order to reduce the sensitivity to noise. The LoG operator (together with other analogous zero-crossing edge extraction methods) is nevertheless affected by two relevant problems: first, one has to exchange noise immunity (a strong lowpass filter is required) with resolution (which is lost in lowpass filtered data), and second, Gaussian filtering causes distortion in the location of edges.

A local skewness estimator, which is a particular polynomial operator, can replace the Laplacian. The image first undergoes a mild Gaussian prefiltering; then, an approximately round mask $\mathcal{M}$ formed by $M \times M$ pixels is used to scan the image.

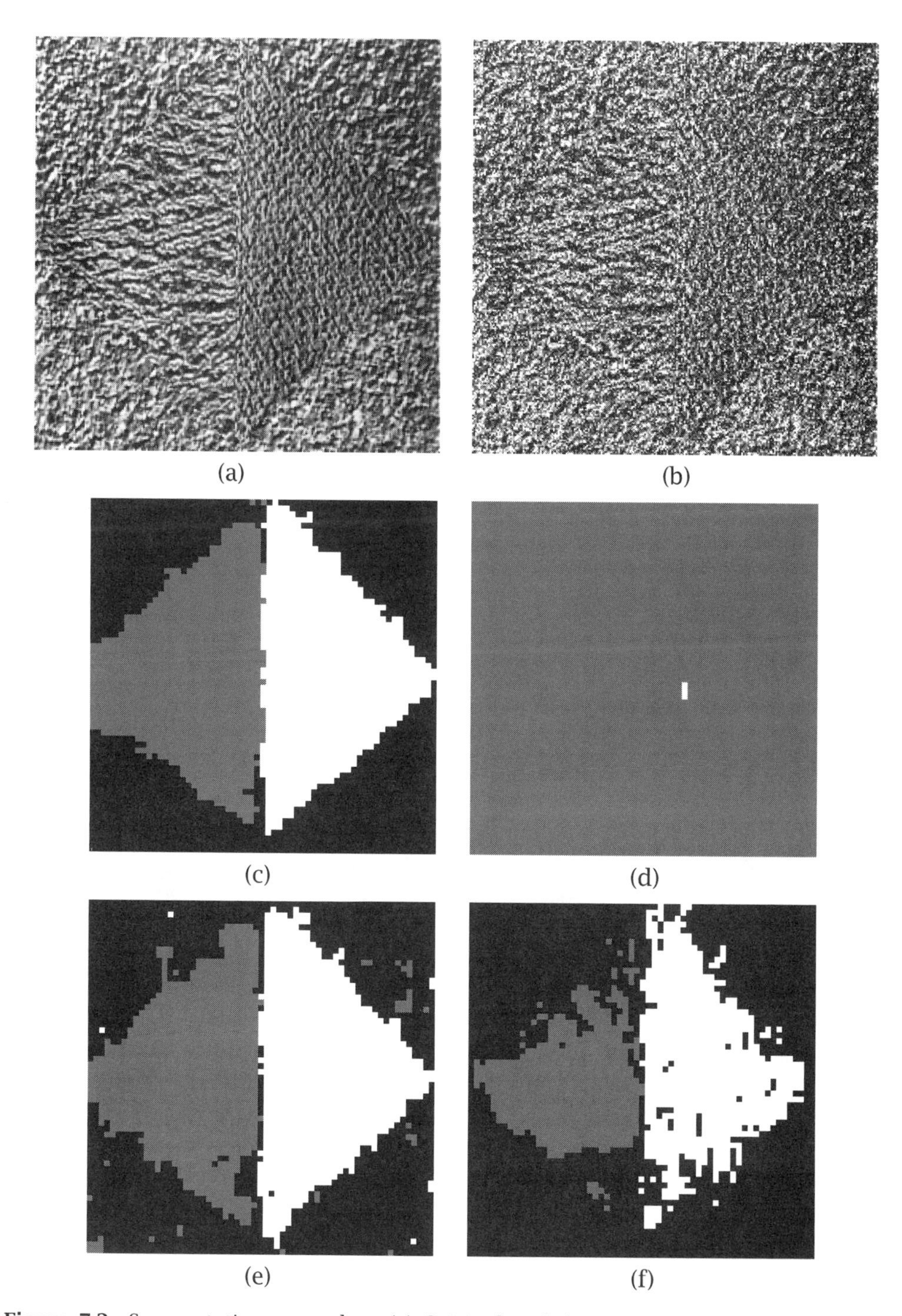

Figure 7.2: Segmentation examples. (a) Original and (b) image with additive Gaussian noise; segmentation of (c) the original image and (d) of the noisy image using quadratic filters; segmentatino of (e) the original image and (f) of the noisy image also using cubic filters.

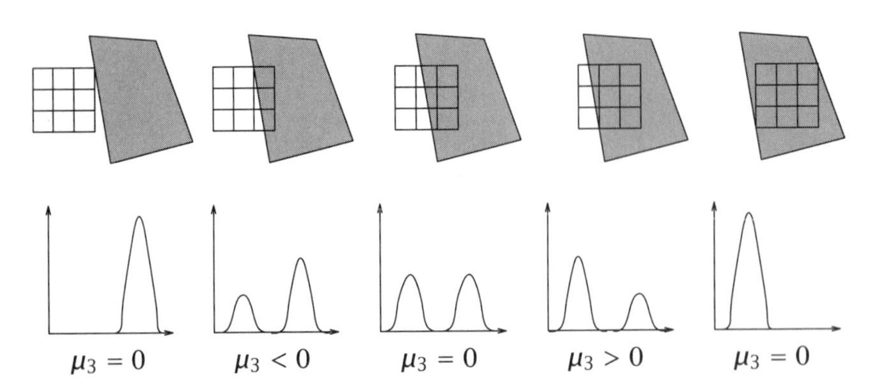

Figure 7.3: Illustration of the behavior of the local skewness when the mask is moved across the edge of an object.

The skewness is estimated in the interior of $\mathcal{M}$:

$$\hat{\mu}_3(\mathbf{n}) = \frac{1}{M^2} \sum_{\mathbf{i} \in I} [x(\mathbf{n}+\mathbf{i}) - \hat{\mu}(\mathbf{n})]^3 , \tag{7.12}$$

where I is a suitable set of indices, and $\hat{\mu}(\mathbf{n})$ is the local average luminance. When a luminance edge is met, the skewness changes its sign, and these zero crossings can be used for edge detection. The change of sign is illustrated in Fig. 7.3. The top row shows an idealized image with a dark object on a bright background, with the analysis mask $\mathcal{M}$ moving toward the interior of the object. In the bottom row the histogram of the pixels inside the mask at each position is plotted; it is seen that the histogram is symmetric ($\mu_3 = 0$) when the mask is completely outside or completely inside the object or when it is located exactly halfway but is asymmetric elsewhere. When the mask is mostly on the background, $\mu_3 < 0$. However, $\mu_3 > 0$ when the mask is mostly on the object. If the curvature of the object border is negligible with respect to the size of $\mathcal{M}$, the center of the mask locates the edge when the sign changes. The same behavior, with opposite sign, results if the object is brighter than the background.

With a Gaussian prefiltering, the skewness-of-Gaussian (SoG) method is obtained, which is intrinsically robust to noise [Ram94]. In fact, most common noise distributions (such as Gaussian, uniform, Laplacian, impulsive) are symmetric and as a consequence tend not to affect the skewness estimation. The SoG operation is controlled by (1) the size of the filtering mask, (2) the size of the mask $\mathcal{M}$ used for the estimation of the skewness, and (3) a threshold needed to reject false zero crossings.

The performance of this polynomial technique is best illustrated by means of an example. Figure 7.4a shows the original test image "Boats," while Fig. 7.4b shows its corrupted version obtained by adding a Gaussian noise with variance $\sigma^2 = 500$. The edges as extracted by the SoG method are shown in Figs. 7.4c and 7.4d for the original and the corrupted images respectively. As a comparison

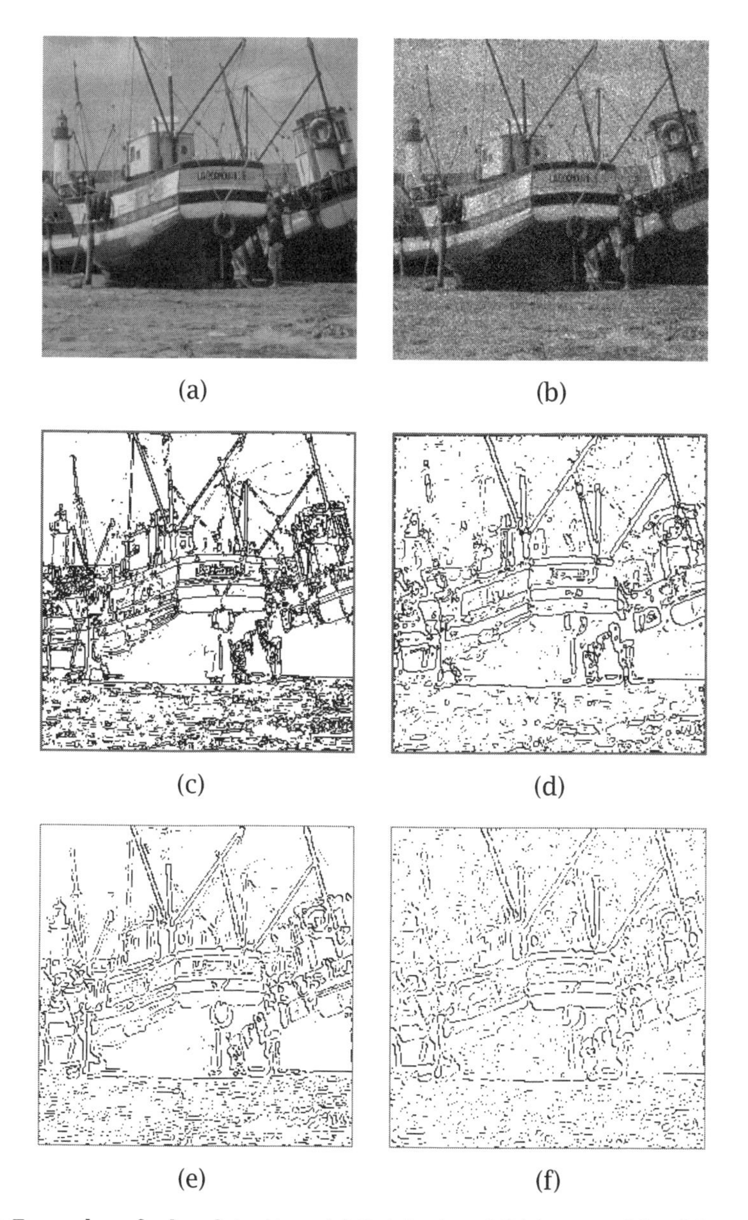

Figure 7.4: Examples of edge detection. (a) Original and (b) image with additive Gaussian noise; edges obtained from (c) the original and (d) the noisy image using the SoG operator; edges from (e) the original and (f) the noisy image using the LoG operator.

the outputs of the LoG operator are also reported (Figs. 7.4e and 7.4f). Threshold values were set subjectively to obtain the best perceptual quality in the various cases. It can be seen that the best quality is obtained by the SoG operator; as a result of its robustness to noise, the amount of Gaussian prefiltering can be small, and this reduces the distortions on small details.

7.4 Applications of Rational Filters

We mentioned in Sec. 7.2.3 that adaptation procedures can be used to determine the coefficients of rational filters. However, this approach has been followed in the literature only for simple 1-D operators. A possible alternative for the design of suitable rational filters for image processing to achieve a desired result is to employ some heuristic criteria and compose simple polynomials into a more complex function. This section gives an overview of the applications of rational operators in the fields of noise smoothing for image data (with various noise distributions), of image interpolation, and of contrast enhancement.

7.4.1 Detail-Preserving Noise Smoothing

A rational filter can be very effective in removing both short-tailed and medium-tailed noise corrupting an image [Ram95, Ram96b]. For the sake of simplicity, we refer to a one-dimensional operator, but its extension to two dimensions is straightforward. The filter can be formulated as

$$y(n) = \frac{x(n-1) + x(n+1)}{S(\mathbf{x}) + A} + x(n)\left[1 - \frac{2}{S(\mathbf{x}) + A}\right], \tag{7.13}$$

where the detail-sensing function $S()$ is defined as

$$S(\mathbf{x}) = k[2x(n) - x(n-1) - x(n+1)]^2$$

and

$$S(\mathbf{x}) = k[x(n-1) - x(n+1)]^2$$

for short-tailed and medium-tailed noise, respectively. In Eq. (7.13), A is a suitably chosen constant. The operator can be expressed in a form similar to that of Eq. (7.8). In the latter case, for example, it becomes

$$y(n) = \frac{x(n-1) + x(n+1) + (A-2)x(n) + kx(n)[x(n-1) - x(n+1)]^2}{k[x(n-1) - x(n+1)]^2 + A}; \tag{7.14}$$

a constant and a quadratic term are recognizable in the denominator, while the numerator consists of the sum of a linear function and a cubic function of the input signal.

We can examine the behavior of these filters for different positive values of the parameter k: If k is very large, $y(n) \simeq x(n)$ and the filter has no effect, while if $k \simeq 0$ and a suitable value is chosen for A, the rational filter becomes a simple linear lowpass filter; for intermediate values of k, the output of the sensor $S(\mathbf{x})$ modulates the response of the filter. Hence, the rational filter can act as an edge-preserving smoother conjugating the noise attenuation capability of a linear lowpass filter and the sensitivity to high frequency details of the edge sensor. Practical tests show that the value of k is not critical; moreover, it is not a function of the processed image but rather of the amount of present noise.

This operator can be applied more than once on the input data to obtain a stronger smoothing action. For p passes, in uniform areas where $S(\mathbf{x})$ is negligible the rational filter is equivalent to a linear lowpass filter of size $(2p+1)$; by defining $w = 1/A$, the coefficients in Eq. (7.13) are $\mathbf{w}_1 = [w, 1-2w, w]$ for $p = 1$, $\mathbf{w}_2 = [w^2, 2w(1-2w), 2w^2 + (1-2w)^2, 2w(1-2w), w^2]$ for $p = 2$, and so on. More important, in the vicinity of a detail the mask of the equivalent filter takes an asymmetric shape, which tends to cover pixels similar to the reference one; hence, the required edge preservation is obtained.

For a 2-D operator, it is sufficient to apply the 1-D operator in a 3×3 mask in the 0°, 45°, 90°, 135° directions; in this way an operator analogous to anisotropic diffusion [Per90] results. In fact, the latter operator was introduced as a scale-space method to perform edge detection at different resolutions, smoothing out details that are considered irrelevant at each scale. However, both anisotropic diffusion (see Chapter 8) and the rational filter perform their smoothing action preferably within a homogeneous region rather than across the borders of different regions. This smoothing action can be conceived as a diffusion of the luminance of the image, with a diffusion constant that varies locally and is an inverse function of the gradient; the higher the gradient, that is, the more an image detail is important, the smaller the diffusion.

Other types of 2-D rational filters can be devised, though, that cannot be represented by the diffusion concept; for example, better results are obtained in general if all the couples of pixels in a 3×3 mask are included in the processing. To formulate the latter filter, let the pixels in the mask be ordered as follows:

$$\mathbf{x} = \begin{bmatrix} x_1 & x_2 & x_3 \\ x_8 & x_0 & x_4 \\ x_7 & x_6 & x_5 \end{bmatrix},$$

where $x_0 = x(m,n)$. The filter expression then becomes

$$y(m,n) = x(m,n) + \sum_{i=1}^{7} \sum_{j=i+1}^{8} \frac{x_i + x_j - 2x_0}{A + k(x_i - x_j)^2}. \tag{7.15}$$

This rational filter outperforms various other nonlinear techniques in terms of MSE and image quality for different noise distributions ranging from uniform to Gaussian and contaminated Gaussian. As an example, Fig. 7.5a shows a test image corrupted with additive contaminated Gaussian noise [Gab94]; after filtering, the image in Fig. 7.5b is obtained.

Based on the rational function approach, other operators have been designed and realized for noise smoothing in image sequences [Coc97], for the filtering of speckle noise in SAR images [Ram97a], and for the attenuation of the blocking effects present in images compressed with DCT-based methods [Mar98]. Combining the advantages of the rational filter and order-statistic operators, median-rational hybrid filters have also been introduced; for the color image filtering problem, a vector rational operation can be performed over the output of three subfilters,

(a) (b)

Figure 7.5: (a) Corrupted test image and (b) its filtered version. (Reproduced with permission from [Ram96b]. © 1996 IEEE.)

such as two vector median subfilters and one center-weighted vector median filter [Khr99a].

7.4.2 Interpolation with Accurate Edge Reproduction

The rational filter paradigm also can be used to design effective operators for image interpolation. A possible approach relies on the observation that any real-world image may be considered to have been obtained from a higher resolution one after lowpass filtering and decimation, with the anti-aliasing lowpass filtering being done explicitly or being produced by the image acquisition system. When an ideal edge is present, this filtering operation symmetrically or asymmetrically modifies the value of the adjacent pixels according to the position of the edge itself in the high-resolution image; after decimation, consequently, an analysis of the values of the pixels of the low-resolution image gives subpixel information on the position of the edge, which hence may be interpolated with higher precision than that obtained with a linear interpolator.

The interpolation is performed by evaluating a nonlinear mean of the pixels adjacent to the pixel $y(m,n)$ to be interpolated. We illustrate the method by considering the 1-D case for simplicity. Figure 7.6a shows the position of the edge in the high-resolution data, Fig. 7.6b shows it in the lowpass-filtered data, and Fig. 7.6c shows it in the decimated data. Given the four consecutive pixels x_1, x_2, x_3, and x_4, an ideal interpolator should yield for the pixel y (which lies between x_2 and x_3) a value similar to that of either x_2 (left) or x_3 (right). As shown in Fig. 7.6d, linear interpolation fails to achieve this result.

A rational interpolator can be introduced, which computes the interpolated sample y as

$$y = \mu x_2 + (1 - \mu) x_3,$$

where

$$\mu = \frac{k(x_3 - x_4)^2 + 1}{k((x_1 - x_2)^2 + (x_3 - x_4)^2) + 2},$$

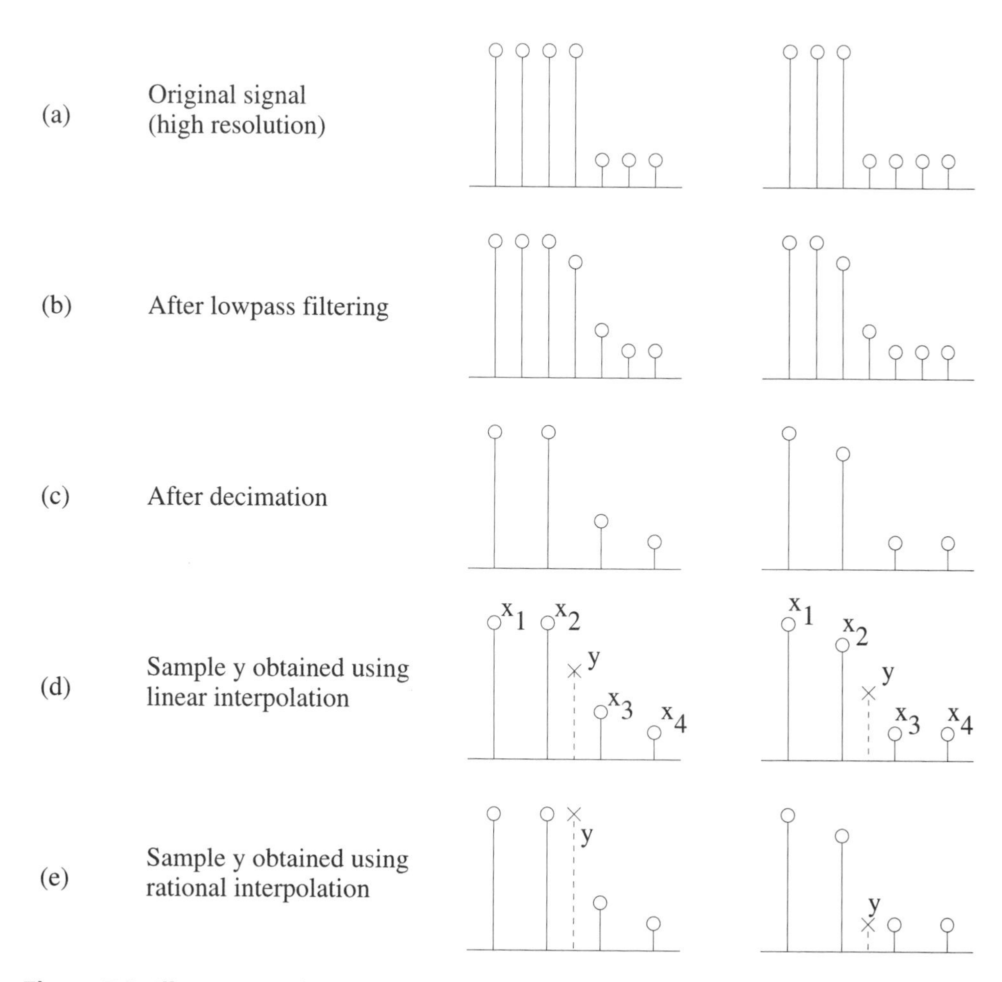

Figure 7.6: Illustration of interpolation in the one-dimensional case. Two possible positions are considered for the edge to be reconstructed.

with k being a user-defined parameter that controls the operator. For $k = 0$ a linear interpolation is obtained, while positive values for k yield the desired edge sensitivity. When the edge is midway between x_2 and x_3, $x_1 - x_2 = x_3 - x_4$, so $\mu = 0.5$ and $y = (x_2 + x_3)/2$, and the filter behaves as a linear one. However, when the edge is positioned asymmetrically, the evaluated differences are no longer equal; for example, if the edge is closer to x_3 (left column in the figure), then $x_1 - x_2 < x_3 - x_4$, so $\mu > 0.5$ and $y \simeq x_2$. In the 2-D case, this method can be applied separately to the rows and the columns of an image. Alternatively, the technique presented by Carrato et al. [Car96] can be used.

Figure 7.7 illustrates the performance of the proposed interpolation method. The original 512×512 image was lowpass filtered with a Gaussian filter, decimated, and interpolated with the rational operator (Fig. 7.7a) and with the cubic convolution technique (Fig. 7.7b) [Key81]. It can be seen that the details (in particular, observe the edges of the hat and the eye) are reconstructed more sharply by the

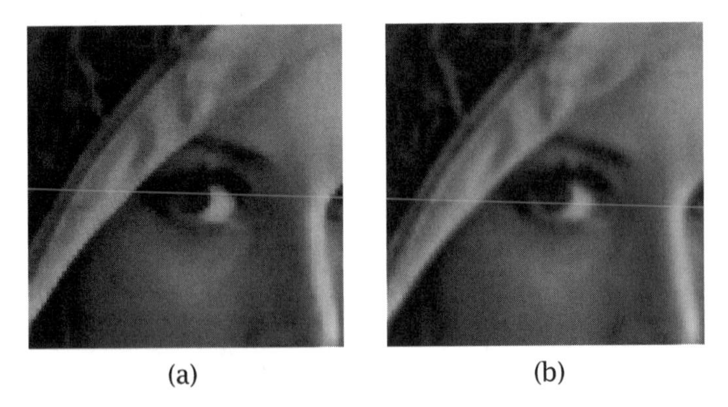

(a) (b)

Figure 7.7: Portion of the interpolated image Lena obtained using (a) the rational operator and (b) cubic convolution.

rational operator.

A similar approach has been used to interpolate images by large factors, for example, to reconstruct a JPEG-coded image using only the dc components of its block DCT [Ram97b], yielding relatively sharp images. Following this approach, it is possible to have images of reasonably good quality at low bit rates using standard coding techniques, while of course the ac components may be transmitted later, possibly on demand, if the actual image is eventually needed.

Another possible application is the interpolation of color images [Khr98]: any rational filter can be used to process multichannel signals by applying it separately to each component, but in general this is not desirable when correlation exists among components because false color may be introduced. Better results are obtained using vector extensions of the operator, such as by extending it to vector form:

$$\mathbf{y} = \frac{w_2\mathbf{x}_2 + w_3\mathbf{x}_3}{w_2 + w_3},$$

with

$$w_2 = 1 + k||\mathbf{x}_2 - \mathbf{x}_4||_p, \quad w_3 = 1 + k||\mathbf{x}_3 - \mathbf{x}_1||_p.$$

Here, $\mathbf{y}$, $\mathbf{x}_1$, $\mathbf{x}_2$, $\mathbf{x}_3$, and $\mathbf{x}_4$ are vector samples (e.g., RGB) and parallel bars indicate the l_1 or l_2 norm.

Finally, it is worth mentioning that rational interpolators have also been used successfully for the restoration of old movies corrupted by scratches and dirt spots [Khr99b]. After localization of the stationary and random defects, a spatial interpolation scheme is used to reconstruct the missing data. The algorithm first checks for the existence of edges so that it can take them into consideration; the edge orientation is estimated and the most convenient data to be used in the reconstruction of the missing pixels are selected. In this way, the edges obtained are free from blockiness and jaggedness.

7.4.3 Contrast Enhancement

In Sec. 7.3.1 we introduced a polynomial technique for contrast enhancement in images that was able to avoid noise amplification. In this subsection a rational filtering approach is suggested that partially retains the noise rejection ability of the polynomial method yet has the advantage of avoiding another problem that typically affects images processed with linear unsharp masking methods: excessive overshoot on sharp details [Ram98]. Wide and abrupt luminance transitions in the input image can produce overshoot effects when linear or cubic unsharp masking methods are used; these are further stressed by the human visual system through the Mach band effect.

The basic unsharp masking scheme is still used here, but a rational function is introduced in the correction path. With such a function, selected details having low and medium sharpness are enhanced but noise amplification is limited, and as a result steep edges, which do not need further emphasis, remain almost unaffected. From a computational viewpoint, it must be stressed that this solution maintains almost the same simplicity as the original linear unsharp masking method.

As was done in Sec. 7.3.1, the output value of each pixel is expressed as the sum of the original value and a correction term $v(m,n)$; here, however, $v(m,n)$ is split into two orthogonal components and expressed as a function of a control signal c:

$$y(m,n) = x(m,n) + \lambda[v_m(m,n)c_m(m,n) + v_n(m,n)c_n(m,n)], \tag{7.16}$$

where

$$\begin{aligned} v_m(m,n) &= 2x(m,n) - x(m,n-1) - x(m,n+1), \\ v_n(m,n) &= 2x(m,n) - x(m-1,n) - x(m+1,n). \end{aligned} \tag{7.17}$$

The two control signals are defined as

$$c_m(m,n) = \frac{g_m(m,n)}{kg_m^2(m,n)+h}, \qquad c_n(m,n) = \frac{g_n(m,n)}{kg_n^2(m,n)+h}, \tag{7.18}$$

where k and h are proper positive factors and g is a measure of the local activity of the signal:

$$\begin{aligned} g_m(m,n) &= [x(m,n+1) - x(m,n-1)]^2, \\ g_n(m,n) &= [x(m+1,n) - x(m-1,n)]^2. \end{aligned} \tag{7.19}$$

Figure 7.8 shows the control term $c(n)$ as a function of $g(n)$ in the 1-D case for a specific choice of k and h ($k = 0.001$, $h = 250$). The presence of a *resonance peak* at the abscissa $g_0 = 500$ can be noticed, this characteristic enables the operator to emphasize details that are represented by low- and medium-amplitude luminance transitions. At the same time, steep and strong edges will yield high values of activity $g(n)$ (larger than g_0) and hence will undergo a more delicate amplification;

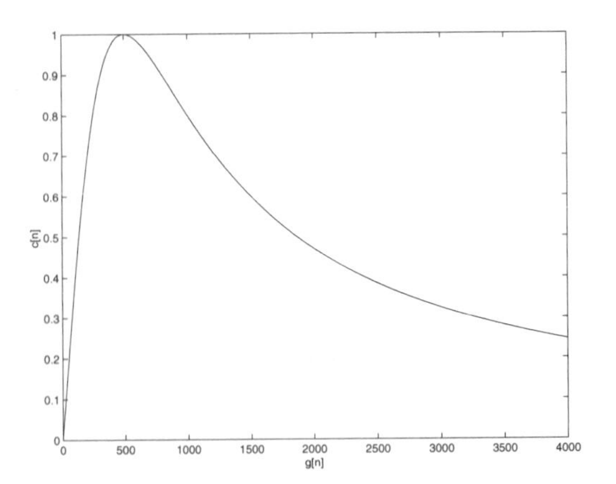

Figure 7.8: Plot of the control function c for $h = 250$, $k = 0.001$. (Reproduced with permission from [Ram98]. © 1998 SPIE.)

in this way, undesired overshoots in the output image will be avoided. At the other end of the diagram, we expect that the noise that is always present in an image will yield small values of $g(n)$ (smaller than g_0), so $c(n)$ will be small, too; hence, the noise itself will not affect the correction signal $v(n)$ in homogeneous areas. By selecting the position g_0 of the resonance, the user can achieve the best balance among these effects.

Different values of g_0 could yield different amplitudes for the peak c_0; this would make the overall strength of the unsharp masking action a function of the selected g_0, hampering the tuning of the operator. To avoid this effect, c should be normalized using the constraint $\sqrt{hk} = 1/2$ so that the peak height is equal to one for any value of g_0. Now the values of h and k that are to be used to place the resonance peak in g_0 are

$$h = \frac{g_0}{2}, \qquad k = \frac{1}{2g_0}. \tag{7.20}$$

In this way, the action of the $c(n)$ term is controlled by a single parameter, that is, g_0. It is interesting to observe that, once g_0 is set, we can freely adjust the intensity of the enhancement using the parameter λ; this choice will not affect the width Δ of the resonance peak.

A simulation example is shown in Fig.7.9. Comparing it to the results in Fig. 7.1, it is clearly seen that the noise sensitivity of the rational operator is still better than that of linear unsharp masking (Fig. 7.1b), even if not as good as that of cubic unsharp masking (Fig. 7.1d); on the other hand, fine details are better amplified, and some excessive overshoots recognizable along the edges of the image processed with the cubic unsharp masking method have disappeared.

(a)

(b)

Figure 7.9: (a) Original test image and (b) result of the rational unsharp masking algorithm. (Reproduced with permission from [Ram98]. © 1998 SPIE.)

7.5 Conclusions and Remaining Issues

Polynomial and rational operators have been introduced in this chapter for a wide spectrum of applications in image processing and analysis. Various simulation results have been included. These results demonstrate that even simple operators can yield effective results.

However, this description clearly indicates that to exploit the possibilities of this family of tools completely a stronger link is needed between theory and practice. It is necessary to formulate design techniques that can overcome the limitations of the heuristic approaches often used as discussed above. To this purpose, mixed competencies are required that should cross-fertilize the research in the mathematics and signal processing worlds; this is presently the most important challenge in this field.

References

[Car96] S. Carrato, G. Ramponi, and S. Marsi. A simple edge-sensitive image interpolation filter. In *Proc. Third IEEE Intl. Conf. on Image Processing*, Vol. III, pp. 711–714 (Lausanne, September 1996).

[Coc97] F. Cocchia, S. Carrato, and G. Ramponi. Design and real-time implementation of a 3-D rational filter for edge preserving smoothing. *IEEE Trans. Consum. Electron.* **43**(4), 1291–1300 (November 1997).

[Gab94] M. Gabbouj and L. Tabus. TUT noisy image database. Tech. Report No. 13, Tampere University of Technology, Finland (December 1994).

[Key81] R. G. Keys. Cubic convolution interpolation for digital image processing. *IEEE Trans. Acoust. Speech Signal Process.* ASSP-**29**(6), 1153–1160 (December 1981).

[Khr98] L. Khriji, F. A. Cheikh, M. Gabbouj, and G. Ramponi. Color image interpolation using vector rational filters. In *Proc. SPIE Conf. on Nonlinear Image Processing IX*, Vol. 3304, pp. 26–29 (1998).

[Khr99a] L. Khriji and M. Gabbouj. Vector median-rational hybrid filters for multichannel image processing. *IEEE Signal Process. Lett.* **6**(7), 186–190 (July 1999).

[Khr99b] L. Khriji, M. Gabbouj, G. Ramponi, and E. D. Ferrandiere. Old movie restoration using rational spatial interpolators. In *Proc. 6th IEEE Intl. Conf. on Electronics, Circuits and Systems* (Cyprus, September 1999).

[Lee94] J. Lee and V. J. Mathews. A stability theorem for bilinear systems. *IEEE Trans. Signal Process.* **41**(7), 1871–1873 (July 1994).

[Leu94] H. Leung and S. Haykin. Detection and estimation using an adaptive rational function filter. *IEEE Trans. Signal Process.* **42**(12), 3366–3376 (December 1994).

[Mak94] A. Makovec and G. Ramponi. Supervised discrimination of noisy textures using third-order moments. In *Proc. COST-229 WG.1+2 Workshop*, pp. 245–250 (Ljubljana, Slovenia, April 1994).

[Mar80] D. Marr and E. Hildreth. Theory of edge detection. *Proc. R. Soc. London B* **207**, 187-217 (1980).

[Mar98] S. Marsi, R. Castagno, and G. Ramponi. A simple algorithm for the reduction of blocking artifacts in images and its implementation. *IEEE Trans. Consum. Electron.* **44**(3), 1062–1070 (August 1998).

[Mat91] V. J. Mathews. Adaptive polynomial filters. *IEEE Sig. Process. Mag.*, pp. 10–26 (July 1991).

[Mat00] V. J. Mathews and G. L. Sicuranza. *Polynomial Signal Processing*, Wiley, New York (2000).

[Mit91] S. K. Mitra, H. Li, I.-S. Lin, and T-H. Yu. A new class of nonlinear filters for image enhancement. In *Proc. Intl. Conf. on Acoustics, Speech, and Signal Processing*, pp. 2525–2528 (Toronto, April 1991).

[Per90] P. Perona and J. Malik. Scale space and edge detection using anisotropic diffusion. *IEEE Trans. Patt. Anal. Machine Intell.* **12**(7), 629–639 (July 1990).

[Ram90] G. Ramponi. Bi-impulse response design of isotropic quadratic filters. *Proc. IEEE* **78**(4), 665–677 (April 1990).

[Ram94] G. Ramponi and S. Carrato. Performance of the Skewness-of-Gaussian (SoG) edge extractor. In *Proc. Seventh European Signal Processing Conf.* Vol. I, pp. 454–457 (Edinburgh, Scotland, September 1994).

[Ram95] G. Ramponi. Detail-preserving filter for noisy images. *Electron. Lett.* **31**(11), 865–866 (25 May 1995).

[Ram96a] G. Ramponi, N. Strobel, S. K. Mitra, and T-H. Yu. Nonlinear unsharp masking methods for image contrast enhancement. *J. Electron. Imaging.* **5**(3), 353–366 (July 1996).

[Ram96b] G. Ramponi. The rational filter for image smoothing. *IEEE Signal Process. Lett.* **3**(3), 63–65 (March 1996).

[Ram97a] G. Ramponi and C. Moloney. Smoothing speckled images using an adaptive rational operator. *IEEE Signal Process. Lett.* **4**(3), 68–71 (March 1997).

[Ram97b] G. Ramponi and S. Carrato. Interpolation of the DC component of coded images using a rational filter. In *Proc. Fourth IEEE Intl. Conf. on Image Processing,* Vol I, pp. 389–392 (Santa Barbara, CA, October 1997).

[Ram98] G. Ramponi and A. Polesel. A rational unsharp masking technique. *J. Electron. Imaging* **7**(2), 333–338 (April 1998).

[Sch80] M. Schetzen. *The Volterra and Wiener Theories of Nonlinear Systems.* Wiley, New York (1980).

[Sch81] M. Schetzen. Nonlinear system modeling based on the Wiener theory. *Proc. IEEE* **69**(12), 1557–1573 (December 1981).

[Sic92] G. L. Sicuranza. Quadratic filters for signal processing. *Proc. IEEE* **80**(8), 1263–1285 (August 1992).

[Vol87] V. Volterra. Sopra le funzioni che dipendono da altre funzioni. *Rendiconti Regia Accademia dei Lincei* (1887).

[Wie58] N. Wiener. *Nonlinear Problems in Random Theory.* Technology Press, M.I.T., and Wiley, New York (1958).

8

Nonlinear Partial Differential Equations in Image Processing

GUILLERMO SAPIRO

Department of Electrical and Computer Engineering
University of Minnesota
Minneapolis, Minnesota

8.1 Introduction

The use of partial differential equations and curvature driven flows in image analysis has become a research topic of increasing interest in the past few years. Let $u_0 : \mathbf{R}^2 \rightarrow \mathbf{R}$ represent a gray-level image, where $u_0(x, y)$ is the gray-level value and $\mathbf{R}$ is the set of all real numbers. Introducing an artificial time t, the image deforms via a partial differential evolution equation (PDE) according to

$$\frac{\partial u}{\partial t} = \mathcal{F}[u(x, y, t)], \tag{8.1}$$

where $u(x, y, t) : \mathbf{R}^2 \times [0, \tau) \rightarrow \mathbf{R}$ is the evolving image, $\mathcal{F} : \mathbf{R} \rightarrow \mathbf{R}$ is an operator that characterizes the given algorithm, and the image u_0 is the initial condition.[1] The solution $u(x, y, t)$ of the differential equation gives the processed image at "scale" t. In the case of vector-valued images, a system of coupled PDEs of the form of Eq. (8.1) is obtained.

[1] $\mathcal{F}$ typically depends on the image and the first and second spatial derivatives.

The same formalism can be applied to planar curves (boundaries of planar shapes), where u is a function from $\mathbf{R}$ to $\mathbf{R}^2$, or to surfaces, functions from $\mathbf{R}^2$ to $\mathbf{R}^3$. In this case, the operator $\mathcal{F}$ must be restricted to the curve, and all isotropic motions can be described as a deformation of the curve or surface in its normal direction, with velocity related to its principal curvature(s). In more formal terms, a flow of the form

$$\frac{\partial u}{\partial t} = \mathcal{F}(\kappa_i)\vec{\mathcal{N}} \tag{8.2}$$

is obtained, where κ_i are the principal curvatures and $\vec{\mathcal{N}}$ is the normal to the curve or surface u. A tangential velocity can be added as well, which may help the analysis but does not affect the geometry of the flow.

Partial differential equations can be obtained from variational problems. Assume a variational approach to an image processing problem formulated as

$$\arg\{\min_u \mathcal{U}(u)\},$$

where $\mathcal{U}$ is a given energy. Let $\mathcal{F}(u)$ denote the corresponding Euler-Lagrange derivative (first variation). Since under general assumptions, a necessary condition for u to be a minimizer of $\mathcal{U}$ is that $\mathcal{F}(u) = 0$, the (local) minima may be computed via the steady state solution of the equation

$$\frac{\partial u}{\partial t} = \mathcal{F}(u),$$

where t is again an "artificial" time marching parameter. PDEs obtained in this way have been used for quite some time in computer vision and image processing, and the literature is large. The classical example is the Dirichlet integral,

$$\mathcal{U}(u) = \int \|\nabla u\|^2(x)\, dx$$

(∇u stands for the gradient of u), which is associated with the linear heat equation (Δ stands for the Laplacian):

$$\frac{\partial u}{\partial t}(t, x) = \Delta u(x).$$

This equation gives birth to the whole *scale-space theory*, which addresses the simultaneous representation of images at multiple scales and levels of detail (see also Chapter 10 in this book). More recently, extensive research is being done on the direct derivation of nonlinear evolution equations, which, in addition, are not necessarily obtained from the energy approaches. This is in fact the case for a number of curvature equations of the form of Eq. (8.2).

Use of partial differential equations and the curve–surface flows in image analysis leads to model images in a continuous domain. This simplifies the formalism, which becomes grid independent and isotropic. The understanding of discrete local nonlinear filters is facilitated when one lets the grid mesh tend to zero and, thanks to an asymptotic expansion, rewrites the discrete filter as a partial differential operator.

Conversely, when the image is represented as a continuous signal, PDEs can be seen as the iteration of local filters with an infinitesimal neighborhood. This interpretation of PDEs allows one to unify and classify a number of the known iterated filters as well as to derive new ones. Actually, Alvarez et al. [Alv93] classified all of the PDEs that satisfy several stability requirements for imaging processing, such as locality and causality. (As pioneered elsewhere [Sap97b], future research might give up the locality requirement.)

An advantage of the PDEs approach is the possibility of achieving high accuracy and stability, with the help of the extensive research available on numerical analysis (this is also addressed in Chapter 10 for the specific case of the implementation of morphological operations). Of course, when considering PDEs for image processing and numerical implementations, we are dealing with derivatives of nonsmooth signals, and the right framework must be defined. The theory of viscosity solutions [Cra92] provides a framework for rigorously employing a partial differential formalism, in spite of the fact that the image may not be smooth enough to give a classical sense to the first and second derivatives involved in the PDE. Last but not the least, this area has quite a unique level of formal analysis, creating the possibility of providing not only successful algorithms but also useful theoretical results, like existence and uniqueness of solutions.

The use of nonlinear PDEs in image processing has become one of the fundamental geometric approaches in this area. Its connections with one of the classical geometric approaches in image processing, mathematical morphology, is fully discussed in in Chapter 10. Additional examples of the use of morphological operators, in their discrete form, are given in Chapter 9.

Two examples of PDE-based image processing algorithms are discussed in this chapter; both are related to the computation of geodesics, although in different spaces and for different applications. The first part of the chapter deals with the use of PDEs for image segmentation. This represents one of the most successful and well established results in the area. The second part discusses the use of PDEs to process multivalued data defined on nonflat manifolds, for example, directional data. This can be considered as the second generation of this PDE-based image processing approach since most of the work reported in the literature so far is for "flat" data. Both topics are representative of the work in this area: they are based on fundamental mathematical concepts while giving practical results that were not possible or were extremely difficult to obtain with other techniques. That is, both topics show a significant interaction between theory and practice, something that has characterized the PDEs framework since its origins.

For additional material on PDEs methods, bibliographies, and some history on the subject, see, for example, Caselles et al. [Cas98], Guichard and Morel [Gui95], and Romeny [Rom94], as well as Chapter 10 in this book.

8.2 Segmentation of Scalar and Multivalued Images

One of the basic problems in image analysis is segmentation. A class of this is object detection, where certain objects in the image are to be singled out. In this case, the image is basically divided into two sets: objects and background. On the other hand, in general image segmentation, the image is partitioned into an unknown number of "uniform" areas. Both problems have been studied since the early days of computer vision and image processing, and different approaches have been proposed (see, for example, [Har85, Mor94, Mum89, Zuc76] and references therein). We first concentrate on object detection and then show how the proposed scheme can be generalized to obtain complete image segmentation.

"Snakes," or active contours, were proposed by Kass et al. [Kas88] to solve the object detection problem. The classical snakes approach is based on deforming an initial contour or surface toward the boundary of the object to be detected. The deformation is obtained by minimizing a global energy designed such that its (local) minima are obtained at the boundary of the object. The energy is basically composed of a term that controls the smoothness of the deforming curve and another that attracts it to the boundary.

Geometric models of deformable contours and surfaces proposed in Caselles et al. [Cas93] and Malladi et al. [Mal95] are based on the theory of curve evolution and geometric flows, which, as pointed out, has gained a large amount of attention from the image analysis community in the past years. These models allow automatic changes in topology when implemented using level-sets [Osh88].

Caselles et al. showed the mathematical relation between these two approaches for two dimensional object detection, proposing the *geodesic active contours* model, [Cas97a], and extended the work to three dimensions based on the theory of minimal surfaces [Cas97b] (see also [Kis95, Sha95, Whi95], for related approaches, especially [Kis95]). For a particular case, the classical energy snakes approach is equivalent to finding a geodesic curve in a Riemannian space with a metric derived from the image. Assuming a level-sets representation of the deforming contour, one can compute this geodesic curve via a geometric flow that is very similar to the one obtained in the curve evolution approaches mentioned above. This flow, however, includes a new term that improves upon those models (for properties and advantages of the model, see [Cas97a, Sap97a]; for additional applications and significant extensions of the geodesic framework, see [Fau98, Lor99, Par97]).

Here we first describe the basic components of this geodesic model for object detection and segmentation in scalar and vector-valued images. Vector-valued images are obtained through imaging modalities in which the data are recorded in a vector fashion, as in color, medical, and LANDSAT applications. In addition, the vector-valued data can be obtained from scale and orientation decompositions often used in texture analysis (see [Sap97a] for the corresponding references). An additional distance based PDE approach for image segmentation, based on morphological watersheds, is formulated in Chapter 10.

8.2.1 Geodesic Active Contours

We now briefly review the active contours [Cas97a], to show their mathematical relation to previous works [Cas93, Kas88, Mal95, Ter88].

Let $C(p) : [0,1] \to \mathbf{R}^2$ be a parametrized planar curve and $u : [0,a] \times [0,b] \to \mathbf{R}^+$ be a given image in which we want to detect the objects boundaries. Let $g_{\text{gray}}(x)$ be a decreasing function such that $g_{\text{gray}} \to 0$ when $x \to \infty$. This is the edge detection function.

Terzopoulos et al. have shown, in one of the most classical and fundamental works in computer vision and image processing [Ter88], that objects can be detected (or segmented out) in an image by letting $C(p)$ deform in such a way that it is smoothly attracted to regions with low values of $g_{\text{gray}}(\cdot)$, that is, to the object's edges. The deformation of C is governed by the minimization of a given energy that combines regularization terms with edge attraction terms. Based on classical dynamical systems principles, it can be shown that minimizing (a version of) the classical energy is basically equivalent to minimizing

$$L_R \triangleq \int_0^{\text{length}} g_{\text{gray}}(\|\nabla u(C)\|)\, \mathrm{d}v,$$

where $\mathrm{d}v$ is the Euclidean arc-length. Therefore, solving the active contours problem is equivalent to finding a path of minimal distance, where distance is given by the modified arc-length $g_{\text{gray}}\, \mathrm{d}v$ (see also Sec. 10.6 for other applications of weighted distances). In other words, the object's boundary is given by a closed geodesic curve.

To find this geodesic curve, we first use the steepest descent method, which gives a local minima of L_R. Then the flow minimizing L_R is given by [Cas97a]

$$\frac{\partial C}{\partial t} = (g_{\text{gray}}\kappa - \nabla g_{\text{gray}} \cdot \vec{\mathcal{N}})\vec{\mathcal{N}},$$

where κ is the Euclidean curvature and $\vec{\mathcal{N}}$ the unit normal.[2] The first component of the right hand side of this equation is regularizing the curve (the classical geometric heat flow), while the second one attracts the curve to the object's boundaries (see also Sec. 10.4 for other curve evolution formulations). Note that both curvature-based and edge attracting velocity terms are present in this geodesic formulation, not just edge stopping terms such as those in watershed-type algorithms; see Chapter 10. These terms are fundamental to segment noisy images and detect objects with height variations in their gradients.

To complete the model, we now introduce the level-set formulation [Osh88] (see also Sec. 10.4). Let us assume that the curve C is parametrized as a level-set of a function $w : [0,a] \times [0,b] \to \mathbf{R}$. That is, C is such that it coincides with the set of points in w such that $w =$ constant (usually zero). Then, the level-set

[2]Note that in contrast with the formulations in Sec. 10.6, this PDE is not an Eikonal equation. On the other hand, starting from a point on the boundary, an open path of minimal weight L_R (open geodesic) from that point can be computed with a type of Eikonal equation, as formulated by Cohen and Kimmel [Coh97].

formulation of the steepest descent method says that solving the above geodesic problem starting from C amounts to searching for the steady state of the evolution equation

$$\frac{\partial w}{\partial t} = \|\nabla w\| \text{div}\left(g_{\text{gray}}(u)\frac{\nabla w}{\|\nabla w\|}\right) = g_{\text{gray}}(u)\kappa\|\nabla w\| + \nabla g_{\text{gray}}(u) \cdot \nabla w,$$

with initial datum $w(0,x) = w_0(x)$ (κ is the Euclidean curvature at the level-sets of w). The minima of L_R are then obtained with the (zero) level-set curve of $w(\infty)$. To increase the speed of convergence, we can add a balloon type force [Coh91], as has been done elsewhere [Cas93, Mal95], and just consider the term $\nu g_{\text{gray}}(u)\|\nabla w\|$, $\nu \in \mathbf{R}^+$, as an extra speed (which minimizes the enclosed area [Coh91]) in the geodesic problem, obtaining

$$\frac{\partial w}{\partial t} = \|\nabla w\| \text{div}\left(g_{\text{gray}}(u)\frac{\nabla w}{\|\nabla w\|}\right) + \nu g_{\text{gray}}(u)\|\nabla w\|. \tag{8.3}$$

Equation (8.3), which is the solution of the geodesic problem (L_R) with an extra area-based speed, constitutes the geodesic active contours and it is an improvement over previous equations for them [Cas93, Mal95] (see [Cas97a, Sap97a]). It can be shown that this PDE has a unique solution (in the viscosity framework) and that under simple conditions, the curve converges to the object's boundaries. That is, in addition to providing state-of-the-art practical results (given below), the formal analysis of this equation is possible. This is quite a unique characteristic of the PDE approach to image processing problems.

This equation, as well as its 3D extension [Cas97b], was independently proposed by Kichenassamy et al. [Kis95] and also by Shah [Sha95] based on a slightly different initial approach (see also [Whi95]). Extensions of the Caselles [Cas93] and Malladi [Mal95] model also were studied by Tek and Kimia [Tek95], motivated by the work of Kimia et al. [Kim95]. The differences between those models have been explained elsewhere [Cas97a, Sap97a].

Figure 8.1 presents an example of this flow. On the left we initialize the curve surrounding all of the objects in the image. Note how the curve splits, and all of the objects are detected at once, on the right side of Fig. 8.1. This is automatically handled by the algorithm, without adding any extra topology tracking procedures.

8.2.2 Geodesic Color Active Contours

We want to extend the geodesic framework for image segmentation presented above to vector-valued data. In general, two different approaches can be adopted to work on vector-valued images. The first approach is to process each plane separately and then to integrate the results of this operation somehow to obtain one unique segmentation for the whole image. The second approach is to integrate the vector information from the very beginning and deform a unique curve based on this information, directly obtaining a unique object segmentation. We adopt the second approach; that is, we integrate the original image information to find

Figure 8.1: Example of geodesic snakes. The original curve surrounds all of the objects (left) and evolves, splitting in the process, to detect all of the objects at once (right).

a unique segmentation directly. The main idea is to define a new Riemannian (metric) space based on information obtained from all of the components in the image at once. More explicitly, edges are computed based on classical results on Riemannian geometry [Kre59]. When the image components are correlated, as in color images, this approach is less sensitive to noise than the combination of scalar gradients obtained from each component [Lee91]. These vector edges are used to define a new metric space in which the geodesic curve is to be computed. The object boundaries are then given by a minimal "color weighted" path.

Vector Edges

We present now the definition of edges in vector-valued images, based on classical Riemannian geometry results [Kre59]. The Riemannian geometry framework for edge detection in multivalued images described below was first suggested by Di Zenzo [DiZ86] and was extended later [Cum91, Lee91, Sap96b]. This approach has a solid theoretical background and constitutes a consistent extension of single-valued gradient computations.

Let $\mathbf{u}(x_1, x_2) : \mathbf{R}^2 \to \mathbf{R}^n$ be a multivalued image with components $u_i(x_1, x_2) : \mathbf{R}^2 \to \mathbf{R}$, $i = 1, 2, \ldots, m$. The value of the image at a given point (x_1^0, x_2^0) is a vector in $\mathbf{R}^n$, and the difference of image values at two points $P = (x_1^0, x_2^0)$ and $Q = (x_1^1, x_2^1)$ is given by $\Delta\mathbf{u} = \mathbf{u}(P) - \mathbf{u}(Q)$. When the (Euclidean) distance $\mathrm{d}(P, Q)$ between P and Q tends to zero, the difference becomes the arc element $\mathrm{d}\mathbf{u} = \sum_{i=1}^{2} (\partial\mathbf{u}/\partial x_i)\,\mathrm{d}x_i$. The quadratic form $\mathrm{d}\mathbf{u}^2$ is called the *first fundamental form* [Kre59]. Although we present now only the Euclidean case, the theory we develop holds for any nonsingular Riemannian metric. For different metrics, either a space transform can be applied to a Euclidean space, if possible, or the metric induced by the given space can be used directly (if it is nonsingular). Using the standard notation of Riemannian geometry [Kre59], we have that $g_{ij} \triangleq (\partial\mathbf{u}/\partial x_i) \cdot (\partial\mathbf{u}/\partial x_j)$, and $\mathrm{d}\mathbf{u}^2 = [\mathrm{d}x_1, \mathrm{d}x_2][g_{ij}][\mathrm{d}x_1, \mathrm{d}x_2]^T$. For a unit vector $\hat{\mathbf{v}} = (\cos\theta, \sin\theta)$, $\mathrm{d}\mathbf{u}^2(\hat{\mathbf{v}})$ indicates the rate of change of the image in the $\hat{\mathbf{v}}$ direction. It is well known that the extrema of the quadratic form are obtained in the directions of the eigenvectors θ_+ and θ_- of the metric tensor $[g_{ij}]$, and the values attained there are the corresponding eigenvalues λ_+ and λ_-. We call θ_+ the *direction of maximal change* and λ_+ the *maximal rate of change*. Similarly, θ_- and λ_- are the *direction of minimal change* and the *minimal rate of change*, respectively.

In contrast with scalar images ($n = 1$), the minimal rate of change λ_- may be different from zero. In the single-valued case, the gradient is perpendicular to the level-sets, and $\lambda_- \equiv 0$. The "strength" of an edge in the multivalued case is not given simply by the rate of maximal change λ_+ but by how λ_+ compares to λ_-. Therefore, a first approximation of edges for vector-valued images, analogous to selecting a function of $\|\nabla u\|$ in the $n = 1$ case, should be a function $f = f(\lambda_+, \lambda_-)$. Selecting $f = f(\lambda_+ - \lambda_-)$ is one choice since for $n = 1$ it reduces to the gradient-based edge detector.

This definition of vector edges was used elsewhere [Sap96b] for color image diffusion and is used here to define vector snakes. It can also be used to define *vector total variation* [Sap97a] (see also [Blo98]).

Vector Snakes

Let $f_{\text{color}}(\mathbf{u}) \triangleq f(\lambda_+, \lambda_-)$ be the edge detector as defined above. The edge stopping function $g_{\text{color}}(\mathbf{u})$ is then defined such that $g_{\text{color}}(\mathbf{u}) \to 0$ when $f \to \max$ (or ∞), as in the scalar case. For example, we can choose $f_{\text{color}}(\mathbf{u}) \triangleq (\lambda_+ - \lambda_-)^{1/p}$ or $f_{\text{color}}(\mathbf{u}) \triangleq \sqrt{\lambda_+}$, $p > 0$, and $g_{\text{color}}(\mathbf{u}) \triangleq 1/(1 + f)$ or $g_{\text{color}}(\mathbf{u}) \triangleq \exp\{-f\}$. The function (metric) $g_{\text{color}}(\mathbf{u})$ defines the Riemannian space on which we compute the geodesic curve. Defining $L_{\text{color}} \triangleq \int_0^{\text{length}} g_{\text{color}}(\mathbf{u})\, dv$, the object detection problem in vector-valued images is then associated with minimizing L_{color}. We therefore have formulated the problem of object segmentation in vector-valued images as a problem of finding a geodesic curve in a Riemannian space defined by a metric induced from the whole vector image.

To minimize L_{color}, that is, the color length, we compute as before the gradient flow. The equations developed for the geodesic active contours are independent of the specific selection of the metric g. Therefore, the same equations hold here. Replacing $g_{\text{gray}}(u)$ by $g_{\text{color}}(\mathbf{u})$ and embedding the evolving curve C in the function $w : \mathbf{R}^2 \to \mathbf{R}$ we obtain the general flow, with additional unit speed, for the *color snakes*:

$$\frac{\partial w}{\partial t} = g_{\text{color}}(\mathbf{u})(\nu + \kappa)\|\nabla w\| + \nabla w \cdot \nabla g_{\text{color}}(\mathbf{u}). \tag{8.4}$$

Recapping, Eq. (8.4) is the modified level-sets flow corresponding to the gradient descent of L_{color}. Its solution (steady state) is a geodesic curve in the Riemannian space defined by the metric $g_{\text{color}}(\lambda_+, \lambda_-)$ of the vector-valued image. This solution gives the boundary of objects in the scene. Note that λ_+ and λ_- can be computed on a smooth image obtained from the vector-valued anisotropic diffusion [Sap96b]. Following Caselles et al. [Cas97a], theoretical results regarding the color active contours can be obtained [Sap97a].

Figure 8.2 shows an example of our vector snakes model. The numerical implementation is based on the algorithm for surface evolution via level-sets developed

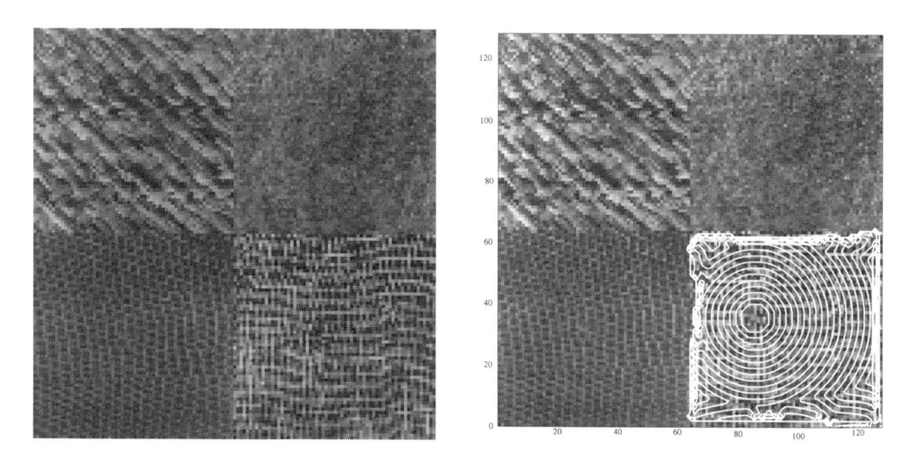

Figure 8.2: Example of the vector snakes for a texture image.

by Osher and Sethian [Osh88]. The original image is filtered with Gabor filters tuned to different frequencies and orientations as proposed by Lee et al. [Lee92] for texture segmentation (see [Sap97a] for additional related references). From this set of frequency-orientation decomposed images, g_{color} is computed accordingly and the vector-valued snakes flow is applied. Four frequencies and four orientations are used, resulting in 16 images.

We should mention that a number of results on color or vector-valued segmentation have been reported in the literature (e.g., [Zhu95]; see [Sap97a] for relevant references and further comparisons). Here we address the geodesic active contours approach with vector-image metrics, a simple and general approach. Other algorithms can be extended to work on vector-valued images as well, following the framework described in this chapter.

8.2.3 Self-Snakes

We now briefly extend the formulation of the geodesic snakes for object detection presented above to a new flow for segmentation-simplification of images. This flow is obtained by deforming each of the image level-sets according to the geodesic snakes. The resulting flow, denoted as *self-snakes*, is closely related to a number of previously reported image processing algorithms based on PDEs, such as anisotropic diffusion [Alv92, Per90] and shock-filters [Osh90], as well as the Mumford-Shah variational approach for image segmentation [Mum89]. The explicit relations are presented in detail elsewhere [Sap96a, Sap97a].

Let us observe again the level-sets flow corresponding to the single-valued geodesic snakes as presented above: $\partial w/\partial t = \|\nabla w\|\text{div}(g_{\text{gray}}(u)(\nabla w/\|\nabla w\|))$. Two functions (maps from $\mathbf{R}^2$ to $\mathbf{R}$) are involved in this flow, the image u and the auxiliary level-sets one, w. Assume now that $w \equiv u$, that is, that the auxiliary

Figure 8.3: Smoothing a portrait of Gauss with the geometric self-snakes (original on the left and enhanced on the right).

level-sets function is the image itself. This equation then becomes

$$\frac{\partial u}{\partial t} = \|\nabla u\| \operatorname{div}\left(g_{\text{gray}}(u)\frac{\nabla u}{\|\nabla u\|}\right). \tag{8.5}$$

A number of interpretations can be given to this equation. First of all, based on the analysis of the geodesic active contours, the flow in Eq. (8.5) indicates that each level-set of the image u moves according to the geodesic active contours flow, being smoothly attracted by the term ∇g_{gray} to boundaries. This gives the name *self-snakes* to the flow. This interpretation also explains why image segmentation is obtained.

Furthermore, Eq. (8.5) can be rewritten as the composition of an anisotropic diffusion term with a shock filter one. This relates the self-snakes, and the topic of active contours in general, with previously developed PDE based algorithms [Alv92, Cat92, Nit92, Osh90, Per90, Rud92, You96] (see [Sap96a, Sap97a]). An example is presented in Fig. 8.3. The same approach can be followed for vector-valued images.

8.2.4 Discussion

We have presented a framework for geometric segmentation of scalar and vector-valued images and have shown that the solution to the deformable contours approach for object detection is given by a geodesic curve in a Riemannian space defined by a metric derived from the data. In the case of vector-valued images, the metric itself is given by a definition of edges based on classical Riemannian geometry (this definition also leads to a novel concept of level-sets in vector-valued images [Chu00]).

The geodesic framework for object detection was then extended to obtain combined anisotropic diffusion and shock filtering in scalar and vector images. This flow simplifies the image and is obtained by deforming each of the image level-sets according to the geodesic flow.

The geodesic active contours provide state-of-the-art object detection and image segmentation results. In addition, they have been extended by Paragios and Deriche for object tracking in video [Par97] and by Faugeras and Keriven for 3D shape reconstruction from stereo [Fau98].

All of the work described so far deals with images defined on the plane, taking values in the general Euclidean space $\mathbf{R}^n$. This is also true for the vast majority of the published work in PDEs based image processing. But what happens if we define the images on more general manifolds or if the images take values not on $\mathbf{R}^n$ but also on general manifolds? We deal with these questions in the rest of this chapter.

8.3 Nonlinear PDEs in General Manifolds: Harmonic Maps and Direction Diffusion

In a number of disciplines, directions provide a fundamental source of information. Examples in the area of computer vision are (2D, 3D, and 4D) gradient directions, optical flow directions, surface normals, principal directions, and color. In the color example, the direction is given by the normalized vector in the color space. Frequently, the available data are corrupted with noise, and thus there is a need for noise removal. In addition, it is often desired to obtain a multiscale-type representation of the direction, similar to those obtained for gray-level images [Koe84, Per90, Per98, Wit83]. Addressing these issues in particular, and diffusion in arbitrary manifolds in general, is the goal of this section. We will see that this is related to geodesic computations as well.

An $\mathbf{R}^n$ direction defined on an image in $\mathbf{R}^2$ is given by a vector $\mathbf{u}(x, y, 0) : \mathbf{R}^2 \to \mathbf{R}^n$ such that the Euclidean norm of $\mathbf{u}(x, y, 0)$ is equal to one; that is, $[\sum_{i=1}^n u_i^2(x, y, 0)]^{1/2} = 1$, where $u_i(x, y, 0) : \mathbf{R}^2 \to \mathbf{R}$ are the components of the vector. The notation can be simplified by considering $\mathbf{u}(x, y, 0) : \mathbf{R}^2 \to S^{n-1}$, where S^{n-1} is the unit ball in $\mathbf{R}^n$. This implicitly includes the unit norm constraint. (Any nonzero vector can be transformed into a direction by normalizing it.) When smoothing the data, or computing a multiscale representation $\mathbf{u}(x, y, t)$ of a direction $\mathbf{u}(x, y, 0)$ (t stands for the scale), it is crucial to maintain the unit norm constraint, which is an intrinsic characteristic of directional data.[3] That is, the smoothed direction $\hat{\mathbf{u}}(x, y, 0) : \mathbf{R}^2 \to \mathbf{R}^n$ must also satisfy $[\sum_{i=1}^n \hat{u}_i^2(x, y, 0)]^{1/2} = 1$, or $\hat{\mathbf{u}}(x, y, 0) : \mathbf{R}^2 \to S^{n-1}$. The same constraint holds for a multiscale representation $\mathbf{u}(x, y, t)$ of the original direction $\mathbf{u}(x, y, 0)$. This is what makes the smoothing of directions different from the smoothing of ordinary vectorial data as performed elsewhere [Sap96b, Whi94]: The smoothing is performed in S^{n-1} instead of $\mathbf{R}^n$.

[3]In this work we do not explicitly address the problem in which the direction smoothing depends on other image attributes (see, for example, [Lin94]); the analysis is done intrinsically to the directional, unit norm data. When other attributes are present, we process them separately or via coupled PDEs (e.g., [Tan00b]). If needed, the unit norm constraint can be relaxed using a framework similar to the one proposed here.

Directions can also be represented by the angle(s) the vector makes with a given coordinate system, denoted in this chapter as *orientation(s)*. In the 2D case, for example, the direction (u_1, u_2) of a vector can be given by the angle θ that this vector makes with the x axis (we consider $\theta \in [0, 2\pi)$): $\theta = \arctan(u_2/u_1)$. There is, of course, a one-to-one map between a direction vector $\mathbf{u}(x, y) : \mathbf{R}^2 \to S^1$ and the angle θ. Using this relation, Perona [Per98] transformed the problem of 2D direction diffusion into a 1D problem of angle or orientation diffusion (see also the comment in Sec. 8.3.2). Perona then proposed PDE based techniques for the isotropic smoothing of 2D orientations (see also [Gra95, Wei96] and the general discussion of these methods in [Per98]).[4] Smoothing orientations instead of directions solves the unit norm constraint but adds a periodicity constraint. Perona showed that a simple heat flow (Laplacian or Gaussian filtering) applied to the $\theta(x, y)$ image, together with special numerical attention, can address this periodicity issue. This approach applies only to small changes in θ, that is, smooth data, thereby disqualifying edges. The straightforward extension of this to S^{n-1} would be to consider $n - 1$ angles and smooth each one of these as a scalar image. The natural coupling is then missing, resulting in a set of decoupled PDEs.

In this work we follow the suggestion in Caselles et al. [Cas00] and directly perform diffusion on the direction space, extending to images representing directions the classical results on diffusion of gray-valued images [Alv92, Koe84, Per90, Rud92, Sap96b, Wei96, Whi95]. That is, from the original unit norm vectorial image $\mathbf{u}(x, y, 0) : \mathbf{R}^2 \to S^{n-1}$ we construct a family of unit norm vectorial images $\mathbf{u}(x, y, t) : \mathbf{R}^2 \times [0, \tau) \to S^{n-1}$ that provides a multiscale representation of directions. The method intrinsically takes care of the normalization constraint, eliminating the need to consider orientations and develop special periodicity preserving numerical approximations. Discontinuities in the directions are also allowed by the algorithm. The approach follows results from the literature on harmonic maps in liquid crystals, and $\mathbf{u}(x, y, t)$ is obtained from a system of coupled partial differential equations that reduces a given (harmonic) energy. Energies giving both isotropic and anisotropic flows are described.

Before presenting the details of the framework for direction diffusion proposed here, let us repeat its main unique characteristics: (1) It includes both isotropic and anisotropic diffusion. (2) It works for directions in any dimension and for general data on nonflat manifolds. (3) It supports nonsmooth data. (4) It is based on a substantial amount of existing theoretical results that help to answer a number of relevant computer vision questions.

---

[4] As Perona pointed out in his work, this is just one example of the diffusion of images representing data beyond flat manifolds. His work has been extended using intrinsic metrics on the manifold [Cha99, Soc98]. Chan and Shen [Cha99] explicitly deal with orientations and present the $\mathcal{L}_1$ norm as well as many additional new features, contributions on discrete formulations, and connections with our approach. The work by Sochen et al. [Soc98] does not deal with orientations or directions. Rudin and Osher [Rud94] also mention the minimization of the $\mathcal{L}_1$ norm of the divergence of the normalized image gradient (curvature of the level-sets); this is done in the framework of image denoising, without addressing the regularization and analysis of directional data or presenting examples. These works do not use the classical and "natural" harmonic maps framework.

8.3.1 The General Problem

Let $\mathbf{u}(x, y, 0) : \mathbf{R}^2 \to S^{n-1}$ be the original image of directions. That is, this is a collection of vectors from $\mathbf{R}^2$ to $\mathbf{R}^n$ such that their unit norm is equal to one; that is, $\|\mathbf{u}(x, y, 0)\| = 1$, where $\|\cdot\|$ indicates Euclidean length. The $u_i(x, y, 0) : \mathbf{R}^2 \to \mathbf{R}$ stands for each of the n components of $\mathbf{u}(x, y, 0)$. We search for a family of images, a multiscale representation, of the form $\mathbf{u}(x, y, t) : \mathbf{R}^2 \times [0, \tau) \to S^{n-1}$, and once again we use $u_i(x, y, t) : \mathbf{R}^2 \to \mathbf{R}$ to represent each of the components of this family. Let us define the *component gradient* ∇u_i as $\nabla u_i \triangleq (\partial u_i/\partial x)\vec{x} + (\partial u_i/\partial y)\vec{y}$, where $\vec{x}$ and $\vec{y}$ are the unit vectors in the x and y directions, respectively. From this, $\|\nabla u_i\| = [(\partial u_i/\partial x)^2 + (\partial u_i/\partial y)^2]^{1/2}$ gives the absolute value of the component gradient. The *component Laplacian* is given by $\Delta u_i = (\partial^2 u_i/\partial x^2) + (\partial^2 u_i/\partial y^2)$. We are also interested in the absolute value of the *image gradient*, given by $\|\nabla \mathbf{u}\| \triangleq [\sum_{i=1}^{n} (\partial u_i/\partial x)^2 + (\partial u_i/\partial y)^2]^{1/2}$.

Having this notation, we are now ready to formulate our framework. The problem of *harmonic maps in liquid crystals* is formulated as the search for the solution to

$$\min_{\mathbf{u}:\mathbf{R}^2 \to S^{n-1}} \iint_{\Omega} \|\nabla \mathbf{u}\|^p \, dx\, dy, \tag{8.6}$$

where Ω stands for the image domain and $p \geq 1$. This variational formulation can be rewritten as $\min_{\mathbf{u}:\mathbf{R}^2 \to \mathbf{R}^n} \iint_{\Omega} \|\nabla \mathbf{u}\|^p \, dx\, dy$, with $\|\mathbf{u}\| = 1$.

This is a particular case of the search for maps $\mathbf{u}$ between Riemannian manifolds (M, g) and (N, h), which are critical points of the *harmonic energy*:

$$E(u) = \int_M \|\nabla_M \mathbf{u}\|^p \mathrm{dvol}M, \tag{8.7}$$

where $\|\nabla_M \mathbf{u}\|$ is the length of the differential in M. In our particular case, M is a domain in $\mathbf{R}^2$ and $N = S^{n-1}$, and $\|\nabla_M \mathbf{u}\|$ reduces to the absolute value of the image gradient. The critical points of Eq. (8.7) are called *p-harmonic maps* (or simply *harmonic maps* for $p = 2$). This is in analogy to the critical points of the Dirichlet energy $\int_\Omega \|\nabla u\|^2$ for real-valued functions u, which are called *harmonic functions.* Note also that the critical points are (generalized) geodesics [Eel78, Eel88].

The general form of the harmonic energy with $p = 2$ has been used successfully, for example, in computer graphics to find smooth maps between two given (triangulated) surfaces (normally a surface and the complex plane) [Eck95, Hak99, Zha99]. In this case, the search is indeed for the critical point, that is, for the harmonic map between the surfaces. This can be done, for example, via finite elements [Hak99]. In our case, the problem is different. We already have a candidate map, the original image of directions $\mathbf{u}(x, y, 0)$, and we want to compute a multiscale or regularized version of it. That is, we are not interested in only the harmonic map between the domain in $\mathbf{R}^2$ and S^{n-1} (the critical point of the energy) but also the process of computing this map via partial differential equations. More specifically, we are interested in the gradient-descent-type flow of the harmonic energy of Eq. (8.7). This is partially motivated by the fact that diffusion equations for gray-valued images can be obtained as gradient descent flows acting on real-valued data (see, for example, [Bla98, Per90, Rud92, You96]). Isotropic diffusion is just the gradient descent of the L_2 norm of the image gradient, while anisotropic diffusion can be interpreted as the gradient descent flow of more robust norms acting on the image gradient.

For the most popular case of $p = 2$, the Euler-Lagrange equation corresponding to Eq. (8.7) is a simple formula based on Δ_M, the Laplace-Beltrami operator of M, and $A_N(\mathbf{u})$, the second fundamental form of N (assumed to be embedded in $\mathbf{R}^k$) evaluated at u (e.g., [Eel78, Str85]): $\Delta_M \mathbf{u} + A_N(\mathbf{u})\langle \nabla_M \mathbf{u}, \nabla_M \mathbf{u}\rangle = 0$. This leads to a gradient descent type of flow, that is,

$$\frac{\partial \mathbf{u}}{\partial t} = \Delta_M \mathbf{u} + A_N(\mathbf{u})\langle \nabla_M \mathbf{u}, \nabla_M \mathbf{u}\rangle. \tag{8.8}$$

In the following sections, we present the gradient descent flows for our particular energy of Eq. (8.6), that is, for M being a domain in $\mathbf{R}^2$ and N equal to S^{n-1}.[5] Below we concentrate on the cases of $p = 2$ (isotropic) and $p = 1$ (anisotropic). The use of $p = 2$ corresponds to the classical heat flow from linear scale-space theory [Koe84, Wit83], while the case $p = 1$ corresponds to the *total variation* flow studied by Rudin et al. [Rud92].

8.3.2 Isotropic Diffusion

It is easy to show that for $p = 2$, the gradient descent flow corresponding to Eq. (8.6) is given by the set of coupled PDEs[6]

$$\frac{\partial u_i}{\partial t} = \Delta u_i + u_i \|\nabla \mathbf{u}\|^2, \qquad 1 \le i \le n. \tag{8.9}$$

This system of coupled PDEs defines the isotropic multiscale representation of $u(x, y, 0)$, which is used as the initial data to solve Eq. (8.9). (Boundary conditions are also added in the case of finite domains.) The first part of Eq. (8.9) comes from the variational form, while the second comes from the constraint. As expected, the first part is decoupled between components u_i and is linear, while the coupling and nonlinearity come from the constraint.

For $p = 2$, we have the following important results from the literature on harmonic maps:

Existence. Existence results for harmonic mappings have already been reported [Eel64] for a particular selection of the target manifold N. Struwe [Str85] showed,

[5] For data such as surface normals and principal directions, M is a surface in 3D and the general flow given by Eq. (8.8) is used. This flow can be implemented using numerical techniques to compute ∇_M and Δ_M on triangulated or implicit surfaces (using [Hak99] and results developed by S. Osher and colleagues at UCLA).

[6] If $n = 2$, that is, if we have 2D directions, then it is easy to show that for $\mathbf{u}(x, y) = (\cos\theta(x, y), \sin\theta(x, y))$, the energy in Eq. (8.6) becomes $\iint_\Omega (\theta_x^2 + \theta_y^2)^{p/2}\,dx\,dy$. For $p = 2$ we then obtain the linear heat flow on θ ($\theta_t = \Delta\theta$) as the corresponding gradient descent flow, as expected from Perona's results in [Per98]. This, of course, is directly derived from the theory of harmonic maps. When the data are not regular, the direction and orientation formulations are not necessarily equivalent.

in one of the classical papers in the area, that for initial data with finite energy [as measured by Eq. (8.7)], M as a two dimensional manifold with $\partial M = \emptyset$ (manifold without boundary), and $N = S^{n-1}$, there is a unique solution to the general gradient descent flow. Moreover, this solution is regular, with the exception of a finite number of isolated points, and the harmonic energy is decreasing in time. If the initial energy is small, the solution is completely regular and converges to a constant value. (The results actually holds for any compact N.) This uniqueness result was later extended to manifolds with smooth $\partial M \neq \emptyset$ and for weak solutions [Fre95]. Recapping, there is a unique weak solution to Eq. (8.9) [weak solutions defined in natural spaces, $H^{1,2}(M,N)$], and the set of possible singularities is finite. These solutions decrease the harmonic energy. The result is not completely true for M with dimension greater than 2, and this was investigated, for example, by Chen [Che89]. Global weak solutions exist, for example, for $N = S^{n-1}$, although there is no uniqueness for the general initial value problem [Cor90]. Results on the regularity of the solution, for a restricted suitable class of weak solutions, to the harmonic flow for high dimensional manifolds M into S^{n-1} have been recently reported [Che95, Fel94]. In this case, it is assumed that the weak solutions hold a number of given energy constraints.

Singularities in 2D. If $N = S^1$ and the initial and boundary conditions are well behaved (smooth, finite energy), then the solution of the harmonic flow is regular. This is the case, for example, for smooth $2D$ image gradients and $2D$ optical flow.

Singularities in 3D. Unfortunately, for $n = 3$ in Eq. (8.9) (that is $N = S^2$, $3D$ vectors), smooth initial data can lead to singularities in finite time [Cha92]. Chang et al. showed examples in which the flow of Eq. (8.9), with initial data $\mathbf{u}(x,y,0) = \mathbf{u}(x,y) \in C^1(D^2, S^2)$ (with D^2 being the unit disk on the plane) and boundary conditions $\mathbf{u}(x,y,t)|_{\partial D^2} = \mathbf{u}|_{\partial D^2}$, develops singularities in finite time. The idea is to use as original data $\mathbf{u}$ a function that covers S^2 more than once. From the point of view of the harmonic energy, the solution is "giving up on" regularity to reduce energy.

Singularities topology. Since singularities can occur, it is then interesting to study them [Bre86, Har97, Qin95]. For example, Brezis et al. studied the value of the harmonic energy when the singularities of the critical point are prescribed (the map is from R^3 to S^2 in this case).[7] Qing characterized the energy at the singularities. A recent review on the singularities of harmonic maps was prepared by Hardt [Har97]. (Singularities for more general energies were studied, for example, by Pismen and Rubinstein [Pis91].) The results reported there can be used to characterize the behavior of the multiscale representation of high dimensional directions, although these results mainly address the shape of the harmonic map, that is, the critical point of the harmonic energy and not the flow. Of course, for

[7]Perona suggested a look at this line of work to analyze the singularities of the orientation diffusion flow.

the case of M being of dimension two, which corresponds to Eq. (8.9), we have Struwe's results mentioned above.

8.3.3 Anisotropic Diffusion

The picture becomes even more interesting for the case $1 \leq p < 2$. Now the gradient descent flow corresponding to Eq. (8.6), in the range $1 < p < 2$ (and formally for $p = 1$), is given by the set of coupled PDEs

$$\frac{\partial u_i}{\partial t} = \operatorname{div}\left(\|\nabla \mathbf{u}\|^{p-2} \nabla u_i\right) + u_i \|\nabla \mathbf{u}\|^p, \qquad 1 \leq i \leq n. \tag{8.10}$$

This system of coupled PDEs defines the anisotropic multiscale representation of $u(x, y, 0)$, which is used as the initial datum to solve Eq. (8.10). In contrast with the isotropic case, now both terms in Eq. (8.10) are nonlinear and include coupled components.

The case of $p \neq 2$ in Eq. (8.7) has been studied less in the literature. When M is a domain in $\mathbf{R}^m$ and $N = S^{n-1}$, the function $\mathbf{v}(X) \triangleq X/\|X\|$, $X \in \mathbf{R}^m$, is a critical point of the energy for $p \in \{2, 3, \ldots, m-1\}$, for $p \in [m-1, m)$ [this interval includes the energy case that leads to Eq. (8.10)], and for $p \in [2, m - 2\sqrt{m-1}]$ [Har97]. For $n = 2$ and $p = 1$, the variational problem has also been investigated by Giaquinta et al. [Gia93], who addressed, among other things, the correct spaces in which to perform the minimization [in the scalar case, $BV(\Omega, \mathbf{R})$ is used] and the existence of minimizers. Of course, we are more interested in the results for the flow given by Eq. (8.10), not just its corresponding energy. Some results exist for $1 < p < \infty$, $p \neq 2$, showing in a number of cases the existence of local solutions that are not smooth. To the best of our knowledge, the case of $1 \leq p < 2$, and in particular $p = 1$, has not been fully studied for the evolution equation.

8.3.4 Examples

Although advanced specialized numerical techniques to solve Eq. (8.6) and its corresponding gradient descent flow have been developed (e.g., [Alo91]), as a first approximation we can basically use the algorithms developed for scalar isotropic and anisotropic diffusion without the unit norm constraint (e.g., [Rud92]) to implement Eqs. (8.9) and (8.10) [Coh87]. Although these equations preserve the unit norm, numerical errors might violate the constraint. Therefore, between every two steps of the numerical implementation of these equations we add a renormalization step [Coh87].

A number of techniques exist for visualizing vectors: (1) Arrows indicating the vector direction are very illustrative but can be used only for sparse images; they are not very informative for dense data such as gradients or optical flow. (2) The HSV color mapping (applied to orientation) is useful for visualizing whole images of directions while also being able to illustrate details such as small noise. (3) Line integral convolution (LIC) [Cab93] is based on locally integrating at each pixel, in the directions given by the directional data, the values of a random image. The

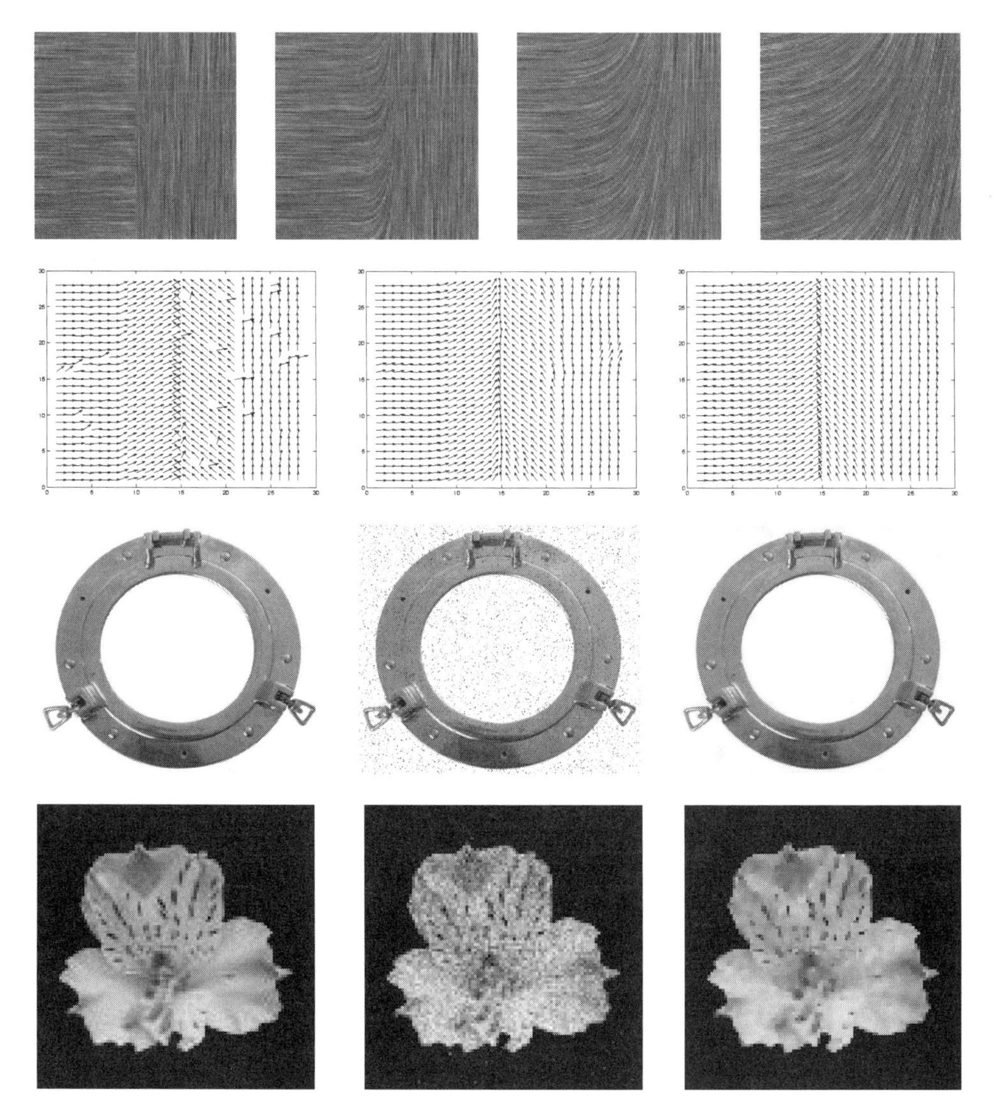

Figure 8.4: Examples illustrating the ideas of direction diffusion (see text for details). (From [Tan00a]. © 2000 Kluwer Academic Publishers.)

LIC technique gives the general form of the flow, while the color map is useful to detect small noise in the direction (orientation) image.

Figure 8.4 (also see color insert) shows a number of simple examples to illustrate the general ideas introduced in this chapter, as well as examples for color image denoising. The top row shows, using LIC, first an image with two regions having different (2D) orientations (original), followed by the results of isotropic diffusion for 200, 2000, and 8000 iterations (scale-space). Note how the edge in the directional data is being smoothed out. The horizontal and vertical directions are being smoothed out to converge to the diagonal average.

The second row shows the result of removing noise in the directional data. The original noisy image is shown first, followed by the results with isotropic and anisotropic smoothing. Note how the anisotropic flow gets rid of the noise (outliers) while preserving the rest of the data, while the isotropic flow also affects the data itself while removing the noise. Note that since the discrete theory developed by Perona [Per98] applies only to small changes in orientation, theoretically it cannot be applied to the images we have seen so far, all of which contain sharp discontinuities in the directional data (and the theory is only isotropic).

We can use the directions diffusion framework to process color images. That is, we separate the color direction from the color magnitude and use the harmonic flows to smooth the direction (chromaticity), relying on standard edge preserving scalar filters to smooth the magnitude (brightness). This gives very good color image denoising results, as shown in the bottom two rows of Fig. 8.4, which show, from left to right, the original, noisy, and reconstructed images, reproduced here in gray-levels only. See Tang et al. [Tan00b] for additional examples, comparisons with the literature, and details. Other directional filters for color denoising, based on vectorial median filtering, can be found elsewhere (e.g., [Cas00, Tra93, Tra96]).

8.3.5 Discussion

We observe that the theory of harmonic maps provides a fundamental framework for directional diffusion in particular and diffusion on general manifolds in general. This framework opens a whole new research area in PDE-based image processing, both in the theoretical and practical arenas. For example, from the theoretical point of view, we need to perform a complete analysis of the harmonic energy and gradient descent flow for $p = 1$, the anisotropic case. On a more practical level, it would be interesting to include the harmonic energy as a regularization term in more general variational problems, such as the general optical flow framework, and to use the theory of harmonic maps for other image processing problems, such as denoising data defined on 3D surfaces.

Acknowledgments

The original work on geodesic snakes was developed with Profs. V. Caselles and R. Kimmel. The direction diffusion work is the result of collaboration with Prof. Caselles and B. Tang. The work on vector edges was jointly developed with Prof. D. Ringach. D. H. Chung helped with many of the images in this chapter. We thank Profs. R. Kohn, K. Rubinstein, S. Osher, T. Chan, J. Shen, L. Vese, D. Heeger, V. Interrante, and P. Perona, and B. Cabral and C. Leedom for their comments. This work was partially supported by Office of Naval Research grant ONR-N00014-97-1-0509, the Office of Naval Research Young Investigator Award, the Presidential Early Career Awards for Scientists and Engineers (PECASE), a National Science Foundation CAREER Award, the National Science Foundation Learning and Intelligent Systems (LIS) Program, and NSF-IRI-9306155 (Geometry Driven Diffusion).

References

[Alo91] F. Alouges. An energy decreasing algorithm for harmonic maps. In *Nematics* (J. M. Coron et al., eds.), pp. 1-13. NATO ASI Series, Kluwer, Dordrecht, The Netherlands (1991).

[Alv92] L. Alvarez, P. L. Lions, and J. M. Morel. Image selective smoothing and edge detection by nonlinear diffusion. *SIAM J. Numer. Anal.* **29**, 845–866 (1992).

[Alv93] L. Alvarez, F. Guichard, P. L. Lions, and J. M. Morel. Axioms and fundamental equations of image processing. *Arch. Rational Mechan.* **123**, 199-257 (1993).

[Bla98] M. Black, G. Sapiro, D. Marimont, and D. Heeger. Robust anisotropic diffusion. *IEEE Trans. Image Process.* **7**(3), 421–432 (March 1998).

[Blo98] P. Blomgren and T. F. Chan. Color TV: Total variation methods for restoration of vector valued images, *IEEE Trans. Image Process.* **7**(3), 304–309 (March 1998).

[Bre86] H. Brezis, J. M. Coron, and E. H. Lieb. Harmonic maps with defects. *Commun. in Mathemat. Phys.* **107**, 649–705 (1986).

[Cab93] B. Cabral and C. Leedom. Imaging vector fields using line integral convolution. In *Computer Graphics* (Proc. SIGGRAPH) (1993).

[Cas93] V. Caselles, F. Catte, T. Coll, F. Dibos. A geometric model for active contours. *Numerische Mathematik* **66**, 1–31 (1993).

[Cas97a] V. Caselles, R. Kimmel, and G. Sapiro. Geodesic active contours. *Intl. J. Comput. Vision* **22**(1), 61–79 (1997).

[Cas97b] V. Caselles, R. Kimmel, G. Sapiro, and C. Sbert. Minimal surfaces based object segmentation. *IEEE Trans. Patt. Anal. Machine Intell.* **19**(4), 394-398 (1997).

[Cas98] V. Caselles, J. M. Morel, G. Sapiro, and A. Tannenbaum, Eds. Special Issue on Partial Differential Equations and Geometry-Driven Diffusion in Image Processing and Analysis. *IEEE Trans. Image Process.* **7**(3), 421–432 (March 1998).

[Cas00] V. Caselles, G. Sapiro, and D. H. Chung. Vector median filters, morphology, and PDEs: Theoretical connections. *J. Mathematical Imaging and Vision* **12**, 108–120 (2000).

[Cat92] F. Catte, P.-L. Lions, J.-M. Morel, and T. Coll. Image selective smoothing and edge detection by nonlinear diffusion. *SIAM J. Numer. Anal.* **29**, 182–193 (1992).

[Cha92] K. C. Chang, W. Y. Ding, and R. Ye. Finite-time blow-up of the heat flow of harmonic maps from surfaces. *J. Differential Geom.* **36**, 507–515 (1992).

[Cha99] T. Chan and J. Shen. Variational restoration of non-flat image features: Models and algorithms. University of California at Los Angeles, CAM-TR (May 1999).

[Che89] Y. Chen. The weak solutions of the evolution problems of harmonic maps. *Math. Z.* **201**, 69–74 (1989).

[Che95] Y. Chen, J. Li, and F. H. Lin. Partial regularity for weak heat flows into spheres. *Commun. Pure Appl. Mathemat.* **48**, 429–448 (1995).

[Chu00] D. H. Chung and G. Sapiro. On the level lines and geometry of vector valued images. *IEEE Signal Process. Lett.*, to appear.

[Coh87] R. Cohen, R. M. Hardt, D. Kinderlehrer, S. Y. Lin, and M. Luskin. Minimum energy configurations for liquid crystals: computational results. In *Theory and Applications of Liquid Crystals*, J. L. Ericksen and D. Kinderlehrer, Eds. pp. 99–121, Springer-Verlag, New York (1987).

[Coh91] L. D. Cohen. On active contour models and balloons. *Comput. Vision, Graph. Image Process.: Image Understanding* **53**, 211–218 (1991).

[Coh97] L. D. Cohen and R. Kimmel. Global minimum for active contours models. *Intl. J. Comput. Vision* **24**, 57–78 (1997).

[Cor90] J. M. Coron. Nonuniqueness for the heat flow of harmonic maps. *Ann. Inst. H. Poincaré, Analyse Non Linéaire* **7**(4), 335–344 (1990).

[Cra92] M. G. Crandall, H. Ishii, and P. L. Lions. User's guide to viscosity solutions of second order partial linear differential equations. *Bull. Amer. Mathemat. Soc.* **27**, 1–67 (1992).

[Cum91] A. Cumani. Edge detection in multispectral images. *Comput. Vision Graph. Image Process.* **53**, 40–51 (1991).

[DiZ86] S. Di Zenzo. A note on the gradient of a multi-image. *Comput. Vision Graph. Image Process.* **33**, 116–125 (1986).

[Eck95] M. Eck, T. DeRose, T. Duchamp, H. Hoppe, M. Lounsbery, and W. Stuetzle. Multiresolution analysis of arbitrary meshes. In *Computer Graphics* (Proc. SIGGRAPH), pp. 173–182 (1995).

[Eel78] J. Eells and L. Lemarie. A report on harmonic maps, *Bull. London Mathemat. Soc.* **10**(1), 1–68 (1978).

[Eel88] J. Eells and L. Lemarie. Another report on harmonic maps, *Bull. London Mathemat. Soc.* **20**(5), 385–524 (1988).

[Eel64] J. Eells and J. H. Sampson. Harmonic mappings of Riemannian manifolds, *Amer. J. Mathemat.* **86**, 109–160 (1964).

[Fau98] O. D. Faugeras and R. Keriven. Variational principles, surface evolution, PDEs, level-set methods, and the stereo problem. *IEEE Trans. Image Process.* **7**(3), 336–344 (March 1998).

[Fel94] M. Feldman. Partial regularity for harmonic maps of evolutions into spheres. *Commun. Partial Differential Eqs.* **19**, 761–790 (1994).

[Fre95] A. Freire. Uniqueness for the harmonic map flow in two dimensions. *Cal. Var.* **3**, 95–105 (1995).

[Gia93] M. Giaquinta, G. Modica, and J. Soucek. Variational problems for maps of bounded variation with values in S^1. *Cal. Var.* **1**, 87–121 (1993).

[Gra95] G. H. Granlund and H. Knuttson, *Signal Processing for Computer Vision*, Kluwer, Boston, MA (1995).

[Gui95] F. Guichard and J. M. Morel, *Introduction to Partial Differential Equations in Image Processing.* Tutorial Notes, IEEE Intl. Conf. Image Processing (Washington, DC, October 1995).

[Hak99] S. Haker, S. Angenent, A. Tannenbaum, R. Kikinis, G. Sapiro, and M. Halle. Conformal surface parametrization for texture mapping. University of Minnesota IMA Preprint Series No. 1611 (April 1999).

[Har85] R. M. Haralick and L. G. Shapiro. Image segmentation techniques. *Comput. Vision Graphics Image Process.* **29**, 100–132 (1985).

[Har87] R. M. Hardt and F. H. Lin. Mappings minimizing the L^p norm of the gradient. *Commun. Pure Appl. Mathemat.* **40**, 555–588 (1987).

[Har97] R. M. Hardt. Singularities of harmonic maps. *Bull. Amer. Mathemat. Soc.* **34**(1), 15–34 (1997).

[Kas88] M. Kass, A. Witkin, and D. Terzopoulos. Snakes: Active contour models. *Intl. J. Comput. Vision* **1**, pp. 321-331, 1988.

[Kim95] B. B. Kimia, A. Tannenbaum, and S. W. Zucker. Shapes, shocks, and deformations, I. *Intl. J. Comput. Vision* **15**, 189–224 (1995).

[Kis95] S. Kichenassamy, A. Kumar, P. Olver, A. Tannenbaum, and A. Yezzi. Gradient flows and geometric active contour models. In *Proc. IEEE Intl. Conf. Computer Vision*, pp. 810–815 (Cambridge, June 1995).

[Koe84] J. J. Koenderink. The structure of images. *Biolog. Cybernet.* **50**, 363–370 (1984).

[Kre59] E. Kreyszig. *Differential Geometry.* University of Toronto Press (1959).

[Lee91] H.-C. Lee and D. R. Cok. Detecting boundaries in a vector field. *IEEE Trans. Signal Process.* **39**, 1181–1194 (1991).

[Lee92] T. S. Lee, D. Mumford, and A. L. Yuille. Texture segmentation by minimizing vector-valued energy functionals: The coupled-membrane model. *Proc. European Conference on Computer Vision '92*, In Lecture Notes in Computer Science No. 588, pp. 165–173, Springer-Verlag, New York (1992).

[Lin94] T. Lindeberg. *Scale-Space Theory in Computer Vision.* Kluwer, Dordrecht, The Netherlands (1994).

[Lor99] L. M. Lorigo, O. Faugeras, W. E. L. Grimson, R. Keriven, and R. Kikinis. Segmentation of bone in clinical knee MRI using texture-based geodesic active contours. In *Medical Image Computing and Computer-Assisted Intervention, MICCAI '98*, pp. 1195–1204, Springer (1998).

[Mal95] R. Malladi, J. A. Sethian, and B. C. Vemuri. Shape modeling with front propagation: A level set approach. *IEEE Trans. Patt. Anal. Machine Intell.* **17**, 158–175 (1995).

[Mor94] J.-M. Morel and S. Solimini. *Variational Methods in Image Segmentation.* Birkhauser, Boston (1994).

[Mum89] D. Mumford and J. Shah. Optimal approximations by piecewise smooth functions and variational problems. *Commun. Pure Appl. Mathemat.* **42**, 577–685 (1989).

[Nit92] M. Nitzberg and T. Shiota. Nonlinear image filtering with edge and corner enhancement. *IEEE Trans. Patt. Anal. Machine Intell.* **14**, 826–833 (1992).

[Osh88] S. J. Osher and J. A. Sethian. Fronts propagation with curvature dependent speed: Algorithms based on Hamilton-Jacobi formulations. *J. Computa. Phys.* **79**, 12–49 (1988).

[Osh90] S. Osher and L. I. Rudin. Feature-oriented image enhancement using shock filters. *SIAM J. Numer. Anal.* **27**, 919–940 (1990).

[Par97] N. Paragios and R. Deriche. A PDE-based level-set approach for detection and tracking of moving objects. INRIA Tech. Report No. 3173, Sophia-Antipolis (May 1997).

[Per90] P. Perona and J. Malik. Scale-space and edge detection using anisotropic diffusion. *IEEE Trans. Patt. Anal. Machine Intell.* **12**, 629–639 (1990).

[Per98] P. Perona. Orientation diffusion. *IEEE Trans. Image Process.* **7**(3), 457–467 (March 1998).

[Pis91] L. M. Pismen and J. Rubinstein. Dynamics of defects. In *Nematics*, J. M. coron et al. eds. pp. 303–326, NATO ASI Series, Kluwer, Dordrecht, The Netherlands (1991).

[Qin95] J. Qing. On singularities of the heat flow for harmonic maps from surfaces into spheres. *Commun. Anal. Geom.* **3**(2), 297–315 (1995).

[Rom94] B. Romeny (ed.). *Geometry Driven Diffusion in Computer Vision.* Kluwer, Dordrecht, The Netherlands (1994).

[Rud92] L. I. Rudin, S. Osher, and E. Fatemi. Nonlinear total variation based noise removal algorithms. *Physica D* **60**, 259–268 (1992).

[Rud94] L. I. Rudin and S. Osher. Total variation based image restoration with free local constraints. In *Proc. IEEE Intl. Conf. Image Processing.* Vol I, pp. 31–35 (Austin, Texas, 1994).

[Sap96a] G. Sapiro. From active contours to anisotropic diffusion: connections between the basic PDE's in image processing. In *Proc. IEEE Intl. Conf. on Image Processing*, pp. 477–480 (Lausanne, Switzerland, 1996).

[Sap96b] G. Sapiro and D. Ringach. Anisotropic diffusion of multivalued images with applications to color filtering. *IEEE Trans. Image Process.* **5**, 1582–1586 (1996).

[Sap97a] G. Sapiro. Color snakes. *Comput. Vision Image Understanding* **68**(2), 247–253 (1997).

[Sap97b] G. Sapiro and V. Caselles. Histogram modification via differential equations. *J. Differential Eqs.* **135**(2), 238–268 (1997).

[Sha95] J. Shah. Recovery of shapes by evolution of zero-crossings. Tech. Report, Northeastern Univ. Math. Dept., Boston, MA (1995).

[Soc98] N. Sochen, R. Kimmel, and R. Malladi. A general framework for low-level vision. *IEEE Trans. Image Process.* **7**(3), 310–318 (1998).

[Str85] M. Struwe. On the evolution of harmonic mappings of Riemannian surfaces. *Comment. Math. Helvetici* **60**, 558–581 (1985).

[Tan00a] B. Tang, G. Sapiro, and V. Caselles. Diffusion of general data on non-flat manifolds via harmonic maps theory: The direction diffusion case. *Intl. J. Comput. Vision* **36**, 149–161 (2000).

[Tan00b] B. Tang, G. Sapiro, and V. Caselles. Color image enhancement via chromaticity diffusion. *IEEE Trans. Image Process.*, to be published.

[Tek95] H. Tek and B. B. Kimia. Image segmentation by reaction-diffusion bubbles. In *Proc. IEEE Intl. Conf. Computer Vision*, pp. 156–162 (Cambridge, June 1995).

[Ter88] D. Terzopoulos, A. Witkin, and M. Kass. Constraints on deformable models: recovering 3D shape and nonrigid motions. *Artific. Intell.* **36**, 91–123 (1988).

[Tra93] P. E. Trahanias and A. N. Venetsanopoulos. Vector directional filters: a new class of multichannel image processing filters. *IEEE Trans. Image Process.* **2**, 528–534 (1993).

[Tra96] P. E. Trahanias, D. Karakos, and A. N. Venetsanopoulos. Directional processing of color images: Theory and experimental results. *IEEE Trans. Image Process.* **5**(6), 868–880 (June 1996).

[Wei96] J. Weickert. Foundations and applications of nonlinear anisotropic diffusion filtering. *Zeitscgr. Angewandte Math. Mechan.* **76**, pp. 283-286, 1996.

[Whi94] R. T. Whitaker and G. Gerig. Vector-valued diffusion. In *Geometry Driven Diffusion in Computer Vision* (B. ter Haar Romeny, ed.), pp. 93–134. Kluwer, Boston (1994).

[Whi95] R. T. Whitaker. Algorithms for implicit deformable models. In *Proc. IEEE Intl. Conf. on Computer Vision* (Cambridge, June 1995).

[Wit83] A. P. Witkin. Scale-space filtering. *Intl. Joint Conf. Artificial Intelligence*, pp. 1019–1021 (Karlsruhe, Germany, 1983).

[You96] Y. L. You, W. Xu, A. Tannenbaum, and M. Kaveh. Behavioral analysis of anisotropic diffusion in image processing. *IEEE Trans. Image Process.* **5**(11), 1539–1553 (November 1996).

[Zha99] D. Zhang and M. Hebert. Harmonic maps and their applications in surface matching. In *Proc. IEEE Computer Vision Pattern Recognition* (Colorado, June 1999).

[Zhu95] S. C. Zhu, T. S. Lee, and A. L. Yuille. Region competition: Unifying snakes, region growing, energy/Bayes/MDL for multi-band image segmentation. In *Proc. IEEE Intl. Conf. on Computer Vision*, pp. 416–423 (Cambridge, MA, June 1995).

[Zuc76] S. W. Zucker. Region growing: Childhood and adolescence (Survey). *Comput. Vision Graph. Image Process.* **5**, 382–399 (1976).

9

Region-Based Filtering of Images and Video Sequences: A Morphological Viewpoint

PHILIPPE SALEMBIER

Department of Signal Theory and Communications
Universitat Politècnica de Catalunya
Barcelona, Spain

9.1 Introduction

Data and signal modeling for images and video sequences is experiencing important developments. Part of this evolution is due to the need to support a large number of new multimedia services. Traditionally, digital images were represented as rectangular arrays of pixels, and digital video was seen as a continuous flow of digital images. New multimedia applications and services imply a representation that is closer to the real world or, at least, that takes into account part of the process that has created the digital information. Content-based compression and indexing are two typical examples of applications for which new modeling strategies and processing tools are necessary.

For content-based image or video compression, the representation based on an array of pixels is not appropriate if one wants to be able to act on objects in the image, to encode differently the areas of interest, or to assign different behaviors to the entities represented in the image. In these applications, the notion of object is essential. As a consequence, the data modeling has to be modified and, for example, has to include regions of arbitrary shapes to represent objects.

Content-based indexing applications are also facing the same kind of challenges. For instance, the video representation based on a flow of frames is inadequate for a large number of video indexing applications. Among the large set of functionalities involved in a retrieval application, consider, for example, browsing. The browsing functionality should go far beyond the "fast forward" and "fast reverse" allowed by VCRs. One would like to have access to a table of contents of the video and be able to jump from one item to another. This kind of functionality implies at least a structuring of the video in terms of individual shots and scenes. Of course, indexing and retrieval involve also a structuring of the data in terms of objects, regions, semantic notions, etc.

In both of these examples, the data modeling has to take into account part of the creation process: an image is created by projection of a visual scene composed of 3D objects onto a 2D plane. Modeling the image in terms of regions is an attempt to know the projection of the 3D object boundaries in the 2D plane. Video shots detection also aims at finding what has been done during the video editing process and where boundaries between elementary components have been introduced. In both cases, the notion of region turns out to be central in the modeling process. Note that regions may be spatial connected components but also temporal or spatiotemporal connected components in the case of video.

Besides the modeling issue, it has to be recognized that most image processing tools are not suited to region-based representations. For example, the vast majority of low level processing tools such as filters are very closely related to the classical pixel-based representation of signals. Typical examples include linear convolution with an impulse response, the median filter, morphological operators based on erosion and dilation with a structuring element, etc. In all cases, the processing strategy consists of modifying the values of individual pixels by a function of the pixel values in a local window.

Early examples of region-based processing can be found in the literature in the field of segmentation. For example, the classical split and merge algorithm [Hor74] first defines a set of elementary regions (the split process) and then interacts directly on these regions, allowing them to merge under certain conditions.

Recently, a set of morphological filtering tools called connected operators has received much attention. Connected operators are region-based filtering tools because they do not modify individual pixel values but directly act on the connected components of the space where the image is constant, the so-called *flat zones*. Intuitively, connected operators can remove boundaries between flat zones but cannot add new boundaries nor shift existing boundaries. The related literature rapidly grows and involves theoretical studies (Chapter 10 herein and [Cre95, Hei97a, Mat97, Mey98a, Mey98b, Ron98, Ser93, Ser98]), algorithm developments [Bre96, Gom99, Sal98, Sal00, Vin93a, Vin93b], and applications [Cre97, Sal92, Sal95, Vil98]. The goals of this chapter are (1) to provide an introduction to connected operators for gray level images and video sequences and (2) to discuss the techniques and algorithms that up to now have been the most successful within the framework of practical applications.

The organization of the chapter is as follows: Section 9.2 introduces the notation and highlights the main drawbacks of classical filtering strategies. Section 9.3 presents the basic notions related to connected operators and discusses some early examples of connected operators. In practice, the two most successful strategies to define connected operators are based either on reconstruction processes or on tree representations, discussed in Sec. 9.4 and 9.5, respectively. Finally, conclusions are given in Sec. 9.6.

9.2 Classical Filtering Approaches

In this section, we define the notation to be used herein and review some of the basic properties of interest in this chapter [Ser82, Ser88]. We deal exclusively with discrete images $f(n)$ or video sequences $f_t(n)$, where n denotes the pixel or space coordinate (a vector in the case of 2D images) and t the time instant in the case of a video sequence. In the lattice of gray level functions, an image f is said to be smaller than an image g if and only if

$$f \leq g \Longleftrightarrow \forall n, \qquad f(n) \leq g(n). \tag{9.1}$$

An operator ψ acting on an input f is said to be:

- *increasing*: $\forall f, g, \quad f \leq g \Rightarrow \psi(f) \leq \psi(g)$
  (The order relationship between images is preserved by the filtering.)
- *idempotent*: $\forall f, \qquad \psi(\psi(f)) = \psi(f)$
  (Iteration of the filtering process is not needed)
- *extensive*: $\forall f, \qquad f \leq \psi(f)$
  (The output image is always brighter (greater) than the input image.)
- *anti-extensive*: $\forall f, \qquad \psi(f) \leq f$
  (The output image is always darker (smaller) than the input image.)
- *a morphological filter*: if it is increasing and idempotent
- *an opening*: if it is an anti-extensive morphological filter
- *a closing*: if it is an extensive morphological filter
- *self-dual*: $\forall f, \qquad \psi(f) = -\psi(-f)$
  (The operator processes in a symmetric fashion bright and dark components.)

Almost all filtering techniques commonly used in image processing are defined by a computation rule and a specific signal $h(n)$ that may be called the impulse response, the window, or the structuring element. Let us review these three classical cases:

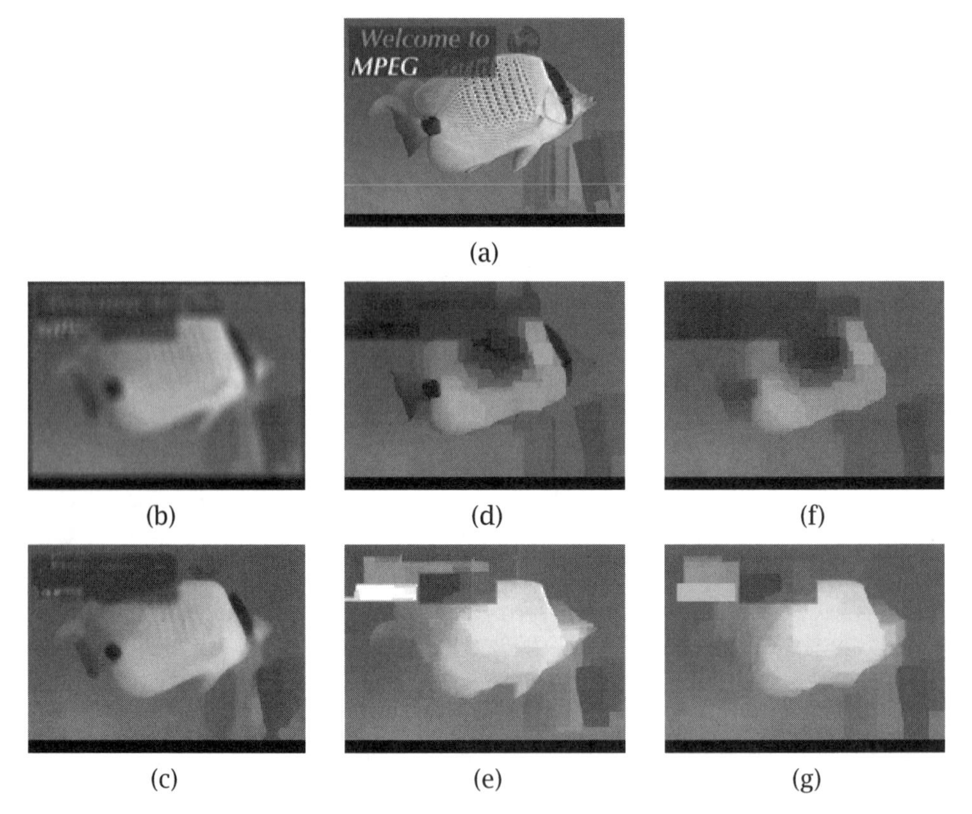

Figure 9.1: Examples of classical filtering strategies: (a) original image, and results of its being filtered using (b) low-pass filter (7×7 average), (c) median (5×5), (d) opening (5×5), (e) closing (5×5), (f) opening followed by a closing (5×5), and (g) closing followed by an opening (5×5).

- Linear convolution and impulse response: the output of any linear translation-invariant system is given by $\psi_h(f)(n) = \sum_{k=-\infty}^{\infty} h(k)f(n-k)$. The impulse response $h(n)$ defines the properties of the filter. Linear filters do not have very interesting lattice properties. They are simply self-dual. In practice, this means that dark and bright components are processed in a symmetrical way. Moreover, in the particular case of *ideal* filters, they are idempotent (but also unstable). Examples of linear filtering results are shown in Fig. 9.1b. The original image, shown in Fig. 9.1a, represents a fish with a written message. As can be seen, most of the details of the original image are attenuated by the filter (average of size 7×7). However, details are not really removed but simply blurred. The characteristics of the blurring are directly related to the extension and shape of the impulse response.

- Median filter and window: The output of a median filter over a window W is defined by $\psi_W(f)(n) = \text{MEDIAN}_{k\in W}\{f(n-k)\}$. Here also the basic properties of the filter are defined by its window. The median filter is increasing and self-dual but not idempotent. An example is shown in Fig. 9.1c. Here, small details are actually removed (for example the texture of the fish). The major drawback of this filtering strategy is that every region tends to have a round shape after filtering. This effect is due to the shape of the window combined with the median processing.

- Morphological erosion–dilation and structuring elements: morphological dilation by a structuring element $h(n)$ is defined in a way similar to the convolution: $\delta_h(f)(n) = \bigvee_{k=-\infty}^{\infty}[h(k) + f(n-k)]$, where $\bigvee$ denotes the supremum (or maximum in the discrete case). The erosion is given by $\varepsilon_h(f)(n) = \bigwedge_{k=-\infty}^{\infty}[h(k) - f(n+k)]$, where $\bigwedge$ denotes the infimum (or minimum in the discrete case).

  Both operators are increasing. In practice, they are seldom used on their own because they do not preserve the position of contours. For example, the dilation enlarges the size of bright components and decreases the size of dark components by displacing their contours. However, they provide a simplification effect: a dilation (erosion) removes dark (bright) components that do not fit within the structuring element. Based on these two primitives, morphological opening and closing can be constructed.

  The opening is given by $\gamma_h(f) = \delta_h(\varepsilon_h(f))$ and the closing by $\varphi_h(f) = \varepsilon_h(\delta_h(f))$. These operators are morphological filters (that is, increasing and idempotent at the same time). Moreover, the opening is antiextensive (it removes bright components), whereas the closing is extensive (it removes dark components). The simplification effect is similar to the one obtained with the erosion and the dilation but contours are not displaced. Processing results are shown in Figs. 9.1d and 9.1e. In the case of opening (closing) with a square structuring element of size 5×5, small bright (dark) components have been removed. As can be seen, the contours remain sharp and centered on their original position. However, the shapes of the components that have not been removed are not perfectly preserved. In both examples, square shapes are clearly visible in the output image. This is due to the square shape of the structuring element.

  If both bright and dark components have to be removed, the classical approach is to concatenate an opening and a closing. Figures 9.1f and 9.1g illustrate the fact that compositions such as $\gamma(\varphi(.))$ or its dual $\varphi(\gamma(.))$[1] can be used to remove bright and dark details of the image, but they are not self-dual [$\varphi\gamma(f) \neq \gamma\varphi(f) = -\varphi\gamma(-f)$]. Moreover, the distortion introduced by the square structuring element is clearly visible in both examples.

Once a processing strategy has been selected (linear convolution, median, morphological operator, etc.), the filter design consists of carefully choosing a specific signal $h(n)$, which may be the impulse response, the window, or the structuring element. While most people would say that this is the heart of the filter design, our point here is to highlight that, in the framework of image processing, the use of

[1] It can be shown that $\gamma(\varphi(f)) = -\varphi(\gamma(-f))$

the signal $h(n)$ has some drawbacks. In all of the examples of Fig. 9.1, $h(n)$ was not related at all with the input signal, and its shape clearly introduced distortions in the output images. The visual effect of the distortion depends on the specific filter, but for a large range of applications requiring high precision on contours, none of these filtering strategies is acceptable.

To reduce the distortion, one possible solution is to adapt $h(n)$ to the local structures of the input signal. This solution may improve the results but still remains unacceptable in many circumstances (it still makes use of a signal that is independent of the input signal). An attractive solution to this problem is provided by connected operators. Most connected operators used in practice rely on a completely different filtering strategy: the filtering is done without using any specific signal such as an impulse response, a window, or a structuring element. In fact, the structures of the input signal are used to act on the signal itself. As a result, no distortion related to a priori selected signals is introduced in the output.

9.3 Connected Operators

9.3.1 Definitions and Basic Properties

Gray level connected operators act by merging flat zones. They cannot create new contours, and as a result, they cannot introduce in the output image a structure that is not present in the input image. Furthermore, they cannot modify the position of existing boundaries between regions and, therefore, have very good contour preservation properties.

Gray level connected operators as originally defined [Ser93] rely on the notion of partition of flat zones. A partition is a set of nonoverlapping connected components or regions that fills the entire space. We assume that the connectivity is defined on the digital grid by a translation-invariant, reflexive, transitive and symmetric relation.[2] Typical examples are the 4- and 8-connectivity. Let us denote by $\mathcal{P}$ a partition and by $\mathcal{P}(n)$ the region that contains pixel at position n. A partial order relationship among partitions can be created: $\mathcal{P}_1$ "is finer than" $\mathcal{P}_2$ (written as $\mathcal{P}_1 \sqsubseteq \mathcal{P}_2$) if $\forall n, \mathcal{P}_1(n) \subseteq \mathcal{P}_2(n)$.

It can be shown that the set of flat zones of an image f is a partition of the space, $\mathcal{P}_f$. Based on these notions, connected operators are defined as follows.

Definition 9.1. Connected operators: A gray level operator ψ is connected if the partition of flat zones of its input f is always finer than the partition of flat zones of its output, that is:

$$\mathcal{P}_f \sqsubseteq \mathcal{P}_{\psi(f)}, \qquad \forall f.$$

This definition clearly highlights the region-based processing of the operator since it states that regions of the output partition are created by union of regions of the

[2]In the context of connected operators, several studies have been carried out on the definition of less usual connectivities; the reader is referred to [Cre96, Hei97b, Ron98, Ser88, Ser98] for more details on this issue.

input partition. An alternative (and equivalent) definition of connected operators was introduced by Meyer [Mey98a] that enhances the role of the boundaries between regions and turns out to be very useful in deriving the notion of leveling (see Sec. 9.4.2).

Definition 9.2. Connected operators: A gray level operator ψ is connected if $\forall f$ input image and $\forall n, n'$ neighboring pixels,

$$\psi(f)(n) \neq \psi(f)(n') \Longrightarrow f(n) \neq f(n').$$

This definition simply states that if two neighboring pixels of the output image have two different gray level values, they also have two different gray level values in the input image; in other words, the operator cannot create new boundaries.

New connected operators can be derived from the combination of primitive connected operators. In particular, the following properties give a few construction rules:

Proposition 9.1. *Properties of connected operators:*

- *If ψ is a connected operator, its dual ψ^*, defined by $\psi^*(f) = -\psi(-f)$, is also connected.*
- *If ψ_1, ψ_2 are connected operators, $\psi_2\psi_1$ is also connected.*
- *If $\{\psi_i\}$ are connected operators, their supremum $\bigvee_i \psi_i$ and infimum $\bigwedge_i \psi_i$ are connected.*

9.3.2 Early Examples of Connected Operators

The first connected operator reported in the literature is known as *binary opening by reconstruction* [Kle76]. This operator eliminates the connected components that would be totally removed by an erosion with a given structuring element and leaves the other components unchanged. This filtering approach was proposed because it offers the advantage of simplifying the image (some components are removed) as well as preserving the contour information (the components that are not removed are perfectly preserved).

It can be shown that the process is increasing, idempotent, and antiextensive. It is therefore an opening. Moreover, it was called "by reconstruction" because of the algorithm used for its implementation. The operator is connected in the sense of Definition 9.1, assuming that the partition corresponding to a binary image is composed of the connected components of the set and of the background. If X is the original binary image, the first step is to compute an erosion with a structuring element B, $\varepsilon_B(X)$ of size k, $\varepsilon_B(X)$. This erosion is used to "mark" the connected components that should be preserved. In fact, the final result is obtained by progressively dilating the erosion inside the mask defined by the original image. This defines a series of images Y_k:

1. $Y_0 = \varepsilon_{B_k}(X)$.

2. $Y_k = \delta_C(Y_{k-1}) \bigcap X$, where C is a binary structuring element defining the connectivity, for example, the elementary square of 3×3 for the 8-connectivity and the elementary cross for the 4-connectivity.

3. Iterate Step 2 until idempotence.

The first gray level connected operator was obtained by a transposition of the previous approach to the lattice of gray level functions [Ser88, Vin93b]. It is known as an *opening by reconstruction of erosions.* The reconstruction process is iterative and creates a series of images g_k:

1. $g_0 = \varepsilon_h(f)$, where f is the input image and h a structuring element.

2. $g_k = \delta_C(g_{k-1}) \bigwedge f$, where C is a flat structuring element defining the connectivity, for example, the elementary square of 3×3 for the 8-connectivity and the elementary cross for the 4-connectivity.

3. Iterate Step 2 until idempotence.

It has been shown [Ser93] that this operator is connected. Intuitively, the erosion acts as a simplification step by removing small bright components. The reconstruction process restores the contours of the components that have not been completely removed by the erosion.

There are several ways to construct connected operators, and many new connected operators have been recently introduced. From a practical viewpoint, the most successful strategies rely either on reconstruction processes or on region-tree pruning. Operators resulting from these two strategies are discussed in Secs. 9.4 and 9.5. Note that the development of connected operators is still in an early stage, and sometimes it is difficult to study the theoretical properties of the operators, or it may even be difficult to define an operator without relying on a specific implementation. However, the connected operators discussed subsequently were selected because they have been used successfully in practical applications.

9.4 Connected Operators Based on Reconstruction Processes

9.4.1 Antiextensive Reconstruction and Connected Operators

The Antiextensive Reconstruction Process

The most classical way to construct connected operators is to use an antiextensive reconstruction process, defined as follows:

Definition 9.3. Antiextensive reconstruction: If f and g are two images (respectively called the "reference" and the "marker" image), the antiextensive reconstruction $\rho^{\downarrow}(g|f)$ of g *under* f is given by

$$g_k = \delta_C(g_{k-1}) \bigwedge f \quad \text{for } k > 0, \tag{9.2a}$$

and

$$\rho^{\downarrow}(g|f) = \lim_{k \to \infty} g_k, \tag{9.2b}$$

where $g_0 = g$ and δ_C is the dilation with the flat structuring element defining the connectivity (3x3 square or cross).

It can be shown that the series g_k always converges and the limit always exists. Depending on which image is considered as being the input, the reconstruction operator has the following properties:

1. If the reference f is considered as being the input and the marker g is assumed to be fixed, the operator $\rho^{\downarrow}(g|.)$ is increasing, antiextensive, and idempotent. It is therefore an opening.
2. If the marker g is considered as being the input and the reference f is assumed to be fixed, the operator $\rho^{\downarrow}(.|f)$ is an idempotent dilation. The operator is increasing but neither antiextensive nor extensive.

Of course, by duality, an extensive reconstruction may be defined:

Definition 9.4. Extensive reconstruction: If f and g are two images (respectively called the "reference" and the "marker" image), the extensive reconstruction $\rho^{\uparrow}(g|f)$ of g *above* f is given by

$$g_k = \varepsilon_C(g_{k-1}) \bigvee f \tag{9.3a}$$

and

$$\rho^{\uparrow}(g|f) = \lim_{k \to \infty} g_k, \tag{9.3b}$$

where $g_0 = g$ and ε_C is the erosion with the flat structuring element defining the connectivity (3×3 square or cross).

Note that Eqs. (9.2a) and (9.3a) define theoretically the reconstruction processes but do not provide efficient implementations. Indeed, the number of iterations is generally fairly high. Assume that the original image f and the marker image g are known and that we want to compute the reconstruction $\rho^{\downarrow}(g|f)$. The most efficient reconstruction algorithms rely on the definition of a clever scanning of the image and are implemented by first-in-first-out (FIFO) queues. A review of the most popular reconstruction algorithms can be found elsewhere [Vin93b]. Here, we describe a simple but efficient one: the basic idea of the algorithm is to start from the regional maxima of the marker image g and to propagate them *under* the original image f. The algorithm works in two steps; the first corresponds to the *initialization* of the queue and the second performs the *propagation*.

- The *initialization* consists of putting in the queue the location of pixels that are on the boundary of the regional maxima of the marker image. Regional maxima are the set of connected components where the image has a constant gray level value and such that every pixel in the neighborhood of the regional maxima strictly has a lower value. Algorithms to compute regional maxima can be found elsewhere [Vin89].

- The *propagation* extracts the first pixel n from the queue (note that n is a pixel of the marker image g). It then assigns to each of its neighbors n' that have a strictly lower gray level value than $g(n)$ (that is, $g(n') < g(n)$) the minimum between the gray level value of the pixel at location n and the gray level value of the pixel of the original image at the same location as n' (that is $g(n') = g(n) \bigwedge f(n')$). Finally, the pixel at location n' is introduced into the queue. This propagation process has to be carried on until idempotence, that is, until the queue is empty. The algorithm is very efficient because the image pixels are processed only once.

In practice, useful connected operators are obtained by considering that f is the input image and one has to derive somehow the marker image g. Most of the time, the marker is itself a transformation $\phi(f)$ of the input image f. As a result, most connected operators ψ obtained by reconstruction can be written as

$$\begin{aligned} \psi(f) &= \rho^{\downarrow}(\phi(f)|f) \text{ (antiextensive operator) or} \\ \psi(f) &= \rho^{\uparrow}(\phi(f)|f) \text{ (extensive operator).} \end{aligned} \tag{9.4}$$

In the following, a few examples are discussed.

Size Filtering

The simplest size-oriented connected operator is obtained by using as the marker image $\phi(f)$ the result of an erosion with a structuring element h. It is the opening by reconstruction of erosion[3]:

$$\psi(f) = \rho^{\downarrow}\left(\varepsilon_h(f)|f\right). \tag{9.5}$$

It can be demonstrated that this operator is an opening [note that this result is not a direct consequence of the reconstruction properties stated in Sec. 9.4.1 because now g is function of f in $\rho^{\downarrow}(g|f)$]. By duality, the closing by reconstruction is given by

$$\psi^*(f) = \rho^{\uparrow}\left(\delta_h(f)|f\right). \tag{9.6}$$

An example of opening by reconstruction of erosion is shown in Fig. 9.2a. In this example, the original signal f has 11 maxima. The marker signal g is created by an erosion with a flat structuring element which eliminates the narrowest maxima. Only five maxima are preserved after erosion. Finally, the marker is reconstructed. In the reconstruction, only the five maxima that were present after erosion are

[3]Note that it can be demonstrated that the same operator is obtained by changing the erosion ε_{h_k} by an opening γ_{h_k} with the same structuring element h_k.

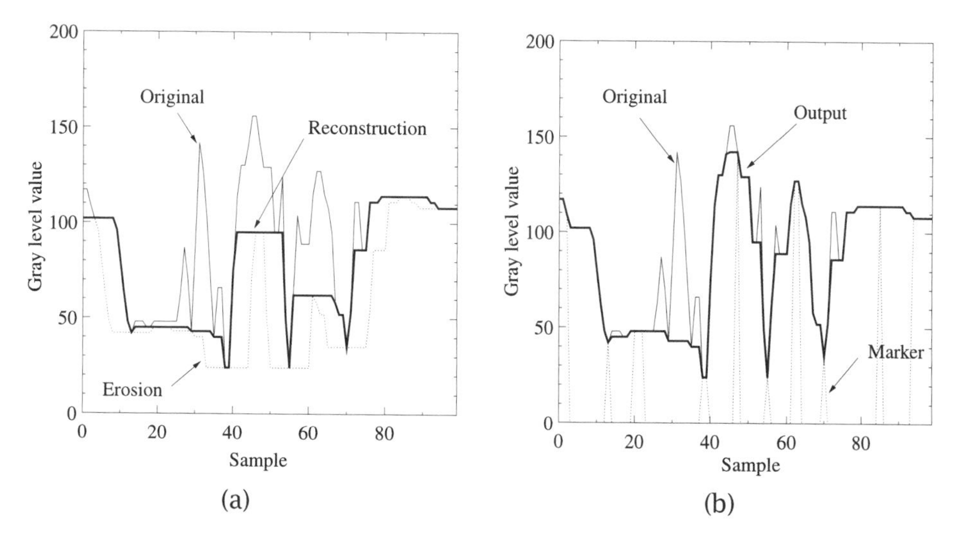

Figure 9.2: Size-oriented connected operators: (a) opening by reconstruction of erosion, and (b) the new marker indicating where the first reconstruction has not been active, plus the second reconstruction.

visible, and narrow maxima have been eliminated. Moreover, the transitions of the reconstructed signal correspond precisely to the transitions of the original signal.

As can be seen, the simplification effect, that is, the elimination of narrow maxima, is almost perfectly done. However, the preservation effect can be criticized: although the maxima contours are well preserved, their shapes and heights are distorted. To reduce this distortion, a new connected operator can be built on top of the first connected operator given by Eq. (9.5). Let us construct a new marker image $m(n)$ indicating the pixels where the reconstruction has not been active, that is, where the final result is different from the erosion:

$$m(n) = \begin{cases} f(n) & \text{for } \rho^{\downarrow}\left(\varepsilon_{h_k}(f) \mid f\right)(n) = \varepsilon_{h_k}(f)(n), \\ 0 & \text{otherwise.} \end{cases} \tag{9.7}$$

This marker image is illustrated in Fig. 9.2b. As can be seen, it is equal to 0 except for the five maxima that were present after erosion and also for the local minima. At those locations, the gray level values of the original image $f(n)$ are assigned to the marker image. Finally, the second connected operator is created by the reconstruction of the marker, m under f:

$$\psi(f) = \rho^{\downarrow}(m|f). \tag{9.8}$$

This operator is also an opening by reconstruction. The final result is shown in Fig. 9.2b. The five maxima are better preserved than with the first opening by reconstruction, whereas the remaining maxima are perfectly removed. Note that in this example, the first opening by reconstruction is used to define a marker of the interesting maxima and the second reconstruction simplifies the image. The

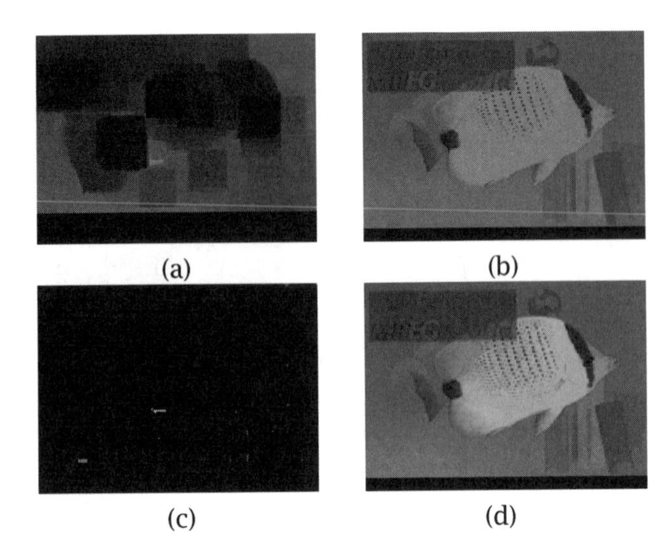

Figure 9.3: Size filtering with opening by reconstruction: (a) erosion of the original image of Fig. 9.1a by a flat structuring element of size 10×10, (b) reconstruction of the erosion, (c) marker indicating where the first reconstruction has not been active [Eq. (9.7)], and (d) second reconstruction.

difference between both reconstructions is also clearly visible in the examples of Fig. 9.3. The first opening by reconstruction removes small bright details of the image: the text in the upper left corner. The fish is a large element and is not removed. It is indeed visible after the first opening by reconstruction (Fig. 9.3b) but its gray level values are not well preserved. This drawback is avoided by using the second reconstruction. Finally, without giving all of the detailed equations and illustrations, let us mention that by duality, closings by reconstruction can be defined. They have the same effect as the openings but on dark components.

Contrast Filtering

The previous section considered size simplification. A contrast simplification can be obtained by substituting the erosion in Eq. (9.5) by a subtraction of a constant c from the original image f. The first reconstruction is therefore

$$\phi(f) = \rho^{\downarrow}(f - c|f). \tag{9.9}$$

This operator, known as the λ-max operator, is connected, increasing, and antiextensive but not idempotent. Its effect is illustrated in Fig. 9.4a. As can be seen, the maxima of small contrast are removed, and the contours of the maxima of high contrast are well preserved. However, the height of the remaining maxima are not well preserved. As in the previous section, this drawback can be removed if a second reconstruction process is used. This second reconstruction process is exactly the same as the previous one defined by Eq. (9.7) (except that now, $m(n) = f(n)$ for $\rho^{\downarrow}(f - c|f)(n) = f(n) - c$) and Eq. (9.8). This second connected operator is an opening, called a dynamic opening [Gri92].

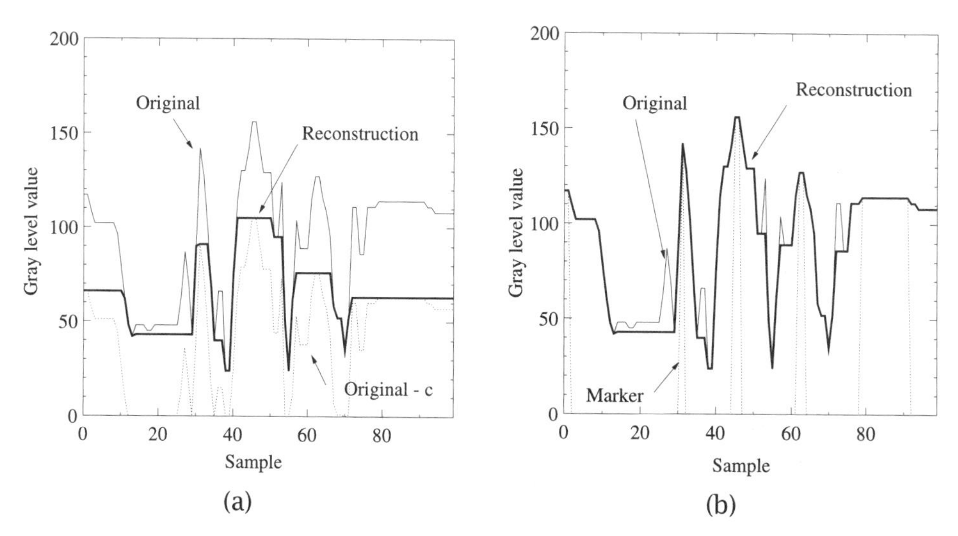

Figure 9.4: Contrast-oriented connected operators: (a) reconstruction of $f - c$ and (b) new marker indicating where the first reconstruction has not been active plus the second reconstruction.

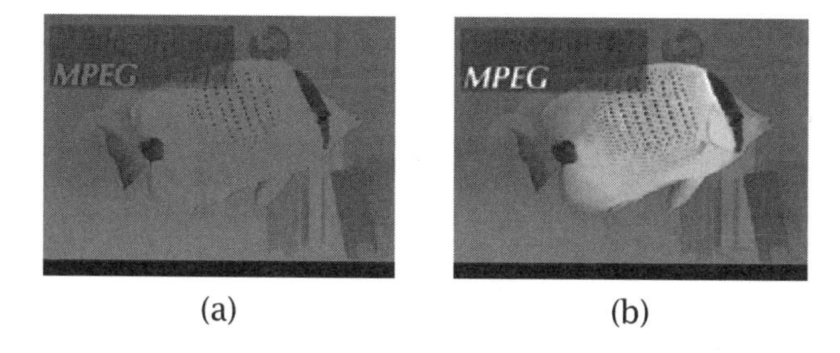

Figure 9.5: Contrast filtering: (a) λ-max operator and (b) dynamic opening.

The visual effect of these operators is illustrated in Fig. 9.5. Both operators have the same simplification effect: they remove maxima of contrast c lower than 100 gray level values. However, the λ-max operator produces an output image of low contrast, even for the preserved maxima. By contrast, the dynamic opening successfully restores the retained maxima.

9.4.2 Self-Dual Reconstruction and Levelings

The connected operators discussed in the previous section were either antiextensive or extensive. In practice, they allow the simplification of either bright or dark image components. For some applications, this behavior is a drawback, and one would like to simplify in a symmetrical way all components. From the theoretical viewpoint, this means that the filter has to be self-dual, that is, $\psi(f) = -\psi(-f)$.

With the aim of constructing self-dual connected operators, the concept of levelings was proposed by Meyer [Mey98a] by adding some restrictions into Definition 9.2:

Definition 9.5. Leveling: The operator ψ is a leveling if $\forall n, n'$ neighboring pixels,

$$\psi(f)(n) > \psi(f)(n') \Rightarrow f(n) \geq \psi(f)(n) \text{ and } \psi(f)(n') \geq f(n').$$

This definition not only states that if a transition exists in the output image, it was already present in the original image (Definition 9.2) but also that (1) the sense of gray level variation between pixels at locations n and n' has to be preserved and (2) the variation $\|\psi(f)(n) - \psi(f)(n')\|$ is bounded by the original variation $\|f(n) - f(n')\|$.

The theoretical properties of levelings were studied elsewhere [Mey98a, Mey98b], and was shown in particular that

- Any opening or closing by reconstruction is a leveling.
- If ψ_1, ψ_2 are levelings, $\psi_2\psi_1$ is also a leveling.
- If $\{\psi_i\}$ are levelings, their supremum $\bigvee_i \psi_i$ and their infimum $\bigwedge_i \psi_i$ are levelings.

The most popular technique to create levelings relies on the following self-dual reconstruction process:

Definition 9.6. Self-dual reconstruction: If f and g are two images (respectively called the "reference" and the "marker" image), the self-dual reconstruction $\rho^{\updownarrow}(g|f)$ of g *with respect to* f is given by

$$\begin{aligned} g_k &= \varepsilon_C(g_{k-1}) \bigvee [\delta_C(g_{k-1}) \bigwedge f] \\ &= \delta_C(g_{k-1}) \bigwedge [\varepsilon_C(g_{k-1}) \bigvee f] \text{ (equivalent expression)} \end{aligned} \quad (9.10a)$$

and

$$\rho^{\updownarrow}(g|f) = \lim_{k \to \infty} g_k, \quad (9.10b)$$

where $g_0 = g$ and δ_C and ε_C are respectively the dilation and the erosion with the flat structuring element defining the connectivity (3×3 square or cross).

An example of self-dual reconstruction is shown in Fig. 9.6. In this example, the marker image is constant everywhere except for two points that mark a maximum and a minimum of the reference image. After reconstruction, the output has only one maximum and one minimum. As can be seen, the self-dual reconstruction is the antiextensive reconstruction of Eq. (9.3) for the pixels where $g(n) < f(n)$ and the extensive reconstruction of Eq. (9.4) for the pixels where $f(n) < g(n)$.

As in the case of antiextensive reconstruction, Eq. (9.10a) does not define an efficient implementation of the reconstruction process. In fact, an efficient implementation of the self-dual reconstruction can be obtained by a combination of the strategies used for antiextensive and extensive reconstruction processes. The *initialization* step consists of putting in the FIFO queue

- the boundary pixels of marker maxima when the marker is smaller than the reference and

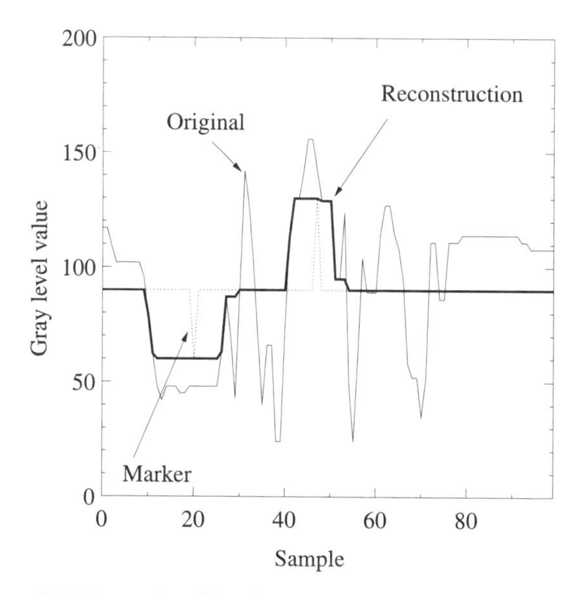

Figure 9.6: Example of leveling with self-dual reconstruction.

- the boundary pixels of marker minima when the marker is greater than the reference.

The *propagation* step is done in a fashion similar to the one described for antiextensive reconstruction: the antiextensive propagation is used when the marker is below the reference and the extensive propagation is used when the marker is above the reference.

In practice, the self-dual reconstruction is used to restore the contour information after a simplification performed by an operator that is neither extensive nor antiextensive. A typical example is an alternating sequential filter: $g = \varphi_{h_k}\gamma_{h_k}\varphi_{h_{k-1}}\gamma_{h_{k-1}}\ldots\varphi_{h_1}\gamma_{h_1}(f)$, where φ_{h_k} and γ_{h_k} are respectively a closing and an opening with a structuring element h_k of size k. These examples are illustrated in Figs. 9.7a and 9.7b. Note in particular the simplification effect that deals with both maxima and minima and how the distortion of the contours introduced by the alternating sequential filter is removed by the reconstruction. However, from a theoretical viewpoint, the operator $\rho^{\dagger}(\varphi_{h_k}\gamma_{h_k}\ldots\varphi_{h_1}\gamma_{h_1}(f)|f)$ is not self-dual because the alternating sequential filter itself is not self-dual. To create an operator that is self-dual, we must create a marker image that also is self-dual. Figures 9.7c and 9.7d show an example in which the marker image was created by a median filter (which is self-dual). This kind of result can be extended to any linear filter, and the self-dual reconstruction can be considered as a general tool that restores the contour information after a filtering process. In other words, the reconstruction allows us to create a connected version $\rho^{\dagger}(\psi(f)|f)$ of any filter $\psi(f)$.

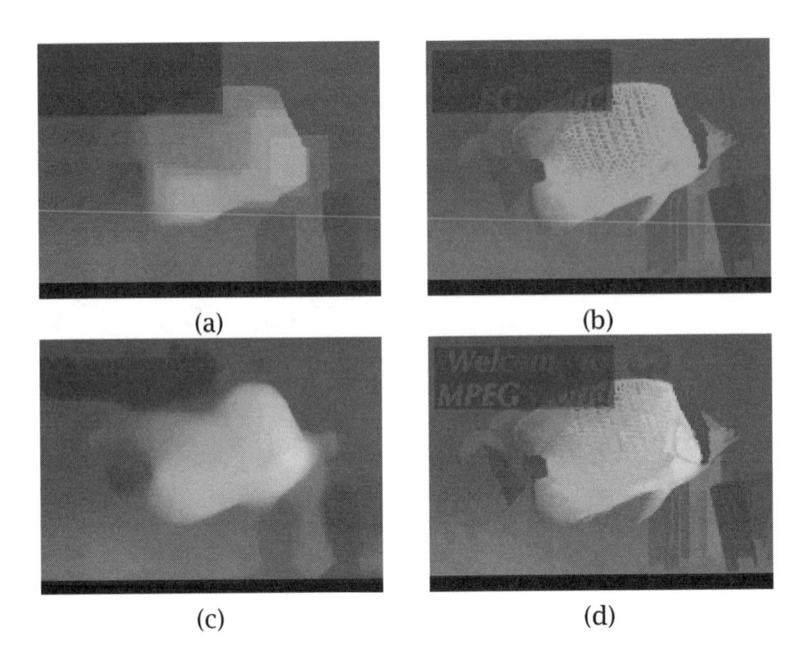

Figure 9.7: Size filtering with leveling: (a) alternating sequential filtering of the original image of Fig. 9.1a by a flat structuring of size 7×7, (b) self-dual reconstruction of the alternating sequential filtered image, (c) median filtering with a window of size 15×15, and (d) self-dual reconstruction of the median filtered image.

9.5 Connected Operators Based on Region-Tree Pruning

9.5.1 Tree Representations and Connected Operators

The reconstruction strategies discussed in Sec. 9.4 can be viewed as tools that work on a pixel-based representation of the image and that provide a way to create connected operators. In this section, we present a different approach: the first step of the filtering process is to construct a region-based representation of the image, then the simplification effect is obtained by direct manipulation of the new representation. The approach may be considered to be conceptually more complex than the reconstruction; however, it provides more flexibility in the choice of the simplification criterion.

Two region-based representations are discussed subsequently: the max-tree-min-tree [Sal98] and the binary partition tree [Sal00]. The first leads to antiextensive connected operators, whereas the second is the basis for self-dual connected operators. Let us first discuss these two region-based representations.

Max-Tree and Min-Tree

The first tree representation, called a max-tree [Sal98], enhances the maxima of the signal. Each node $\mathcal{N}_k$ in the tree structure represents a connected component

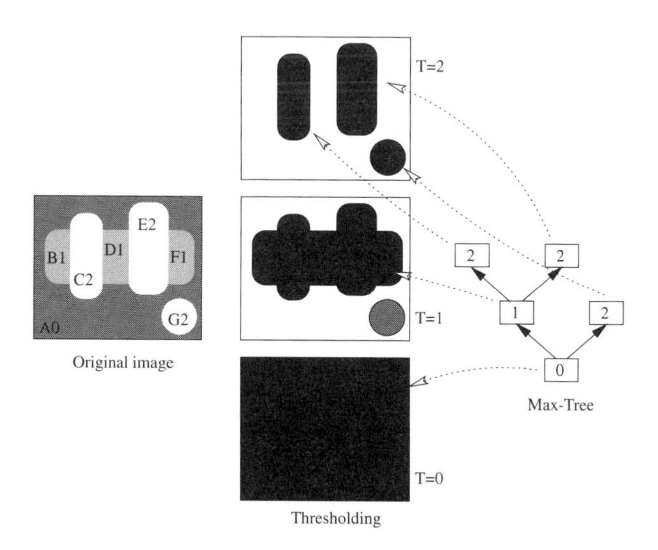

Figure 9.8: Max-tree representation of images.

of the space that is extracted by the following thresholding process: for a given threshold T, consider the set X of pixels that have a gray level value larger than T and the set of Y pixels that have a gray level value equal to T,

$$\begin{aligned} X &= \{n, \text{ such that } f(n) \geq T\}, \\ Y &= \{n, \text{ such that } f(n) = T\}. \end{aligned} \tag{9.11}$$

The tree nodes $\mathcal{N}_k$ represent the connected components of X such that $X \cap Y \neq \emptyset$. A simple example of a max-tree is shown in Fig. 9.8. The original image is made of seven flat zones identified by a letter {A,...,G}. The number following each letter defines the gray level value of the flat zone. The binary images X resulting from the thresholding with $0 \leq T \leq 2$ are shown in the center of the figure. The max-tree is given on the right side. It is composed of five nodes that represent the connected components, shown in black. The number inside each square represents the threshold value where the component was extracted. Finally, the links in the tree represent the inclusion relationships among the connected components following the threshold values. Note that when the threshold is set to $T = 1$, the circular component does not create a connected component that is represented in the tree because none of its pixels has a gray level value equal to 1. However, the circle itself is obtained when $T = 2$. The three regional maxima are represented by three leaves, and the tree root represents the entire support of the image.

A more complex case is shown in Fig. 9.9. In this example, the max-tree and its dual representation, the min-tree, are shown. The former highlights the maxima of the image (leaves of the tree), whereas the latter highlights the minima. These trees are region-based representations of the image that are used subsequently to define connected operators. Computation of the max-tree can be done efficiently,

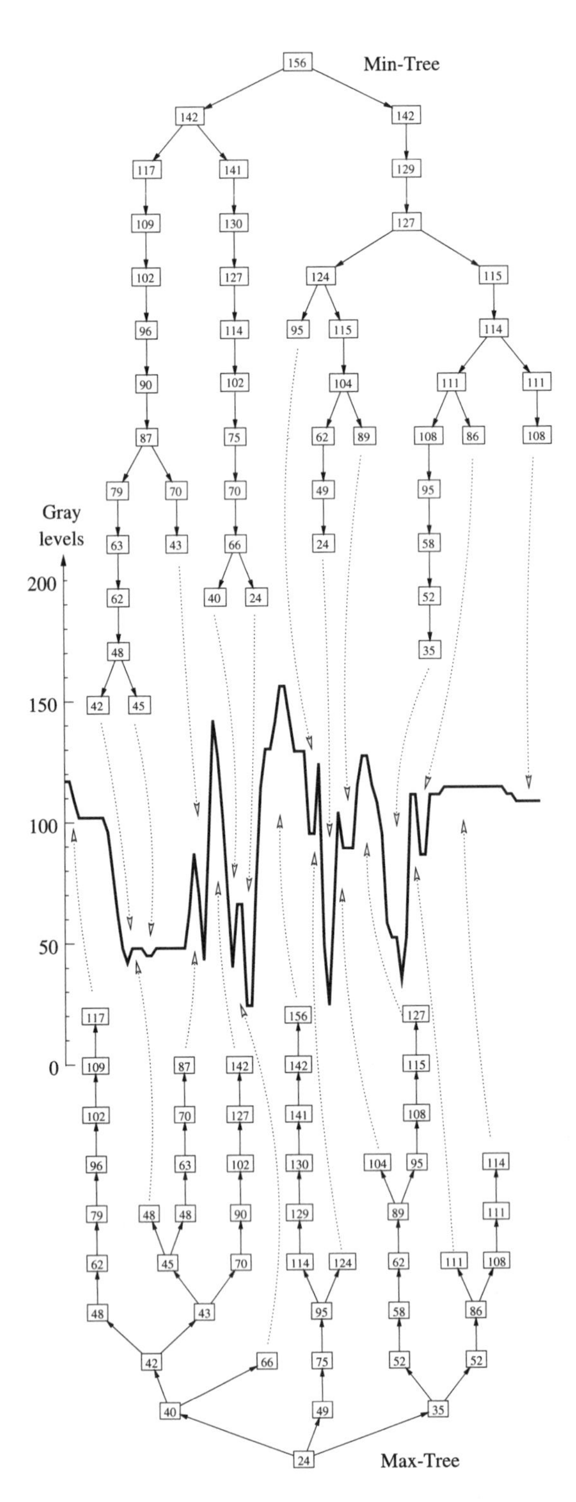

Figure 9.9: Example of a max-tree and min-tree representation.

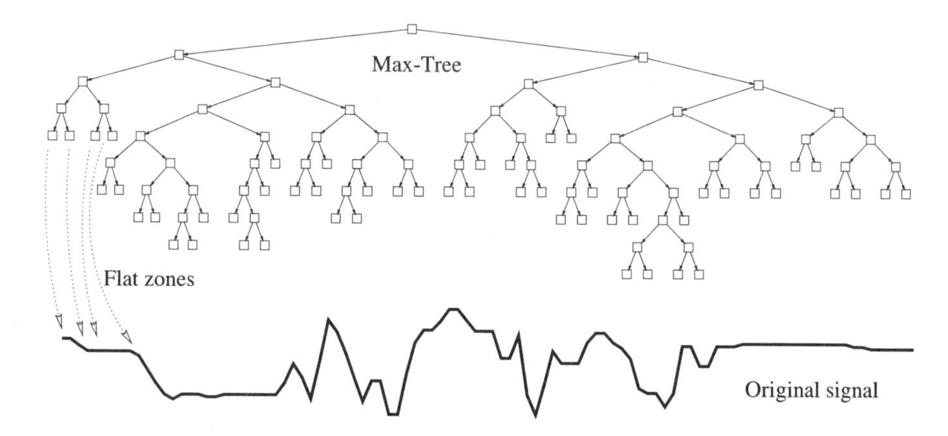

Figure 9.10: Example of binary partition tree representation.

and the appendix describes an efficient algorithm for this purpose.

Binary Partition Tree

The second example of region-based representations of images is the binary partition tree [Sal00], representing a set of regions that can be obtained from an initial partition that we assume to be the partition of flat zones. An example is shown in Fig. 9.10. The leaves of the tree represent the flat zones of the original signal. The remaining nodes of the tree represent regions that are obtained by merging the regions represented by the children. As in the cases of the max-tree and min-tree, the root node represents the entire image support. As can be seen, the tree represents a fairly large set of regions at different scales. Large regions appear close to the root, whereas small details can be found at lower levels. This representation should be considered as a compromise between representation accuracy and processing efficiency. Indeed, all possible mergings of regions belonging to the initial partition are not represented in the tree. Only the most "likely" or "useful" merging steps are represented in the binary partition tree. The connectivity encoded in the tree structure is binary in the sense that a region is explicitly connected to its sibling (since their union is a connected component represented by the father), but the remaining connections between regions of the original partition are not represented in the tree. Therefore, the tree encodes only part of the neighborhood relationships between the regions of the initial partition. However, as will be seen, the main advantage of the tree representation is that it allows the fast implementation of sophisticated processing techniques.

The binary partition tree should be created in such a way that the most "interesting" or "useful" regions are represented. This issue can be application dependent. However, a possible solution, suitable for a large number of cases, is to create the tree by keeping track of the merging steps performed by a segmentation algorithm based on region merging (see [Gar98, Mor86], for example). In the following, this information is called the *merging sequence*. Starting from the partition of flat

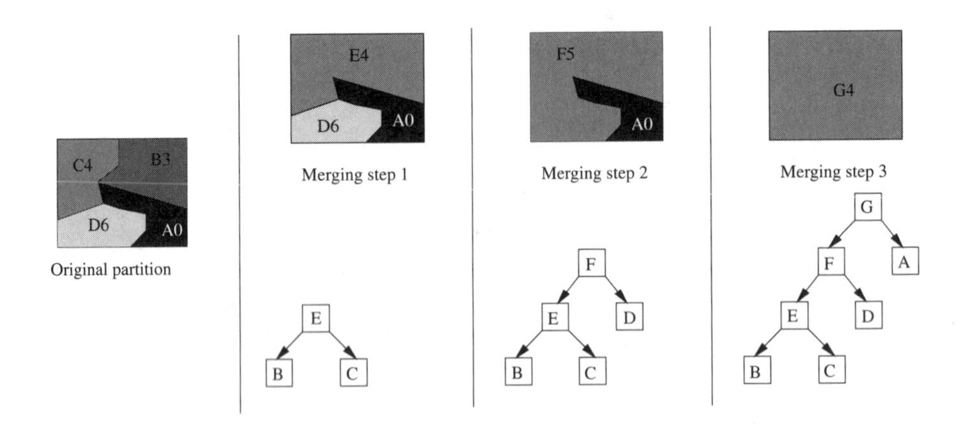

Figure 9.11: Example of binary partition tree creation with a region merging algorithm.

zones (in principle, any other precomputed partition may be used), the algorithm merges neighboring regions following a homogeneity criterion until a single region is obtained. An example is shown in Fig. 9.11. The original partition involves four regions. The regions are indicated by a letter, and the number indicates the gray level value of the flat zone. The algorithm merges the four regions in three steps. In the first step, the pair of most similar regions, B and C, are merged to create region E, which is then merged with region D to create region F. Finally, region F is merged with region A, and this creates region G, corresponding to the region of support of the whole image. In this example, the merging sequence that progressively defines the binary partition tree is $(B,C)|(E,D)|(F,A)$.

To completely define the merging algorithm, one has to specify the region merging order and the region model, that is, the model used to represent the union of two regions. Note that the merging process is carried on until one single region (the image region of support) is obtained. To create the binary partition trees used to illustrate the processing examples discussed in this chapter, we used a merging algorithm following the color homogeneity criterion described by Garrido et al. [Gar98]. Let us define the merging order $O(R_1,R_2)$ and the region model M_R:

- **Merging order:** At each step the algorithm looks for the pair of most similar neighboring regions. The similarity between regions R_1 and R_2 is defined by the following expression:

$$O(R_1,R_2) = N_1||M_{R_1} - M_{R_1\cup R_2}||_2 + N_2||M_{R_2} - M_{R_1\cup R_2}||_2, \tag{9.12}$$

  where N_1 and N_2 are the numbers of pixels of regions R_1 and R_2, $||.||_2$ denotes the $\mathcal{L}_2$ norm, and M_R represents the model for region R. It consists of three constant values describing the YUV components. The value of this merging order, compared to other classical criteria, is discussed elsewhere [Gar98].

- **Region model:** As mentioned, each region is modeled by a constant YUV value. M_R is therefore a vector of three components. During the merging

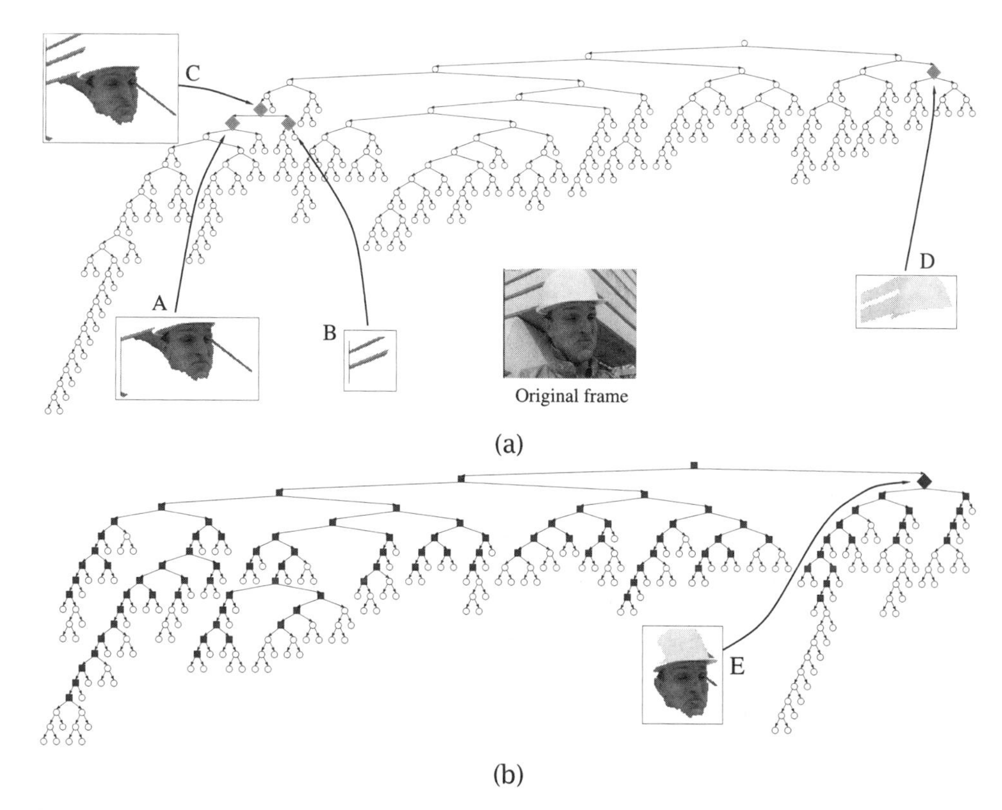

Figure 9.12: Examples of creation of binary partition tree: (a) using the color homogeneity criterion alone and (b) using both the color and the motion homogeneity criteria.

process, the YUV components of the union of two regions, R_1 and R_2, are computed as follows [Gar98]:

$$\begin{aligned} &\text{if } N_1 < N_2 \Rightarrow M_{R_1 \cup R_2} = M_{R_2}, \\ &\text{if } N_2 < N_1 \Rightarrow M_{R_1 \cup R_2} = M_{R_1}, \\ &\text{if } N_1 = N_2 \Rightarrow M_{R_1 \cup R_2} = (M_{R_1} + M_{R_2})/2. \end{aligned} \tag{9.13}$$

As can be seen, if $N_1 \neq N_2$, the model of the union of two regions is equal to the model of the largest region.

It should be noted that the homogeneity criterion does not have to be restricted to color. For example, if the image for which we create the binary partition tree belongs to a sequence of images, motion information should also be used to generate the tree: In the first stage, regions are merged using a color homogeneity criterion, whereas a motion homogeneity criterion is used in the second stage. Figure 9.12 shows an example for the video sequence *Foreman*. In Fig. 9.12a, the binary partition tree has been constructed exclusively with the color homogeneity criterion described above. In this case, it is not possible to concentrate the information about the foreground object (head and shoulder regions of the foreman) within a single subtree. For example, the face appears mainly in the subtree

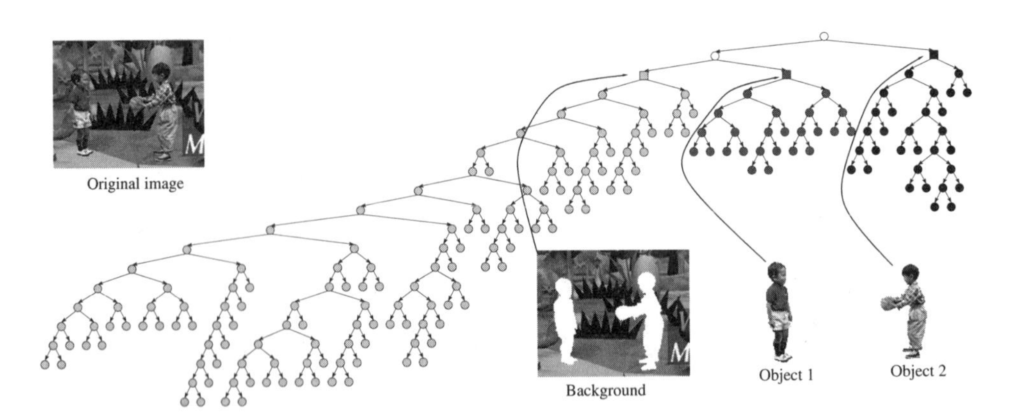

Figure 9.13: Example of partition tree creation with restriction imposed by object masks.

hanging from region A, whereas the helmet regions are located below region D. In practice, the nodes that are close to the root have no clear meaning because they are not homogeneous in color. Figure 9.12b presents an example of binary partition tree created with color and motion criteria. The nodes appearing in the lower part of the tree as white circles correspond to the color criterion, whereas the dark squares correspond to a motion criterion. The motion criterion is formally the same as the color criterion except that the YUV color distance is replaced by the YUV displaced frame difference. As can be seen, the process starts with the color criterion as in Fig. 9.12a and then when a given peak signal to noise ratio (PSNR) is reached, it changes to the motion criterion. Using motion information, the face and the helmet now appear as a single region E.

Additional information from previous processing or detection algorithms can also be used to generate the tree in a more robust way. For instance, a mask of an object included in the image can be used to impose constraints on the merging algorithm in such a way that the object itself is represented with only one node in the tree. Typical examples of such algorithms are face, skin, character, or foreground object detection. An example is illustrated in Fig. 9.13. Assume, for example, that the original video sequence *Children* has been analyzed so that the masks of the two foreground objects (children) are available. If the merging algorithm is constrained to merge regions within each mask before dealing with the remaining regions, the region of support of each mask is represented as a single node in the resulting binary partition tree. In the figure, the nodes corresponding to the background and the two foreground objects are represented by squares. The three subtrees further decompose each object into elementary regions.

As can be seen, the construction of a binary partition tree is fairly more complex than the creation of a max-tree or a min-tree. However, binary partition trees offer more flexibility because one can choose the homogeneity criterion through the proper selection of the region model and the merging order. Furthermore, if the functions defining the region model and the merging order are self-dual, the tree itself is self-dual. The same binary partition tree can be used to represent f

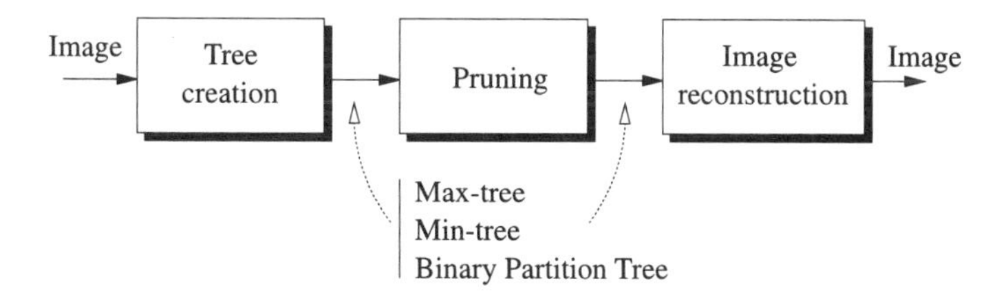

Figure 9.14: Connected operators based on tree representations.

and $-f$. The example of Fig. 9.9 clearly shows that the structures of a max-tree and its dual, the min-tree, are quite different. The binary partition tree representation is appropriate to derive self-dual connected operators, whereas the max-tree (min-tree) is adequate for antiextensive (extensive) connected operators. Note that in all cases, the trees are hierarchical region-based representations. They encode a large set of regions and partitions that can be derived for the flat zone partition of the original image without adding new contours.

Filtering Strategy

Once the tree representation has been created, the filtering strategy consists of *pruning* the tree and reconstructing an image from the pruned tree. The global processing strategy is illustrated in Fig. 9.14. The simplification effect of the filter is done by pruning because the idea is to eliminate the image components that are represented by the leaves and branches of the tree. The nature of these components depends on the tree. In the case of max-trees (min-trees), the components that may be eliminated are regional maxima (minima), whereas the elements that may be simplified in the case of binary partition trees are unions of the most similar flat zones. The simplification itself is governed by a criterion that may involve simple notions such as size or contrast or more complex ones such as texture, motion, or even semantic criteria.

One of the values of the tree representations is that the set of possible merging steps is fixed (the possible merging steps are represented by the tree branches). As a result, sophisticated simplification (pruning) strategies may be designed. A typical example of such strategy deals with nonincreasing simplification criteria.

Mathematically, a criterion C assessed on a region R is said to be increasing if the following property holds:

$$\forall R_1 \subseteq R_2 \Rightarrow C(R_1) \leq C(R_2). \tag{9.14}$$

Assume that all nodes corresponding to regions in which the criterion value is lower than a given threshold should be removed by merging. If the criterion is increasing, the pruning strategy is straightforward: merge all nodes that should be removed. It is indeed a pruning strategy since the increasingness of the criterion guarantees that if a node has to be removed, all of its descendants also have to be removed. An example of a binary partition tree with an increasing decision

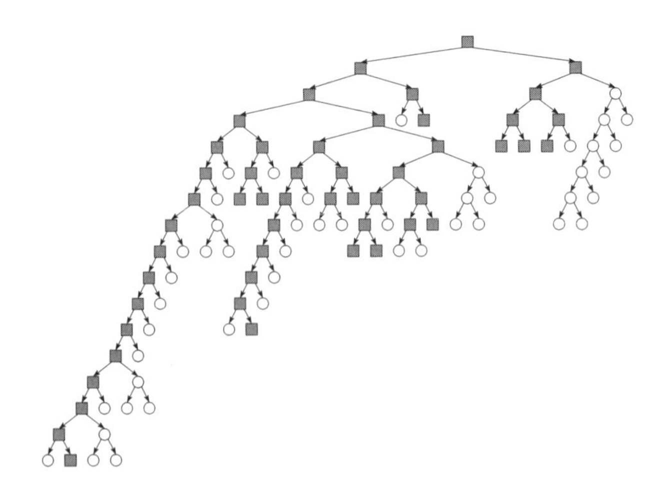

Figure 9.15: Example of an increasing criterion (size). If a node has to be removed, all of its descendants also have to be removed. The gray squares are nodes to be preserved, while the white circles are nodes to be removed.

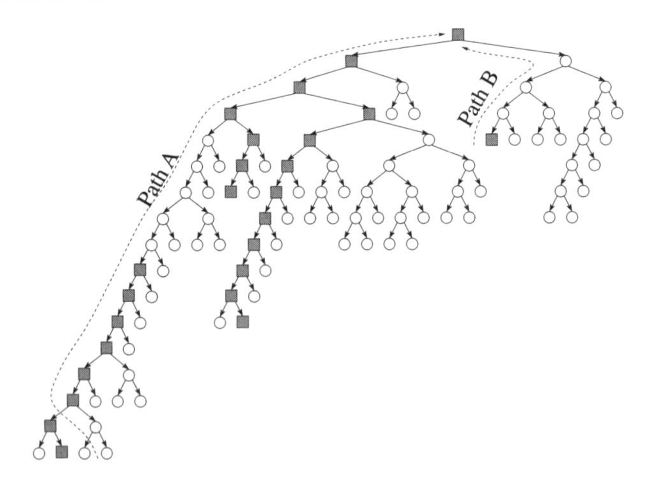

Figure 9.16: Example of a nonincreasing criterion (perimeter). No relation exists between the decisions among descendants (see decisions along path A or path B). The gray squares are nodes to be preserved, while the white circles are nodes to be removed.

criterion is shown in Fig. 9.15. The criterion used to create this example is the size, measured as the number of pixels belonging to the region, which is indeed increasing. Note that this example involves a binary partition tree, but the same issue also applies to max-tree and min-tree representations.

If the criterion is not increasing, the pruning strategy is not straightforward since the descendants of a node to be removed need not necessarily be removed. An example of such a criterion is the region perimeter. Figure 9.16 illustrates this case. If we follow either path A or path B in the figure, we see that there are some oscillations of the remove-preserve decisions. In practice, the non-increasingness of the criterion implies a lack of robustness of the operator. For example, similar

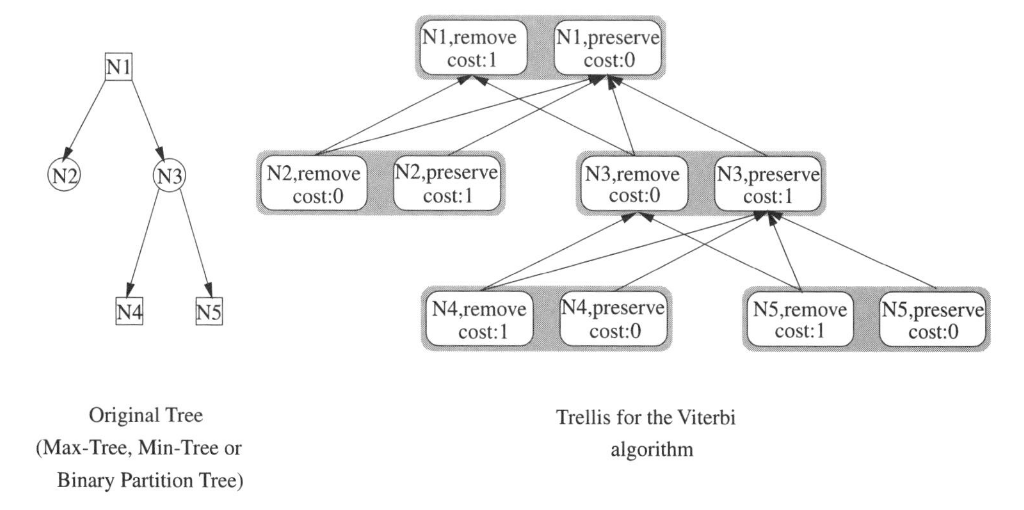

Figure 9.17: Creation of the trellis structure for the Viterbi algorithm. A circular (square) node on the tree indicates that the criterion value states that the node has to be removed (preserved). The trellis on which the Viterbi algorithm is run duplicates the structure of the tree and defines a *preserve state* and a *remove state* for each node of the tree. Paths from *remove* states to child *preserve* states are forbidden so that the decisions are increasing.

images may produce quite different results, or small modifications of the criterion threshold can result in drastic changes in the output.

A possible solution to the nonincreasingness of the criterion consists of applying a transformation on the set of decisions. The transformation should create a set of increasing decisions while preserving as much as possible the decisions defined by the criterion. This problem may be viewed as a dynamic programming issue that can be efficiently solved with the Viterbi algorithm.

The dynamic programming algorithm is explained and illustrated here assuming that the tree is binary. The extension to N-ary trees is straightforward, and the example of binary tree is used only to simplify the explanation and the corresponding notations. An example of a trellis on which the Viterbi algorithm [Vit79] is applied is illustrated in Fig. 9.17. The trellis has the same structure as the tree except that two trellis states, *preserve* state $\mathcal{N}_k^{\mathrm{P}}$ and *remove* state $\mathcal{N}_k^{\mathrm{R}}$, correspond to each node $\mathcal{N}_k$ of the tree. The two states of each child node are connected to the two states of its parent. However, to avoid nonincreasing decisions, the *preserve* state of a child is not connected to the *remove* state of its parent. As a result, the trellis structure guarantees that if a node has to be removed, its children also have to be removed. The cost associated with each state is used to compute the number of modifications the algorithm has to do to create an increasing set of decisions. If the criterion value states that the node of the tree has to be removed, the cost associated with the *remove* state is equal to zero (no modification) and the cost associated with the *preserve* state is equal to one (one modification). Similarly, if the criterion value states that the node has to be preserved, the cost of the *remove*

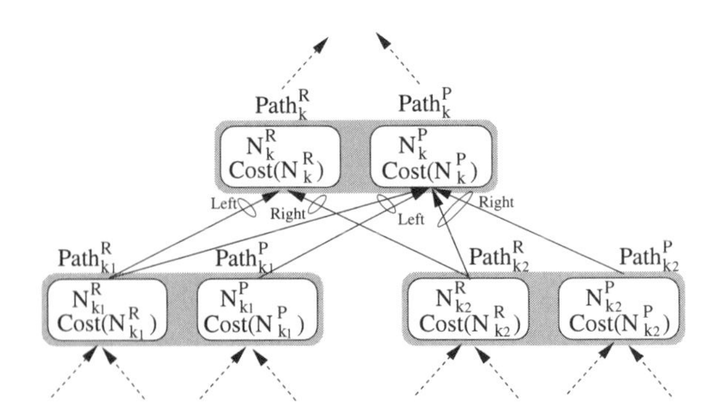

Figure 9.18: Example of the path and cost for the Viterbi algorithm [see Eqs. (9.17), (9.18), and (9.19)].

state is equal to one and the cost of the *preserve* state is equal to zero.[4] The cost values appearing in the figure assume that nodes $\mathcal{N}_1$, $\mathcal{N}_4$, and $\mathcal{N}_5$ should be preserved and that $\mathcal{N}_2$ and $\mathcal{N}_3$ should be removed. The goal of the Viterbi algorithm is to define the set of decisions such that

$$\min \sum_k \text{cost}(\mathcal{N}_k) \text{ such that the decisions are increasing.} \tag{9.15}$$

To find the optimum set of decisions, a set of paths going from all leaf nodes to the root node is created. For each node, the path can go through either the *preserve* or the *remove* state of the trellis. The Viterbi algorithm is used to find the paths that minimize the global cost at the root node. Note that the trellis structure itself guarantees that this optimum decision is increasing. The optimization is achieved in a bottom-up iterative fashion. For each node, it is possible to define the optimum paths ending at the *preserve* state and at the *remove* state:

- Let us consider a node $\mathcal{N}_k$ and its *preserve* state $\mathcal{N}_k^P$. A path path$_k$ is a continuous set of transitions between nodes $(\mathcal{N}_\alpha \to \mathcal{N}_\beta)$ defined in the trellis:

$$\text{path}_k = (\mathcal{N}_\alpha \to \mathcal{N}_\beta) \cup (\mathcal{N}_\beta \to \mathcal{N}_\gamma) \cup \ldots \cup (\mathcal{N}_\psi \to \mathcal{N}_k). \tag{9.16}$$

The path path_k^P starting from a leaf node and ending at that state is composed of *two* subpaths[5]: the first one, $\text{path}_k^{P,\text{left}}$, comes from the left child and the second one, $\text{path}_k^{P,\text{right}}$, from the right child (see Fig. 9.18). In both cases, the path can emerge either from the *preserve* or from the *remove* state of the child nodes. If $\mathcal{N}_{k_1}$ and $\mathcal{N}_{k_2}$ are respectively the left and right child nodes of $\mathcal{N}_k$, we have

$$\text{path}_k^{P,\text{left}} = \text{path}_{k_1}^R \bigcup (\mathcal{N}_{k_1}^R \to \mathcal{N}_k^P) \quad \text{or} \quad \text{path}_{k_1}^P \bigcup (\mathcal{N}_{k_1}^P \to \mathcal{N}_k^P),$$

[4] Although some modifications may be much more severe than others, the cost choice has no strong effect on the final result. This issue of cost selection is similar to the hard versus soft decision of the Viterbi algorithm in the context of digital communications [Vit79].

[5] In the general case of an N-ary tree, the number of incoming paths may be arbitrary.

$$\text{path}_k^{\text{P,Right}} = \text{path}_{k_2}^{\text{R}} \bigcup (\mathcal{N}_{k_2}^{\text{R}} \to \mathcal{N}_k^{\text{P}}) \quad \text{or} \quad \text{path}_{k_2}^{\text{P}} \bigcup (\mathcal{N}_{k_2}^{\text{P}} \to \mathcal{N}_k^{\text{P}}),$$

$$\text{path}_k^{\text{P}} = \text{path}_k^{\text{P,left}} \bigcup \text{path}_k^{\text{P,right}}. \tag{9.17}$$

The cost of a path is equal to the sum of the costs of its individual state transitions. Therefore, the optimum path (path of lowest cost) for each child can be easily selected.

$$\text{If cost } (\text{path}_{k_1}^{\text{R}}) < \text{cost}(\text{path}_{k_1}^{\text{P}}),$$

$$\text{then} \begin{cases} \text{path}_k^{\text{P,left}} = \text{path}_{k_1}^{\text{R}} \cup (\mathcal{N}_{k_1}^{\text{R}} \to \mathcal{N}_k^{\text{P}}), \\ \text{cost}(\text{path}_k^{\text{P,left}}) = \text{cost } (\text{path}_{k_1}^{\text{R}}); \end{cases}$$

$$\text{otherwise} \begin{cases} \text{path}_k^{\text{P,left}} = \text{path}_{k_1}^{\text{P}} \cup (\mathcal{N}_{k_1}^{\text{P}} \to \mathcal{N}_k^{\text{P}}), \\ \text{cost}(\text{path}_k^{\text{P,left}}) = \text{cost } (\text{path}_{k_1}^{\text{P}}). \end{cases}$$

$$\text{If cost } (\text{path}_{k_2}^{\text{R}}) < \text{cost}(\text{path}_{k_2}^{\text{P}}),$$

$$\text{then} \begin{cases} \text{path}_k^{\text{P,right}} = \text{path}_{k_2}^{\text{R}} \cup (\mathcal{N}_{k_2}^{\text{R}} \to \mathcal{N}_k^{\text{P}}), \\ \text{cost}(\text{path}_k^{\text{P,right}}) = \text{cost } (\text{path}_{k_2}^{\text{R}}); \end{cases}$$

$$\text{otherwise} \begin{cases} \text{path}_k^{\text{P,right}} = \text{path}_{k_2}^{\text{P}} \cup (\mathcal{N}_{k_2}^{\text{P}} \to \mathcal{N}_k^{\text{P}}), \\ \text{cost}(\text{path}_k^{\text{P,right}}) = \text{cost } (\text{path}_{k_2}^{\text{P}}). \end{cases}$$

$$\text{cost}(\text{path}_k^{\text{P}}) = \text{cost}(\text{path}_k^{\text{P,left}}) + \text{cost}(\text{path}_k^{\text{P,right}}) + \text{cost}(\mathcal{N}_k^{P}). \tag{9.18}$$

- In the case of the *remove* state $\mathcal{N}_k^{\text{R}}$, the two subpaths can come only from the *remove* states of the children, so no selection has to be done. The path and its cost are constructed as follows:

$$\text{path}_k^{\text{R,left}} = \text{path}_{k_1}^{\text{R}} \cup (\mathcal{N}_{k_1}^{\text{R}} \to \mathcal{N}_k^{\text{R}}),$$

$$\text{path}_k^{\text{R,right}} = \text{path}_{k_2}^{\text{R}} \cup (\mathcal{N}_{k_2}^{\text{R}} \to \mathcal{N}_k^{\text{R}}),$$

$$\text{path}_k^{\text{R}} = \text{path}_k^{\text{R,left}} \cup \text{path}_k^{\text{R,right}},$$

$$\text{cost } (\text{path}_k^{\text{R}}) = \text{cost } (\text{path}_{k_1}^{\text{R}}) + \text{cost } (\text{path}_{k_2}^{\text{R}}) + \text{cost } (\mathcal{N}_k^{\text{R}}). \tag{9.19}$$

This procedure is iterated in a bottom-up fashion until the root node is reached. One path of minimum cost ends at the *preserve* state of the root node and the other ends at the *remove* state of the root node. From these two paths, the one of minimum cost is selected. This path connects the root node to all leaves, and the states it goes through define the final decisions. By construction, these decisions are increasing, and they are as close as possible to the original decisions.

A complete example of optimization is shown in Fig. 9.19. The original tree involves five nodes. As before, the *preserve* decisions are shown by a square, whereas the *remove* decisions are indicated by a circle. As can be seen, the original tree does not correspond to a set of increasing decisions because $\mathcal{N}_3$ should

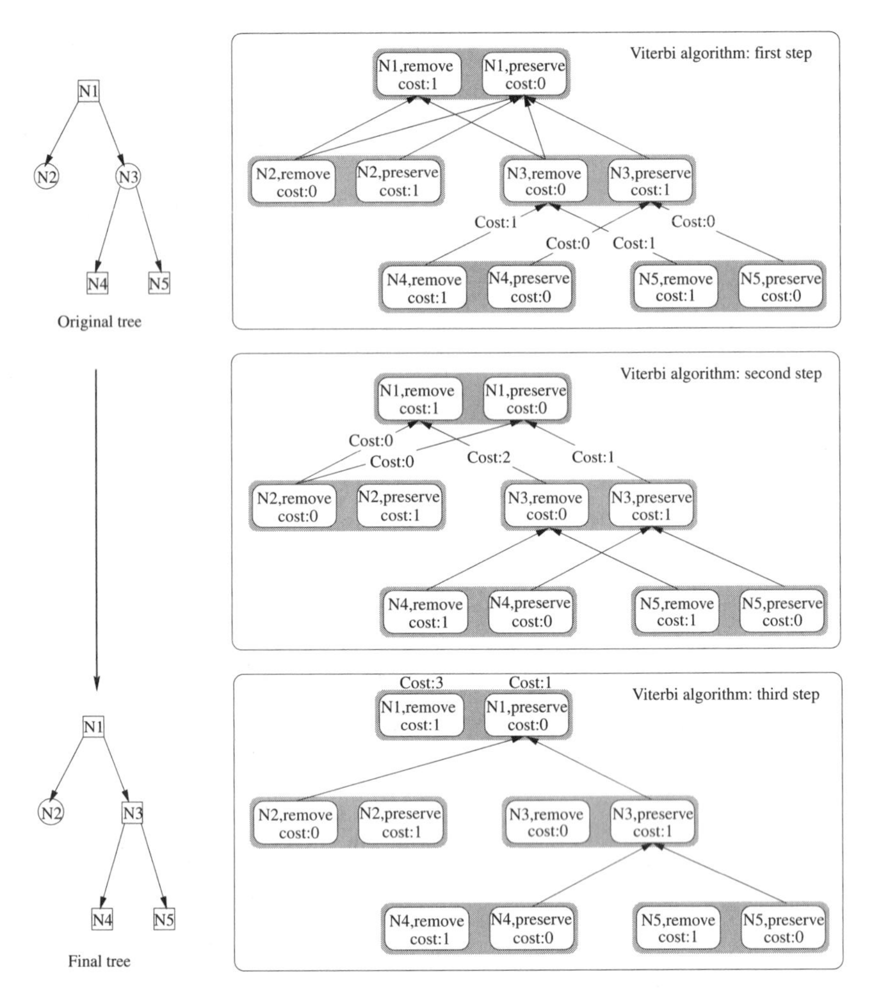

Figure 9.19: Example of the optimum decisions by the Viterbi algorithm. The squares indicate nodes to be preserved, while the circles indicate nodes to be removed.

be removed but $\mathcal{N}_4$ and $\mathcal{N}_5$ should be preserved. The algorithm is initialized by creating the trellis and by populating the states by their respective cost (see Fig. 9.17). Then, the first step of the algorithm consists in selecting the paths that go from states $\mathcal{N}_4^R$, $\mathcal{N}_4^P$, $\mathcal{N}_5^R$, $\mathcal{N}_5^P$ to states $\mathcal{N}_3^R$, $\mathcal{N}_3^P$. The corresponding trellis is shown in the upper part of Fig. 9.19, together with the corresponding costs of the four surviving paths. The second step iterates the procedure between states $\mathcal{N}_2^R$, $\mathcal{N}_2^P$, $\mathcal{N}_3^R$, $\mathcal{N}_3^P$ and states $\mathcal{N}_1^R$, $\mathcal{N}_1^P$. Here again, only four paths survive, indicated in the central diagram of the figure. The final step consists of selecting the path of lowest cost that terminates at the root states. In this example, the path ending at the *remove* state of the root node ($\mathcal{N}_1^R$) has a cost of 3, whereas the path ending at the *preserve* state ($\mathcal{N}_1^P$) has a cost of 1. This last path is taken since it corresponds to an increasing set of decisions and involves just one modification of the original decisions. To find the optimum increasing decisions, one has to track back up the

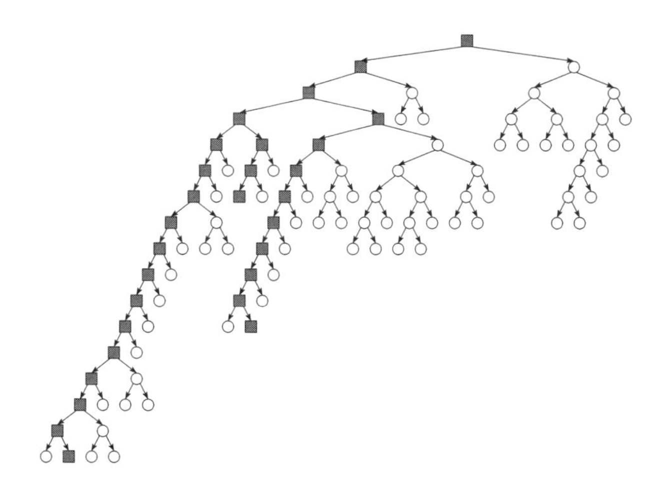

Figure 9.20: Set of increasing decisions resulting from the use of the Viterbi algorithm on the original tree of Fig. 9.16. Five decisions along path A and one decision along path B have been modified. The gray squares are nodes to be preserved, while the white circles are nodes to be removed.

selected path from the root to all leaves. In our example, we see that the paths hit the following states: $\mathcal{N}_1^P$, $\mathcal{N}_2^R$, $\mathcal{N}_3^P$, $\mathcal{N}_4^P$ and $\mathcal{N}_5^P$. The bottom diagram of Fig. 9.19 shows the final path together with the modified tree. As can be seen, the only modification has been to change the decision of node $\mathcal{N}_3$ and the resulting set of decisions is increasing.

A complete example of decision modification is shown in Fig. 9.20. The original tree corresponds to the one shown in Fig. 9.16. The Viterbi algorithm has to modify five decisions along path A and one decision along path B (see Fig. 9.16) to get the optimum set of increasing decisions.

To summarize this section, let us say that any pruning strategy can be applied directly on the tree if the decision criterion is increasing (size is a typical example). In the case of a nonincreasing criterion such as the perimeter, the Viterbi algorithm can be used to modify the smallest number of decisions so that increasingness is obtained. These modifications define a pruning strategy. Once the pruning has been performed, it defines an output partition, and each region is filled with a constant value. In the case of a max-tree (min-tree), the constant value is equal to the minimum (maximum) gray level value of the original pixels belonging to the region. As a result, the operator is antiextensive (extensive). In the case of a binary partition tree, the goal is to define a self-dual operator. Therefore, each region of the output partition has to be filled by a self-dual model, such as the mean or the median of the original pixels belonging to the region.

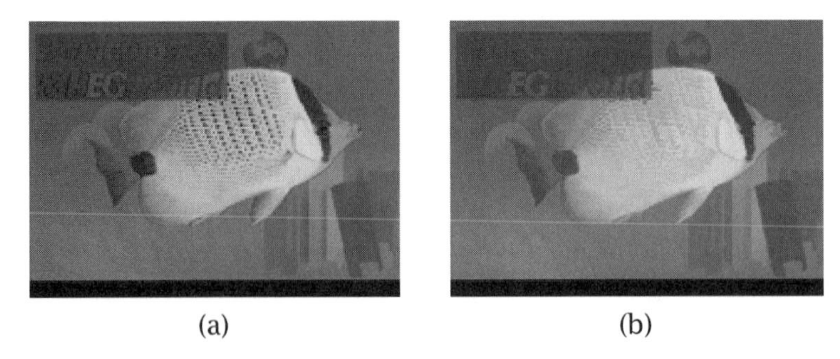

(a) (b)

Figure 9.21: Area filtering examples: (a) area opening γ^{area} and (b) area opening followed by area closing, $\varphi^{\text{area}}\gamma^{\text{area}}$.

9.5.2 Examples of Connected Operators Based on Tree Representations

Increasing Criterion ⇒ Direct Pruning

The first example deals with situations in which the criterion is increasing. In this case, the comparison of the criterion value with a threshold directly defines a pruning strategy. A typical example is the area opening [Vin93a]. One of the possible implementations of the area opening consists of creating a max-tree and measuring the area (the number of pixels) $\mathcal{A}_k$ contained in each node $\mathcal{N}_k$. If the area $\mathcal{A}_k$ is smaller than a threshold $\mathcal{T}_{\mathcal{A}}$, the node is removed. The area criterion is increasing, so the Viterbi algorithm discussed in the previous section does not have to be used. It can be shown that the area opening is equal to the supremum of all possible openings by a connected structuring element involving $\mathcal{T}_{\mathcal{A}}$ pixels. The simplification effect of the area opening is illustrated in Fig. 9.21a. As expected, the operator removes small bright components of the image. If this simplified image is processed by the dual operator, the area closing, small dark components are also removed (see Fig. 9.21b).

Using the same strategy, a large number of connected operators can be obtained. For example, if the criterion is the volume, $\sum_{n \in R} f(n)$ (also increasing), the resulting operator is the volumic opening [Mey97]. The reader is referred to Salembier and Garrido [Sal00] to see examples of this situation involving a binary partition tree.

Nonincreasing Criterion ⇒ Modification of the Decision (Viterbi Algorithm) and Pruning

This situation is illustrated here by a motion-oriented connected operator [Sal98]. Denote by $f_t(n)$ an image sequence, where n represents the coordinates of the pixels and t the time instant. The goal of the connected operator is to eliminate the image components that do not undergo a given motion. The first step is therefore to define the motion model that gives, for example, the displacement field at each position $\Delta(n)$. The field can be constant Δ if one wants to extract all objects following a translation, but in general the displacement can depend on the spatial position n to deal with more complex motion models such as affine or quadratic models.

The sequence processing is performed as follows: each frame is transformed into its corresponding max-tree representation, and each node $\mathcal{N}_k$ is analyzed. To check whether the pixels contained in a given node $\mathcal{N}_k$ are moving in accordance with the motion field $\Delta(n)$, a simple solution consists in computing the mean displaced frame difference (DFD) of this region with the previous frame:

$$\mathrm{DFD}_{f_t}^{f_{t-1}}(\mathcal{N}_k) = \frac{\sum_{n \in \mathcal{N}_k} |f_t(n) - f_{t-1}(n - \Delta(n))|}{\sum_{n \in \mathcal{N}_k} 1}. \tag{9.20}$$

In practice, however, it is not very reliable to assess the motion on the basis of only two frames. The criterion should include a reasonable memory of past decisions. This idea can be easily introduced in the criterion by adding a recursive term. Two mean DFDs are measured: one between the current frame f_t and the previous frame f_{t-1} and a second one between the current frame and the previous filtered frame $\psi(f_{t-1})$ (ψ denotes the connected operator). The motion criterion is finally defined as

$$\mathrm{motion}(\mathcal{N}_k) = \alpha \mathrm{DFD}_{f_t}^{f_{t-1}}(\mathcal{N}_k) + (1 - \alpha)\mathrm{DFD}_{f_t}^{\psi(f_{t-1})}(\mathcal{N}_k), \tag{9.21}$$

with $0 \le \alpha \le 1$. If α is equal to 1, the criterion is memoryless, whereas low values of α allow the introduction of an important recursive component into the decision process. In a way similar to all recursive filtering schemes, the selection of a proper value for α depends on the application: if one wants to detect very rapidly any changes in motion, the criterion should be mainly memoryless ($\alpha \approx 1$), whereas if a more reliable decision involving the observation of a larger number of frames is necessary, then the system should rely heavily on the recursive part ($0 \le \alpha \ll 1$).

The motion criterion described by Eqs. (9.20) and (9.21) deals with one set of motion parameters. Objects that do not follow the given motion produce a high DFD and should be removed. The criterion is not increasing, so the Viterbi algorithm has to be used. This motion-oriented pruning strategy can be used on max-tree, min-tree, or binary partition tree representation.

A motion filtering example relying on a max-tree representation is shown in Fig. 9.22. The objective of the operator is to remove all moving objects. The motion model is defined by $\Delta(n) = (0,0), \forall n$. In this sequence, all objects are still except the ballerina behind the two speakers and the speaker on the left side who is speaking. The application of the connected operator $\psi(f)$ described previously removes all bright moving objects (Fig. 9.22b). The application of the dual operator $\psi^*(f) = -\psi(-f)$ removes all dark moving objects (Fig. 9.22c). The residue (that is, the difference with the original image) presented in Fig. 9.22d shows what has been removed by the operator. As can be seen, the operator has very precisely extracted the ballerina and the moving details of the speaker's face.

Figure 9.22: Example of a motion connected operator preserving fixed objects: (a) original frame f, (b) motion connected operator ψ, (c) dual operator $\psi^*\psi(f)$, and (d) residue, $f - \psi^*\psi(f)$.

The motion connected operator can potentially be used for a large set of applications. It permits, in particular, different ways of handling the motion information. Indeed, motion information generally is measured without knowing anything about the image structure. Connected operators take a different viewpoint by making decisions on the basis of the analysis of flat zones. By using motion connected operators, we can "inverse" the classical approach to motion and, for example, analyze simplified sequences in which objects are following a known motion. Various connected operators involving nonincreasing criteria such as entropy, simplicity, and perimeter can be found in the literature [Sal98, Sal00].

9.5.3 Pruning Strategies Involving a Global Optimization Under Constraint

In this section, we illustrate a more complex pruning strategy involving a global optimization under constraint. To fix the notations, let us denote by C the criterion that has to be optimized (we assume, without loss of generality, that the criterion has to be minimized) and by $\mathcal{K}$ the constraint. The problem is to minimize the criterion C with the restriction that the constraint $\mathcal{K}$ is below a given threshold $\mathcal{T}_{\mathcal{K}}$. Moreover, we assume that both the criterion and the constraint are additive over the regions represented by the nodes $\mathcal{N}_k$: $C = \sum_{\mathcal{N}_k} C(\mathcal{N}_k)$ and

$\mathcal{K} = \sum_{\mathcal{N}_k} \mathcal{K}(\mathcal{N}_k)$. The problem is therefore to define a pruning strategy such that the resulting partition is composed of nodes $\mathcal{N}_i$

$$\min \sum_{\mathcal{N}_i} C(\mathcal{N}_i), \qquad \text{with} \sum_{\mathcal{N}_i} \mathcal{K}(\mathcal{N}_i) \leq \mathcal{T}_{\mathcal{K}}. \tag{9.22}$$

It has been shown [Sho88] that this problem can be reformulated as the minimization of the Lagrangian: $C + \lambda\mathcal{K}$, where λ is the so-called Lagrange parameter. Both problems have the same solution if we find λ^* such that $\mathcal{K}$ is equal (or very close) to the constraint threshold $\mathcal{T}_{\mathcal{K}}$. Therefore, the problem consists of pruning the tree to find a set of nodes that create a partition

$$\min \left(\sum_{\mathcal{N}_i} C(\mathcal{N}_i) + \lambda^* \sum_{\mathcal{N}_i} \mathcal{K}(\mathcal{N}_i) \right). \tag{9.23}$$

Assume, in a first step, that the optimum λ^* is known. In this case, the pruning is done by a bottom-up analysis of the tree. If the Lagrangian value corresponding to a given node $\mathcal{N}_0$ is smaller than the sum of the Lagrangians of the children nodes $\mathcal{N}_i$, then the children are pruned:

$$\text{If } C(\mathcal{N}_0) + \lambda^* \mathcal{K}(\mathcal{N}_0) < \sum_{\mathcal{N}_i} C(\mathcal{N}_i) + \lambda^* \sum_{\mathcal{N}_i} \mathcal{K}(\mathcal{N}_i), \text{ prune the children nodes } \mathcal{N}_i. \tag{9.24}$$

This procedure is iterated up to the root node. In practice, of course, the optimum λ^* parameter is not known, and the previous bottom-up analysis of the tree is embedded in a loop that searches for the best λ parameter. The computation of the optimum λ parameter can be done with a gradient search algorithm. The bottom-up analysis itself is not expensive in terms of computation since the algorithm simply has to perform a comparison of Lagrangians for all nodes of the tree. The parts of the algorithm that might be expensive are the computation of the criterion and the constraint values associated with the regions. Note, however, that this computation has to be done once. Finally, the theoretical properties depend mainly on the criterion and on the constraint. In any case, the operator is connected and self-dual.

This type of pruning strategy is illustrated by two examples relying on a binary partition tree representation. In the first example, the goal of the connected operator is to simplify the input image by minimizing the number of flat zones of the output image: $C_1 = \sum_{\mathcal{N}_k} 1$. In the second example, the criterion is to minimize the total length of the contours of the flat zones: $C_2 = \sum_{\mathcal{N}_k} \text{perimeter}(\mathcal{N}_k)$. In both cases, the criterion has no meaning if there is no constraint because the algorithm would prune all nodes. The constraint we use is to force the output image to be a faithful approximation of the input image: the squared error between the input and the output images $\mathcal{K} = \sum_{\mathcal{N}_k} \sum_{n \in \mathcal{N}_k} [\psi(f)(n) - f(n)]^2$ is constrained to be below a given quality threshold. In the examples shown in Fig. 9.23, the squared error is constrained to be of at least 31 dB. Figure 9.23a shows the output image when the criterion is the number of flat zones. The image is visually a

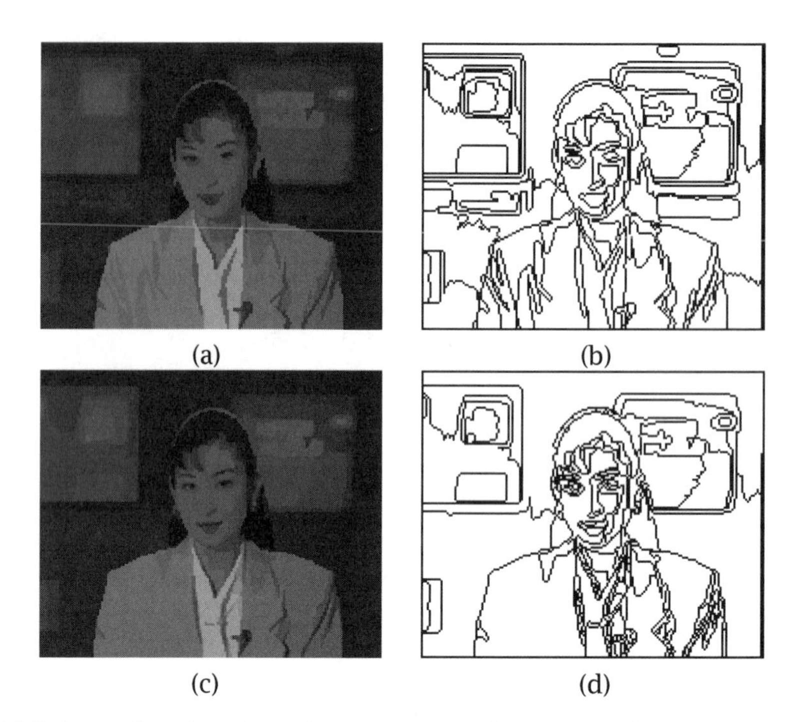

Figure 9.23: Examples of optimization strategies under a squared error constraint of 31 dB: (a) minimization of the number of flat zones and (b) contours of the flat zones (the number of flat zones is 87 and the perimeter length is 4491); (c) minimization of the total perimeter length and (d) contours of the flat zones (the number of flat zones is 219 and the perimeter length is 3684).

good approximation of the original image, but it involves a much lower number of flat zones: the original image is composed of 14335 flat zones, whereas only 87 flat zones are present in the filtered image. The second criterion is illustrated in Fig. 9.23c. The approximation provided by this image is of the same quality as the previous one (squared error of 31 dB). However, the characteristics of its flat zones are quite different. The total length of the perimeter of its flat zones is equal to 3684 pixels, whereas the example of Fig. 9.23a involves a total perimeter length of 4491 pixels. The reduction of perimeter length is obtained at the expense of a drastic increase of the number of flat zones: 219 instead of 87. Figures 9.23b and 9.23d show the flat zone contours. As can be seen, the flat zone contours are more complex in the first example, but the number of flat zones is higher in the second one.

This kind of strategy can be applied for a large number of criteria and constraints. Note that without defining a tree structure such as a max-tree, min-tree, or binary partition tree, it would be extremely difficult to implement these kinds of connected operators.

9.6 Conclusions

This chapter has presented and discussed a region-based processing technique involving connected operators. There is currently an interest in defining processing tools that do not act on the pixel level but on a region level. Connected operators are examples of such tools that come from mathematical morphology.

Connected operators are operators that process the image by merging flat zones. As a result, they cannot introduce any contours or move existing ones. The two most popular approaches to create connected operators have been reviewed. The first one works on a pixel-based representation of the image and involves a reconstruction process. The operator involves first a simplification step based on a "classical" operator (such as morphological open, close, low-pass filter, median filter, etc.) and then a reconstruction process. Three kinds of reconstruction processes have been analyzed: antiextensive, extensive, and self-dual. The goal of the reconstruction process is to restore the contour information after the simplification. In fact, the reconstruction can be seen as a way to create a connected version of an arbitrary operator. Note that the simplification effect is defined and limited by the first step. The examples we have shown involved simplification in terms of size or contrast.

The second strategy to create connected operators involves three steps: In the first step, a region-based representation of the input image is constructed. Three examples have been discussed: max-tree, min-tree, and binary partition tree. In the second step, the simplification is obtained by pruning the tree, and in the third step, the output image is constructed from the pruned tree. The tree creation defines the set of regions that the pruning strategy can use to create the final partition. It represents a compromise between flexibility and efficiency: on the one hand, not all possible mergings of flat zones are represented in the tree, but on the other hand, once the tree has been defined, complex pruning strategies can be defined. In particular, it is possible to deal in a robust way with nonincreasing criteria. Criteria involving the notions of area, motion, and optimization under a quality constraint have been demonstrated.

Appendix: Fast Implementation of the Max-Tree Creation

This section describes an example of fast implementation of the max-tree creation. This implementation is valid for the classical case in which the connected components are defined by 4- or 8-connectivity. The implementation relies on the use of a hierarchical FIFO queue, that is, a set of first-in-first-out queues in which each individual queue is assigned to a particular gray level value h. These queues are used to define an appropriate scanning and processing order of the pixels. To create the max-tree, the following three queue functions are necessary:

- **hqueue-add(h,p)**: Add the pixel p (of gray level h) in the queue of priority h.
- **hqueue-first(h)**: Extract the first available pixel in the queue of priority h.
- **hqueue-empty(h)**: Return "TRUE" if queue of priority h is empty.

We also make use of the following notations: **number-nodes(h)** defines the number of nodes $\mathcal{N}_h^k$ at level h. Its values are initialized to zero at the beginning of the tree construction. ORI(p) denotes the original gray level value of pixel p and STATUS(p) stores the information of the pixel status: the pixel can be "Not-analyzed," "In-the-queue," or assigned to the node k of level h. In this last case, STATUS(p) $= k$. As can be seen, pixel p belongs to the node $\mathcal{N}_{\mathrm{ORI}(p)}^{\mathrm{STATUS}(p)}$. The max-tree definition given in Sec. 9.5.1 implies a severe redundancy in the node content. Indeed, a child node represents a connected component of the space that is totally included in its father node. For practical implementation, this is inefficient in terms of memory requirements. The algorithm that is described here avoids this redundancy by storing in each node $\mathcal{N}_k$ only the pixel locations that belong to $\mathcal{N}_k$ and not to its father. In practice, if one wants to access to all of the pixel locations belonging to $\mathcal{N}_k$, the node and all of its descendants have to be scanned.

The max-tree creation relies on a simple recursive flooding procedure. The STATUS is initialized to "Not-analyzed," and one of the pixels of lowest gray level value $h_{\min}$ is put in the queue. The tree is created by calling flood($h_{\min}$). The flooding procedure flood(h) is precisely described in Fig. 9.24. It has two basic steps: the first actually performs the propagation and the updating of the STATUS, whereas the second defines the father–child relationships. The execution time is typically of less than one second for 256×256 images of 256 gray levels on a 200 MHz Pentium system.

```
flood(h)                                              /* Flood gray level h */
                                                      /* Step 1:  propagation */
    while not hqueue-empty(h)
        p ← hqueue-first(h)
        STATUS(p) ← number-nodes(h)                   /* Process p */
        for every neighbor q of p                     /* 4 or 8 connectivity */
            if STATUS(q) == ''Not-analyzed''
                hqueue-add(ORI(q), q)
                STATUS(q) ← ''In-the-queue''
                node-at-level(ORI(p)) ← true
                if (ORI(q) > ORI(p))                  /* A child at level q found */
                    m = ORI(q)
                    repeat                            /* Flood the child */
                        m ← flood(m)
                    until m = h
    number-nodes(h) ← number-nodes(h) + 1
                                                      /* Step 2:  define the father */
    m ← h - 1
    while m ≥ 0 and node-at-level(m) = false          /* Look for the father */
        m ← m - 1
    if m ≥ 0                                          /* Assign the father */
        i ← number-nodes(h) - 1
        j ← number-nodes(m)
        father of N_h^i ← node N_m^j
    else
        N_h^i has no father (N_h^i is root node)
    node-at-level(h) ← false
    return m
```

Figure 9.24: Flooding procedure for the max-tree creation.

References

[Bre96] E. Breen and R. Jones. An attribute-based approach to mathematical morphology. In *Proc. International Symposium on Mathematical Morphology* (P. Maragos, R.W. Schafer, and M.A. Butt, eds.), pp. 41–48. Kluwer, Amsterdam (1996).

[Cre95] J. Crespo, J. Serra, and R.W. Schafer. Theoretical aspects of morphological filters by reconstruction. *Signal Process.* **47**(2), 201–225 (1995).

[Cre96] J. Crespo. Space connectivity and translation-invariance. In *Proc. International Symposium on Mathematical Morphology* (P. Maragos, R.W. Schafer, and M.A. Butt, eds.), pp. 118–126. Kluwer, Amsterdam (1996).

[Cre97] J. Crespo, R.W. Shafer, J. Serra, C. Gratin, and F. Meyer. A flat zone approach: a general low-level region merging segmentation method. *Signal Process.* **62**(1), 37–60 (October 1997).

[Gar98] L. Garrido, P. Salembier, and D. Garcia. Extensive operators in partition lattices for image sequence analysis. *Signal Process.* **66**(2), 157–180 (April 1998).

[Gom99] C. Gomila and F. Meyer. Levelings in vector space. In *Proc. IEEE Intl. Conf. on Image Processing* (Kobe, Japan, October 1999).

[Gri92] M. Grimaud. A new measure of contrast: the dynamics. In *Visual Communications and Image Processing '92* (Serra, Gader, and Dougherty eds.). Proc. SPIE Vol. 1769, pp. 292–305 (1992).

[Hei97a] H. Heijmans. Connected morphological operators and filters for binary images. In *Proc. IEEE International Conference on Image Processing* **2**, pp. 211–214 (Santa Barbara, CA 1997).

[Hei97b] H. Heijmans. Connected morphological operators for binary images. Tech. Report PNA-R9708. CWI, Amsterdam, The Netherlands (1997).

[Hor74] S. L. Horowitz and T. Pavlidis. Picture segmentation by a directed split-and-merge procedure. In *Proc. Second International Joint Conference on Pattern Recognition*, pp. 424–433 (1974).

[Kle76] J.C. Klein. Conception et réalisation d'une unité logique pour l'analyse quantitative d'images. Ph.D. thesis, Nancy University, France (1976).

[Mat97] G. Matheron. Les nivellements. Tech. Report, Paris School of Mines, Center for Mathematical Morphology, France (1997).

[Mey97] F. Meyer, A. Oliveras, P. Salembier, and C. Vachier. Morphological tools for segmentation: connected filters and watersheds. *Ann. Télécommun.* **52**(7-8), 366–379 (July-August 1997).

[Mey98a] F. Meyer. From connected operators to levelings. In *Proc. Fourth International Symposium on Mathematical Morphology*, pp. 191–198. Kluwer, Amsterdam (1998)

[Mey98b] F. Meyer. The levelings. In *Proc. Fourth International Symposium on Mathematical Morphology*, pp. 199–206. Kluwer, Amsterdam (1998).

[Mor86] O. Morris, M. Lee, and A. Constantinidies. Graph theory for image analysis: an approach based on the shortest spanning tree. *IEE Proc., F* **133**(2), 146–152 (April 1986).

[Ron98] C. Ronse. Set-theoretical algebraic approaches to connectivity in continuous or digital spaces. *J. Mathemat. Imaging Vision* **8**, pp.41–58 (1998).

[Sal92] P. Salembier and M. Kunt. Size-sensitive multiresolution decomposition of images with rank order based filters. *Signal Process.* **27**(2), 205–241 (May 1992).

[Sal95] P. Salembier and J. Serra. Flat zones filtering, connected operators and filters by reconstruction. *IEEE Trans. Image Process.*, **3**(8), 1153–1160 (August 1995).

[Sal98] P. Salembier, A. Oliveras, and L. Garrido. Anti-extensive connected operators for image and sequence processing. *IEEE Trans. Image Process.* **7**(4), 555–570 (April 1998).

[Sal00] P. Salembier and L. Garrido. Binary partition tree as an efficient representation for image processing, segmentation and information retrieval. *IEEE Trans. Image Process.* **9**(4), 561–576 (April 2000).

[Ser82] J. Serra. *Image Analysis and Mathematical Morphology.* Academic Press (1982).

[Ser88] J. Serra. *Image Analysis and Mathematical Morphology, Vol II: Theoretical Advances.* Academic Press (1988).

[Ser93] J. Serra and P. Salembier. Connected operators and pyramids. In *Image Algebra and Mathematical Morphology*, Proc. SPIE Vol. 2030, pp. 65–76 (1993).

[Ser98] J. Serra. Connectivity on complete lattices. *J. Mathemat. Imaging Vision* **9**, 231–251 (1998).

[Sho88] Y. Shoham and A. Gersho. Efficient bit allocation for an arbitrary set of quantizers. *IEEE Trans. Acoust. Speech Signal Process.* **36**, 1445–1453 (September 1988).

[Vil98] V. Vilaplana and F. Marqués. Face segmentation using connected operators. In *Proc. of the Fourth International Symposium on Mathematical Morphology*, pp. 207–214, Kluwer, Amsterdam (1998).

[Vin89] L. Vincent. Graphs and mathematical morphology. *Signal Process.* **16**(4), 365–388 (April 1989).

[Vin93a] L. Vincent. Grayscale area openings and closings, their efficient implementation and applications. In *Proc. First Workshop Mathematical Morphology and Its Applications to Signal Processing* (J. Serra and P. Salembier, eds.), pp. 22–27. UPC, Barcelona (1993).

[Vin93b] L. Vincent. Morphological gray scale reconstruction in image analysis: Applications and efficients algorithms. *IEEE Trans. Image Process.* **2**(2), 176–201 (April 1993).

[Vit79] A.J. Viterbi and J.K. Omura. *Principles of Digital Communications and Coding*. McGraw-Hill, New York (1979).

10

Differential Morphology

PETROS MARAGOS

Department of Electrical and Computer Engineering
National Technical University of Athens
Athens, Greece

10.1 Introduction

Morphological image processing has been based traditionally on modeling images as sets or as points in a complete lattice of functions and viewing morphological image transformations as set or lattice operators. Thus, so far, the two classic approaches to analyzing or designing the deterministic systems of mathematical morphology have been (1) geometry, by viewing them as image set transformations in Euclidean spaces, and (2) algebra, to analyze their properties using set or lattice theory. Geometry was used mainly for intuitive understanding, and algebra was restricted to the space domain. Despite their limitations, these approaches have produced a powerful and self-contained broad collection of nonlinear image analysis concepts, operators, and algorithms. In parallel with these directions, there is a recently growing part of morphological image processing that is based on ideas from differential calculus and dynamical systems. It combines some early ideas on using simple morphololological operations to obtain signal gradients with some recent ideas on using differential equations to model nonlinear multiscale processes or distance propagation in images. In this chapter we present a unified view of the various interrelated ideas in this area and develop some systems analysis tools in both the space and a (slope) transform domain.

The main tools of morphological image processing are a broad class of nonlinear image operators, of which the two most fundamental are dilation and erosion.

The space domain in which images are defined can be either continuous, $\mathbf{E} = \mathbf{R}^2$, or discrete, $\mathbf{E} = \mathbf{Z}^2$. For a binary image represented by a set $S \subseteq \mathbf{E}$, its morphological dilation by another planar set B, denoted by $S \oplus B$, and its erosion, denoted by $S \ominus B$, are the Minkowski set operations

$$S \oplus B \triangleq \{\mathbf{x} + \mathbf{b} : \mathbf{x} \in S, \mathbf{b} \in B\}, \tag{10.1}$$

$$S \ominus B \triangleq \{\mathbf{x} : B_{+\mathbf{x}} \subseteq S\}, \tag{10.2}$$

where $\mathbf{x} = (x, y)$ denotes points on the plane and $B_{+\mathbf{x}} \triangleq \{\mathbf{x}+\mathbf{b} : \mathbf{b} \in B\}$ denotes set translation by a vector. The corresponding signal operations are the Minkowski dilation and erosion of an image function $f : \mathbf{E} \to \overline{\mathbf{R}}$ by another (structuring) function g:

$$f \oplus g(\mathbf{x}) \triangleq \bigvee_{\mathbf{y} \in \mathbf{E}} f(\mathbf{y}) + g(\mathbf{x} - \mathbf{y}), \tag{10.3}$$

$$f \ominus g(\mathbf{x}) \triangleq \bigwedge_{\mathbf{y} \in \mathbf{E}} f(\mathbf{y}) - g(\mathbf{y} - \mathbf{x}), \tag{10.4}$$

where $\vee$ and $\wedge$ denote supremum and infimum. The signal range is a subset of $\overline{\mathbf{R}} = \mathbf{R} \cup \{-\infty, +\infty\}$. The scalar addition in $\overline{\mathbf{R}}$ is like addition in $\mathbf{R}$ extended by the rules $r \pm \infty = \pm\infty$, $\forall r \in \mathbf{R}$ and $(+\infty) + (-\infty) = -\infty$. In convex analysis [Roc70] and optimization, the nonlinear operation $\oplus$ is called *supremal convolution*, and an operation closely related to $\ominus$ is the *infimal convolution*

$$f \oplus' g(\mathbf{x}) \triangleq \bigwedge_{\mathbf{y} \in \mathbf{E}} f(\mathbf{y}) +' g(\mathbf{x} - \mathbf{y}) \tag{10.5}$$

where $+'$ is like the extended addition in $\overline{\mathbf{R}}$ except that $(+\infty) +' (-\infty) = +\infty$. A simple case of the signal dilation and erosion results when g is *flat*, that is, equal to 0 over its support set B and $-\infty$ elsewhere. Then the weighted dilation and erosion of f by g reduce to the flat dilation and erosion of f by B:

$$f \oplus B(\mathbf{x}) \triangleq \bigvee_{\mathbf{y} \in B} f(\mathbf{x} - \mathbf{y}), \qquad f \ominus B(\mathbf{x}) \triangleq \bigwedge_{\mathbf{y} \in B} f(\mathbf{x} + \mathbf{y}).$$

Additionally, a wide variety of (simple or complex) parallel and/or serial interconnections of the basic morphological operations, called generally "morphological systems," have found a broad range of applications in image processing and computer vision; examples include problems in nonlinear filtering, noise suppression, contrast enhancement, geometric feature detection, skeletonization, multiscale analysis, size distributions, segmentation, and shape recognition (see [Ser82, Ser88, Mar90, Hei94, Mar98] for broad surveys and more references).

Among the very few early connections between morphology and calculus were the *morphological gradients*. Specifically, given a function $f : \mathbf{R}^d \to \mathbf{R}$, with $d = 1, 2$, its isotropic morphological *sup-derivative* at a point x is defined by

$$\mathcal{M}f(x) \triangleq \lim_{r \downarrow 0} \frac{f \oplus rB(x) - f(x)}{r}, \tag{10.6}$$

where $rB = \{rb : b \in B\}$ is a d-dimensional disk B scaled to radius r. The derivative $\mathcal{M}$ has been used in morphological image analysis for edge detection. It actually becomes equal to $\|\nabla f\|$ when f is differentiable.

A more recent application area in which calculus-based ideas have been used in morphology is that of *multiscale image analysis.* Detecting features, motion, and objects as well as modeling many other information extraction tasks in image processing and computer vision has necessitated the analysis of image signals at multiple scales. Following the initial formulation of multiscale image analysis using Gaussian convolutions by Marr and his co-workers [Mar82] were two other important developments, the continuous Gaussian scale-space by Witkin [Wit83] and the observation by Koenderink [Koe84] that this scale-space can be modeled via the heat diffusion partial differential equation (PDE). Specifically, if

$$u(x,y,t) = \iint_{\mathbf{R}^2} f(x-v, y-w) G_t(v,w) \mathrm{d}v \mathrm{d}w \tag{10.7}$$

is the multiscale linear convolution of an original image signal $f(x,y)$ with a Gaussian function $G_t(x,y) = \exp[-(x^2+y^2)/4t]/4\pi t$ whose variance (= $2t$) is proportional to scale t, then the scale-space function u can be generated from the isotropic and homogeneous heat diffusion PDE[1]:

$$u_t = \nabla^2 u = u_{xx} + u_{yy}, \tag{10.8}$$

with initial condition $u(x,y,0) = f(x,y)$. The popularity of this approach is due to its linearity and its relation to the heat PDE, about which much is known from physics and mathematics. The big disadvantage of the Gaussian scale-space approach is the fact that linear smoothers blur and shift important image features, for example, edges. There is, however, a variety of *nonlinear* smoothing filters that can smooth while preserving important image features and can provide a multiscale image ensemble, that is, a nonlinear scale-space. A large class of such filters consists of the standard morphological *openings* and *closings* (which are serial compositions of erosions and dilations) in a multiscale formulation [Che89, Mar89, Mat75, Ser82] and their lattice extensions, of which the most notable are the *reconstruction* openings and closings [Sal95, Vin93]. These reconstruction filters, starting from a reference image f consisting of several parts and a marker g (initial seed) inside some of these parts, can reconstruct whole objects with exact preservation of their boundaries and edges. In this reconstruction process they simplify the original image by completely eliminating smaller objects inside which the marker cannot fit. A detailed discussion of reconstruction filters and their applications can be found in Chapter 9.

Until recently the vast majority of implementations of multiscale morphological filtering had been discrete. In 1992 three teams of researchers independently published nonlinear PDEs that model the continuous multiscale morphological

[1] Notation for PDEs: $u_t = \partial u/\partial t$, $u_x = \partial u/\partial x$, $u_y = \partial u/\partial y$, $\nabla u = (u_x, u_y)$, $\mathrm{div}\,((v,w)) = \nabla \cdot (v,w) = v_x + w_y$.

scale-space. Specifically, Alvarez et al. [Alv92, Alv93] obtained PDEs for multiscale flat dilation and erosion by compact convex structuring sets as part of their general work on developing PDE-based models for multiscale image processing that satisfy certain axiomatic principles. Brockett and Maragos [Bro92, Bro94] developed nonlinear PDEs that model multiscale morphological dilation, erosion, opening, and closing by compact support structuring elements that are either convex sets or concave functions and may have nonsmooth boundaries. Their work was based on the semigroup structure of the multiscale dilation and erosion operators and the use of morphological derivatives to deal with the development of shocks. Boomgaard and Smeulders [Boo92, Boo94] obtained PDEs for multiscale dilation and erosion by studying the propagation of the boundaries of 2D sets and signal graphs under multiscale dilation and erosion. Their work applies to convex structuring elements whose boundaries contain no linear segments, are smooth, and possess a unique normal at each point.

To illustrate the basic idea behind morphological PDEs, we consider a 1D example, for which we define the multiscale flat dilation and erosion of a 1D signal $f(x)$ by the set $[-t, t]$ as the scale-space functions

$$\delta(x,t) = \bigvee_{|y|\leq t} f(x-y), \qquad \varepsilon(x,t) = \bigwedge_{|y|\leq t} f(x+y).$$

An example is shown in Fig. 10.1. The PDEs generating these multiscale flat dilations and erosions are

$$\begin{aligned} \delta_t &= |\delta_x|, \qquad \varepsilon_t = -|\varepsilon_x|, \\ \delta(x,0) &= \varepsilon(x,0) = f(x). \end{aligned} \tag{10.9}$$

In parallel to the development of the above ideas, there have been some advances in the field of differential geometry for evolving curves or surfaces using level set methods. Specifically, Osher and Sethian [Osh88, Set96] have developed PDEs of the Hamilton-Jacobi type to model the propagation of curves embedded as level curves (isoheight contours) of functions evolving in scale-space. Further, to solve these PDEs they developed robust numerical algorithms based on stable and shock-capturing schemes that had been formulated to solve similar shock-producing nonlinear wave PDEs [Lax73]. Kimia et al. [Kim90, Tek98] have applied and extended these curve evolution ideas to shape analysis in computer vision. Arehart et al. [Are93] and Sapiro et al. [Sap93] implemented continuous-scale morphological dilations and erosions using the numerical algorithms of curve evolution to solve the PDEs for multiscale dilation and erosion.

Multiscale dilations and erosions of binary images can also be obtained via distance transforms, that is, the distance function from the original image set. Discrete distance transforms can be implemented quickly via 2D min-sum difference equations, which are special cases of recursive erosions, developed by Rosenfeld and Pfaltz [Ros66] and Borgefors [Bor86]. Using Huygen's construction, the boundaries of multiscale dilations–erosions by disks can also be viewed as the wavefronts of a wave initiating from the original image boundary and propagating

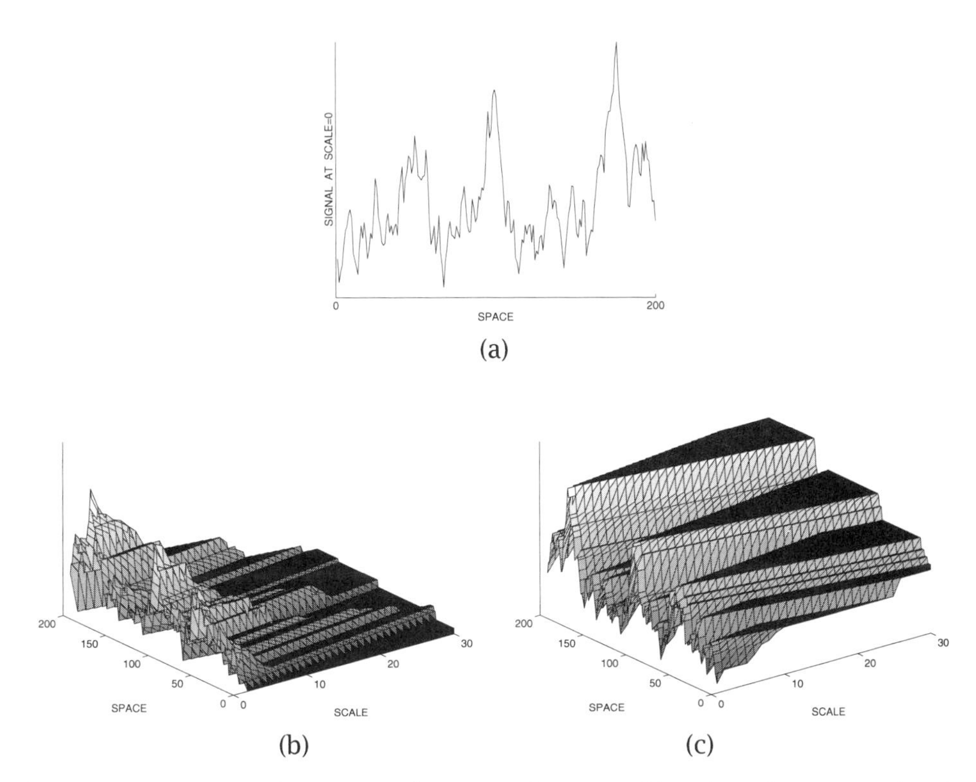

Figure 10.1: (a) Original 1D signal $f(x)$ at scale $t = 0$, (b) multiscale erosion $\varepsilon(x,t) = f \ominus tB(x)$, and (c) multiscale dilation $\delta(x,t) = f \oplus tB(x)$ of $f(x)$ by a set $B = [-1,1]$ for scales $t = [0,30]$.

with constant normal speed in a homogeneous medium [Blu73]. This idea can also be extended to heterogeneous media by using a weighted distance function, in which the weights are inversely proportional to the propagation speeds [Lev70]. In geometrical optics, the distance wavefronts are obtained from the isolevel contours of the solution of the eikonal PDE. This ubiquitous PDE has been applied to solving various problems in image analysis and computer vision such as shape-from-shading, gridless halftoning, and image segmentation.

In this chapter we discuss some close relationships between the morphological derivatives, the PDEs for multiscale morphology, the eikonal PDE of optics, and the difference equations used to implement distance transforms. The unifying theme is a collection of nonlinear differential–difference equations modeling the scale or space dynamics of morphological systems. We call this area *differential morphology*. Whereas classical morphological image processing is based on set and lattice theory, differential morphology offers calculus-based tools and some exciting connections to the physics of wave propagation. Some additional material on nonlinear PDEs and curve evolution as applied to image processing can be found in Chapter 8.

We also present analysis tools for the nonlinear systems used in differential morphology, which have many similarities with the tools used to analyze linear differential schemes. These tools apply either to the space domain or to a new transform domain, the *slope domain*. To understand their behavior in the slope domain, we discuss some nonlinear signal transforms, called *slope transforms*, whose properties and applications to morphological systems have some interesting conceptual similarities with Fourier transforms and their application to linear systems.

The chapter is organized as follows. We begin in Sec. 10.2 by presenting analytic methods for 2D morphological systems both in the spatial domain, using their impulse response and generating 2D nonlinear difference equations, as well as in the slope domain, using slope transforms. In Sec. 10.3 we discuss the basic PDEs for multiscale dilations and erosions, refine them using morphological derivatives, give their slope domain interpretation, describe PDEs for opening filters, and outline numerical algorithms for their implementation. Section 10.4 summarizes the main ideas from curve evolution as they apply to differential morphology. Section 10.5 deals with distance transforms for binary images and the analysis of their computation methods based on min-sum difference equations and slope filters. Finally, the eikonal PDE, its solution via weighted distance transforms, and some of its applications to image processing are discussed in Sec. 10.6.

10.2 2D Morphological Systems and Slope Transforms

10.2.1 2D Morphological Systems

A 2D signal operator or system Ψ is generally called

dilation if $\Psi(\bigvee_i f_i) = \bigvee_i \Psi(f_i)$ for any signal collection $\{f_i\}$;

erosion if $\Psi(\bigwedge_i f_i) = \bigwedge_i \Psi(f_i)$;

shift-invariant if $\Psi[f(\mathbf{x}-\mathbf{y})] = \Psi(f)(\mathbf{x}-\mathbf{y})$ for any signal $f(\mathbf{x})$ and shift $\mathbf{y}$;

translation-invariant if $\Psi[c + f(\mathbf{x}-\mathbf{y})] = c + \Psi(f)(\mathbf{x}-\mathbf{y})$ for any $f, \mathbf{y}$ and any constant c.

Of particular interest in this chapter are operators $\mathcal{E}$ that are *erosion and translation-invariant (ETI)* systems. Such systems are shift-invariant and obey an infimum-of-sums superposition:

$$\mathcal{E}\left[\bigwedge_i c_i + f_i(\mathbf{x})\right] = \bigwedge_i c_i + \mathcal{E}[f_i(\mathbf{x})]. \tag{10.10}$$

Similarly, dilation and translation-invariant (DTI) systems are shift-invariant and obey a supremum-of-sums superposition as in Eq. (10.10) but with $\bigwedge$ replaced by $\bigvee$. Two elementary signals useful for analyzing such systems are the *zero impulse* $\xi(\mathbf{x})$ and the *zero step* $\zeta(\mathbf{x})$:

$$\xi(\mathbf{x}) \triangleq \begin{cases} 0 & \text{for } \mathbf{x} = \mathbf{0}, \\ -\infty & \text{for } \mathbf{x} \neq \mathbf{0}, \end{cases} \qquad \zeta(\mathbf{x}) \triangleq \begin{cases} 0 & \text{for } \mathbf{x} \geq 0, \\ -\infty & \text{for } \mathbf{x} < 0, \end{cases}$$

where $\mathbf{x} = (x, y) \geq 0$ means that both x and y are greater than or equal to 0. Occasionally we shall refer to ξ as an "upper" impulse and to its negated version $-\xi$ as a "lower" impulse. A signal can be represented as a sup or inf of weighted upper or lower impulses; for example,

$$f(\mathbf{x}) = \bigwedge_{\mathbf{y}} f(\mathbf{y}) +' [-\xi(\mathbf{x} - \mathbf{y})]. \tag{10.11}$$

If we define the lower impulse response of an ETI system $\mathcal{E}$ as its output $g_\wedge = \mathcal{E}(-\xi)$ when the input is the zero lower impulse, we find that the system's action is equivalent to the infimal convolution of the input with its lower impulse response [Mar94]:

$$\mathcal{E} \text{ is ETI} \iff \mathcal{E}(f) = f \oplus' g_\wedge, \quad g_\wedge \triangleq \mathcal{E}(-\xi).$$

Similarly, a system $\mathcal{D}$ is DTI iff $\mathcal{D}(f) = f \oplus g_\vee$, where $g_\vee \triangleq \mathcal{D}(\xi)$ is the system's upper impulse response. Thus, DTI and ETI systems are uniquely determined in the spatial domain by their impulse responses, which also control their causality and stability [Mar94].

To create a transform domain for morphological systems, we first note that the planes $f(\mathbf{x}) = \mathbf{s} \cdot \mathbf{x} + c$ are either eigenfunctions of a DTI system $\mathcal{D}$ or an ETI system $\mathcal{E}$ since

$$\mathcal{D}[\mathbf{s} \cdot \mathbf{x} + c] = \mathbf{s} \cdot \mathbf{x} + c + G_\vee(\mathbf{s}),$$
$$\mathcal{E}[\mathbf{s} \cdot \mathbf{x} + c] = \mathbf{s} \cdot \mathbf{x} + c + G_\wedge(\mathbf{s}),$$

where $\mathbf{s} \cdot \mathbf{x} \triangleq s_1 x + s_2 y$ for $\mathbf{s} = (s_1, s_2)$ and $\mathbf{x} = (x, y)$ in $\mathbf{R}^2$ and where

$$G_\vee(\mathbf{s}) \triangleq \bigvee_{\mathbf{x}} g_\vee(\mathbf{x}) - \mathbf{s} \cdot \mathbf{x}, \qquad G_\wedge(\mathbf{s}) \triangleq \bigwedge_{\mathbf{x}} g_\wedge(\mathbf{x}) - \mathbf{s} \cdot \mathbf{x}$$

are the corresponding eigenvalues, called, respectively, the upper and lower *slope responses* of the DTI and ETI systems. They measure the amount of shift in the intercept of the input hyperplanes with slope vector $\mathbf{s}$. They are also conceptually similar to the frequency response of linear systems.

10.2.2 Slope Transforms

Viewing the 2D slope response as a signal transform with variable slope vector, we define for any 2D signal $f(\mathbf{x})$ its upper slope transform as the 2D function $F_\vee : \mathbf{R}^2 \to \overline{\mathbf{R}}$, defined by

$$F_\vee(\mathbf{s}) \triangleq \bigvee_{\mathbf{x} \in \mathbf{R}^2} f(\mathbf{x}) - \mathbf{s} \cdot \mathbf{x}, \tag{10.12}$$

and as its lower slope transform[2] the function

$$F_\wedge(\mathbf{s}) \triangleq \bigwedge_{\mathbf{x} \in \mathbf{R}^2} f(\mathbf{x}) - \mathbf{s} \cdot \mathbf{x}. \tag{10.13}$$

[2]In convex analysis [Roc70], given a convex function h there uniquely corresponds another convex function $h^*(\mathbf{s}) = \bigvee_{\mathbf{x}} \mathbf{s} \cdot \mathbf{x} - h(\mathbf{x})$, called the *Fenchel conjugate* of h. The lower slope transform of h and its conjugate function are closely related since $h^*(\mathbf{s}) = -H_\wedge(\mathbf{s})$.

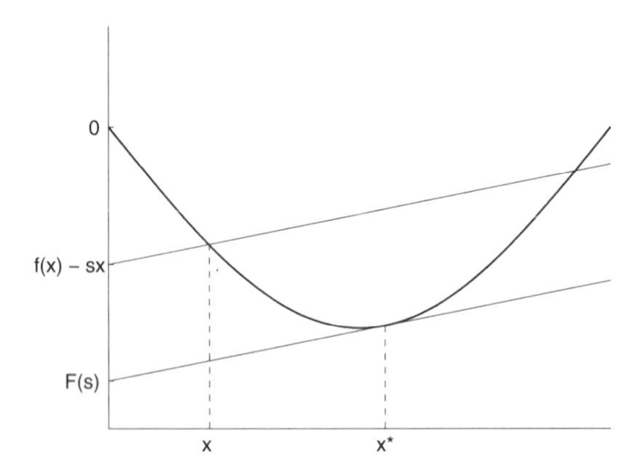

Figure 10.2: Convex signal f, its tangent with slope s, and a line parallel to the tangent.

As shown in Fig. 10.2 for a 1D signal $f(x)$, $f(x) - sx$ is the intercept of a line with slope s passing from the point $(x, f(x))$ on the signal's graph. Hence, for each s, the lower slope transform of f is the minimum value of this intercept. For *differentiable*[3] 1D signals, this minimum occurs when the above line becomes a tangent; for 2D signals the tangent line becomes a tangent plane. Examples of lower slope transforms are shown in Fig. 10.3.

In general, a 2D signal $f(\mathbf{x})$ is covered from above by all planes $F_\vee(\mathbf{s}) + \mathbf{s} \cdot \mathbf{x}$ whose infimum creates an *upper envelope*,

$$\hat{f}(x) \triangleq \bigwedge_{\mathbf{s} \in \mathbf{R}^2} F_\vee(\mathbf{s}) + \mathbf{s} \cdot \mathbf{x}, \tag{10.14}$$

and from below by planes $F_\wedge(\mathbf{s}) + \mathbf{s} \cdot \mathbf{x}$ whose supremum creates the *lower envelope*

$$\check{f}(x) \triangleq \bigvee_{\mathbf{s} \in \mathbf{R}^2} F_\wedge(\mathbf{s}) + \mathbf{s} \cdot \mathbf{x}. \tag{10.15}$$

We view the signal envelopes $\hat{f}(x)$ and $\check{f}(x)$ as the "inverse" upper and lower slope transforms of $f(\mathbf{x})$, respectively. Examples are shown in Fig. 10.4. The upper (lower) slope transform is always a convex (concave) function. Similarly, the upper (lower) envelope created by the "inverse" upper (lower) slope transform is always a concave (convex) function. Further, for any signal f,

$$\check{f} \leq f \leq \hat{f}. \tag{10.16}$$

[3]For differentiable signals, the maximization or minimization of the intercept $f(\mathbf{x}) - \mathbf{s} \cdot \mathbf{x}$ involved in both slope transforms can also be done, for a fixed $\mathbf{s}$, by finding its value at the stationary point $\mathbf{x}^*$ such that $\nabla f(\mathbf{x}^*) = \mathbf{s}$. This extreme value of the intercept (as a function of the slope $\mathbf{s}$) is the *Legendre transform* [Cou62] of the signal f. If f is convex (or concave) and has an invertible gradient, its Legendre transform is single-valued and equal to the lower (or upper) transform; otherwise, the Legendre transform is multivalued. This possibly multivalued transform has been defined as a "slope transform" [Dor94] and its properties are similar to those of the upper-lower slope transforms, but there are also some important differences [Mar94].

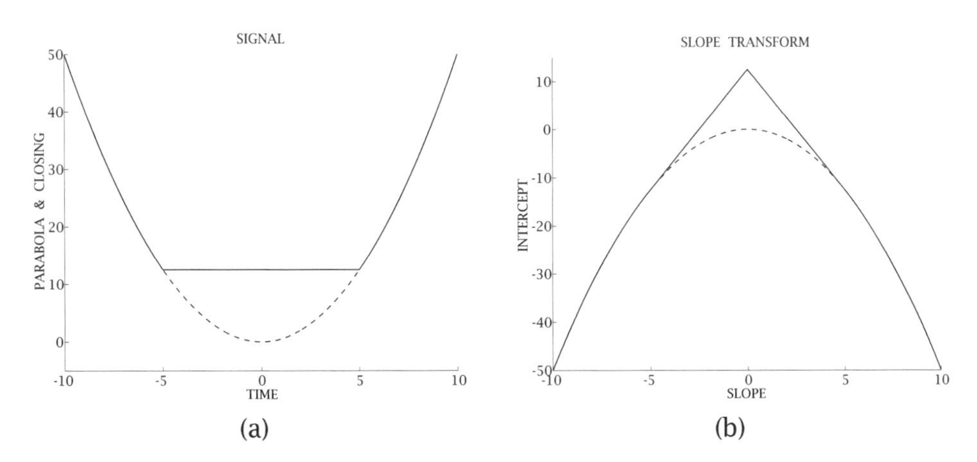

Figure 10.3: (a) Convex parabola signal $f(x) = x^2/2$ (dashed line) and its morphological closing (solid line) by a flat structuring element $[-5, 5]$. (b) Lower slope transform $F_{\wedge}(s) = -s^2/2$ of the parabola (dashed line) and of its closing (solid line).

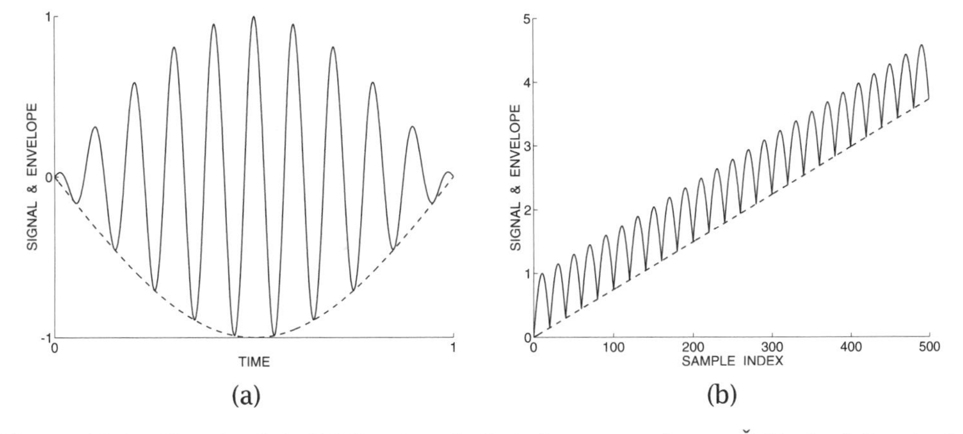

Figure 10.4: Signals f (solid lines) and their lower envelopes $\check{f}$ (dashed lines) obtained via the composition of the lower slope transform and its inverse. (a) Cosine whose amplitude has been modulated by a slower cosine pulse. (b) Impulse response f of a discrete ETI system generated by the min-sum difference equation $f(n) = \min[-\xi(n), \bigwedge_{1\leq k\leq 20} f(n-k) + a_k]$, where $a_k = \sin(\pi k/21)$.

Tables 10.1 and 10.2 list several properties and examples of the 2D upper slope transform. The most striking is that supremal convolution in the time-space domain corresponds to addition in the slope domain. Note the analogy with linear systems, in which linearly convolving two signals in space corresponds to multiplying their Fourier transforms. Very similar properties also hold for the 2D lower slope transform, the only differences being the interchange of suprema with infima, concave with convex, and the supremal $\oplus$ with the infimal convolution $\oplus'$.

Table 10.1: Properties of 2D Upper Slope Transform

Signal $f(\mathbf{x})$ [a]	Transform $F_\vee(\mathbf{s})$ [b]
$\bigvee_i c_i + f_i(\mathbf{x})$	$\bigvee_i c_i + F_i(\mathbf{s})$
$f(\mathbf{x}-\mathbf{x}_0)$	$F(\mathbf{s}) - \mathbf{s}\cdot\mathbf{x}_0$
$f(\mathbf{x}) + \mathbf{s}_0\cdot\mathbf{x}$	$F(\mathbf{s}-\mathbf{s}_0)$
$f(r\mathbf{x}),\quad r\in\mathbf{R}$	$F(\mathbf{s}/r)$
$rf(\mathbf{x}),\quad r>0$	$rF(\mathbf{s}/r)$
$(f\oplus g)(\mathbf{x})$	$(F+G)(\mathbf{s})$
$f(\mathbf{x})\le g(\mathbf{x})\quad\forall\mathbf{x}$	$F(\mathbf{s})\le G(\mathbf{s})\quad\forall\mathbf{s}$
$g(\mathbf{x}) = \begin{cases} f(\mathbf{x}), & \Vert\mathbf{x}\Vert_p \le r \\ -\infty, & \Vert\mathbf{x}\Vert_p > r \end{cases}$	$G(\mathbf{s}) = F(\mathbf{s}) \oplus' r\Vert\mathbf{s}\Vert_q,\quad \frac{1}{p}+\frac{1}{q}=1$
[a] $\mathbf{x}=(x,y)\in\mathbf{R}^2$	[b] $\mathbf{s}=(s_1,s_2)\in\mathbf{R}^2$

Table 10.2: Examples of 2D Upper Slope Transforms

Signal $f(\mathbf{x})$	Transform $F_\vee(\mathbf{s})$
$\mathbf{s}_0\cdot\mathbf{x}$	$-\xi(\mathbf{s}-\mathbf{s}_0)$
$\mathbf{s}_0\cdot\mathbf{x}+\zeta(\mathbf{x})$	$-\zeta(\mathbf{s}-\mathbf{s}_0)$
$\xi(\mathbf{x}-\mathbf{x}_0)$	$-\mathbf{s}\cdot\mathbf{x}_0$
$\zeta(\mathbf{x}-\mathbf{x}_0)$	$-\mathbf{s}\cdot\mathbf{x}_0 - \zeta(\mathbf{s})$
$\begin{cases} 0, & \Vert\mathbf{x}\Vert_p\le r, \\ -\infty, & \Vert\mathbf{x}\Vert_p > r, \end{cases}\quad p\ge 1$	$r\Vert\mathbf{s}\Vert_q,\quad \frac{1}{p}+\frac{1}{q}=1$
$-s_0\Vert\mathbf{x}\Vert_p,\quad s_0>0$	$\begin{cases} 0, & \Vert\mathbf{s}\Vert_q\le s_0 \\ +\infty, & \Vert\mathbf{s}\Vert_q > s_0 \end{cases}$
$\sqrt{1-x^2-y^2},\quad x^2+y^2\le 1$	$\sqrt{1+s_1^2+s_2^2}$
$-(x^2+y^2)/2$	$(s_1^2+s_2^2)/2$
$-(\vert x\vert^p+\vert y\vert^p)/p,\quad p>1$	$(\vert s_1\vert^q+\vert s_2\vert^q)/q$

Given a discrete-domain 2D signal $f(i,j)$, we define its lower slope transform as follows:

$$F_\wedge(s_1,s_2) = \bigwedge_{i=-\infty}^{\infty}\bigwedge_{j=-\infty}^{\infty} f(i,j) - (is_1 + js_2), \tag{10.17}$$

and likewise for its upper slope transform using $\bigvee$. The properties of these slope transforms for signals defined on the discrete plane are almost identical to the ones for signals defined on $\mathbf{R}^2$ (see [Mar94] for details).

A more general treatment of upper and lower slope transforms on complete lattices can be found in Heijmans and Maragos [Hei97].

10.2.3 Min-Sum Difference Equations and Discrete Slope Filters

The space dynamics of a large class of 2D discrete ETI systems can be described by the following general 2D min-sum difference equation:

$$u(i,j) = \left(\bigwedge_{(k,\ell)\in M_o} a_{k\ell} + u(i-k, j-\ell) \right) \wedge \left(\bigwedge_{(k,\ell)\in M_i} b_{k\ell} + f(i-k, j-\ell) \right), \tag{10.18}$$

which we view as a 2D discrete nonlinear system, mapping the input signal f to the output u. The masks M_o, M_i are pixel coordinate sets that determine which output and input samples will be added with constant weights to form the current output sample. Similarly, the dynamics of DTI systems can be described by max-sum difference equations as in Eq. (10.18) but with $\bigwedge$ replaced by $\bigvee$.

As explained by Dudgeon and Mersereau [Dud84], for 2D linear difference equations the recursive computability of Eq. (10.18) depends on (1) the shape of the output mask $M_o = \{(k,\ell) : a_{k\ell} < +\infty\}$ determining which past output samples are involved in the recursion, (2) the boundary conditions, that is, the locations and values of the output samples $u(i,j)$ that are prespecified as initial conditions, and (3) the scanning order in which the output samples should be computed. We assume boundary conditions of value $+\infty$ and of a shape (dependent on M_o and the scanning order) appropriate so that the difference equation is an ETI system recursively computable. Obviously, $(0,0) \notin M_o$. The nonrecursive part of Eq. (10.18) represents an infimal convolution of the input array $f(i,j)$ with the 2D finite-support structuring function $b(i,j) = b_{ij}$, which is well understood. Thus, we henceforth focus only on the recursive version of Eq. (10.18) by setting $b_{k\ell} = +\infty$, except for $b_{00} = 0$. This yields the autoregressive 2D min-sum difference equation,

$$u(i,j) = \left(\bigwedge_{(k,\ell)\in M_o} a_{k\ell} + u(i-k, j-\ell) \right) \wedge f(i,j). \tag{10.19}$$

If $g = \mathcal{E}(-\xi)$ is the impulse response of the corresponding ETI system $\mathcal{E} : f \mapsto u$, then $u = f \oplus' g$. Finding a closed-formula expression for g is generally not possible. However, we can first find the slope response G and then, via the inverse lower slope transform, find the impulse response g or its envelope $\breve{g}$. (For notational simplicity, we have dropped here the subscript $\wedge$ from g and G.) Let us consider the 2D finite-support signal of the mask coefficients,

$$a(i,j) \triangleq \begin{cases} a_{ij} & \text{for } (i,j) \in M_o, \\ +\infty & \text{otherwise,} \end{cases} \tag{10.20}$$

and its lower slope transform,

$$A_\wedge(s_1, s_2) = \bigwedge_{(i,j)\in M_o} a_{ij} - is_1 - js_2. \tag{10.21}$$

Then, rewriting Eq. (10.19) as

$$u = (u \oplus' a) \wedge f \tag{10.22}$$

and applying lower slope transforms to both sides yields

$$U_\wedge(\mathbf{s}) = [U_\wedge(\mathbf{s}) + A_\wedge(\mathbf{s})] \wedge F_\wedge(\mathbf{s}). \tag{10.23}$$

Since $U_\wedge(\mathbf{s}) = G(\mathbf{s}) + F_\wedge(\mathbf{s})$, assuming that $F_\wedge(\mathbf{s})$ is finite yields

$$G(\mathbf{s}) = \min[0, G(\mathbf{s}) + A_\wedge(\mathbf{s})]. \tag{10.24}$$

The largest nontrivial solution of Eq. (10.24) is [Mar96]

$$G(\mathbf{s}) = \begin{cases} 0 & \text{for } \mathbf{s} \in P, \\ -\infty & \text{for } \mathbf{s} \notin P, \end{cases} \tag{10.25}$$

where P is the convex planar region,

$$P = \{(s_1, s_2) : is_1 + js_2 \leq a_{ij} \;\forall (i,j) \in M_o\}. \tag{10.26}$$

Thus, the system acts as an *ideal-cutoff* spatial slope filter, passing all input lower slope vectors $\mathbf{s}$ in the planar region P unchanged but rejecting the rest. The inverse slope transform on G yields the lower envelope $\breve{g}$ of the impulse response g. Over short-scale periods, g has the shape induced by the sequence $\{a_{ij}\}$, but over scales much longer than the size of the output coefficient mask M_o, g behaves like its lower envelope $\breve{g}$ (see the 1D example of Fig. 10.4b). Together G and $\breve{g}$ can describe the long-scale dynamics of the system. In addition, if $g = \breve{g}$, then the above analysis is also exact for the short-scale behavior.

As an example, let $M_o = \{(0,1),(0,1)\}$ and consider the min-sum difference equation,

$$u(i,j) = \min\left[u(i-1,j) + a_{10},\; u(i,j-1) + a_{01},\; f(i,j)\right], \tag{10.27}$$

which can model the forward pass of the sequential computation of the cityblock distance transform. Assuming boundary conditions $u(i,j) = +\infty$ if $i < 0$ or $j < 0$ and a bottom-left to top-right scanning order, the impulse response (found by induction) and slope response (shown in Fig. 10.5a for $a_{10} = a_{01} = 1$) are

$$\begin{aligned} g(i,j) &= a_{10}i + a_{01}j - \zeta(i,j), \\ G(s_1,s_2) &= \zeta(a_{10} - s_1, a_{01} - s_2). \end{aligned} \tag{10.28}$$

Thus, this system acts as a 2D lowpass spatial slope filter, passing all input lower slopes $s_1 \leq a_{10}$ and $s_2 \leq a_{01}$, but rejecting the rest. In this case, $g = \breve{g}$ is convex. This example demonstrates that the impulse response of ETI systems described by min-sum difference equations with a recursive part has an infinite support.

10.3 PDEs for Morphological Image Analysis

10.3.1 PDEs Generating Dilations and Erosions

Let $k : \mathbf{R}^2 \to \overline{\mathbf{R}}$ be a unit-scale upper-semicontinuous concave structuring function, to be used as the kernel for morphological dilations and erosions. Scaling both its

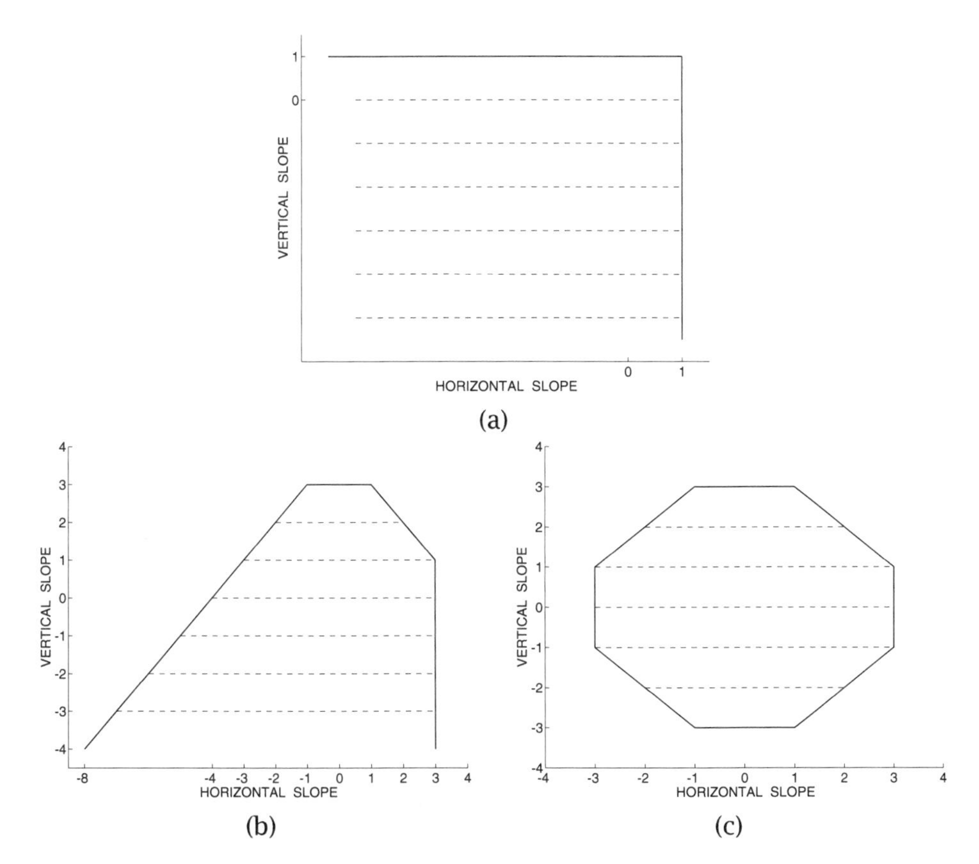

Figure 10.5: Regions of support of binary slope responses of discrete ETI systems, representing (a) the forward pass of the cityblock distance transform, (b) the forward pass of the chamfer (3,4) distance transform, and (c) the chamfer (3,4) distance transform.

values and its support by a scale parameter $t \geq 0$ yields a parameterized family of multiscale structuring functions:

$$k_t(x, y) \triangleq \begin{cases} tk(x/t, y/t) & \text{for } t > 0, \\ \xi(0, 0) & \text{for } t = 0, \end{cases} \tag{10.29}$$

which satisfies the semigroup property $k_s \oplus k_t = k_{s+t}$. Using k_t in place of g as the kernel in the basic morphological operations leads to defining the *multiscale* dilation and erosion of $f : \mathbf{R}^2 \to \mathbf{R}$ by k_t as the scale-space functions

$$\delta(x, y, t) \triangleq f \oplus k_t(x, y), \qquad \varepsilon(x, y, t) \triangleq f \ominus k_t(x, y), \tag{10.30}$$

where $\delta(x, y, 0) = \varepsilon(x, y, 0) = f(x, y)$.

In practice, a useful class of functions k consists of flat structuring functions,

$$k(x, y) = \begin{cases} 0 & \text{for } (x, y) \in B, \\ -\infty & \text{for } (x, y) \notin B, \end{cases} \tag{10.31}$$

that are the $0/-\infty$ indicator functions of compact convex planar sets B. The general PDE generating the multiscale flat dilations of f by a general compact convex symmetric B is [Alv93, Bro94, Hei97]

$$\delta_t = \mathrm{sptf}(B)(\delta_x, \delta_y), \tag{10.32}$$

where sptf(B) is the support function of B,

$$\mathrm{sptf}(B)(x,y) \triangleq \bigvee_{(a,b)\in B} ax + by. \tag{10.33}$$

Useful cases of structuring sets B are obtained by the unit balls

$$B_p = \{(x,y) \in \mathbf{R}^2 : \|(x,y)\|_p \leq 1\}$$

of the metrics induced by the $\mathcal{L}_p$ norms $\|\cdot\|_p$, for $p = 1, 2, \ldots, \infty$. The PDEs generating the multiscale flat dilations of f by B_p for three special cases of norms $\|\cdot\|_p$ are as follows:

$$B = \text{rhombus } (p = 1) \Longrightarrow \delta_t = \max(|\delta_x|, |\delta_y|) = \|\nabla\delta\|_\infty, \tag{10.34}$$

$$B = \text{disk } (p = 2) \Longrightarrow \delta_t = \sqrt{(\delta_x)^2 + (\delta_y)^2} = \|\nabla\delta\|_2, \tag{10.35}$$

$$B = \text{square } (p = \infty) \Longrightarrow \delta_t = |\delta_x| + |\delta_y| = \|\nabla\delta\|_1, \tag{10.36}$$

with $\delta(x,y,0) = f(x,y)$. The corresponding PDEs generating mutliscale flat erosions are

$$B = \text{rhombus} \Longrightarrow \varepsilon_t = -\|\nabla\varepsilon\|_\infty, \tag{10.37}$$

$$B = \text{disk} \Longrightarrow \varepsilon_t = -\|\nabla\varepsilon\|_2, \tag{10.38}$$

$$B = \text{square} \Longrightarrow \varepsilon_t = -\|\nabla\varepsilon\|_1, \tag{10.39}$$

with $\varepsilon(x,y,0) = f(x,y)$.

These simple but nonlinear PDEs are satisfied at points where the data are smooth, that is, the partial derivatives exist. However, even if the initial image or signal f is smooth, at finite scales $t > 0$ the above dilation or erosion evolution may create discontinuities in the derivatives, called *shocks*, which then continue propagating in scale-space. Thus, the multiscale dilations δ or erosions ε are *weak solutions* of the corresponding PDEs, in the sense put forth by Lax [Lax73]. Ways to deal with these shocks include replacing standard derivatives with morphological derivatives [Bro94] or replacing the PDEs with differential inclusions [Mat93]. For example, let

$$\mathcal{M}_x f(x,y) \triangleq \lim_{r\downarrow 0} \frac{\sup\{f(x+v,y) : |v| \leq r\} - f(x,y)}{r}$$

be the sup-derivative of f along the x-direction. If the right $[f_x(x+,y)]$ and left $[f_x(x-,y)]$ derivatives of f along the x-direction exist, then

$$\mathcal{M}_x f(x,y) = \max[0, f_x(x+,y), -f_x(x-,y)]. \tag{10.40}$$

The sup-derivative $\mathcal{M}_y f$ along the y-direction can be treated similarly. Then, a generalized PDE generating flat dilations by a compact convex symmetric B is

$$\delta_t = \mathrm{sptf}(B)(\mathcal{M}_x \delta, \mathcal{M}_y \delta). \tag{10.41}$$

This new PDE can handle discontinuities (i.e., shocks) in the partial derivatives of δ, provided that its left and right derivatives exist everywhere.

The above PDEs for dilations–erosions of graylevel images by flat structuring elements directly apply to binary images, since flat dilations–erosions commute with thresholding, and hence when the graylevel image is dilated–eroded, each of its thresholded versions representing a binary image is simultaneously dilated–eroded by the same element and at the same scale. However, this is not the case with graylevel structuring functions. We provide two examples of PDEs generating multiscale dilations by graylevel structuring functions: If k is the compact-support spherical function

$$k(x,y) = \begin{cases} \sqrt{1+x^2+y^2} & \text{for } x^2+y^2 \le 1, \\ -\infty & \text{for } x^2+y^2 > 1, \end{cases} \tag{10.42}$$

the dilation PDE becomes

$$\delta_t = \sqrt{1+(\delta_x)^2+(\delta_y)^2}. \tag{10.43}$$

For the infinite-support parabolic structuring function

$$k(x,y) = -r(x^2+y^2), \quad r > 0, \tag{10.44}$$

the dilation PDE becomes

$$\delta_t = [(\delta_x)^2 + (\delta_y)^2]/4r. \tag{10.45}$$

10.3.2 Slope Transforms and Dilation PDEs

All of the above dilation (and erosion) PDEs can be unified using slope transforms. Specifically, let the unit-scale kernel $k(x,y)$ be a general upper-semicontinuous concave function and consider its upper slope transform,

$$K_\vee(s_1,s_2) \triangleq \bigvee_{(x,y)\in\mathbf{R}^2} k(x,y) - (s_1 x + s_2 y). \tag{10.46}$$

Then, as discussed elsewhere [Hei97, Mat93], the PDE generating multiscale signal dilations by k is

$$\delta_t = K_\vee(\delta_x, \delta_y). \tag{10.47}$$

Thus, the rate of change of δ in the scale (t) direction is equal to the upper slope transform of the structuring function evaluated at the spatial gradient of δ. Similarly, the PDE generating the multiscale erosion by k is

$$\varepsilon_t = -K_\vee(\varepsilon_x, \varepsilon_y). \tag{10.48}$$

All of the dilation and erosion PDEs examined are special cases of Hamilton-Jacobi equations, which are of paramount importance in physics. Such equations usually do not admit classic (i.e., everywhere differentiable) solutions. Viscosity solutions of Hamilton-Jacobi PDEs have been extensively studied by Crandall et al. [Cra92]. Based on the theory of viscosity solutions, Heijmans and Maragos [Hei97] have shown via slope transforms that the multiscale dilation by a general upper-semicontinuous concave function is the viscosity solution of the Hamilton-Jacobi dilation PDE of Eq. (10.47).

10.3.3 Numerical Algorithm for Dilation PDEs

The PDEs generating flat dilation and erosion by disks are special cases of Hamilton-Jacobi PDEs of the type

$$\phi_t = \beta \|\nabla \phi\|_2,$$

where $\phi = \phi(x, y, t)$ and where $\beta = \beta(x, y)$ has constant sign for all (x, y). An efficient algorithm for numerically solving such PDEs for applications of curve evolution has been developed by Osher and Sethian [Osh88] by adapting the technology of conservative monotone discretization schemes for shock-producing PDEs of hyperbolic conservation laws [Lax73]. The main steps of such a first-order algorithm are

$$\begin{aligned}
\Phi_{i,j}^n &= \text{approximation of } \phi(i\Delta x, j\Delta y, n\Delta t) \text{ on a grid,} \\
V_{ij} &= \beta(i\Delta x, j\Delta y), \\
D_{+x}\Phi_{i,j}^n &= (\Phi_{i+1,j}^n - \Phi_{i,j}^n)/\Delta x, \quad D_{-x}\Phi_{i,j}^n = (\Phi_{i,j}^n - \Phi_{i-1,j}^n)/\Delta x, \\
D_{+y}\Phi_{i,j}^n &= (\Phi_{i,j+1}^n - \Phi_{i,j}^n)/\Delta y, \quad D_{-y}\Phi_{i,j}^n = (\Phi_{i,j}^n - \Phi_{i,j-1}^n)/\Delta y, \\
H &= [\max(0, D_{-x}\Phi_{i,j}^n)]^2 + [\min(0, D_{+x}\Phi_{i,j}^n)]^2 + [\max(0, D_{-y}\Phi_{i,j}^n)]^2 \\
&\quad + [\min(0, D_{+y}\Phi_{i,j}^n)]^2, \\
\Phi_{i,j}^{n+1} &= \Phi_{i,j}^n + \Delta t V_{ij}\sqrt{H}, \qquad n = 0, 1, 2, \ldots, (T_{max}/\Delta t),
\end{aligned} \tag{10.49}$$

where $T_{\max}$ is the maximum time (or scale) of interest, Δx, and Δy are the spatial grid spacings, and Δt is the time (scale) step. For stability, the space/time steps must satisfy $(\Delta t/\Delta x + \Delta t/\Delta y)V_{ij} \leq 0.5$.

By choosing fine grids (and possibly higher-order terms), an arbitrarily low error (between signal values on the continuous plane and the discrete grid) can be achieved in implementing morphological operations involving disks as structuring elements. This is a significant advantage of the PDE approach, as observed elsewhere [Are93, Sap93]. Thus, curve evolution provides a geometrically better implementation of multiscale morphological operations with the disk-shaped structuring element. Figure 10.6 shows the results of a simulation to compare the traditional dilation of digital images via discrete max-sum convolution of the image by digital approximations to disks (e.g., squares) versus a dilation that is the solution $\delta(x, y, t)$ of the dilation PDE numerically solved using algorithm (10.49). Comparing the graylevel images to their binary versions (from thresholding at

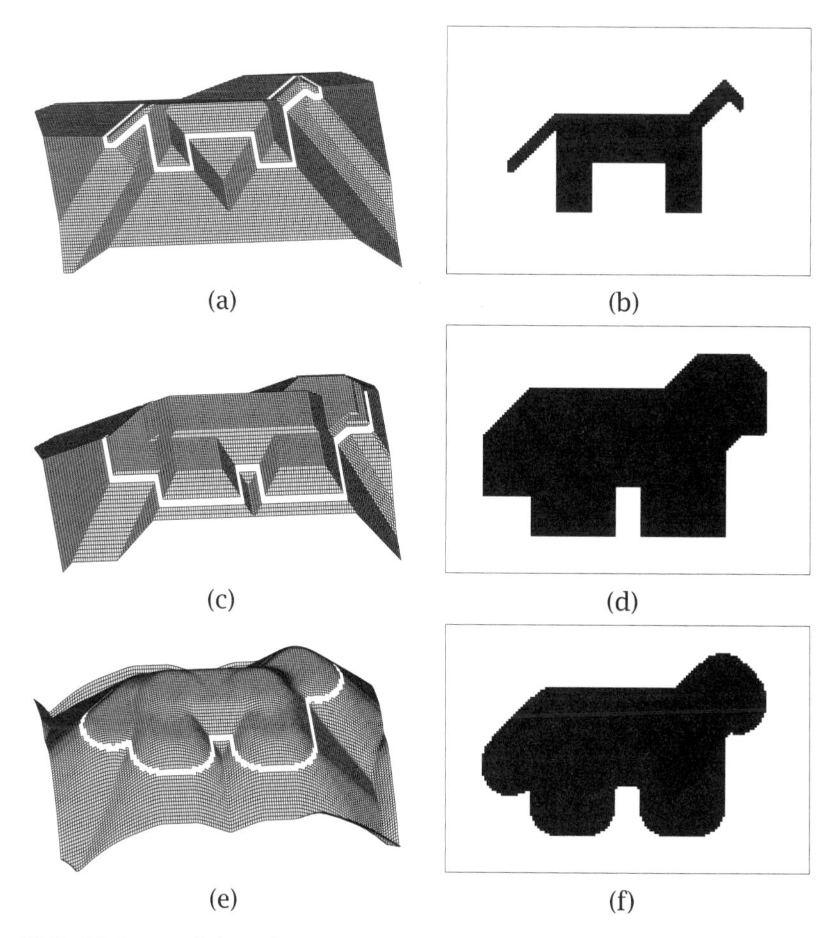

Figure 10.6: (a) Original digital graylevel image f and its contour at level $= 0$. (b) Binary image S from thresholding f at level $= 0$. (c) Flat dilation $f \oplus B$ of f by a discrete disk, that is, a square of $(2t+1)\times(2t+1)$ pixels, with $t = 5$. (d) Binary image $S \oplus B$ from thresholding $f \oplus B$ at level $= 0$. (e) Dilation of f by running the PDE $\partial\delta/\partial t = \|\nabla\delta\|_2$ for scales $t \in [0, 5]$ with initial condition $\delta(x, y, 0) = f(x, y)$. (f) Binary image from thresholding the image in (e) at level $= 0$.

level $= 0$), it is evident that the PDE approach to multiscale dilations can give much better approximations to Euclidean disks and hence avoid the abrupt shape discretization inherent in modeling digital multiscale dilations using discrete disks.

10.3.4 PDEs Generating Openings and Closings

Let $u(x, y, t) = [f(x, y) \ominus tB] \oplus tB$ be the multiscale flat opening of an image f by the disk B. This standard opening can be generated at any scale $r > 0$ by running the following PDE [Alv93]

$$u_t = -\max(\text{sgn}(r - t), 0)\, \|\nabla u\|_2 + \max(\text{sgn}(t - r), 0)\, \|\nabla u\|_2, \qquad (10.50)$$

from time $t = 0$ until time $t = 2r$ with initial condition $u(x, y, 0) = f(x, y)$, where $\text{sgn}(\cdot)$ denotes the signum function. This PDE has time-dependent switching coefficients that make it act as an erosion PDE during $t \in [0, r]$ but as a dilation PDE during $t \in [r, 2r]$. The discontinuities that this PDE exhibits at the instants it switches can be dealt with by making appropriate changes to the time scale, as suggested by Alvarez et al. [Alv93].

Recently, the reconstruction openings have found many more applications than the standard openings in a large variety of problems. We next present a nonlinear PDE, introduced by Maragos and Meyer [Mar99], that can model and generate openings and closings by reconstruction. Consider a 2D reference signal $f(x, y)$ and a marker signal $g(x, y)$. If $g \leq f$ everywhere and we start iteratively growing g via incremental flat dilations with an infinitesimally small disk ΔtB but without ever growing the result above the graph of f, then in the limit we shall have produced the reconstruction opening of f (with respect to the marker g). The infinitesimal generator of this signal evolution can be modeled via a dilation PDE that has a mechanism to stop the growth whenever the intermediate result attempts to create a function larger than f. Specifically, let $u(x, y, t)$ represent the evolutions of f with initial value $u_0(x, y) = u(x, y, 0) = g(x, y)$. Then, u is a weak solution of the PDE

$$\begin{aligned} u_t(x, y, t) &= \|\nabla u(x, y, t)\| \text{sgn}[f(x, y) - u(x, y, t)], \\ u(x, y, 0) &= g(x, y). \end{aligned} \tag{10.51}$$

This PDE models a *conditional dilation* that grows the intermediate result as long as it does not exceed f. In the limit we obtain the final result $u_\infty(x, y) = \lim_{t\to\infty} u(x, y, t)$. The mapping $u_0 \mapsto u_\infty$ is the *reconstruction opening* filter.

If in the PDE of Eq. (10.51) we reverse the order between f and g (i.e., assume that $g \geq f$), then the positive growth (dilation) of g is replaced with negative growth (erosion) because now $\text{sgn}(f - u) \leq 0$. This negative growth stops when the intermediate result attempts to become smaller than f; in the limit we obtain the *reconstruction closing* of f with respect to the marker g.

The following shock-capturing and entropy-satisfying numerical algorithm has been used to solve the PDE of Eq. (10.51) [Mar99]. Let $U_{i,j}^n$ be the approximation of $u(x, y, t)$ on a computational grid $(i\Delta x, j\Delta y, n\Delta t)$. We then approximate PDE Eq. (10.51) by the following 2D nonlinear difference equation:

$$\begin{aligned} U_{i,j}^{n+1} = U_{i,j}^n - \Delta t\, [\cdots \\ (S_{i,j}^n)^+ \sqrt{((D_{-x}U_{i,j}^n)^+)^2 + ((D_{+x}U_{i,j}^n)^-)^2 + ((D_{-y}U_{i,j}^n)^+)^2 + ((D_{+y}U_{i,j}^n)^-)^2} + \\ (S_{i,j}^n)^- \sqrt{((D_{+x}U_{i,j}^n)^+)^2 + ((D_{-x}U_{i,j}^n)^-)^2 + ((D_{+y}U_{i,j}^n)^+)^2 + ((D_{-y}U_{i,j}^n)^-)^2}\, \Big], \end{aligned} \tag{10.52}$$

where $S_{i,j}^n = \text{sgn}\left(f(i\Delta x, j\Delta y) - U_{i,j}^n\right)$ and we denote $r^+ \triangleq \max(r, 0)$, $r^- \triangleq \min(r, 0)$. For stability, $(\Delta t/\Delta x + \Delta t/\Delta y) \leq 0.5$ is required. Also, a sign consistency is enforced at each iteration: $\text{sgn}(U^n - f) = \text{sgn}(g - f)$. Examples of simulating this numerical algorithm are shown in Fig. 10.7.

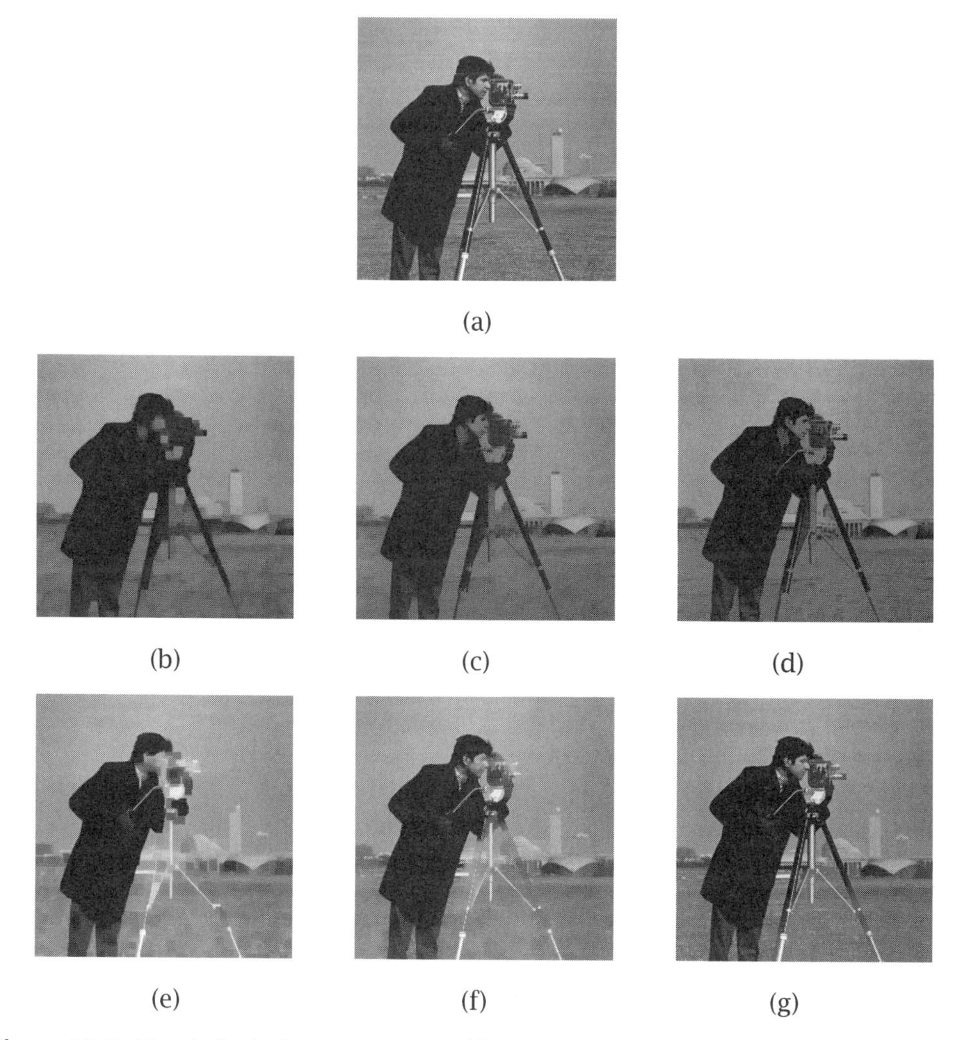

Figure 10.7: Morphological reconstruction filtering generated by the PDE Eq. (10.51) using two markers. (a) Reference image Cameraman, 256×256 pixels. Second row: (b) the marker ($t = 0$) was a standard opening by a 7×7 square of pixels, (c) the evolution at $t = 20\Delta t$, and (d) the final reconstruction opening (after convergence). Third row: (e) the marker ($t = 0$) was a standard closing by a square of 7×7 pixels, (f) the evolution at $t = 20\Delta t$, and (g) the final reconstruction closing (after convergence). ($\Delta x = \Delta y = 1$, $\Delta t = 0.25$.)

What happens if we now use the PDE of Eq. (10.51) when there is no specific order between f and g? In such a case, the PDE has a sign-varying coefficient $\text{sgn}(f - u)$ with spatiotemporal dependence, which controls the instantaneous growth and stops it whenever $f = u$. (Of course, there also is no growth at stationary points, where $u_x = u_y = 0$.) The control mechanism is of a switching type: For each t, at points (x, y) where $u(x, y, t) < f(x, y)$, it acts as a dilation PDE, whereas if $u(x, y, t) > f(x, y)$, it acts as an erosion PDE and reverses the direction of propagation. The final result $u_\infty(x,y) = \lim_{t\to\infty} u(x,y,t)$ is equal to the output from a general class of morphological filters, called *levelings*, which were introduced by Meyer [Mey98], have many useful properties, and contain as special cases the reconstruction openings and closings.

10.4 Curve Evolution

Consider at time $t = 0$ an initial simple, smooth, closed planar curve $\gamma(0)$ that is propagated along its normal vector field at speed V for $t > 0$. Let this evolving curve (front) $\gamma(t)$ be represented by its position vector $\vec{C}(p,t) = (x(p,t), y(p,t))$ and be parameterized by $p \in [0, J]$ so that it has its interior on the left in the direction of increasing p and $\vec{C}(0,t) = \vec{C}(J,t)$. The curvature along the curve is

$$\kappa = \kappa(p,t) \triangleq \frac{y_{pp}x_p - y_p x_{pp}}{(x_p^2 + y_p^2)^{3/2}}. \tag{10.53}$$

A general front propagation law (flow) is

$$\frac{\partial \vec{C}(p,t)}{\partial t} = V\vec{N}(p,t), \tag{10.54}$$

with initial condition $\gamma(0) = \{\vec{C}(p,0) : p \in J\}$, where $\vec{N}(p,t)$ is the instantaneous unit *outward normal* vector at points on the evolving curve and $V = \vec{C}_t \cdot \vec{N}$ is the *normal speed*, with $\vec{C}_t = \partial\vec{C}/\partial t$. This speed may depend on local geometrical information such as the curvature, global image properties, or other factors independent of the curve. If $V = 1$ or $V = -1$, then $\gamma(t)$ is the boundary of the dilation or erosion of the initial curve $\gamma(0)$ by a disk of radius t. In general, if B is an arbitrary compact, convex, symmetric planar set of unit scale and if we dilate the initial curve $\gamma(0)$ with tB and set the new curve $\gamma(t)$ equal to the outward boundary of $\gamma(0) \oplus tB$, then this curve evolution can also be generated by the PDE of Eq. (10.54) using a speed [Are93, Sap93]

$$V = \mathrm{sptf}(B)(\vec{N}), \tag{10.55}$$

where $\mathrm{sptf}(B)$ is the support function of B.

Another important speed model has been studied extensively by Osher and Sethian [Osh88, Set96] for general evolution of interfaces and by Kimia et al. [Kim90] for shape analysis in computer vision:

$$V = 1 - \varepsilon\kappa, \qquad \varepsilon \geq 0. \tag{10.56}$$

As analyzed by Sethian [Set96], when $V = 1$ the front's curvature will develop singularities, and the front will develop corners (i.e., the curve derivatives will develop shocks—discontinuities) at finite time if the initial curvature is anywhere negative. Two ways to continue the front beyond the corners are as follows: (1) If the front is viewed as a geometric curve, then each point is advanced along the

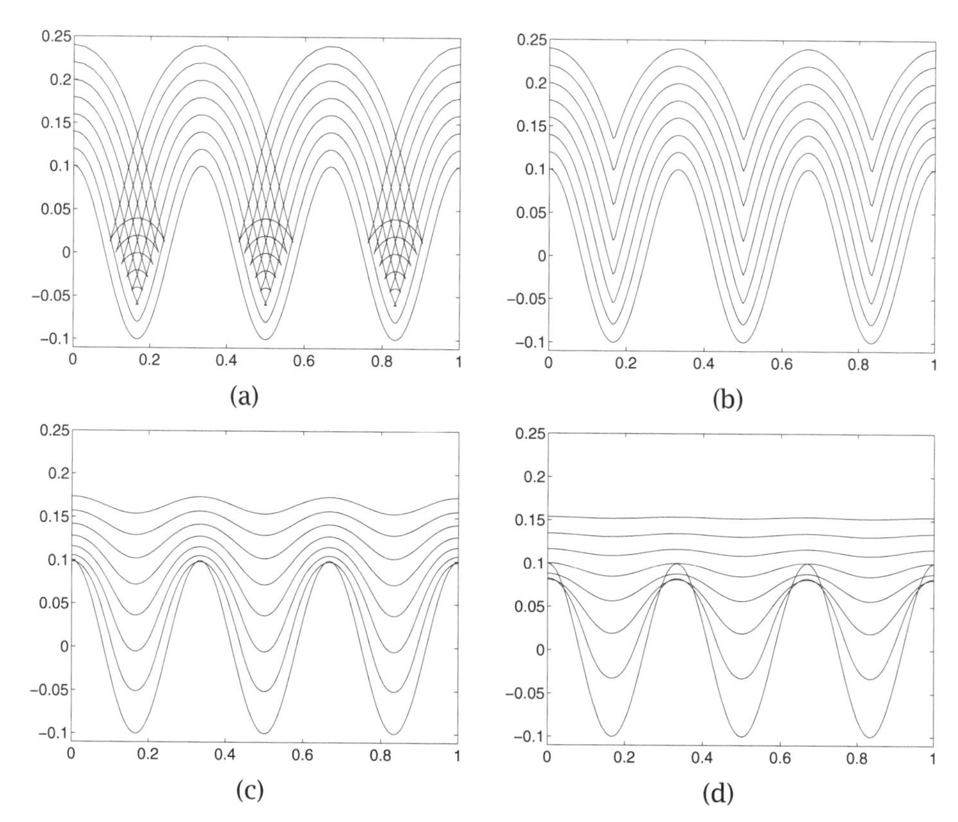

Figure 10.8: Evolution of the curve (signal graph) $(-p, \cos(6\pi p)/10)$, $p \in [0, 1]$. Evolved curves are plotted from $t = 0$ to $t = 0.14$ at increments of 0.02. The numerical simulation for (b), (c), and (d) is based on the Osher and Sethian algorithm with $\Delta x = 0.005$ and Δt chosen small enough for stability. (a) $V = 1$, "swallowtail" weak solution. (b) $V = 1$, entropy weak solution with $\Delta t = 0.002$. (c) $V = 1 - 0.05\kappa$ with $\Delta t = 0.0002$. (d) $V = 1 - 0.1\kappa$ with $\Delta t = 0.0001$.

normal by a distance t, and hence a "swallowtail" is formed beyond the corners by allowing the front to pass through itself. 2) If the front is viewed as the boundary separating two regions, an *entropy condition* is imposed to disallow the front to pass through itself. In other words, if the front is a propagating flame, then "once a particle is burnt it stays burnt" [Set96]. The same idea has also been used to model grassfire propagation leading to the medial axis of a shape [Blu73]. It is equivalent to using Huygen's principle to construct the front as the set of points at distance t from the initial front. This can also be obtained from multiscale dilations of the initial front by disks of radii $t > 0$. Both the swallowtail and the entropy solutions are weak solutions. The examples in Fig. 10.8 show that, when $\varepsilon > 0$, motion with curvature-dependent speed has a smoothing effect. Further, the limit of the solution for the $V = 1 - \varepsilon\kappa$ case as $\varepsilon \downarrow 0$ is the entropy solution for the $V = 1$ case [Set96].

To overcome the topological problem of splitting and merging and numerical problems with the Lagrangian formulation of Eq. (10.54), an Eulerian formulation was proposed by Osher and Sethian [Osh88] in which the original curve $\gamma(0)$ is first embedded in the surface of an arbitrary 2D Lipschitz continuous function $\phi_0(x,y)$ as its level set (contour line) at zero level. For example, we can select $\phi_0(x,y)$ to be equal to the signed distance function from the boundary of $\gamma(0)$, positive (negative) in the interior (exterior) of $\gamma(0)$. Then, the evolving planar curve is embedded as the zero-level set of an evolving space-time function $\phi(x,y,t)$:

$$\gamma(t) = \{(x,y) : \phi(x,y,t) = 0\}, \tag{10.57}$$

$$\gamma(0) = \{(x,y) : \phi(x,y,0) = \phi_0(x,y) = 0\}. \tag{10.58}$$

Geometrical properties of the evolving curve can be obtained from spatial derivatives of the level function. Thus, at any point on the front the curvature and outward normal of the level sets can be found from ϕ:

$$\vec{N} = -\frac{\nabla\phi}{\|\nabla\phi\|}, \quad \kappa = -\text{div}\left(\frac{\nabla\phi}{\|\nabla\phi\|}\right). \tag{10.59}$$

The curve evolution PDE of Eq. (10.54) induces a PDE generating its level function:

$$\begin{aligned} \phi_t &= V\|\nabla\phi\|, \\ \phi(x,y,0) &= \phi_0(x,y). \end{aligned} \tag{10.60}$$

If $V = 1$, the above function evolution PDE is identical to the flat circular dilation PDE of Eq. (10.35) by equating scale with time. Thus, we can view this specific dilation PDE as a special case of the general function evolution PDE of Eq. (10.60) in which all level sets expand in a homogeneous medium with $V = 1$. Propagation in a heterogeneous medium with $V = V(x,y) > 0$ leads later to the eikonal PDE.

10.5 Distance Transforms

10.5.1 Distance Transforms and Wave Propagation

For binary images, the distance transform is a compact way to represent their multiscale dilations and erosions by convex polygonal structuring elements whose shape depends upon the norm used to measure distances. Specifically, a binary image can be divided into the foreground set $S \subseteq \mathbf{R}^2$ and the background set $S^c = \{(x,y) : (x,y) \notin S\}$. For shape analysis of an image object S, it is often more useful to consider its inner distance transform by using S as the domain to measure distances from its background. However, for the applications discussed herein, we need to view S as a region marker or a source emanating a wave that will propagate away from it into the domain of S^c. Thus, we define the *outer distance transform* of a set S with respect to the metric induced by some norm $\|\cdot\|_p$, $p = 1, 2, \ldots, \infty$, as the distance function:

$$\mathrm{D}_p(S)(x,y) \triangleq \bigwedge_{(v,w)\in S} \|(x-v, y-w)\|_p. \tag{10.61}$$

If B_p is the unit ball induced by the norm $\| \cdot \|_p$, thresholding the distance transform at level $r > 0$ and obtaining the corresponding level set yields the morphological dilation of S by the ball B_p at scale r:

$$S \oplus rB_p = \{(x, y) : \mathrm{D}_p(S)(x, y) \geq r\}. \tag{10.62}$$

The boundaries of these dilations are the wavefronts of the distance propagation. Multiscale erosions of S can be obtained from the outer distance transform of S^c.

In addition to being a compact representation for multiscale erosions and dilations, the distance transform has found many applications in image analysis and computer vision. Examples include smoothing, skeletonization, size distributions, shape description, object detection and recognition, segmentation, and path finding [Blu73, Bor86, Nac96, Pre93, Ros66, Ros68, Ver91, Vin91b]. Thus, many algorithms have been developed for its computation.

Using Huygen's construction, the boundaries of multiscale dilations-erosions by disks can also be viewed as the wavefronts of a wave initiating from the original image boundary and propagating with constant normal speed, that is, in a homogeneous medium. Thus, the distance function has a minimum time-of-arrival interpretation [Blu73], and its isolevel contours coincide with those of the wave phase function. Points at which these wavefronts intersect and extinguish themselves (according to Blum's grassfire propagation principle) are the points of the Euclidean skeleton axis of S [Blu73]. Overall, the Euclidean distance function $\mathrm{D}_2(S)$ is the weak solution of the following nonlinear PDE:

$$\begin{aligned} \|\nabla E(x, y)\|_2 &= 1, \quad (x, y) \in S^c, \\ E(x, y) &= 0, \quad (x, y) \in \partial S. \end{aligned} \tag{10.63}$$

This is a special case of the eikonal PDE $\|\nabla E(x, y)\|_2 = \eta(x, y)$ that corresponds to wave propagation in heterogeneous media and whose solution E is a weighted distance function, whose weights $\eta(x, y)$ are inversely proportional to the varying propagation speed [Lev70, Rou92, Ver90].

10.5.2 Distance Transforms as Infimal Convolutions and Slope Filters

If we consider the $0/\infty$ indicator function of S,

$$\mathrm{I}(S)(x, y) \triangleq \begin{cases} 0 & \text{for } (x, y) \in S, \\ +\infty & \text{for } (x, y) \notin S, \end{cases} \tag{10.64}$$

and the L_p norm structuring function,

$$g_p(x, y) = \|(x, y)\|_p, \tag{10.65}$$

it follows that the distance transform can be obtained from the infimal convolution of the indicator function of the set with the norm function:

$$\mathrm{D}_p(S) = \mathrm{I}(S) \oplus' g_p. \tag{10.66}$$

Further, since the relative ordering of distance values does not change if we raise them to a positive power $m > 0$, it follows that we can obtain powers of the distance function by convolving with the respective powers of the norm function:

$$[\mathrm{D}_p(S)]^m = \mathrm{I}(S) \oplus' (g_p)^m. \tag{10.67}$$

The infimal convolution in Eq. (10.66) is equivalent to passing the input signal, that is, the set's indicator function, through an ETI system with slope response [Mar96]

$$G_\wedge(\mathbf{s}) = \begin{cases} 0 & \text{for } \|\mathbf{s}\|_q \leq 1, \\ -\infty & \text{for } \|\mathbf{s}\|_q > 1, \end{cases} \tag{10.68}$$

where q is the conjugate exponent of p $(1/p + 1/q = 1)$. That is, the distance transform is the output of an ideal-cutoff slope-selective filter that rejects all input planes whose slope vector falls outside the unit ball with respect to the $\|\cdot\|_q$ norm but passes all the rest unchanged.

10.5.3 Euclidean Distance Transforms of Binary Images and Approximations

To obtain isotropic distance propagation, we want to employ the Euclidean distance transform, that is, using the norm $\|\cdot\|_2$ in Eq. (10.61), since it gives multiscale morphology with the disk as the structuring element. However, computing the Euclidean distance transform of discrete images has a significant computational complexity. Thus, various techniques have been used to obtain an approximate or the exact Euclidean distance transform at a lower complexity. Four types of approaches that deal with this problem are as follows:

(1) Discrete metrics on grids that yield approximations to the Euclidean distance. Their early theory was developed by Rosenfeld and Pfaltz [Ros66, Ros68], based on either sequential or parallel operations. This was followed later by a generalization developed by Borgefors [Bor86] and based on *chamfer* metrics that yielded improved approximations to the Euclidean distance.

(2) Fast algorithmic techniques that can obtain the exact Euclidean distances by operating on complex data structures (e.g., [Dan80, Vin91a]).

(3) Infimal convolutions of binary images with a parabolic structuring function, which yield the exact squared Euclidean distance transform [Boo92, Hua94, Ste80]. This follows from Eq. (10.67) by using $m = 2$ with the Euclidean norm $(p = 2)$:

$$[\mathrm{D}_2(S)]^2 = \mathrm{I}(S) \oplus' (g_2)^2. \tag{10.69}$$

The kernel in the above infimal convolution is a convex parabola $[g_2(x, y)]^2 = \|(x, y)\|_2^2 = x^2 + y^2$. Note that the above result holds for images and kernels defined both on the continuous and on the discrete plane. Of course, convolution of the image with an infinite-extent kernel is not possible, and hence truncation of the parabola is used, which incurs an approximation error. The complexity of this convolution approach can be reduced significantly by using dimensional decomposition of the 2D parabolic structuring function by expressing it either as the dilation of two 1D quadratic structuring functions [Boo92] or as the dilation of several 3×3 kernels that yields a truncation of the 2D parabola [Hua94, Shi91].

(4) Efficient numerical algorithms for solving the nonlinear PDE (10.63) that yield arbitrarily close approximations to the Euclidean distance function.

Approach (4) yields the best approximations and will be discussed later. Of the other three approaches, (1) and (3) are more general than (2), have significant theoretical structure, and can be used with even the simplest data structures, such as rectangularly or hexagonally sampled image signals. Next we elaborate on approach (1), which has been studied the most.

10.5.4 Chamfer Distance Transforms

The general chamfer distance transform is obtained by propagating local distance steps within a small neighborhood. For each such neighborhood the distance steps form a mask of weights that is infimally convolved with the image. For a 3×3-pixel neighborhood, if a and b are the horizontal and diagonal distance steps, respectively, the outer (a, b) chamfer distance transform of a planar set S can also be obtained directly from the general definition in Eq. (10.61) by replacing the general L_p norm $\|\cdot\|_p$ with the (a, b) chamfer norm

$$\|(x, y)\|_{a,b} \triangleq \max(|x|, |y|)a + \min(|x|, |y|)(b - a). \tag{10.70}$$

The unit ball corresponding to this chamfer norm is a symmetric octagon, and the resulting distance transform is

$$\mathrm{D}_{a,b}(S)(x, y) = \bigwedge_{(v,w)\in S} \|(x - v, y - w)\|_{a,b}. \tag{10.71}$$

Note that the above two equations apply to sets S and points (x, y) both in the continuous plane $\mathbf{R}^2$ as well as in the discrete plane $\mathbf{Z}^2$.

For a 3×3-pixel neighborhood, the outer (a, b) chamfer distance transform of a discrete set $S \subseteq \mathbf{Z}^2$ can be obtained via the following sequential computation [Bor86, Ros66]:

$$\begin{aligned} u_n(i, j) = \min\big[&u_n(i-1, j) + a,\ u_n(i, j-1) + a, \\ &u_n(i-1, j-1) + b,\ u_n(i+1, j-1) + b,\ u_{n-1}(i, j)\big]. \end{aligned} \tag{10.72}$$

Starting from $u_0 = \mathrm{I}(S)$ as the $0/\infty$ indicator function of S, two passes $(n = 1, 2)$ of the 2D recursive erosion of Eq. (10.72) suffice to compute the chamfer distance transform of S if S^c is bounded and simply connected. During the first pass the image is scanned from top-left to bottom-right using the four-point nonsymmetric half-plane submask of the 3×3 neighborhood. During the second pass the image is scanned in the reverse direction using the reflected submask of distance steps. The final result $u_2(i, j)$ is the outer (a, b) chamfer distance transform of S evaluated at points of $\mathbf{Z}^2$. An example of the three images, u_0, u_1, and u_2, is shown in Fig. 10.9.

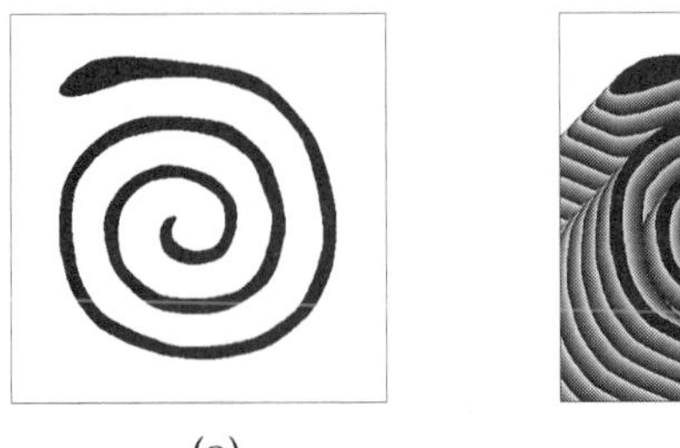

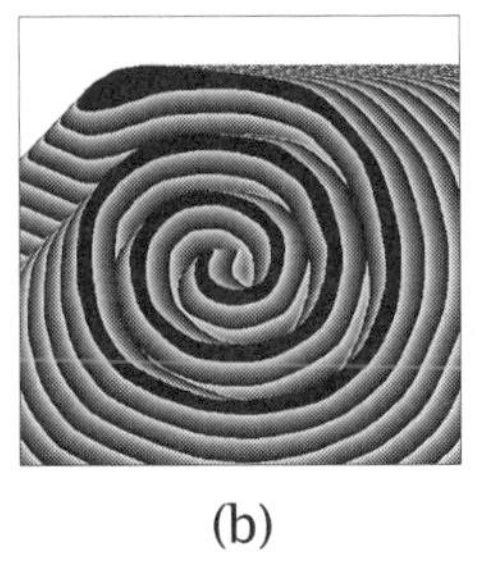

(a) (b) (c)

Figure 10.9: Sequential computation of the chamfer distance transform with optimal distance steps in a 3×3 mask. (a) Original binary image, 450×450 pixels, (b) result after forward scan, and (c) final result after backward scan. [In (b) and (c) the distances are displayed as intensity values modulo a constant.]

Thus, the sequential implementation of the local distance propagation is done via simple recursive min-sum difference equations. We shall show that these equations correspond to ETI systems with infinite impulse responses and binary slope responses. The distance propagation can also be implemented in parallel via nonrecursive min-sum equations, which correspond to ETI systems with finite impulse responses, as explained next.

Sequential Computation and IIR Slope Filters

Consider a 2D min-sum autoregressive difference equation with output mask and coefficients as in Fig. 10.10c:

$$\begin{aligned} u(i,j) = \min[&u(i-1,j)+a,\ u(i,j-1)+a, \\ &u(i-1,j-1)+b,\ u(i+1,j-1)+b,\ f(i,j)], \end{aligned} \tag{10.73}$$

where $f = \mathrm{I}(S)$ is the $0/\infty$ indicator function of a set S representing a discrete binary image. Then consider the following distance transformation of f obtained in two passes: During the forward pass Eq. (10.73) is recursively run over $f(i,j)$ in a bottom-to-top, left-to-right scanning order. The forward pass mapping $f \mapsto u$ is an ETI system with an infinite impulse response (found via induction):

$$g_f(i,j) = \begin{cases} \|(i,j)\|_{a,b} & \text{for } i+j \geq 0,\ j \geq 0, \\ +\infty & \text{otherwise.} \end{cases} \tag{10.74}$$

A truncated version of g_f is shown in Fig. 10.10d. The slope response $G_f(\mathbf{s})$ of this ETI system is equal to the $0/-\infty$ indicator function of the region shown (for $a = 3$, $b = 4$) in Fig. 10.5(b).

During the backward pass a recursion similar to Eq. (10.73) but in the opposite scanning order and using as output mask the reflected version of Fig. 10.10c is run over the previous result $u(i,j)$ to yield a signal $d(i,j)$ that is the final distance transform of $f(i,j)$. The backward pass mapping $u \mapsto d$ is an ETI system with an infinite impulse response $g_b(i,j) = g_f(-i,-j)$ and a slope response $G_b(\mathbf{s}) = G_f(-\mathbf{s})$.

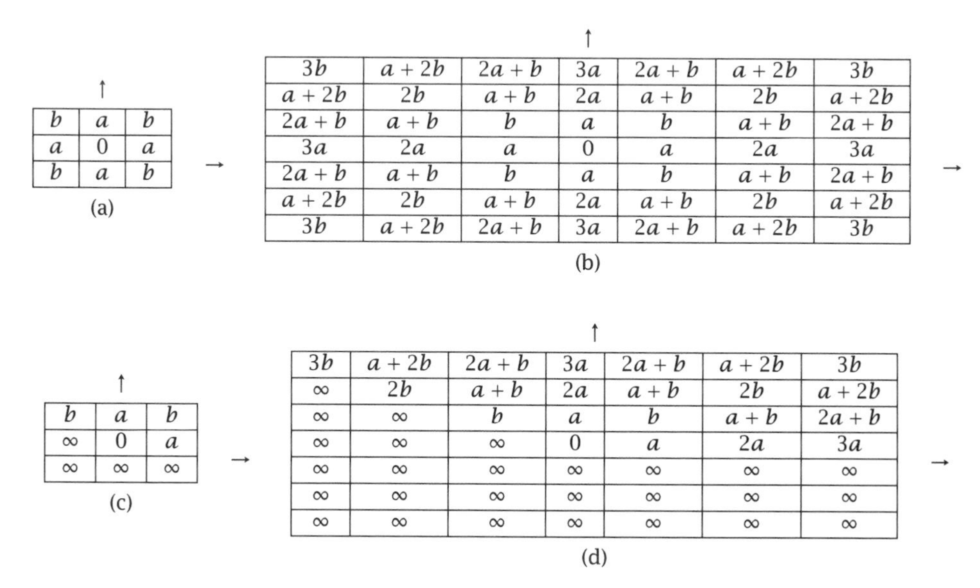

(a)

b	a	b
a	0	a
b	a	b

(b)

$3b$	$a+2b$	$2a+b$	$3a$	$2a+b$	$a+2b$	$3b$
$a+2b$	$2b$	$a+b$	$2a$	$a+b$	$2b$	$a+2b$
$2a+b$	$a+b$	b	a	b	$a+b$	$2a+b$
$3a$	$2a$	a	0	a	$2a$	$3a$
$2a+b$	$a+b$	b	a	b	$a+b$	$2a+b$
$a+2b$	$2b$	$a+b$	$2a$	$a+b$	$2b$	$a+2b$
$3b$	$a+2b$	$2a+b$	$3a$	$2a+b$	$a+2b$	$3b$

(c)

b	a	b
∞	0	a
∞	∞	∞

(d)

$3b$	$a+2b$	$2a+b$	$3a$	$2a+b$	$a+2b$	$3b$
∞	$2b$	$a+b$	$2a$	$a+b$	$2b$	$a+2b$
∞	∞	b	a	b	$a+b$	$2a+b$
∞	∞	∞	0	a	$2a$	$3a$
∞	∞	∞	∞	∞	∞	∞
∞	∞	∞	∞	∞	∞	∞
∞	∞	∞	∞	∞	∞	∞

Figure 10.10: Coefficient masks and impulse responses of ETI systems associated with computing the (a, b) chamfer distance transform. (a) Local distances within the 3×3-pixel unit "disk." (b) Distances from origin by propagating three times the local distances in (a); also equal to a 7×7-pixel central portion of the infinite impulse response of the overall system associated with the distance transform. (c) Coefficient mask for the min-sum difference equation computing the forward pass for the chamfer distance. (d) A 7×7-pixel portion of the infinite impulse response of the system corresponding to the min-sum difference equation computing the forward pass.

Since infimal convolution is an associative operation, the distance transform mapping $f \mapsto d$ is an ETI system with an infinite impulse response $g = g_f \oplus' g_b$:

$$d = (f \oplus' g_f) \oplus' g_b = f \oplus' (g_f \oplus' g_b) = f \oplus' g. \tag{10.75}$$

The overall slope response,

$$G(\mathbf{s}) = G_f(\mathbf{s}) + G_b(\mathbf{s}) = G_f(\mathbf{s}) + G_f(-\mathbf{s}), \tag{10.76}$$

of this distance transform ETI system is the $0/-\infty$ indicator function of a bounded convex region shown in Fig. 10.5c for $a = 3$, $b = 4$. Further, by using induction on (i, j) and symmetry, we find that $g = g_f \oplus' g_b$ is equal to the (a, b) chamfer distance function:

$$g(i, j) = \|(i, j)\|_{a,b}. \tag{10.77}$$

A truncated version of g is shown in Fig. 10.10b. Thus, our analysis has proved using ETI systems theory that the two-pass computation via recursive min-sum difference equations whose coefficients are the local chamfer distances yields the (a, b) chamfer distance transform:

$$[\mathrm{I}(S) \oplus' g_f] \oplus' g_b = \mathrm{D}_{a,b}(S). \tag{10.78}$$

Two special cases are the well-known cityblock and chessboard discrete distances [Ros66]. The cityblock distance transform is obtained using $a = 1$ and $b = +\infty$, that is, using the five-pixel rhombus as the unit "disk." It is an ETI system with impulse response $g(i,j) = |i| + |j|$ and the slope response being the indicator function of the unit square $\{\mathbf{s} : \|\mathbf{s}\|_\infty = 1\}$. Similarly, the chessboard distance transform is obtained using $a = b = 1$. It is an ETI system with impulse response $g(i,j) = \max(|i|,|j|)$ and the slope response being the indicator function of the unit rhombus $\{\mathbf{s} : \|\mathbf{s}\|_1 = 1\}$.

Parallel Computation and FIR Slope Filters

The (a,b) chamfer distance transform can be implemented alternatively using parallel operations. Namely, let

$$g_0(i,j) \triangleq \begin{cases} g(i,j) & \text{for } |i|,|j| \le 1, \\ +\infty & \text{otherwise}, \end{cases} \tag{10.79}$$

be the 3×3-pixel central portion of g defined in Eq. (10.77). It can be shown via induction that the nth-fold infimal convolution of g_0 with itself yields g in the limit:

$$g = \lim_{n\to\infty} \underbrace{(g_0 \oplus' g_0) \dots \oplus' g_0}_{n \text{ times}}. \tag{10.80}$$

Figure 10.10b shows the intermediate result for $n = 3$ iterations. Similar finite decompositions of discrete conical functions into infimal convolutions of smaller kernels have been studied elsewhere [Shi91]. Consider now the nonautoregressive min-sum difference equation

$$d_n(i,j) = \bigwedge_{k=-1}^{1} \bigwedge_{\ell=-1}^{1} g_0(i,j) + d_{n-1}(i-k, j-\ell), \tag{10.81}$$

run iteratively for $n = 1, 2, \dots,$ starting from $d_0 = f$. Each iteration is equivalent to the infimal convolution of the previous result with a finite impulse response equal to g_0. By iterating these local distances to the limit, the final distance transform is obtained: $d = \lim_{n\to\infty} d_n$. In practice, when the input image f has finite support, the number of required iterations is finite and bounded by the image diameter.

Optimal Chamfer Distance Transforms

Selecting the steps a, b under certain constraints leads to an infinite variety of chamfer metrics based on a 3×3 mask. The two well-known and easily computable special cases of the cityblock metric with $(a,b) = (1,\infty)$ and the chessboard metric with $(a,b) = (1,1)$ give the poorest discrete approximations to Euclidean distance (and to multiscale morphology with a disk structuring element), with errors reaching 41.4% and 29.3%, respectively. Using Euclidean steps $(a,b) = (1,\sqrt{2})$ yields a 7.61% maximum error. Thus, a major design goal is to reduce

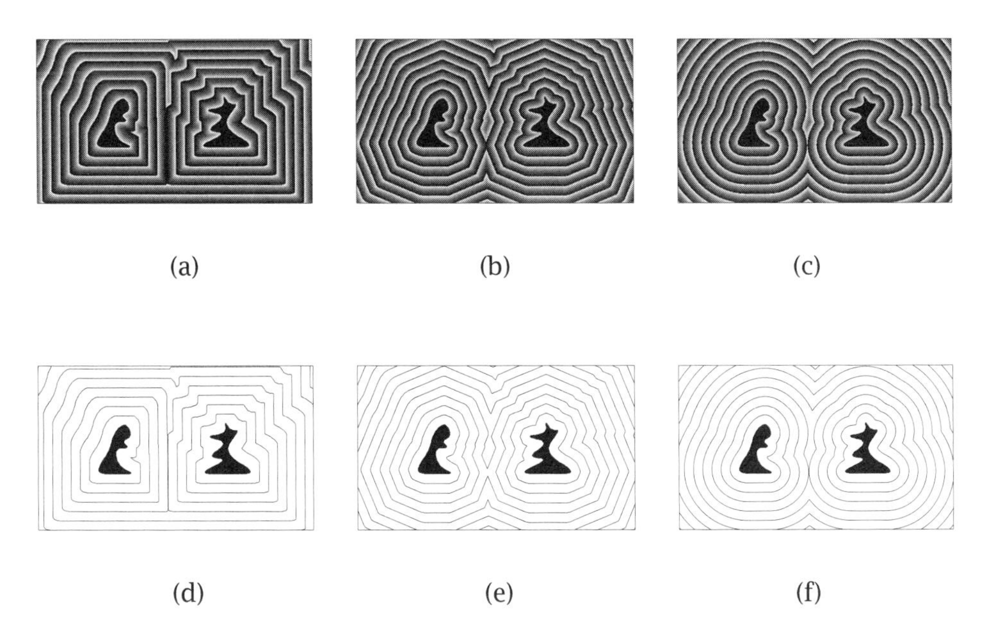

Figure 10.11: Top row: distance transforms of a binary image obtained via (a) a (1,1) chamfer metric, (b) the optimal 3×3 chamfer metric, and (c) curve evolution. These distances are displayed as intensities modulo a constant $h = 20$. Bottom row: the multiscale dilations (at scales $t = nh$, $n = 1, 2, 3, \ldots,$) of the original set (filled black regions) were obtained by thresholding the three distance transforms at isolevel contours whose levels are multiples of h using the following structuring elements: (d) and (e) the unit-scale polygons corresponding to the metrics used in (a) and (b) and (f) the disk. All images have a resolution of 450×600 pixels.

the approximation error between the chamfer distances and the corresponding Euclidean distances [Bor86]. A suitable error criterion is the maximum absolute error (MAE) between a unit chamfer ball and the corresponding unit disk [But98, Ver91]. The optimal steps obtained by Butt and Maragos [But98] for minimizing this MAE are $a = 0.9619$ and $b = 1.3604$, which give a 3.957% maximum error. In practice, for faster implementation, integer-valued distance steps A and B are used, and the computed distance transform is divided by a normalizing constant k, which can be real-valued. We refer to such a metric as $(a, b) = (A, B)/k$. Using two decimal digits for truncating optimal values and optimizing the triple (A, B, k) as done by Butt and Maragos [But98] for the smallest possible error yields $A = 70$, $B = 99$, and $k = 72.77$. The corresponding steps are

$$(a, b) = (70, 99)/72.77,$$

yielding a 3.959% MAE. See Fig. 10.11 for an example. By working as above, optimal steps that yield an even lower error can also be found for chamfer distances with a 5×5 mask or larger neighborhood [But98].

10.6 Eikonal PDE and Distance Propagation

The main postulate of geometrical optics [Bor59] is Fermat's principle of *least time.* Let us assume a 2D (i.e., planar) medium with (possibly space-varying) refractive index $\eta(x, y) = c_0/c(x, y)$, defined as the ratio of the speed c_0 of light in free space divided by its speed $c(x, y)$ in the medium. Given two points A and B in such a medium, the optical path length along a ray trajectory Γ_{AB} (parameterized by ℓ) between points A and B is

$$\text{optical path length} = \int_{\Gamma_{AB}} \eta\left(\Gamma_{AB}(\ell)\right) \, \mathrm{d}\ell = c_0 T(\Gamma_{AB}), \tag{10.82}$$

where $\mathrm{d}\ell$ is the differential length element along this trajectory and $T(\Gamma_{AB})$ is the time required for the light to travel this path. Fermat's principle states that light will choose a path between A and B that minimizes the optical path length.

An alternative viewpoint of geometrical optics is to consider the scalar function $E(x, y)$, called the *eikonal,* whose isolevel contours are normal to the rays. Thus, the eikonal's gradient $\|\nabla E\|$ is parallel to the rays. It can be shown [Bor59] using calculus of variations that Fermat's principle is equivalent to the following PDE:

$$\|\nabla E(x, y)\| = \sqrt{\left(\frac{\partial E}{\partial x}\right)^2 + \left(\frac{\partial E}{\partial y}\right)^2} = \eta(x, y), \tag{10.83}$$

called the *eikonal equation.* Thus, the minimal optical path length between two points located at A and B is

$$E(B) - E(A) = \inf_{\Gamma_{AB}} \int_{\Gamma_{AB}} \eta(\Gamma_{AB}(\ell)) \, \mathrm{d}\ell. \tag{10.84}$$

Assume an optical wave propagating in a 2D medium of index $\eta(x, y)$ at wavelengths much smaller than the image objects, so that ray optics can approximate wave optics. Then, the eikonal E of ray optics is proportional to the phase of the wavefunction. Hence, the isolevel contour lines of E are the wavefronts. Assuming that at time $t = 0$ there is an initial wavefront at a set of source points S_i, we can trace the wavefront propagation using Huygen's *envelope construction*: Namely, if we dilate the points $P = (x, y)$ of the wavefront curve at a certain time t with circles of infinitesimal radius $c(x, y)\,\mathrm{d}t$, the envelope of these circles yields the wavefront at time $t + \mathrm{d}t$. If $T(P)$ is the time required for the wavefront to arrive at P from the sources, then

$$E(P) = c_0 T(P) = \inf_i \left\{ \inf_{\Gamma_{S_i P}} \int_{\Gamma_{S_i P}} \eta\left(\Gamma_{S_i P}(\ell)\right) \, \mathrm{d}\ell + E(S_i) \right\}. \tag{10.85}$$

Thus, we can equate the eikonal $E(x, y)$ to the *weighted distance function* between a point (x, y) and the sources along a path of minimal optical length and also view E as proportional to the wavefront arrival time $T(x, y)$ (see also [Bor59, Kim96, Lev70, Rou92, Ver90]). An example is shown in Fig. 10.12.

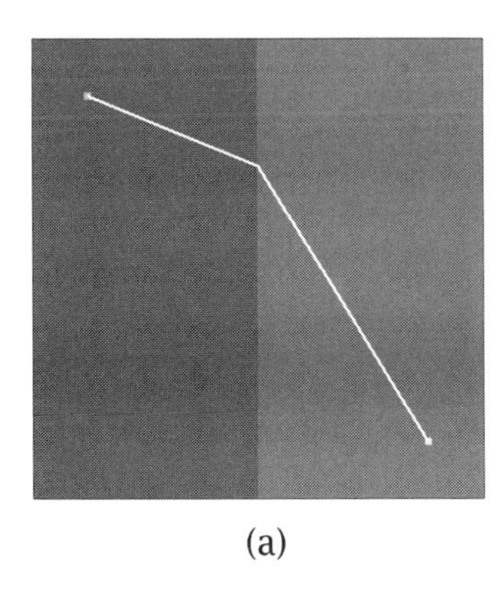

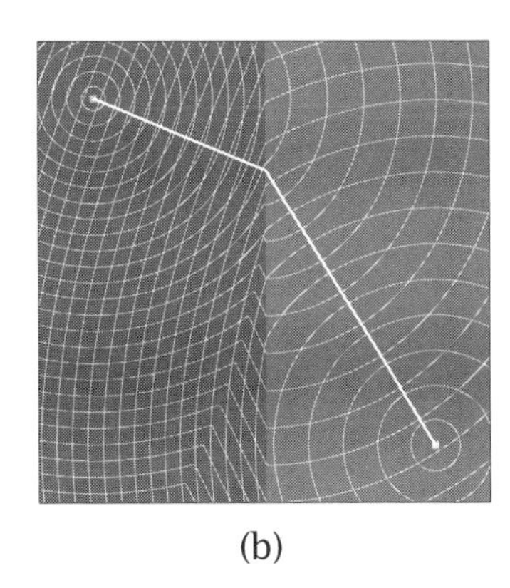

(a) (b)

Figure 10.12: (a) Image of an optical medium consisting of two areas of different refractive indexes (whose ratio is 5/3) and the correct path of the light ray (from Snell's law) between two points. (b) Path found using the weighted distance function (numerically estimated via curve evolution); the thin light contours show the wavefronts propagating from the two source points.

Many tasks for extracting information from visible images have been related to optics and wave propagation via the eikonal PDE. Its solution $E(x, y)$ can provide shape from shading, analog contour-based halftoning, and topographic segmentation of an image by choosing the refractive index field $\eta(x, y)$ to be an appropriate function of the image brightness [Hor86, Kim96, Naj94, Pnu94, Sch83, Ver90]. Further, in the context of curve evolution, the eikonal PDE can be seen as a stationary formulation of the embedding level function evolution PDE of Eq. (10.60) with positive speed $V = \beta(x, y) = \beta_0/\eta(x, y) > 0$. Namely, if

$$T(x, y) = \inf\{t : \phi(x, y, t) = 0\} \tag{10.86}$$

is the minimum time at which the zero-level curve of $\phi(x, y, t)$ crosses (x, y), then it can be shown [Bar90, Fal94, Osh88])

$$\|\nabla T(x, y)\| = \frac{1}{\beta(x, y)}. \tag{10.87}$$

Setting $E = \beta_0 T$ leads to the eikonal.

In short, we can view the solution $E(x, y)$ of the eikonal as a weighted distance transform (WDT) whose values at each pixel give the minimum distance from the light sources weighted by the gray values of the refractive index field. On a computational grid this solution is numerically approximated using discrete WDTs, which can be implemented either via 2D recursive min-sum difference equations or via numerical algorithms of curve evolution. The former implementation employs adaptive 2D recursive erosions and is a stationary approach to solving the eikonal, whereas the latter involves a time-dependent formulation and evolves curves based on a dilation-type PDE at a speed varying according to the gray values. Next we outline these two ways of solving the eikonal PDE and discuss some of its applications.

WDT Based on Chamfer Metrics

Let $f(i,j) \geq 1$ be a sampled nonnegative graylevel image and let us view it as a discrete refractive index field. Also let S be a set of reference points or the "sources" of some wave or the location of the wavefront at time $t = 0$. The discrete WDT finds at each pixel $P = (i,j)$ the smallest sum of values of f over all possible discrete paths connecting P to the sources S. It can also be viewed as a procedure for finding paths of minimal "cost" among nodes of a weighted graph or as discrete dynamic programming. It has been used extensively in image analysis problems such as minimal path finding, weighted distance propagation, and graylevel image skeletonization (for example, [Lev70, Mey92, Rut68, Ver91].

The above discrete WDT can be computed by running a 2D min-sum difference equation like Eq. (10.72) that implements the chamfer distance transform of binary images but with spatially varying coefficients proportional to the gray image values [Rut68, Ver90]:

$$\begin{aligned} u_n(i,j) = \min[&u_n(i-1,j) + af(i,j), \\ &u_n(i,j-1) + af(i,j),\ u_n(i-1,j-1) + bf(i,j), \\ &u_n(i+1,j-1) + bf(i,j),\ u_{n-1}(i,j)], \end{aligned} \tag{10.88}$$

where $u_0 = \mathrm{I}(S)$ is the $0/\infty$ indicator function of the source set S. Starting from u_0, a sequence of functions u_n is iteratively computed by running Eq. (10.88) over the image domain in a forward scan for even n, whereas for odd n an equation similar to Eq. (10.88) but with a reflected coefficient mask is run in a backward scan. In the limit $n \to \infty$ the final WDT u_∞ is obtained. In practice, this limit is reached after a finite number of passes. The number of iterations required for convergence depends on both the sources and the gray values. There are also other, faster implementations using queues (see [Mey92, Ver90]). The final transform is a function of the source set S, the index field, and the norm used for horizontal distances.

The above WDT based on discrete chamfer metrics is a discrete approximate solution of the eikonal PDE. The rationale for such a solution is that, away from the sources, this difference equation mapping $f \mapsto u$ corresponds to

$$\bigvee_{(k,\ell)\in B} \frac{u(i,j) - u(i-k, j-\ell)}{a_{ij}} = f(i,j), \tag{10.89}$$

where B is equal to the union of the output mask and its reflection and a_{ij} are the chamfer steps inside B. The left side of Eq. (10.89) is a weighted discrete approximation to the morphological derivative $\mathcal{M}(-u)$ with horizontal distances weighted by a_{ij}. Thus, since in the continuous case $\mathcal{M}(-u) = \|\nabla u\|$, Eq. (10.89) is an approximation of the eikonal. In fact, as established elsewhere [Mey92], it is possible to recover a digital image u from its half morphological gradient $u - u \ominus B$ using the discrete weighted distance transform if one uses 1-pixel sources in each regional minimum of u.

The constants a and b in Eq. (10.88) are the distance steps by which the planar chamfer distances are propagated within a 3×3 neighborhood. The propagation of the local distances (a, b) starts at the points of sources S and moves with speed $V = \beta(i, j) = \beta_0 / f(i, j)$. If f is a binary image, then the propagation speed is constant and the solution of the above difference equation (after convergence) yields the discrete chamfer distance transform of S. To improve the WDT approximation to the eikonal's solution, one can optimize (a, b) so that the error is minimized between the planar chamfer distances and the Euclidean distances. Using a neighborhood larger than 5×5 can further reduce the approximation error but at the cost of an even slower implementation. However, larger neighborhood masks cannot be used with WDTs because they give erroneous results since the large masks can bridge over a thin line that separates two segmentation regions. Overall, this chamfer metric approach to WDT is fast and easy to implement but due to the required small neighborhoods is not isotropic and cannot achieve high accuracy.

WDT Based on Curve Evolution

In this approach, at time $t = 0$ the boundary of each source is modeled as a curve $\gamma(0)$, which is then propagated with normal speed $V = \beta(x, y) = \beta_0 / \eta(x, y)$. The propagating curve $\gamma(t)$ is embedded as the zero-level curve of a function $\phi(x, y, t)$, where $\phi(x, y, 0) = \phi_0(x, y)$ is the signed (positive in the curve interior) distance from $\gamma(0)$. The function ϕ evolves according to the PDE

$$\phi_t = \beta(x, y) \|\nabla \phi\|, \tag{10.90}$$

which corresponds to curve evolution in a heterogeneous medium with position-dependent speed $\beta > 0$, or equivalently to a successive front dilation by disks with position-varying radii $\beta(x, y)\, dt$. This is a time-dependent formulation of the eikonal PDE [Fal94, Osh88]. It can be solved via Osher and Sethian's numerical algorithm given by Eq. (10.49). The value of the resulting WDT at any pixel (x, y) of the image is the time it takes for the evolving curve to reach this pixel, that is, the smallest t such that $\phi(x, y, t) \geq 0$. This continuous approach to the WDT can achieve subpixel accuracy, as investigated by Kimmel et al. [Kim96]. In the applications of the eikonal PDE examined herein, the global speed function $\beta(x, y)$ is everywhere nonnegative. In such cases the computational complexity of Osher and Sethian's level set algorithm (which can handle sign changes in the speed function) can be significantly reduced by using Sethian's *fast marching* algorithm [Set96], which is designed to solve the corresponding stationary formulation of the eikonal PDE, that is, $\|\nabla T\| = 1/\beta$. There are also other types of numerical algorithms for solving stationary eikonal PDEs; for example, Rouy and Tourin [Rou92] have proposed an efficient iterative algorithm for solving $\|\nabla E\| = \eta$.

All of the numerical image experiments with curve evolution shown herein were produced using an implementation of fast marching based on a simple data structure of two queues. This data structure is explained in [Mar00] and has been used to implement WDTs based on either chamfer metrics or fast marching

Figure 10.13: Eikonal halftoning of the Cameraman image I from the weighted distance transform of the "negative" image $\max(I) - I$. Top row: the light source was at the top left corner, and the WDTs (displayed as intensities modulo a height such that 25 waves exist per image) were obtained via (a) a (1,1) chamfer metric, (b) the optimal 3×3 chamfer metric, and (c) curve evolution. Bottom row: 100 contour lines of the WDTs in the top row give gridless halftoning of the original images.

for applications both with single sources as well as with multiple sources, where triple points develop at the collision of several wavefronts.

Gridless Halftoning via the Eikonal PDE

Inspired by the use in Schröder [Sch83] of the eikonal function's contour lines for visually perceiving an intensity image $I(x, y)$, Verbeek and Verwer [Ver90] and especially Pnueli and Bruckstein [Pnu94] attempted to solve the PDE

$$\|\nabla E(x, y)\| = \text{constant} - I(x, y) \tag{10.91}$$

and create a binary *gridless* halftone version of $I(x, y)$ as the union of the level curves of the eikonal function $E(x, y)$. The larger the intensity value $I(x, y)$, the smaller the local density of these contour lines in the vicinity of (x, y). This eikonal PDE approach to gridless halftoning, which we call *eikonal halftoning*, is indeed very promising and can simulate various artistic effects, as shown in Fig. 10.13, which also shows that the curve evolution WDT gives a smoother halftoning of the image than the WDTs based on chamfer metrics.

Watershed Segmentation via the Eikonal

A powerful morphological approach to image segmentation is the *watershed* algorithm [Mey90, Vin91b], which transforms an image $f(x, y)$ to the crest lines separating adjacent catchment basins that surround regional minima or other "marker" sets of feature points. Najman and Schmitt [Naj94] established that (in the continuous domain and assuming that the image is smooth and has isolated critical points) the continuous watershed is equivalent to finding a skeleton by influence zones with respect to a weighted distance function that uses points in the regional minima of the image as sources and $\|\nabla f\|$ as the field of indices. A similar result was obtained by Meyer [Mey94] for digital images. In Maragos and Butt [Mar00] the eikonal PDE modeling the watershed segmentation of an image-related function f was solved by finding a WDT via the curve evolution PDE of Eq. (10.90) in which the speed β is proportional to $1/\|\nabla f\|$. Further, the results of this new segmentation approach [Mar00] have been compared to the digital watershed algorithm via flooding [Vin91b] and to the eikonal approach solved via a discrete WDT based on chamfer metrics [Mey94, Ver90]. In all three approaches, robust features are extracted first as markers of the regions, and the original image I is transformed to another function f by smoothing via alternating opening-closing by reconstruction, taking the gradient magnitude of the filtered image, and changing (via morphological reconstruction) the homotropy of the gradient image so that its only minima occur at the markers. The segmentation is done on the final outcome f of the above processing.

In the standard digital watershed algorithm via flooding [Mey90, Vin91b], the flooding at each level is achieved by a planar distance propagation that uses the chessboard metric. This kind of distance propagation is not isotropic and could give wrong results, particularly for images with large plateaus. Eikonal segmentation using WDTs based on chamfer metrics improves this situation a little but not entirely. In contrast, for images with large plateaus or regions, segmentation via the eikonal PDE and curve evolution WDT gives results close to ideal. Using a test image that was difficult (because expanding wavefronts meet watershed lines at many angles ranging from being perpendicular to almost parallel), Fig. 10.14 shows that the continuous segmentation approach based on the eikonal PDE and curve evolution outperforms the discrete segmentation results (using either the digital watershed flooding algorithm or chamfer metric WDTs). In real images, which may contain a large variety of region sizes and shapes, the digital watershed flooding algorithm may give results comparable to the eikonal PDE approach. Details on comparing the two segmentation approaches can be found in [Mar00].

10.7 Conclusions

We have provided a unified view and some analytic tools for a recently growing part of morphological image processing that is based on ideas from differential calculus and dynamic systems, including the use of partial differential equations

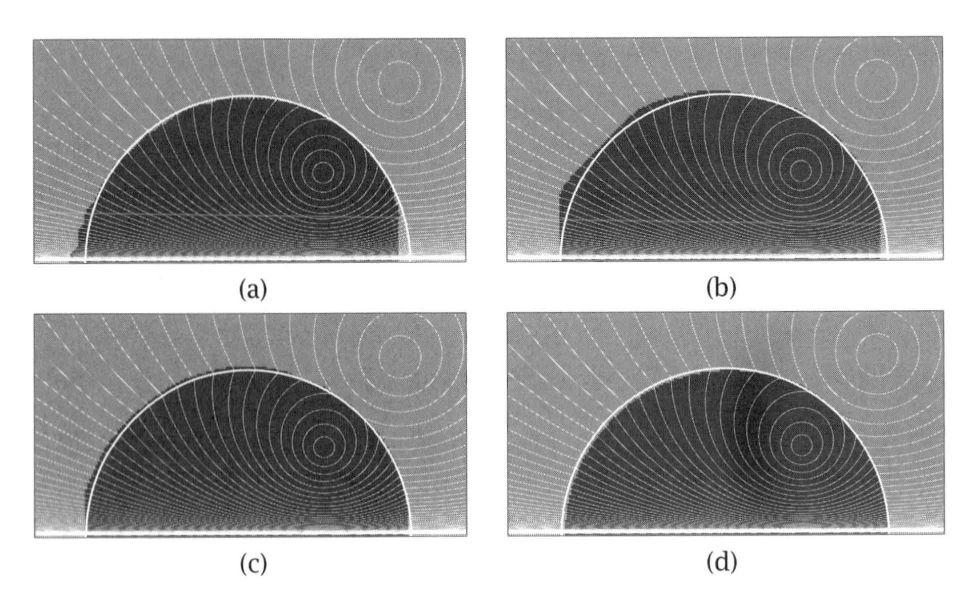

Figure 10.14: Performance of various segmentation algorithms on a test image, 250×400 pixels that is the minimum of two potential functions. Its contour plot (thin bright curves) is superimposed on all segmentation results. Markers are the two source points of the potential functions. The segmentation results are from (a) the digital watershed flooding algorithm and from WDTs based on (b) the optimal 3×3 chamfer metric, (c) the optimal 5×5 chamfer metric, and (d) curve evolution. (The thick bright curve shows the correct segmentation.)

or difference equations to model nonlinear multiscale analysis or distance propagation in images. We have discussed general 2D nonlinear difference equations of the max-sum or min-sum type that model the space dynamics of 2D morphological systems (including the distance computations) and some nonlinear signal transforms, called slope transforms, that can analyze these systems in a transform domain in ways conceptually similar to the application of Fourier transforms to linear systems. We have used these nonlinear difference equations to model discrete distance transforms and relate them to numerical solutions of the eikonal PDE of optics. In this context, distance transforms are shown to be bandpass slope filters. We have also reviewed some nonlinear PDEs that model the evolution of multiscale morphological operators and use morphological derivatives. Related to these morphological PDEs is the area of curve evolution, which employs methods of differential geometry to study the differential equations governing the propagation of time-evolving curves. The PDEs governing multiscale morphology and most cases of curve evolution are of the Hamilton-Jacobi type and are related to the eikonal PDE of optics.

We view the analysis of the multiscale morphological PDEs and of the eikonal PDE solved via weighted distance tranforms as a unified area in nonlinear image processing that we call *differential morphology*, and we have briefly discussed some of its potential applications to image processing.

References

[Alv92] L. Alvarez, F. Guichard, P. L. Lions, and J. M. Morel. Axiomatization et nouveaux operateurs de la morphologie mathematique. *C. R. Acad. Sci. Paris* **315**, Series I, 265–268 (1992).

[Alv93] L. Alvarez, F. Guichard, P. L. Lions, and J. M. Morel. Axioms and fundamental equations of image processing. *Archiv. Rat. Mech.* **123**(3), 199–257 (1993).

[Are93] A. Arehart, L. Vincent and B. Kimia. Mathematical morphology: the Hamilton-Jacobi connection. In *Proc. Intl. Conf. on Computer Vision*, pp. 215–219 (1993).

[Bar90] M. Bardi and M. Falcone. An approximation scheme for the minimum time function. *SIAM J. Control Optimization* **28**, 950–965 (1990).

[Blu73] H. Blum. Biological shape and visual science (part I). *J. Theoret. Biol.* **38**, 205–287 (1973).

[Boo92] R. van den Boomgaard. Mathematical morphology: extensions towards computer vision. Ph.D. thesis, Univ. of Amsterdam, The Netherlands (1992).

[Boo94] R. van den Boomgaard and A. Smeulders. The morphological structure of images: the differential equations of morphological scale-space. *IEEE Trans. Pattern Anal. Mach. Intellig.* **16**, 1101–1113 (November 1994).

[Bor59] M. Born and E. Wolf. *Principles of Optics.* Pergamon Press, Oxford, England (1959; 1987 edition).

[Bor86] G. Borgefors. Distance transformations in digital images. *Comput. Vision Graph. Image Process.* **34**, 344–371 (1986).

[Bro92] R. W. Brockett and P. Maragos. Evolution equations for continuous-scale morphology. In *Proc. IEEE Intl. Conf. on Acoustics, Speech, and Signal Processing* (San Francisco, CA, March 1992).

[Bro94] R. Brockett and P. Maragos. Evolution equations for continuous-scale morphological filtering. *IEEE Trans. Signal Process.* **42**, 3377–3386 (December 1994).

[But98] M. A. Butt and P. Maragos. Optimal design of chamfer distance transforms. *IEEE Trans. Image Process.* **7**, 1477–1484 (October 1998).

[Che89] M. Chen and P. Yan. A multiscaling approach based on morphological filtering. *IEEE Trans. Pattern Anal. Machine Intell.* **11**, 694–700 (July 1989).

[Cou62] R. Courant and D. Hilbert. *Methods of Mathematical Physics*, Wiley, New York (1962).

[Cra92] M. G. Crandall, H. Ishii, and P.-L. Lions. User's guide to viscosity solutions of second order partial differential equations. *Bull. Amer. Math. Soc.* **27**,1–66 (July 1992).

[Dan80] P.-E. Danielsson, Euclidean distance mapping. *Comp. Graph. Image Process.* **14**, 227–248 (1980).

[Dor94] L. Dorst and R. van den Boomgaard. Morphological signal processing and the slope transform. *Signal Process.* **38**, 79–98 (July 1994).

[Dud84] D. E. Dudgeon and R. M. Mersereau. *Multidimensional Digital Signal Processing.* Prentice Hall, Englewood Cliffs, NJ (1984).

[Fal94] M. Falcone, T. Giorgi, and P. Loreti. Level sets of viscocity solutions: some applications to fronts and rendez-vous problems. *SIAM J. Appl. Math.* **54**, 1335–1354 (October 1994).

[Hei94] H. J. A. M. Heijmans. *Morphological Image Operators,* Academic Press, Boston (1994).

[Hei97] H. J. A. M. Heijmans and P. Maragos. Lattice calculus and the morphological slope transform. *Signal Process.* **59**, 17–42 (1997).

[Hor86] B. K. P. Horn. *Robot Vision.* MIT Press, Cambridge, MA (1986).

[Hua94] C. T. Huang and O. R. Mitchell. A euclidean distance transform using grayscale morphology decomposition. *IEEE Trans. Pattern Anal. Machine Intell.* **16**, 443–448 (April 1994).

[Kim90] B. Kimia, A. Tannenbaum, and S. Zucker. Toward a computational theory of shape: an overview. In *Proc. European Conf. on Computer Vision* (France, April 1990).

[Kim96] R. Kimmel, N. Kiryati, and A. M. Bruckstein. Sub-pixel distance maps and weighted distance transforms. *J. Mathemat. Imaging Vision* **6**, 223–233 (1996).

[Koe84] J. J. Koenderink. The structure of images. *Biolog. Cybern.* **50**, 363–370 (1984).

[Lax73] P. D. Lax. *Hyperbolic Systems of Conservation Laws and the Mathematical Theory of Schock Waves.* SIAM, Philadelphia (1973).

[Lev70] G. Levi and U. Montanari. A grey-weighted skeleton. *Inform. Control* **17**, 62–91 (1970).

[Mar89] P. Maragos. Pattern spectrum and multiscale shape representation. *IEEE Trans. Pattern Anal. Machine Intell.* **11**, 701–716 (July 1989).

[Mar94] P. Maragos. Morphological systems: slope transforms and max-min difference and differential equations. *Signal Process.* **38**, 57–77 (July 1994).

[Mar96] P. Maragos. Differential morphology and image processing. *IEEE Trans. Image Process.* **78**, 922–937 (June 1996).

[Mar98] P. Maragos. Morphological signal and image processing. In *The Digital Signal Processing Handbook* (V. Madisetti and D. Williams, eds.) CRC Press, Boca Raton, FL (1998).

[Mar00] P. Maragos and M. A. Butt. Curve evolution, differential morphology, and distance transforms applied to multiscale and eikonal problems. *Fundamenta Informaticae* **41**, 91–129 (2000).

[Mar99] P. Maragos and F. Meyer. Nonlinear PDEs and numerical algorithms for modeling levelings and reconstruction filters. In *Scale-Space Theories in Computer Vision* (Proc. Intl. Conf. Scale-Space'99) pp. 363–374. Lecture Notes in Computer Science Vol. 1682. Springer-Verlag (1999).

[Mar90] P. Maragos and R. W. Schafer. Morphological systems for multidimensional signal processing. *Proc. IEEE* **78**, 690–710 (April 1990).

[Mar82] D. Marr. *Vision.* Freeman, San Francisco (1982).

[Mat75] G. Matheron. *Random Sets and Integral Geometry.* Wiley, New York (1975).

[Mat93] J. Mattioli. Differential relations of morphological operators. In *Proc. Intl. Workshop on Mathematical Morphology and Its Application to Signal Processing* (J. Serra and P. Salembier, eds.). Univ. Polit. Catalunya, Barcelona, Spain (1993).

[Mey92] F. Meyer. Integrals and gradients of images. In *Image Algebra and Morphological Image Processing III.* Proc. SPIE Vol. 1769, pp. 200–211 (1992).

[Mey94] F. Meyer. Topographic distance and watershed lines. *Signal Process.* **38**, 113–125 (July 1994).

[Mey98] F. Meyer. The levelings. In *Mathematical Morphology and Its Applications to Image and Signal Processing* (H. Heijmans and J. Roerdink, eds.), pp. 199–206, Kluwer Acad. Publ. (1998).

[Mey90] F. Meyer and S. Beucher. Morphological segmentation. *J. Visual Commun. Image Representation* **1**(1), 21–45 (1990).

[Nac96] P. F. M. Nacken. Chamfer metrics, the medial axis and mathematical morphology. *J. Mathemat. Imaging Vision* **6**, 235–248 (1996).

[Naj94] L. Najman and M. Schmitt. Watershed of a continuous function. *Signal Process.* **38**, 99–112 (July 1994).

[Osh88] S. Osher and J. Sethian. Fronts propagating with curvature-dependent ppeed: Algorithms based on Hamilton-Jacobi formulations. *J. Comput. Phys.* **79**, 12–49 (1988).

[Pnu94] Y. Pnueli and A. M. Bruckstein. Digi$_D$ürer—a digital engraving system. *Visual Comput.* **10** 277-292 (1994).

[Pre93] F. Preteux. On a distance function approach for gray-level mathematical morphology. In *Mathematical Morphology in Image Processing* (E.R. Dougherty, ed.) Marcel Dekker, New York (1993).

[Roc70] R. T. Rockafellar. *Convex Analysis.* Princeton Univ. Press (1970).

[Ros66] A. Rosenfeld and J. L. Pfaltz. Sequential operations in digital picture processing. *J. ACM* **13**, 471-494 (October 1966).

[Ros68] A. Rosenfeld and J. L. Pfaltz. Distance functions on digital pictures. *Pattern Recog.* **1**, 33-61 (1968).

[Rou92] E. Rouy and A. Tourin. A viscocity solutions approach to shape from shading. *SIAM J. Numer. Anal.* **29**(3), 867-884 (June 1992).

[Rut68] D. Rutovitz. Data structures for operations on digital images. In *Pictorial Pattern Recognition* (G.C. Cheng et al. eds.), pp. 105-133. Thompson, Washington, DC (1968).

[Sal95] P. Salembier and J. Serra. Flat zones filtering, conencted operators, and filters by reconstruction. *IEEE Trans. Image Process.* **4**, 1153-1160 (August 1995).

[Sap93] G. Sapiro, R. Kimmel, D. Shaked, B. Kimia, and A. Bruckstein. Implementing continuous-scale morphology via curve evolution. *Pattern Recog.* **26**(9), 1363-1372 (1993).

[Sch83] M. Schröder. The eikonal equation. *Math. Intelligencer* **1**, 36-37 (1983).

[Ser82] J. Serra. *Image Analysis and Mathematical Morphology*, Academic Press, New York (1982).

[Ser88] J. Serra (ed.). *Image Analysis and Mathematical Morphology*, Vol. 2. Academic Press, New York (1988).

[Set96] J. A. Sethian. *Level Set Methods.* Cambridge Univ. Press (1996).

[Shi91] F. Y.-C. Shih and O. R. Mitchell. Decomposition of gray-scale morphological structuring elements. *Pattern Recog.* **24**, 195-203 (1991).

[Ste80] S. R. Sternberg. Language and Architecture for Parallel Image Processing. In *Pattern Recognition in Practice* (E. S. Gelsema and L. N. Kanal, eds.). North Holland, New York (1980).

[Tek98] H. Tek and B. B. Kimia. Curve evolution, wave propagation, and mathematical morphology. In *Mathematical Morphology and Its Applications to Image and Signal Processing* (H. Heijmans and J. Roerdink, eds.), pp. 115-126, Kluwer, Boston (1998).

[Ver90] P. Verbeek and B. Verwer. Shading from shape, the eikonal equation solved by grey-weighted distance transform. *Pattern Recog. Lett.* **11**, 618–690 (1990).

[Ver91] B. J. H. Verwer. Distance transforms: metrics, algorithms, and applications. Ph.D. thesis, Tech. Univ. of Delft, The Netherlands (1991).

[Vin91a] L. Vincent. Exact Euclidean distance function by chain propagations. In *Proc. Conf. on Computer Vision and Pattern Recognition*, pp. 520–525 (1991).

[Vin91b] L. Vincent and P. Soille. Watershed in digital spaces: an efficient algorithm based on immersion simulations. *IEEE Trans. Pattern Anal. Machine Intell.* **13**, 583–598 (June 1991).

[Vin93] L. Vincent. Morphological grayscale reconstruction in image analysis: applications and efficient algorithms. *IEEE Trans. Image Process.* **2**, 176–201 (April 1993).

[Wit83] A. P. Witkin. Scale-space filtering. *Proc. Intl. Joint Conf. on Artificial Intelligence* (Karlsruhe, Germany, 1983).

11

Coordinate Logic Filters: Theory and Applications in Image Analysis

BASIL G. MERTZIOS AND KONSTANTINOS D. TSIRIKOLIAS

Department of Electrical and Computer Engineering
Democritus University of Thrace
Xanthi, Greece

11.1 Introduction

Logic operations have been used successfully for a variety of image processing tasks with binary images, such as removal of isolated points that usually represent noise, separation of multiple objects, extraction of depth maps and skeletons, shape smoothing, coding, compression, region filling, shape recognition and restoration of shapes [Jam87]. Coordinate logic operations (CLOs) are logic operations (AND, OR, NOT, and XOR and their combinations) among the corresponding binary values of two or more signals or image pixels. In this chapter we present the coordinate logic (CL) filters [Mer93, Mer98, Tsi91, Tsi93], which constitute a novel class of nonlinear digital filters that are based on the execution of CLOs among the pixels of the image [Die71].

CL filters can execute the four basic morphological operations (erosion, dilation, opening, and closing), as well as the key issue of the successive filtering and managing of the residues. Therefore, CL filters are suitable to perform the range of tasks and applications that are executed by morphological filters, achieving the same functionality. These applications include multidimensional filtering (noise removal, lowpass and highpass filtering, image magnification), edge extraction and enhancement, shape description and recognition, region filling, shape smoothing, skeletonization, feature extraction using the pattern spectrum, multiscale shape representation and multiresolution approaches [Mar89], coding, fractal modeling, and special video effects. However, due to their different definition, in practice, CL filters display a slightly different response to gray-level images. In the case of binary signals or images, CL filters coincide with the conventional morphological filters.

Since CL filters use only CLOs, they are simple, fast, and very efficient in various 1D, 2D, or higher-dimensional digital image analysis and pattern recognition applications. The CLOs CAND (coordinate AND) and COR (coordinate OR) are analogous to the operations MIN and MAX, respectively [Nak78], while CXOR (coordinate XOR) does not correspond to any morphological operation. In gray-level images the use of CXOR operation provides efficient edge detection and enhancement since there is no need to compute the difference between the original and filtered images, as is required by morphological filters.

The desired processing in CL filtering is achieved by executing only direct logic operations among the pixels of the given image. Considering the underlying concept of CLOs, the CL filtering may be seen as independent parallel processing of the bit planes that are produced from the decomposition of the given image. Therefore, they are characterized by inherent parallelism and are appropriate for high-speed real-time applications.

CL filters can also be used to execute tasks in image processing that cannot be executed by morphological filters, such as implementing of an approximation of the majority function [Ahm89] by the majority CL filter, fractal transformations and modeling, and design of cellular automata. In fact, a remarkable property of the CL filters is their direct relation with known fractal structures. For example, the CAND and COR operators are generators of Sierpinsky structures. Using CL filters, very simple algorithms for fractal transformation and cellular automata may be designed [Bar88, Mer98], that may be implemented simply using special purpose hardware and architectures based on logic circuits.

CL filters satisfy the defining property of idempotency of morphological filters [Mar87, Mar89, Mar90, Nak78, Ser83, Ser92] for opening and closing and their combinations. However, they are nonincreasing filters since the second defining property of increasingness that characterizes the morphological filters is not generally satisfied. In fact, the CL filters satisfy the defining properties of idempotency, duality, and extensivity of morphological filters [Mar90, Par96, Ser83, Ser92] for opening and closing as well as for their combinations.

The fundamental operations of dilation and erosion that correspond to topological max and min operations on pixel neighborhoods are special cases of the rth ranked order filters and imply a sorting procedure. From this point of view, the morphological operations that in general use combinations of dilation and erosion, overlap with the order-statistic (OS) filters [Mar87] or even may be considered as a class of OS filters [Kim86]. In contrast, CL filters do not involve any kind of sorting, and their CLOs of dilation and erosion result in signal values that may not be included in the initial set of input signal values; therefore, CL filters do not overlap with any class of OS filters.

A methodology has been developed for implementing morphological filters using CLOs. If the image gray levels are mapped to an appropriately selected set of decimal numbers, the CLOs of erosion and dilation act on this specific set exactly as the MIN and MAX operators of morphological filtering do, using very simple hardware structures. The only drawback of this approach is the large binary word lengths required for the assignment of the image gray quantization levels to this specific set of numbers, which is alleviated by exploiting the properties of CLOs and developing decomposition techniques that operate with smaller binary lengths, and with less demanding hardware structures. This approach may be extended for the implementation of any nonlinear filter that falls in the general class of OS filters, using CLOs.

A class of nonlinear filters that is based on Boolean operators is that of the generalized stack (GS) filters [Lin90, Mar87]. The difference between GS and CL filters is that the stack filters operate on signal levels, whereas CL filters operate on binary representations. The definition of GS filters is based on threshold decomposition and Boolean operators. Threshold decomposition maps the 2^n-level input signal (where n is the word length) into $2^n - 1$ binary signals by thresholding the original signal at each of the allowable levels. The set of $2^n - 1$ signals is then filtered by $2^n - 1$ Boolean operators that are constrained to have the stacking property. The multilevel output is finally obtained as the sum of the $2^n - 1$ binary output signals. In contrast, CL filters decompose the signal into n binary signals that operate in parallel and achieve the desired processing by executing only direct logic operations among the binary values of the given signal.

The CL operators are defined and their properties are derived in Sec. 11.2, and Sec. 11.3 presents the basic CL filters that execute the dilation, erosion, opening, and closing operations as well as their filter structures. Section 11.4 reviews the properties of duality for dilation–erosion and opening–closing, of idempotency for opening and closing, and of extensivity for the dilation, erosion, opening, and closing of CL filters. Section 11.5 introduces the proposed scheme for performing morphological operations using CLOs and presents a simple hardware implementation of the basic MIN and MAX operations using this scheme. Section 11.6 covers image analysis and feature extraction applications, including edge extraction, calculation of the pattern spectrum, noise removal, fractal transformation, and the design of cellular automata. Concluding remarks are given in Sec. 11.7.

11.2 Coordinate Logic Operations on Digital Signals

The underlying idea in coordinate logic image processing is the execution of CLOs among gray-level pixels. These operations are executed among the corresponding binary bits of equal position of the considered pixels, without counting the carry bits. The fundamental properties of logic operations also hold for CLOs since they constitute an extension of Boolean algebra [Die71].

For our discussion, we assume that A and B are two decimal numbers with n bits in their binary representations, given by

$$A = [a_1, a_2, \ldots, a_n], \qquad B = [b_1, b_2, \ldots, b_n].$$

The following definitions refer to the coordinate operators that are used by CL filters.

Definition 11.1. The coordinate logic operation $\bullet$ of two numbers A and B in the decimal system, in their binary sequence, is given by

$$C = [c_1, c_2, \ldots, c_n] = A \bullet B, \tag{11.1}$$

where $\bullet$ denotes the CL operation corresponding to the logic operation $\mathbf{o}$. The ith bit c_i is defined as the output of the logic operation $\mathbf{o}$ among the corresponding ith bits a_i and b_i of the operands, that is,

$$c_i = a_i \mathbf{o} b_i, \qquad i = 1, 2, \ldots, n. \tag{11.2}$$

The operation $\mathbf{o}$ may be the logical OR, AND, or NOT, or a function of them. A distinct characteristic of the coordinate operators is that they do not encounter any carry bits as occurs in regular logical operations among binary sequences.

Definition 11.2. The coordinate logic operation CNOT on a number A results in a decimal number C with an n-bit representation:

$$C = [c_1 c_2, \ldots, c_n] = \text{CNOT } A, \tag{11.3}$$

where the ith bit c_i is the logical NOT of the corresponding a_i bit of the number A, that is,

$$c_i = \text{NOT } a_i, \qquad i = 1, 2, \ldots, n. \tag{11.4}$$

It is seen that the CAND and COR operations result in values that are not included in the initial range of the values of the pixels. This characteristic does not constitute a problem in the proposed image processing feature extraction procedure since after each successive filtering, only the population of the remaining pixels is important, not their gray-level values. However, for the other image processing tasks presented in this chapter, this characteristic is either ignored by the combinations of CL operators or has a beneficial effect (e.g., in fractal transformation).

Let A and B be two integer positive decimal numbers and let the CAND and COR operations on the numbers A and B result in the two decimal numbers

$$E = A \,\text{CAND}\, B = [e_1, e_2, \ldots, e_n], \tag{11.5}$$

$$D = A \,\text{COR}\, B = [d_1, d_2, \ldots, d_n]. \tag{11.6}$$

Table 11.1: Fundamental Properties of Coordinate Logic Operations.

Fundamental laws	0 CAND $A = 0$, 0 COR $A = A$ A CAND $A = A$, A COR $A = A$ (idempotence laws) $(2^n - 1)$ CAND $A = A$, $(2^n - 1)$ COR $A = (2^n - 1)$ A CAND CNOT $(A) = 0$, A COR CNOT $(A) = (2^n - 1)$
Commutative laws	A CAND $B = B$ CAND A, A COR $B = B$ COR A
Associative laws	A CAND $(B$ CAND $C) = (A$ CAND $B)$ CAND C A COR $(B$ COR $C) = (A$ COR $B)$ COR C
Distributive laws	CAND $(A,($COR $(B,C)))$ = COR (CAND (A,B), CAND (A,C)) CAND $(A,($CAND $(B,C)))$ = CAND (CAND $(A,B),C)$ COR $(A,($CAND $(B,C)))$ = CAND (COR (A,B), COR (A,C)) COR $(A,($COR $(B,C)))$ = COR (COR $(A,B),C)$
De Morgan's laws	CNOT $(A_1$ CAND A_2 CAND ... CAND $A_n)$ = CNOT (A_1) COR CNOT (A_2)... COR CNOT (A_n) CNOT $(A_1$ COR A_2 COR ... CAND $A_n)$ = CNOT (A_1) CAND CNOT (A_2)... CAND CNOT (A_n).
Absorption laws	A COR $(B$ CAND $A) = A$, A CAND $(B$ COR $A) = A$ $(A$ COR $B)$ CAND $(A$ COR $C) = A$ COR B CAND C

The fundamental properties of logic operations also hold for CLOs since they constitute an extension of Boolean algebra [Die71]. These properties are summarized in Table 11.1.

Theorems 11.1 and 11.2 below provide useful relations among the decimal numbers A, B and the decimal numbers D, E obtained from A, B using the CL operations given in Eqs. (11.5) and (11.6).

Theorem 11.1. *Let E and D be decimal numbers as defined by Eqs. (11.5) and (11.6), respectively. Then*

$$0 \leq E \leq \min(A, B), \tag{11.7}$$

$$\max(A, B) \leq D \leq (2^n - 1), \tag{11.8}$$

where n is the word length. In other words, the COR *and* CAND *of A and B represent a measure of the maximum and of the minimum functions of A and B, respectively.*

Proof. The proof results directly by taking into account the definitions of AND and OR, and from the fact that no carry bits are encountered in the CLOs. Note that $(2^n - 1)$ is the maximum number represented by a word length of n bits. Theorem 11.1 can be extended to the case of more than two numbers. □

Theorem 11.2. *Let E and D be decimal numbers as defined by Eqs. (11.5) and (11.6), respectively. The sum of E and D is equal to the sum of A and B [Mer98], that is,*

$$E + D = (A \text{ COR } B) + (A \text{ CAND } B) = A + B \tag{11.9}$$

Theorem 11.3. *Let A and B be decimal numbers with $A < B$. Then it holds*

$$E = \sum_{i=1}^{n} e_i 2^i = A \text{ CAND } B = A - \sum_{i=1}^{n} \delta(a_i - 1)\delta(b_i)2^i \leq A, \quad (11.10a)$$

$$D = \sum_{i=1}^{n} d_i 2^i = A \text{ COR } B = B + \sum_{i=1}^{n} \delta(a_i - 1)\delta(b_i)2^i \geq B, \quad (11.10b)$$

where $\delta(.)$ is the delta or Dirac function.

Proof. For any bit position i of A, B for which a_i, b_i have different binary values, it holds that $a_i \oplus b_i = 1$ (where $\oplus$ denotes XOR), or equivalently $\delta(a_i - 1)\delta(b_i) = 1$. However, if however a_i, b_i have the same binary values then $\delta(a_i - 1)\delta(b_i) = 0$. Hence, $e_i = a_i - \delta(a_i - 1)\delta(b_i)$ and $d_i = b_i + \delta(a_i - 1)\delta(b_i)$ and resulting in Eqs. (11.10). □

Corollary 11.1. *It follows from Theorem 11.2 that*

$$A - E = D - B = K = \sum_{i=1}^{n} \delta(a_i - 1)\delta(b_i)2^i, \quad (11.11)$$

which means that COR *and* CAND *differ from the* min *and* max *operations by an equal quantity K. Therefore, the application of* COR *and* CAND *instead of the* min *and* max *operations, respectively, to two numbers attenuates or amplifies them by the same quantity K.*

Theorem 11.4. *Let A and B be decimal numbers, with $A < B$. Then*

$$E = A \text{ CAND } B = \min(A, B) = A, \quad (11.12a)$$

$$D = A \text{ COR } B = \max(A, B) = B, \quad (11.12b)$$

holds if and only if for each $a_i = 1$ the corresponding $b_i = 1$, or equivalently $\delta(a_i - 1)\delta(b_i - 1) = 1$.

Proof. If $b_i = 1$ for all bit positions at which $a_i = 1$, then $d_i = 1$ and $e_i = 1$. If $a_i = 0$ and $b_i = 1$, then $d_i = 1$ and $e_i = 0$. If $a_i = 0$ and $b_i = 0$, then $d_i = 0$ and $e_i = 0$. Therefore, it results that in all cases we have $d_i = b_i$ and $e_i = a_i$. □

Theorem 11.5. *Let A_i, $i = 1, 2, \ldots, N$, be decimal numbers. Then*

$$0 \leq E = A_1 \text{ CAND } A_2 \ldots \text{ CAND } A_N = \text{ CAND } \{A_i, i = 1, 2, \ldots, N\} \leq \min\{A_i\}, \quad (11.13a)$$

$$\max\{A_i\} \leq D = A_1 \text{ COR } A_2 \ldots \text{ COR } A_N = \text{ COR } \{A_i, i = 1, 2, \ldots, N\} \leq (2^n - 1). \quad (11.13b)$$

In Eq. (11.13a) note that $E = \min\{A_i\}$ if and only if all A_i have ones at all the bits where $\min\{A_i\}$ has ones, while in Eq. (11.13b) $D = 2^n - 1$ if and only if all A_i have zeros at all the bits where $\max\{A_i\}$ has zeros.

Proof. It can be proved easily by extending Theorems 11.1 and 11.4 in the case of CLOs among more than two numbers. □

Corollary 11.2. *Let A and B be decimal numbers with $A < B$. If for each $a_i = 1$ the corresponding $b_i = 0$, that is, if $\delta(a_i)\delta(b_i - 1) = 1$, then*

$$E = A \,\text{CAND}\, B = 0, \qquad D = A \,\text{COR}\, B = A + B \le 2^n - 1. \tag{11.14}$$

Proof. If $b_i = 0$ for all bit positions at which $a_i = 1$, then $d_i = 0$, and therefore $E = 0$. Then, using Theorems 11.2 and 11.1, it follows that $D = A + B \le 2^n - 1$. □

Corollary 11.3. *Let A and B be decimal numbers with $E = A \,\text{CAND}\, B$ and $D = A \,\text{COR}\, B$. Then the following propositions hold:*

1. *If $E > 0$, then $D < A + B$.*
2. *If the summation of A and B is $A + B > 2^n - 1$, then $E > 0$.*

Proof. 1. It follows from Eq. (11.9) that $A+B-D = E > 0$, that is, that $D < A+B$.

2. If $A+B > 2^n - 1$, from Eq. (11.9) it follows that $E = A+B-D > 2^n - 1 - D > 0$. □

11.3 Derivation of the Coordinate Logic Filters

We denote a two-dimensional digital signal by a 2D set $G = \{g(i,j),\ i = 1,2,\ldots, M,\ j = 1,2,\ldots,N\}$, where M and N are the finite dimensions of the signal in the horizontal and vertical directions, respectively. The corresponding filtered output signal is

$$F = \{f(i,j),\ i = 1,2,\ldots,M,\ j = 1,2,\ldots,N\}. \tag{11.15}$$

In what follows, for simplicity we denote the input and output images as G and F, respectively.

At this point we derive four basic CLOs: dilation, erosion, opening, and closing. First, we decompose the given gray-level image G into a set of binary images $S_k = \{s_k(i,j);\ i = 1,2,\ldots,M,\ j = 1,2,\ldots,N\}$, $k = 0,1,\ldots,n-1$, according to the decomposition of the (i,j) pixel as follows:

$$g(i,j) = \sum_{k=0}^{n-1} s_k(i,j)2^k, \quad i = 1,2,\ldots,M, \quad j = 1,2,\ldots,N, \tag{11.16}$$

where $s_k(i,j)$, $k = 0,1,\ldots,n-1$ are the binary components of the decimal pixel values $g(i,j)$, $i = 1,2,\ldots,M$, $j = 1,2,\ldots,N$.

11.3.1 Coordinate Logic Dilation

CL dilation of the image G by the structuring element B is denoted by $G_B^{\mathrm{D}}(g(i,j))$, or G_B^{D} for simplicity, and is defined by

$$\begin{aligned} F = G_B^{\mathrm{D}} &= \mathrm{COR}\, g(i,j) \in B \\ &= \sum_{k=0}^{n-1} (s_k(i,j))_B^{\mathrm{D}} 2^k, \quad i = 1,2,\ldots,M, \quad j = 1,2,\ldots,N, \end{aligned} \tag{11.17}$$

where $(s_k(i,j))_B^{\mathrm{D}}$ denotes the dilation operation on the binary value $s_k(i,j)$ by the structuring element B, given by $(s_k(i,j))_B^{\mathrm{D}} = \mathrm{OR}\,(s_k(i,j)) \in B$. Moreover, note that $f(i,j)$, $i = 1,2,\ldots,M$, $j = 1,2,\ldots,N$, is in the range

$$(\max g(i,j) \in B) \leq f(i,j) \leq 2^n - 1. \tag{11.18}$$

11.3.2 Coordinate Logic Erosion

CL erosion of the image G by the structuring element B is denoted by G_B^{E} and is defined by

$$\begin{aligned} F = G_B^{\mathrm{E}} = \mathrm{CAND}\, g(i,j) \in B = \sum_{k=0}^{n-1} (s_k(i,j))_B^{E} 2^k, \\ i = 1,2,\ldots,M, \quad j = 1,2,\ldots,N, \end{aligned} \tag{11.19}$$

where $(s_k(i,j))_B^{\mathrm{E}}$ denotes the erosion operation on the binary value $s_k(i,j)$ by the structuring element B, given by $(s_k(i,j))_B^{\mathrm{E}} = \mathrm{AND}\,(s_k(i,j)) \in B$. Moreover, note that $f(i,j)$, $i = 1,2,\ldots,M$, $j = 1,2,\ldots,N$, is in the range

$$0 \leq f(i,j) \leq \min(g(i,j) \in B). \tag{11.20}$$

11.3.3 Coordinate Logic Opening and Closing

An erosion operation followed by a dilation operation tends to remove small objects and projections without changing the overall size of an object. The resulting operation is called CL opening and is denoted by $\left(G_B^{\mathrm{E}}\right)_B^{\mathrm{D}}$.

A dilation followed by an erosion tends to fill holes and concavities, without changing the overall size of an object. The resulting operation is called CL closing and is denoted by $\left(G_B^{\mathrm{D}}\right)_B^{\mathrm{E}}$.

11.3.4 Filter Structures

CL erosion and dilation operations have a functionality similar to those of the corresponding morphological operations. Specifically, CL erosion tends to remove small objects and small projections, whereas dilation tends to fill holes and concavities in objects, that is, the operations have complementary smoothing actions, sharing the characteristic of either decreasing or increasing the size of an object. The output ranges of CL erosion and dilation satisfy Eqs. (11.18) and (11.20), respectively.

Various filter structures may be considered, depending on the size of the edge, the location of the origin, and the type of CLOs and on whether the origin is taken into account. Suggestively, one characteristic rhombus structuring element B is described by

$$\begin{array}{ccc} & * & \\ * & [*] & * \\ & * & \end{array}$$

where the asterisk inside the square brackets denotes the location of the origin (i, j) and the surrounding asterisks denote the pixels in the structuring element. The structure of a 2D CL filter corresponding to this structuring element is given by

$$f(i,j) = g(i-1,j) \mathbf{\ o\ } g(i,j-1) \mathbf{\ o\ } g(i,j) \mathbf{\ o\ } g(i+1,j) \mathbf{\ o\ } g(i,j+1). \quad (11.21)$$

Using the filter structure of Eq. (11.21), the erosion of the image G using CL filters is given by

$$f(i,j) = g(i-1,j) \text{ CAND } g(i,j-1) \text{ CAND } g(i,j) \text{ CAND } g(i+1,j) \text{ CAND } g(i,j+1). \quad (11.22)$$

Since the new state of each pixel depends only on the present state of that pixel and those of its neighbors, the new state for every pixel in the filtered image can be computed independently and simultaneously. In a cellular context, the "neighborhood" of a given cell is defined as the extent of the structure of the CL filter when the origin of the structure is centered on that cell.

11.4 Properties of Coordinate Logic Filters

This section discusses the most fundamental properties of the operations of CL erosion, dilation, opening, and closing. These properties hold for both binary and gray-level signals and images.

The duality properties for CL dilation and erosion, which allow for the interchange of the functionality of the image and its complementary image, are respectively as follows:

$$G_B^{\mathrm{D}} = \text{NOT}\,(\text{NOT}\,G)_B^{\mathrm{E}}, \quad (11.23)$$

$$G_B^{\mathrm{E}} = \text{NOT}\,(\text{NOT}\,G)_B^{\mathrm{D}}. \quad (11.24)$$

The proof of Eqs. (11.23) and (11.24) results from the application of the De Morgan laws in coordinate logic [Die71]. These properties can also be extended to derive the duality properties for CL opening and closing, which are, respectively,

$$(G_B^{\mathrm{E}})_B^{\mathrm{D}} = \text{NOT}\left((\text{NOT}\,G)_B^{\mathrm{D}}\right)_B^{E} \quad (11.25)$$

$$(G_B^{\mathrm{D}})_B^{\mathrm{E}} = \text{NOT}\left((\text{NOT}\,G)_B^{\mathrm{E}}\right)_B^{\mathrm{D}}. \quad (11.26)$$

Proof. Applying CL dilation on G_B^{E}, we obtain from Eq. (11.24)

$$(G_B^{\mathrm{E}})_B^{\mathrm{D}} = \left(\mathrm{NOT}\,(\,\mathrm{NOT}\,G)_B^{\mathrm{D}}\right)_B^{\mathrm{D}} = \mathrm{NOT}\,\left((\,\mathrm{NOT}\,G)_B^{\mathrm{D}}\right)_B^{\mathrm{E}}, \tag{11.27}$$

where the property of Eq. (11.23) was applied on $\mathrm{NOT}\,(\,\mathrm{NOT}\,G)_B^{\mathrm{D}}$. This proves the property of Eq. (11.25). □

The property of Eq. (11.26) may be similarly proved.

The idempotence properties for CL opening and closing, respectively, are as follows:

$$(G_B^{\mathrm{E}})_B^{\mathrm{D}} = \left(\left((G_B^{\mathrm{E}})_B^{\mathrm{D}}\right)_B^{\mathrm{E}}\right)_B^{\mathrm{D}} \tag{11.28}$$

$$(G_B^{\mathrm{D}})_B^{\mathrm{E}} = \left(\left((G_B^{\mathrm{D}})_B^{\mathrm{E}}\right)_B^{\mathrm{D}}\right)_B^{\mathrm{E}}. \tag{11.29}$$

Proof. The CLOs and morphological operators coincide for binary images $S_k = \{s_k(i,j),\ i = 1,2,\ldots,M,\ j = 1,2,\ldots,N\}$, $k = 0,1,2,\ldots,n-1$; thus, the idempotence property holds for each of the binary images S_k. Therefore, for opening we have

$$\left((s_k)_B^{\mathrm{E}}\right)_B^{\mathrm{D}} = \left(\left(\left((s_k)_B^{\mathrm{E}}\right)_B^{\mathrm{D}}\right)_B^{\mathrm{E}}\right)_B^{\mathrm{D}}. \tag{11.30}$$

From Eq. (11.30) we may write

$$\sum_{k=0}^{n-1}\left[\left((s_k)_B^{\mathrm{E}}\right)_B^{\mathrm{D}}\right]2^k = \sum_{k=0}^{n-1}\left[\left(\left(\left((s_k)_B^{\mathrm{E}}\right)_B^{\mathrm{D}}\right)_B^{\mathrm{E}}\right)_B^{\mathrm{D}}\right]2^k, \tag{11.31}$$

which is Eq. (11.28) in its decomposed form. Therefore, the CL opening operation has the idempotence property. □

The CL closing property of idempotence expressed by Eq. (11.29) may be similarly proved.

We now turn to the extensivity for CL dilation and CL erosion, CL opening, and CL closing. The dilation and erosion operations are related by

$$G_B^{\mathrm{E}} \le G \le G_B^{\mathrm{D}}. \tag{11.32}$$

CL opening is an antiextensive operation, while CL closing is extensive:

$$(G_B^{\mathrm{E}})_B^{\mathrm{D}} \le G, \quad (G_B^{\mathrm{D}})_B^{\mathrm{E}} \ge G. \tag{11.33}$$

11.5 Morphological Filtering Using Coordinate Logic Operations on Quantized Images

In this section a technique is proposed that enables the CL filters to behave exactly like the morphological ones. This is achieved by quantizing the image pixel intensities in fewer gray levels than the original image, with the new image intensities taking only specific decimal values defined below. The exact number of new levels allowed depends on the particular image application and on the available word length for data representation. This approach is explained after Definition 11.3 and Theorem 11.6, which are the basis of the subsequent development.

Definition 11.3. The set of quantized decimal numbers $\hat{A}_n$ is given by

$$\hat{A}_n = \{A_i : A_i = 2^i - 1, \quad i = 0, 1, 2, \ldots, n\}, \tag{11.34}$$

where n is the length of the binary word that stores the decimal numbers representing the pixel intensities. The cardinality $|\hat{A}_n|$ of set $\hat{A}_n$ equals $n + 1$; that is, there are only $n + 1$ such specially quantized numbers.

Theorem 11.6. *Let $A_1, A_2, \ldots, A_m \in \hat{A}_n$, that is, they are numbers taking one of the values $0, 1, 3, 7, \ldots, 2^n$. For the binary representations of $A_i \in \hat{A}_n$ the following then holds:*

$$E = A_1 \,\text{CAND}\, A_2 \,\text{CAND}, \ldots, \text{CAND}\, A_m = \min\{A_1, A_2, \ldots, A_m\}, \tag{11.35}$$

$$D = A_1 \,\text{COR}\, A_2 \,\text{COR}, \ldots, \text{COR}\, A_m = \max\{A_1, A_2, \ldots, A_m\}. \tag{11.36}$$

Proof. Let a_{ij}, $i = 1, \ldots, m$, $j = 1, \ldots, n$ be the jth bit of the number A_i and let e_j, d_j be the jth bits of the numbers E and D, respectively. It is easy to verify then that

$$e_j = a_{1j} \,\text{AND}\, a_{2j} \ldots \text{AND}\, a_{mj} = \prod_{i=1}^{m} [1 - \delta(a_{ij})], \tag{11.37}$$

$$d_j = a_{1j} \,\text{OR}\, a_{2j} \ldots \text{OR}\, a_{mj} = 1 - \prod_{i=1}^{m} \delta(a_{ij}). \tag{11.38}$$

According to Eq. (11.37), if $a_{ij} = 1$, $\forall i = 1, 2, \ldots, m$, then $e_j = 1$; however, if there is at least one a_{ij}, $i = 1, 2, \ldots, m$, such that $a_j = 0$, then $e_j = 0$. Similarly, from Eq. (11.38), if $a_{ij} = 0$, $\forall i = 1, 2, \ldots, m$, then $d_j = 0$, but if there is at least one a_{ij} $i = 1, 2, \ldots, m$, such that $a_{ij} = 1$, then $d_j = 1$. Using Eq. (11.37), E may be written in the form

$$E = \sum_{j=1}^{n} e_j 2^{j-1} = \sum_{j=1}^{n} \left[\prod_{i=1}^{m} [1 - \delta(a_{ij})] \right] 2^{j-1}. \tag{11.39}$$

Since $\prod_{i=1}^{m} [1 - \delta(a_{ij})] = 1$ only when every ith bit of A_j equals 1, then E is equal to the number A_j having the smallest number of nonzero bits, that is $E = \min\{A_1, A_2, \ldots, A_m\}$. Similarly, using Eq. (11.38), D may be written in the form

$$D = \sum_{j=1}^{n} d_j 2^{j-1} = \sum_{j=1}^{n} \left[1 - \prod_{i=1}^{m} \delta(a_{ij}) \right] 2^{j-1}. \tag{11.40}$$

Since $1 - \prod_{i=1}^{m} \delta(a_{ij}) = 0$ only when every jth bit of A_i equals 0, then D is equal to the number A_i having the greatest number of nonzero bits. That is $D = \max\{A_1, A_2, \ldots, A_m\}$. □

Theorem 11.6 also shows that when CLOs are applied to a specific set of decimal numbers belonging to the set $\hat{A}_n$, the resulting numbers here E and D always belong within the given set, that is $E, D \in \hat{A}_n$. This property is not valid when the CLOs are applied to decimal numbers $A_1, A_2, \ldots, A_m \notin \hat{A}_n$ since then E and D satisfy Eqs. (11.35) and (11.36). This means that, in general, E and D do not belong within the initial set of processed numbers $A_1, A_2, \ldots, A_m$.

Since CAND and COR operators, when operating on numbers belonging to $\hat{A}_n$, are identical to the MIN and MAX operators, respectively, all of the CL filters defined in this section are identical to the corresponding morphological ones, and the increasing property is valid for CL filters as well. Based on Theorem 11.6, we may devise a method for performing morphological operations and filtering using only CLOs, which are faster and simpler in their hardware implementation.

Suppose that n-bit words are used to represent 2^n image intensity levels. If we wish to represent the image intensities via the set of numbers $A_1, A_2, \ldots A_{k+1} \in \hat{A}_n$, $k \leq n$, then we should quantize the image intensities in $k+1$ levels. This is achieved by using a look-up table of size $2^n(k+1)$ bits, which assigns each initial image intensity to one of the $k+1$ levels. Although in this preprocessing stage this quantization constitutes a coarser image level representation than the original one, it is sufficient to use a small number $k \leq 8$ of quantization levels for a variety of image processing applications, such as image segmentation and feature extraction. Note that reduction of the number of gray levels is a usual practice in image feature extraction applications.

For the case in which exact implementation of the morphological filters using CLOs is desired, we apply a decomposition procedure to the given image so that the total number of bit planes is organized in subsets that each comprising $n+1$ bit planes. The final results are then derived by appropriate parallel composition of the partial results.

11.6 Image Analysis and Pattern Recognition Applications

In this section a number of typical image analysis and pattern recognition applications using the CL filters are reviewed. These applications may be grouped into two classes:

1. Those that are based on the implementation of morphological operations using CLOs, and

2. novel techniques that are not based or related to morphology and are based on the exploitation of the properties, the inherent structure and the characteristics of the CLOs.

11.6.1 Edge Extraction

Edge extraction in an image G can be achieved with CL filters using the same approach adopted with morphological filters, with the eroded image G_B^{E} subtracted from the original image G, so that the edge detector is $G - G_B^{\mathrm{E}}$. Alternative combinations may be used to obtain the edge estimators, such as $G_B^{\mathrm{D}} - G$, or $G_B^{\mathrm{D}} - G_B^{\mathrm{E}}$ with the latter providing a more symmetric treatment between the image and its background. Edges in different orientations can be obtained by using a properly oriented 1D structuring element. The size of the structuring element controls the thickness of the edge markers. Among the variety of the CL-based edge detectors, an efficient one that corresponds to the morphological $G_B^{\mathrm{D}} - G_B^{\mathrm{E}}$ edge detector and gives very similar results is the following:

$$\left[\left(G_B^{\mathrm{D}} \text{ CXOR } G\right) - \left(G_B^{\mathrm{E}} \text{ CXOR } G\right)\right], \tag{11.41}$$

where A CXOR B represents a measure of the difference between A and B.

A novel approach for edge extraction and enhancement is based on the direct application of appropriate CL filters to the original image, without using arithmetic subtraction between images. Such an edge detector CL filter requires only b CL operators, where b is the number of pixels belonging to the structuring element B. For the CL filter corresponding to the rhombus structuring element given by Eq. (11.21) with $b = 5$, the resulting edge extractor is given by

$$\begin{aligned} f(i,j) = {} & g(i,j) \text{ CAND } \left[\text{CNOT } [g(i-1,j) \right. \\ & \left. \text{CAND } g(i+1,j) \text{ CAND } g(i,j+1) \text{ CAND } g(i,j-1)]\right]. \end{aligned} \tag{11.42}$$

In the same class of edge detector CL filters are those that use CXOR operations. An example of such a simple CL filter based on the same rhombus structuring element is given by

$$\begin{aligned} f(i,j) = {} & [g(i-1,j) \text{ CXOR } g(i,j+1)] \\ & \text{COR } g(i,j) \text{ COR } [g(i+1,j) \text{ CXOR } g(i-1,j)]. \end{aligned} \tag{11.43}$$

In the edge extractor CL filter of Eq. (11.43), the CL operation CXOR detects the existence of a difference in a specific direction, while the CL operation COR detects the existence of a difference in an active neighborhood. The performance of the proposed CL edge detectors can be improved by using threshold values and appropriate structuring elements that depend on the specific application. Figure 11.1a shows an original image G_1 and 11.1b shows the edges of G_1 using the CL edge extractor of Eq. (11.41).

11.6.2 The Pattern Spectrum, Multiscale Shape Representation, and Multiresolution Approaches

An efficient analysis tool for extracting quantitative description of geometrical structures from signals in the mathematical morphology framework is based on

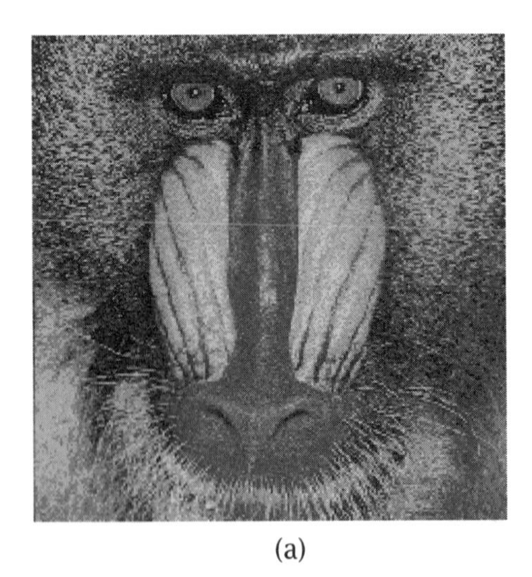

(a)

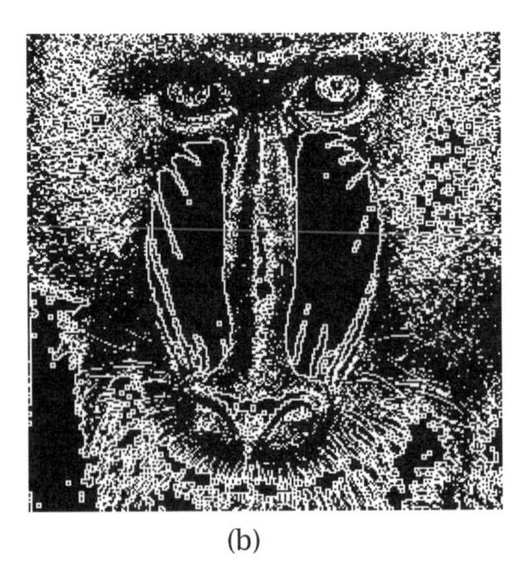

(b)

Figure 11.1: (a) Original image G_1 and (b) the edges of G_1 obtained with the CL filter of Eq. (11.41).

the theory of granulometries [Mat75]. It uses a family of transformations $G_{nB}^*(X)$ that may represent any of the operations of erosion, dilation, opening, or closing of object X with a structuring element nB. The size distribution is a function representing a mapping from $G_{nB}^*(X)$ to a quantitative measure $\mathrm{M}[G_{nB}^*(X)]$. The most basic measure of the transformed image $G_{nB}^*(X)$ is its area, that is, the number of activated pixels in the image. The *pattern spectrum*, or *pecstrum*, is a shape–size descriptor that detects and quantifies critical scales in an image object, where scale is defined as the smallest varying size of a shape pattern [Mar89]. The pecstrum is based on the granulometry and is well suited for use in low-level computer vision tasks, including the ability to operate on data gathered from multidimensional sensors. Corresponding to the morphological filters is the digital version $P(n)$ of the pecstrum function. Applying the structuring element $(n + 1)B$ using CLOs results in the following [Mer98]:

$$\begin{aligned} P(n) &= \frac{M[G_{nB}^* - G_{(n+1)B}^*]}{M[G]} \\ &= \frac{\sum_{i=1}^{N}\sum_{i=1}^{N} F\left(g_{nB}^*(i,j) - g_{(n+1)B}^*(i,j)\right)}{\sum_{i=1}^{n}\sum_{i=1}^{n} F\left(g(i,j)\right)}, \quad n = 0,1,2,\ldots, \end{aligned} \tag{11.44}$$

where G_{nB}^* denotes any of the CLOs of erosion, dilation, opening, or closing of the image G by the structuring element nB, while $g(i,j)$ and $g_{nB}^*(i,j)$ denote the integer gray-level pixels of the initial image and transformed image. Also, considering that the measure $\mathrm{M}[\cdot]$ is the area of the shape, then $F(x) = 1$ if $x > 0$, and $F(x) = 0$ if $x = 0$. The pecstrum sequence may be efficiently implemented on cellular logic computer architectures and provides compact, quantitative information of shape and structure that has a direct physical interpretation. For the special case in which G^*_{nB} represents the multiscale opening, $P(n)$ equals the rejected area, normalized to the whole area of the image.

Figure 11.2 shows an original gray-level image G_2 representing a cell. The images that result from using CL filters for the opening of G_2 by square-shaped structuring elements of increasing size B, $2B$, $3B$, and $4B$, where $B = 3 \times 3$, are shown in Figs. 11.2b through 11.2e. The remaining figure parts show the residues that result from successively subtracting the images shown in Figs. 11.2a–e. The application of Eq. (11.44) produces the pecstrum sequence $P(1) = 0.003$, $P(2) = 0.002$, $P(3) = 0.04$, $P(4) = 0.03$, which denote the rejected areas normalized to the whole area of the image and are extracted using a measure of the number of activated pixels in the residues.

The pecstrum sequence computed with CL filters may also be used for a multiscale shape representation (MSR) scheme using opening CL filters [Che89, Mar89]. Moreover, nonlinear multiresolution representations, the morphological pyramids, that are based on opening-closing (OC) CL filters and are related to skeleton transforms, such as the medial axis transform [Har87, Mar90], have been developed for image analysis and compression. The OC filters provide nonlinear signal smoothing by eliminating impulses whose spatial widths are smaller than the scale, while preserving edges. The multiresolution procedure is a multistage process in which an OC filter at each stage filters the input image, producing spatial image features of various resolutions. The filtered image from the current stage becomes the input of the next stage, and the difference between the input image and the output image is calculated. The structuring element used in the OC filter of the next stage has a larger size than the one used in the current stage. For an original image G_0, its N-level multiresolution pyramid $\{L_1, L_2, \ldots, L_N\}$ is described by

$$\begin{aligned} L_n &= G^*_{(n-1)B} - G^*_{nB} \\ &= \mathrm{OC}_{n-1}(G^*_{(n-2)B}) - \mathrm{OC}_n(G^*_{(n-1)B}), \\ &\qquad 1 \le n < N, \qquad L_N = G^*_{(N-1)B}, \end{aligned} \tag{11.45}$$

where $\mathrm{OC}_n(X)$ is an OC filter for the nth stage and G^*_{nB} represents the OC operation, using a set-structuring element nB. The L_N is analogous to a lowpass filtered image in the sense that it loses fine details. The first N decomposed subimages $\{L_1, L_2, \ldots, L_N\}$ are called feature images since each contains fine image features of known resolution. They are implemented with CL OC filters and may be used for the reconstruction of the original image G. The multiresolution process may be implemented in a way similar to Fig. 11.2 but with OC operations instead of just openings.

11.6.3 Noise Removal—The Majority Coordinate Logic Filter

Efficient filters for noise removal may be built using CLOs and the underlying Boolean operations. In binary signals, the majority function is a simple predicate, whose output has the same value (1 or 0) as the majority of the population b of the input units [Ahm89]. If b is odd, the majority is $(b + 1)/2$ terms or more,

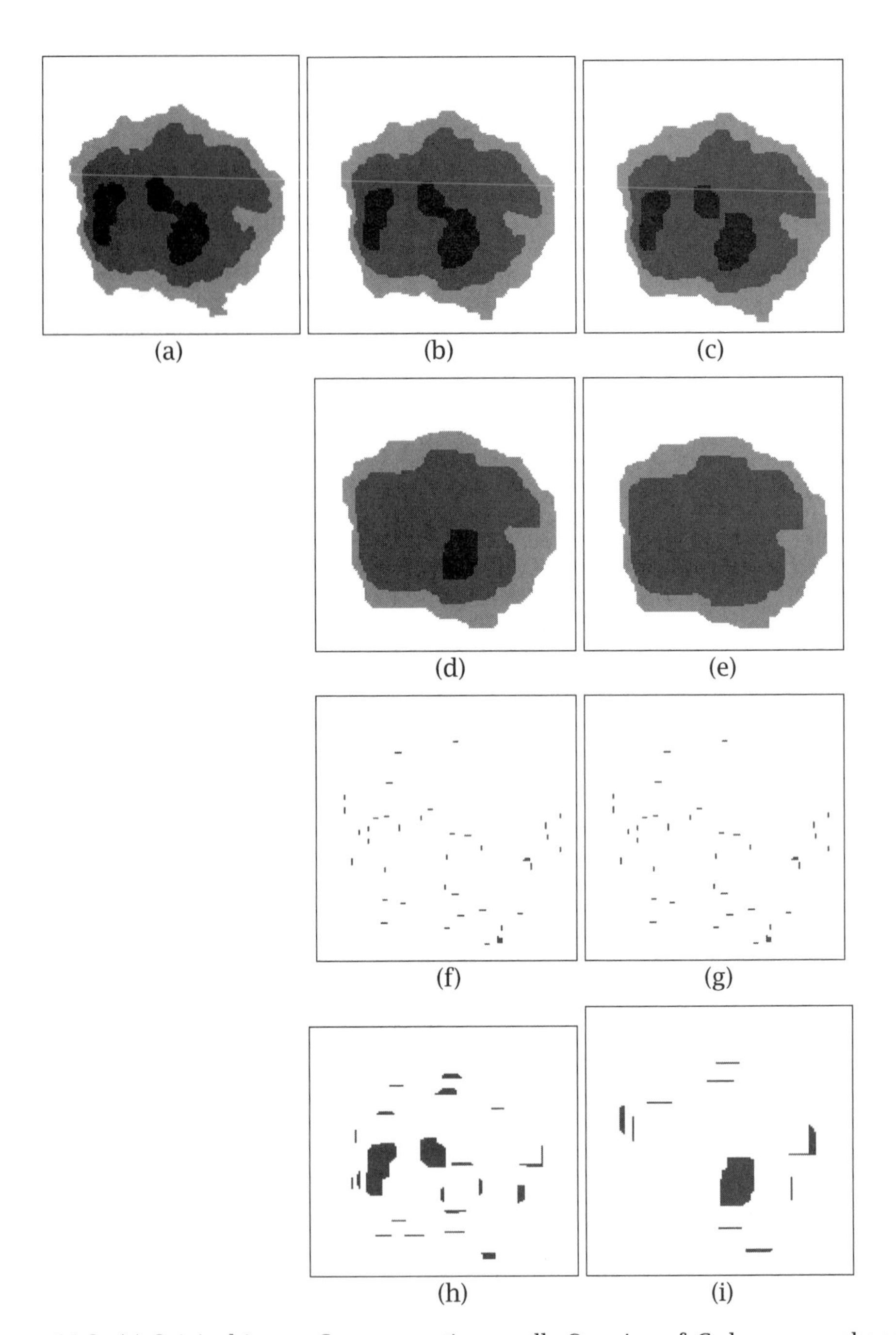

Figure 11.2: (a) Original image G_2 representing a cell. Opening of G_2 by square-shaped structuring elements of increasing sizes: (b) B, (c) $2B$, (d) $3B$ and (e) $4B$, where B is of size 3×3. (f)–(i) The residues that result from successively subtracting the images (a) through (e).

while if b is even the majority is $b/2$ terms or more. If no pixel value in the specific neighborhood satisfies the majority criterion, the output of the function is 0. However, in practice, for this case one usually sets the initial pixel value $f(i,j)$ equal to $g(i,j)$, instead of 0. In general, the output of the regular majority function for gray-level pixels is the pixel value that has the majority in this neighborhood.

The majority coordinate logic (MCL) filter implements the majority function at each level of the signal in the specific active neighborhood by selecting the n bits having the majority in this neighborhood in each of the n binary images. The MCL filter results by checking all possible combinations of the b pixels of the structuring element at all n binary levels, taken $\lfloor (b+1)/2 \rfloor$ at a time, where $\lfloor x \rfloor$ denotes the integer part of the positive number x. Thus, the MCL filter quickly and easily implements an approximation of the exact majority function applied on an active neighborhood containing b pixels, which is suitable for additive noise removal. Considering the filter structure of Eq. (11.21) with $b = 5$, the output $f(i,j)$ of the MCL filter that implements the majority function at each binary level of the signal is formed by selecting at each binary level those bits that correspond to at least $(5+1)/2 = 3$ binary pixels of equal value. The specific MCL filter that results from checking 10 possible combinations of 5 objects, taken 3 at a time, is as follows:

$$
\begin{aligned}
f(i,j) = {} & [g(i,j) \text{ CAND } g(i+1,j) \text{ CAND } g(i,j+1)] \\
& \text{COR } [g(i,j) \text{ CAND } g(i+1,j) \text{ CAND } g(i-1,j)] \\
& \text{COR } [g(i,j) \text{ CAND } g(i+1,j) \text{ CAND } g(i,j-1)] \\
& \text{COR } [g(i,j) \text{ CAND } g(i,j+1) \text{ CAND } g(i-1,j)] \\
& \text{COR } [g(i,j) \text{ CAND } g(i,j+1) \text{ CAND } g(i,j-1)] \\
& \text{COR } [g(i,j) \text{ CAND } g(i-1,j) \text{ CAND } g(i,j-1)] \\
& \text{COR } [g(i+1,j) \text{ CAND } g(i,j+1) \text{ CAND } g(i-1,j)] \\
& \text{COR } [g(i+1,j) \text{ CAND } g(i,j+1) \text{ CAND } g(i,j-1)] \\
& \text{COR } [g(i+1,j) \text{ CAND } g(i-1,j) \text{ CAND } g(i,j-1)] \\
& \text{COR } [g(i,j-1) \text{ CAND } g(i-1,j) \text{ CAND } g(i+1,j)] \\
& \text{COR } [g(i,j-1) \text{ CAND } g(i-1,j) \text{ CAND } g(i,j)] \\
& \text{COR } [g(i,j+1) \text{ CAND } g(i-1,j) \text{ CAND } g(i,j-1)]. \qquad (11.46)
\end{aligned}
$$

For binary images, the CL filter becomes simpler, since the CL operations are reduced to simple logic operations. Comparisons with median and weighted mean filters, which are the most prominent filters for impulse noise removal, have been done using signal-to-noise ratio (SNR) criteria. Specifically, Fig. 11.3b shows the application of the filter of Eq. (11.46) to the noisy image G_3 of Fig. 11.3a. The SNR [defined as $20\log(\text{signal})/(\text{noise})$] of the "salt-and-pepper noise" image of Fig. 11.3a is 22.5 dB and results from contaminating the original image with 10% impulse noise of random values.

Figures 11.3c and 11.3d show the results of applying a 4-neighbor median and a 4-neighbor weighted mean filter, respectively. The mean square errors using

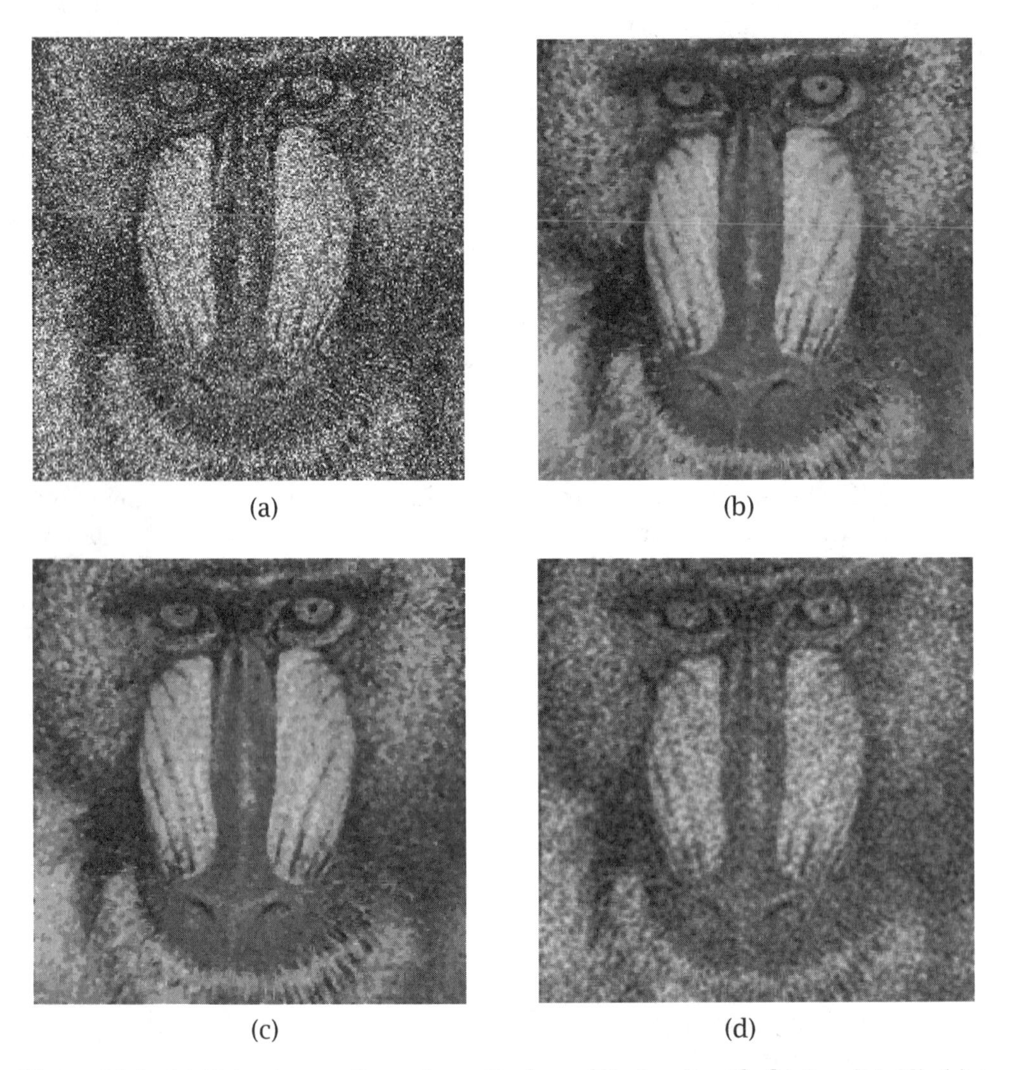

Figure 11.3: (a) Noisy image G_3, and results from filtering it with (b) Eq. (11.46), (c) a 4-neighbor median filter, and (d) a 4-neighbor weighted mean filter.

the MCL, the median, and the weighted mean filter are 5.98×10^{-3}, 5.23×10^{-3}, and 2.94×10^{-2}, respectively. The corresponding percentages of improvement of the images, in the sense that $x\%$ noisy pixels have been removed and replaced correctly, are 40%, 42%, and 5%. Thus, the efficiency of CL filters is comparable with that of median filters.

CL filters can achieve fast image magnification (zoom) of an $N \times M$ image G in one or two directions by the integer scale factor K. The associated spread image $\hat{G}$ has dimensions $(NK) \times (MK)$ and its empty areas are "zeros" representing background. The typical CL filter that may be used for the image magnification task of G has a square-shaped structuring element of size $K \times K$ and uses the COR operation, implementing a dilation on the image $\hat{G} = \hat{g}(i,j)$. Using a CL dilation filter for $K = 3$ magnification of G, the pixels $f(i,j)$ of the resulting magnified image F are given by

$$\begin{aligned} f(i,j) = {} & \hat{g}(i-1,j-1) \text{ COR } \hat{g}(i-1,j) \text{ COR } \hat{g}(i-1,j+1) \\ & \text{COR } \hat{g}(i,j-1) \text{ COR } \hat{g}(i,j) \text{ COR } \hat{g}(i,j+1) \\ & \text{COR } \hat{g}(i+1,j-1) \text{ COR } \hat{g}(i+1,j) \text{ COR } \hat{g}(i+1,j+1). \end{aligned} \quad (11.47)$$

11.6.4 Fractal Transformations and Modeling

Peitgen et al. [Pei92] presented an algorithm to create the Sierpinski triangle, which is claimed to be the shortest possible program, since all of the information needed is included in one line of code [of the form "IF (xAND$(y-x)$) = 0 THEN PSET ($x+158-.5*y, y+30$)" in pseudocode BASIC]. This program actually computes a color coding of the Pascal triangle based on the divisibility of the numbers of the triangle and constitutes an *a posteriori* construction, determined by the coordinates (x, y) of the cell. The algorithm is based on the binary coding of the coordinates, where white and black cells are associated with even and odd numbers, respectively. It transforms the mathematical background of Pascal's triangle by computing the white or black color of an arbitrary cell in a manner that avoids running all the rows above it. The comparison of the binary expression is the AND operation. The successive transformations have been done by shearing the top of the triangle a bit to the right and centering it at $x = 158$.

The CL filters can be used to construct quickly and easily various forms of fractals. An additional advantage of using them for fractals is the insight one gains into the relationship between direct logic operations and known fractal structures. We wish to emphasize this relationship, which applies to both artificial and natural fractal structures. The applications of this remarkable property of CL filters are not confined to the construction of fractals but also extend to fractal transformations, that is, to the transformation of an image G to a fractal encoded form F. The goal of this transformation is that, by describing the transformed image as a fractal, new features and descriptions may be extracted. The transformed image obviously has a much higher order of self-similarity than the original one, thus making the transformed image attractive for fractal compression. Since the proposed fractal transformation is nonlinear, reconstructing the original image from the transformed one is a very difficult task, achieved using the properties of CL filters only at the expense of a big computational load and a very small error tolerance.

The proposed fractal construction and modeling based on CL filters is faster and simpler than Sierpinski's gasket and does not presuppose the mathematics underlying it; rather it reveals them unexpectedly and rediscovers a direct and conceptually simple alternative to the shortest algorithm [Pei92]. The logical basis of the proposed CL fractal construction system is related to modeling formal systems in terms of fractal geometry. We present two characteristic CL based fractals: (a) the Sierpinsky triangles [Bar88] and (b) the shell creation. The Sierpinsky triangle may be considered as a graphical mapping of the basic CLOs (CAND, COR) in the Euclidean space, that is, the plotting of the initial coordinates (i, j) at the position $(i, i \bullet j)$, where $\bullet$ denotes the CAND or the COR. Algorithm 11.1 below gives the pseudocode in BASIC for one specific form of the Sierpinsky triangle.

Algorithm 11.1. *Sierpinsky Triangle Using CLOs*

```
For i = 1 to N
  For j = 1 to M
  PSET (i, i COR j), c
  Next j
Next i
```

In Algorithm 11.1, we set $N = M = 2^m$ (i.e., they are powers of 2), while c represents the drawing color in the general case. The command [PSET$(i, i\,\text{COR}\ j), c$] is used to set the pixel value c to the position $(i, i\,\text{COR}\ j)$, that is, $g(i, i\,\text{COR}\ j) = c$. Figure 11.4a shows the result of executing Algorithm 11.1. Alternative forms of Sierpinsky triangles may be derived by using CAND operations or successive applications of COR and CAND.

The shell construction algorithm shows how we can build complex natural structures using coordinate logic operations. Here the basic structure is a circle; the position and color of each new circle generated are determined by CXOR and its radian by the factor j/s, where j and s are variables (Figure 11.4b; also see color insert).

Algorithm 11.2. *Shell Construction Using CLOs*

```
For i = 1 to N
  For j = 1 to M
  Circle(i XOR j, j + t), j/s, x
  Next j
Next i
  N = 500, M = 50, s = 3, t = 100, x = i XOR j
```

11.6.5 Cellular Automata

Using very simple rules we can design cellular automata for the construction of various types of surfaces and materials. Important applications include the development of cellular automata for surface-material modeling and generation, for modeling of natural shapes and structures, and for understanding the organization of atoms in order to produce structures and materials using nanotechnology. The proposed CL-based rules for designing cellular automata that follow are applied m times on the image G_4, which is a 2D white noise signal representing the initial condition.

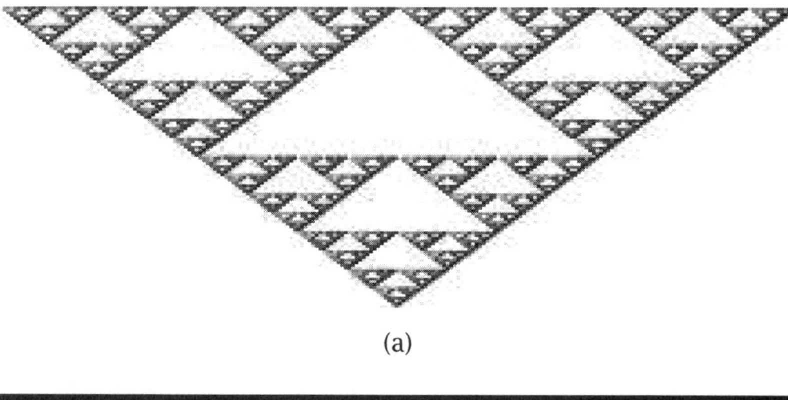

(a)

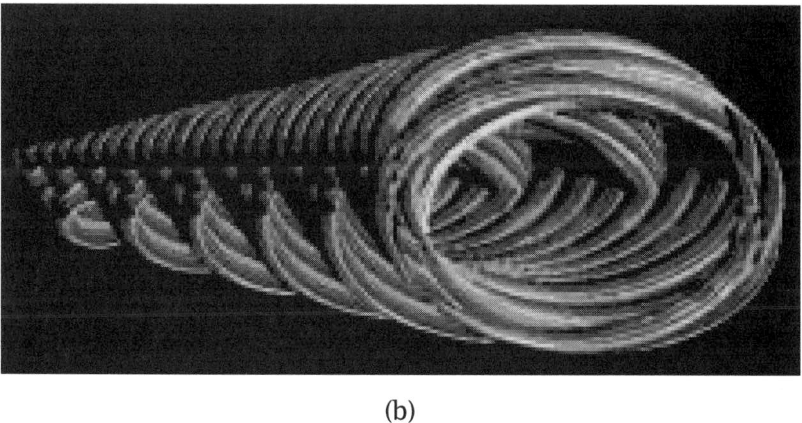

(b)

Figure 11.4: Construction of (a) the Sierpinsky triangle, based on Algorithm 11.1, and (b) the fractal form "shell," based on Algorithm 11.2.

Rule 1 For $k = 1$ to m
  For $i = 1$ to N, $j = 1$ to M
  $g(i, j) = g(i + 1, j)$ CAND $g(i, j + 1)$
  If $g(i, j) = 0$, then $g(i, j) = g(i, j)$
Next k

Rule 2 For $k = 1$ to m
  For $i = 1$ to N, $j = 1$ to M
  $g(i, j) = g(i+1, j)$ CAND $g(i, j+1)$ CAND $g(i, j)$
  If $g(i, j) = 0$, then $g(i, j) = g(i, j)$
Next k

Rule 3 For $k = 1$ to m
  $g(i, j) = g(i, j)$ CAND $g(i$ COR $j, i+1)$ CAND $g(i$ COR $j, j+1)$
    COR $(g(i$ COR $j, i - 1)$ CAND $g(i$ COR $j, j - 1))$
  If $g(i, j) = 0$, then $g(i, j) = g(i, j)$
  If $g(i, j) = 2^n - 1$, then $g(i, j) = g(i, j)$
Next k

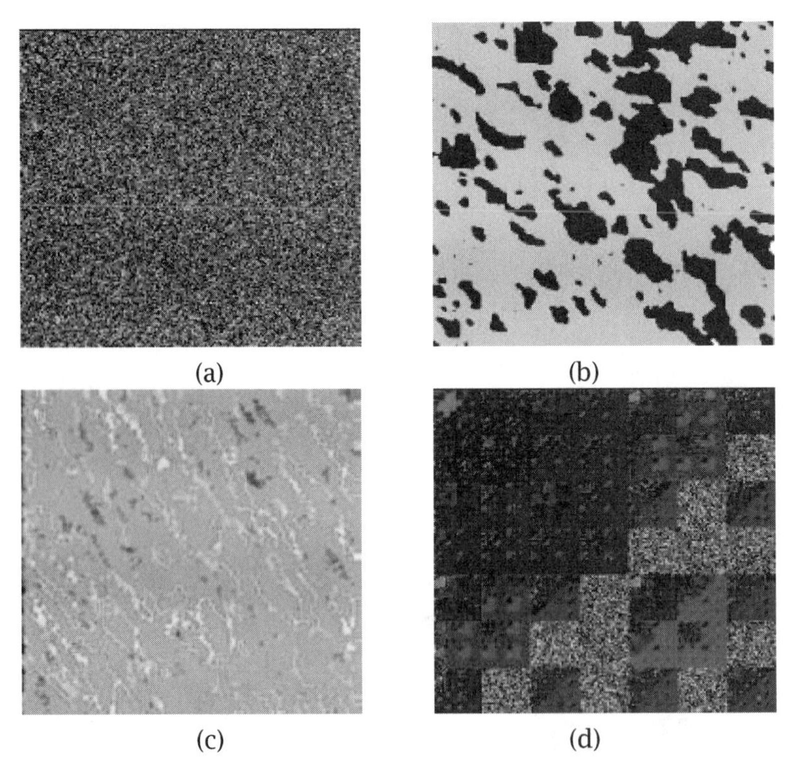

Figure 11.5: (a) Initial image G_4, a 2D white noise signal, and results from applying CL Rules (b) 1, (c) 2, and (d) 3 for $m = 4$ iterations.

In Rule 3, n is the word length. Figure 11.5 (also see color insert) shows the surfaces that result from applying these rules to the initial image G_4 for $m = 4$ iterations.

11.7 Concluding Remarks

This chapter has presented the coordinate logic filters, a new family of nonlinear digital filters based on the execution of coordinate logic operations. The key issues in coordinate logic analysis of images are the method of fast successive filtering and managment of the residues. Image processing and analysis using CL filters is achieved by executing only direct coordinate logic operations among the pixels of a given image. CL filters can be applied to both binary and gray-level images and are efficient in various 1D and 2D processing applications, such as opening, closing, region filling, shape smoothing, image magnification, and noise removal, as well as in feature extraction tasks, such as edge detection, calculation of the pattern spectrum, multiscale shape representation, multiresolution analysis, fractal transformations, and the design and modeling of cellular automata.

CL filters provide a tool for processing gray-level images as a set of binary images. Therefore, CL based decomposition of such images can be used to extend existing processing and analysis techniques (e.g., [Spi97]) for binary images to gray-level ones. These tasks are an objective of future work.

CL filters may be implemented more easily and quickly than order-statistic and other nonlinear digital filters by using logic circuits or cellular automata, and they are characterized by inherent parallelism.

CL filters are functionally similar to morphological filters. In general, using them instead of morphological filters for the development of specific image analysis tools does not lead to better performance. Their critical advantage is simplicity and high-speed implementation since they use only Boolean logic operations, whereas morphological filters require repetitive and time-consuming sorting operations. In certain cases, such as in noise removal, in which the performance of the CL filters is comparable with that of median filters, CL filters show better performance than morphological ones. CL filters also can implement certain applications, such as fractal transformations, that cannot be implemented by morphological ones.

More sophisticated CL filters can be designed, depending on the specific requirements of the application, using selected bit planes or applying different structuring elements at each bit level. Moreover, processing and compression of gray-level images can be done by running the CL operations on the "more significant" bit planes, ignoring "less significant" bit planes that might correspond to noise.

References

[Ahm89] S. Ahmad and G. Tesauro. Scaling and generalization in neural networks. In *Advances in Neural Information Processing Systems I* (D. S. Touretzky, ed.), pp. 160–168 (1989).

[Bar88] M. F. Barnsley. *Fractals Everywhere.* Academic Press, New York (1988).

[Che89] M. Chen and P. Yan. A multiscaling approach based on morphological filtering. *IEEE Trans. Pattern Anal. Machine Intell.* PAMI-**11**(7), 694–700 (July 1989).

[Die71] D. L. Dietmeyer. *Logic Design of Digital Systems.* Allyn and Bacon Inc, Boston, MA (1971).

[Har87] R. M. Haralick, C. Lin, J. S. J. Lee, and X. Zhuang. Multiresolution morphology. In *Proc. First Intl. Conf. on Computer Vision*, pp. 516–520 (1987).

[Jam87] M. James. *Pattern Recognition.* BSP Professional Books, Oxford (1987).

[Lin90] J. H. Lin and E. J. Coyle. Minimum mean absolute error estimation over the class of generalized stack filters. *IEEE Trans. Acoust. Speech Signal Process.* ASSP-**38**(4), 663–678 (April 1990).

[Kim86] V. Kim and L. Yaroslawskii. Rank algorithms for picture processing. *Comput. Vision Graph. Image Process.* **35**, 234–258 (1986).

[Mar87] P. Maragos and R. W. Schafer. Morphological filters, Part II: Their relations to median, order statistic and stack filters. *IEEE Trans. Acoust. Speech, Signal Process.* ASSP-**35**(8), 1170–1184 (August 1987).

[Mar89] P. Maragos. Pattern spectrum and multiscale shape representation. *IEEE Trans. Pattern Anal. Machine Intell.* PAMI-**11**(7), 701–716 (July 1989).

[Mar90] P. Maragos and R. W. Schafer. Morphological systems for multidimensional signal processing. *Proc IEEE* **78**(4), 690–710 (April 1990).

[Mat75] G. Matheron. *Random Sets and Integral Geometry.* Wiley, New York (1975).

[Mer93] B. G. Mertzios and K. Tsirikolias. Coordinate logic filters in image processing. In *Proc. IEEE Winter Workshop on Nonlinear Digital Signal Processing.* pp. 11-51–11-56 (Tampere, Finland 1993).

[Mer98] B. G. Mertzios and K. Tsirikolias. Coordinate logic filters and their applications in image processing and pattern recognition. *Circ. Syst. Signal Process.* **17**(4), 517–538 (1998).

[Nak78] Y. Nakagawa and A. Rosenfeld. A note on the use of local min and max operations in digital picture processing. *IEEE Trans. Syst. Man Cybern.* SMC-**8**(8), 632–635 (August 1978).

[Par96] J. R. Parker. *Algorithms for Image Processing and Computer Vision.* Wiley, New York (1996).

[Pei92] H.-O. Peitgen, H. Jurgens, and D. Saupe. *Chaos and Fractals: New Frontiers of Science.* Springer Verlag, Berlin (1992).

[Ser83] J. Serra. *Image Analysis and Morphological Filters.* Academic Press, New York (1983).

[Ser92] J. Serra and L. Vincent. An overview of morphological filtering. *Circ. Syst. Signal Process.* **11**(1), 47–108 (1992).

[Spi97] I. M. Spiliotis and B. G. Mertzios. Fast algorithms for basic processing and analysis operations on block represented binary images. *Pattern Recog. Lett.* **17**, 1437–1450 (February 1997).

[Tsi91] K. Tsirikolias and B. G. Mertzios. Logic filters in image processing. In *Proc. Intl. Conf. on Digital Signal Processing,* pp. 285–287 (1991).

[Tsi93] K. Tsirikolias and B. G. Mertzios. Edge extraction and enhancement using coordinate logic filters. In *Proc. Intl. Conf. on Image Processing: Theory and Applications,* pp. 251–254 (1993).

12

Nonlinear Filters Based on Fuzzy Models

FABRIZIO RUSSO

Departimento di Elettrotecnica Elettronica Informatica
Università degli Studi di Trieste
Trieste, Italy

12.1 Introduction

The concept of fuzzy sets was introduced by Zadeh in 1965 as a mathematical tool able to model the concept of partial membership [Zad65]. Only in the mid-1980s, however, did fuzzy technology emerge as an effective problem-solving resource for engineering applications. Due to the successful development of fuzzy controllers, techniques based on fuzzy rules became very attractive, and the number of research and application areas grew rapidly. In 1992 special fuzzy systems entered the area of nonlinear filtering of noisy images. Since then, the number of different methods has been progressively increasing.

Unlike classical techniques that adopt a YES–NO approach, fuzzy filters use a gradual approach based on degrees of certainty. Hence, fuzzy filters are very well suited to address the uncertainty that typically occurs when both noise cancellation and detail preservation must be achieved. It is worth pointing out that the role (i.e., the degree of significance) played by fuzzy models is highly variable for the different methods. Some approaches use fuzzy sets to extend the formal framework of existing filters. Many other techniques, however, exploit the power and innovation of fuzzy reasoning to process information in a less traditional way. These approaches represent the subject of this chapter.

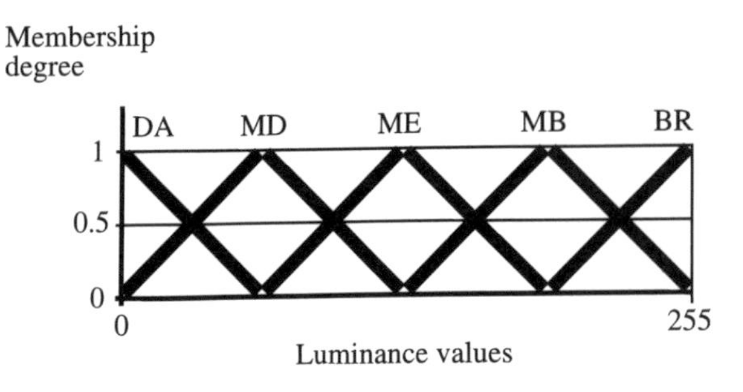

Figure 12.1: Example of fuzzy sets.

Regarding the processing architecture, such techniques can be grouped into two main classes: (1) indirect and (2) direct approaches. Indirect approaches typically adopt the basic structure of a weighted mean filter and use fuzzy models to evaluate the corresponding weights. Direct approaches adopt special fuzzy systems that directly yield the result. Such techniques include fuzzy inference ruled by else-action (FIRE) filters and evolutionary neural fuzzy (ENF) filters. After a brief introduction to fuzzy models, these families of nonlinear filters are described in detail.

12.2 Fuzzy Models

12.2.1 Fuzzy Sets

A fuzzy set is a generalization of a classical ("crisp") set based on the concept of partial membership. A fuzzy set F defined on U (universe of discourse) is represented as a set of ordered pairs [Zad65]:

$$F = \{(u, \mu_F(u)) | u \in U\}, \tag{12.1}$$

where $\mu_F(u)$ is the membership function that maps U to the real interval $[0,1]$, that is, $\mu_F(u) : U \to [0,1]$. For each element u $\in U$, the function $\mu_F(u)$ yields the degree of membership of u to the fuzzy set F ($0 \leqslant \mu_F(u) \leqslant 1$). This degree ranges from zero (no membership) to unity (full membership) according to the particular choice of fuzzy set shape. Very popular choices are triangular and trapezoidal shapes because they reduce the computational burden. More sophisticated choices such as bell-shaped fuzzy sets can also be adopted, depending on the specific application. Fuzzy sets are often identified by linguistic labels. As an example, Fig. 12.1 shows a possible definition of five fuzzy sets dealing with the pixel luminance of a digitized image: dark (DA), medium dark (MD), medium (ME), medium bright (MB), and bright (BR).

Some useful definitions concerning fuzzy sets are briefly reported [Rus99a]. The *crossover point* of a fuzzy set F is the element u_c with membership degree $\mu_F(u_c) = 0.5$. The *support* of a fuzzy set F is the crisp set $S(F)$ formed by the elements having nonzero degree of membership $S(F) = \{u \in U | \mu_F(u) > 0\}$. A fuzzy *singleton* is a fuzzy set whose support is a single element u with $\mu_F(u) = 1$. The *complement* of a fuzzy set F is the fuzzy set $\bar{F}$ described by the membership function $\mu_{\bar{F}}(u) = 1 - \mu_F(u)$.

Let F_1 and F_2 be two fuzzy sets on U. The union $F_{\text{un}} = F_1 \cup F_2$ (often expressed as "F_1 OR F_2") is described by the membership function: $\mu_{F_{\text{un}}}(u) = \max(\mu_{F_1}(u), \mu_{F_2}(u))$. The intersection $F_{\text{int}} = F_1 \cap F_2$ (often expressed as "F_1 AND F_2") is described by the membership function: $\mu_{F_{\text{int}}}(u) = \min(\mu_{F_1}(u), \mu_{F_2}(u))$. Finally, we can observe that a fuzzy relation R between sets U and V is a fuzzy set characterized by a membership function μ_R that maps the product space $U \times V$ to the real interval $[0, 1]$, expressed by

$$R = \{((u, v), \mu_R(u, v)) \,|\, (u, v) \in U \times V\}. \tag{12.2}$$

12.2.2 Fuzzy Systems

It is well known that fuzzy rules efficiently process data by mimicking human decision making [Kos92]. A fuzzy rule typically includes a group of antecedent clauses that define conditions and a consequent clause that defines the corresponding action; for example, "if x is dark and y is medium, then z is bright." A fuzzy system is a nonlinear system formed by a set of fuzzy rules (rule base) and an appropriate inference mechanism. The set of rules represents the knowledge base of the fuzzy system, and the inference mechanism numerically processes the knowledge base to yield the result. A very important class of fuzzy systems is constituted by systems that map a set of scalar inputs to one scalar output. As an example, let us consider a typical fuzzy system that maps N input variables $u_1, u_2, \ldots, u_N$ to one output variable v_0 by means of M fuzzy rules [Rus99a]:

Rule 1. IF $(u_1, F_{1,1})$ AND $(u_2, F_{2,1})$ AND...AND $(u_N, F_{N,1})$ THEN (v_0, G_1)
Rule 2. IF $(u_1, F_{1,2})$ AND $(u_2, F_{2,2})$ AND...AND $(u_N, F_{N,2})$ THEN (v_0, G_2)
...
Rule j. IF $(u_1, F_{1,j})$ AND $(u_2, F_{2,j})$ AND...AND $(u_N, F_{N,j})$ THEN (v_0, G_j)
...
Rule M. IF $(u_1, F_{1,M})$ AND $(u_2, F_{2,M})$ AND...AND $(u_N, F_{N,M})$ THEN (v_0, G_M)

Here $F_{i,j}$ $(1 \le i \le N,\ 1 \le j \le M)$ formally identifies the (antecedent) fuzzy set associated with the ith input variable in the jth rule, and G_j is the (consequent) fuzzy set associated with the output variable in the same rule. Different inference schemes to evaluate the output of the fuzzy system are available in the literature. From a conceptual point of view, the following steps are involved:

1. evaluation of the activation of each rule,
2. evaluation of the effect on the corresponding action, and
3. combination of the effects produced by all fuzzy rules and evaluation of the resulting scalar value.

The degree of activation of a fuzzy rule measures how much this rule is satisfied by the set of inputs. This degree ranges from zero (no activation) to unity (full activation). Let us focus on a method that is very attractive because it is computationally efficient. Let λ_j be the degree of activation of the jth rule. This degree is evaluated as follows:

$$\lambda_j = \min_i \left\{\mu_{F_{i,j}}(u_i)\right\}. \tag{12.3}$$

Notice that λ_j is determined by the group of antecedent clauses that define conditions about the inputs. Hence, only antecedent fuzzy sets are considered in Eq. (12.3). Now, let us consider the consequence of this activation on the action defined by the consequent clause "THEN (v_0, G_j)." A new fuzzy set G'_j is generated, whose membership function is defined by

$$\mu_{G'_j}(u) = \lambda_j \mu_{G_j}(u). \tag{12.4}$$

We can observe that the fuzzy set G'_j is a scaled version of the fuzzy set G_j, that is, G'_j has the same shape as G_j (this scheme is usually called "correlation-product inference" [Kos92]). The output v_0 is thus evaluated by the following relationships:

$$v_0 = \frac{\sum_{j=1}^{M} w_j v_j}{\sum_{j=1}^{M} w_j}, \tag{12.5}$$

$$w_j = \int_V \mu_{G'_j}(v)\, dv, \quad j = 1, \ldots, M, \tag{12.6}$$

$$v_j = \frac{\int_V v \mu_{G'_j}(v)\, dv}{w_j}, \quad j = 1, \ldots, M, \tag{12.7}$$

where w_j can be interpreted as the "weight" of the fuzzy set G'_j and v_j is the "centroid" of the same fuzzy set. All the consequent fuzzy sets $G'_1, G'_2, \ldots, G'_M$, are defined on V. If V is discrete, the summation symbol should replace the integral symbol in Eqs. (12.6) and (12.7).

12.2.3 Fuzzy Aggregation Connectives

The minimum operator is a very common choice for combining the antecedent clauses in a fuzzy rule. More sophisticated aggregation connectives are available in the literature. Basically, they include intersection connectives, union connectives, and compensative connectives. These connectives are (possibly nonlinear) functions that map a set of degrees of membership $\mu_1, \mu_2, \ldots, \mu_n$ to the real interval $[0, 1]$.

If an aggregation scheme ranging from minimum to zero (i.e., something more pessimistic than the minimum) is desired, we can choose the following class of intersection connectives [Yag80]:

$$y_I(\mu_1, \mu_2, ..., \mu_n) = 1 - \min\left\{1, \left(\sum_{i=1}^{n} (1 - \mu_i)^p\right)^{1/p}\right\}. \tag{12.8}$$

Likewise, for an aggregation scheme ranging from maximum to unity (that is, something more optimistic than the maximum), we can adopt the following class of union aggregators:

$$y_U(\mu_1, \mu_2, ..., \mu_n) = \min\left\{1, \left(\sum_{i=1}^{n} \mu_i^p\right)^{1/p}\right\}. \tag{12.9}$$

Finally, for an aggregation scheme ranging from minimum to maximum, we can choose the generalized mean connective [Dyc84]:

$$y_M(\mu_1, \mu_2, ..., \mu_n; w_1, w_2, ..., w_n) = \left(\sum_{i=1}^{n} w_i \mu_i^p\right)^{1/p}, \tag{12.10}$$

where $\sum_{i=1}^{n} w_i = 1$. Indeed, this connective yields all values between minimum and maximum by varying the parameter p between $-\infty$ and $+\infty$.

Finally, we can resort to hybrid connectives to combine outputs of union and intesection aggregators. The combination can be performed by using an additive or a multiplicative model, respectively, as follows:

$$y_H = (1 - \gamma)(y_I) + \gamma(y_U), \tag{12.11}$$

$$y_H = (y_I)^{1-\gamma}(y_U)^{\gamma}. \tag{12.12}$$

The degree of compensation between the union and intersection components depends on the value of the parameter γ $(0 \le \gamma \le 1)$.

12.3 Fuzzy Weighted Mean (FWM) Filters

These methods belong to the class of indirect approaches mentioned in Sec. 12.1. FWM filters adopt fuzzy sets or fuzzy systems to evaluate the weights of a weighted linear filter that, in turn, yields the output. To describe these and other nonlinear techniques, we shall adopt a common mathematical notation. Suppose we deal with digitized images having L gray levels. Let $x(\mathbf{n})$ be the pixel luminance at location $\mathbf{n} = [n_1, n_2]$ in the input image and let $y(\mathbf{n})$ be the corresponding pixel luminance in the output image $(0 \le x(\mathbf{n}) \le L - 1, 0 \le y(\mathbf{n}) \le L - 1)$. Let $W(n) = \{x_i(\mathbf{n}); i = 0, \ldots, N\}$ be the set of pixel values which belong to a window around $x(\mathbf{n})$, where $x_0 = x(\mathbf{n})$. Finally, let $\Delta x_i(\mathbf{n}) = x_i(\mathbf{n}) - x(\mathbf{n})$ be the luminance difference between the neighbor x_i and the central pixel x.

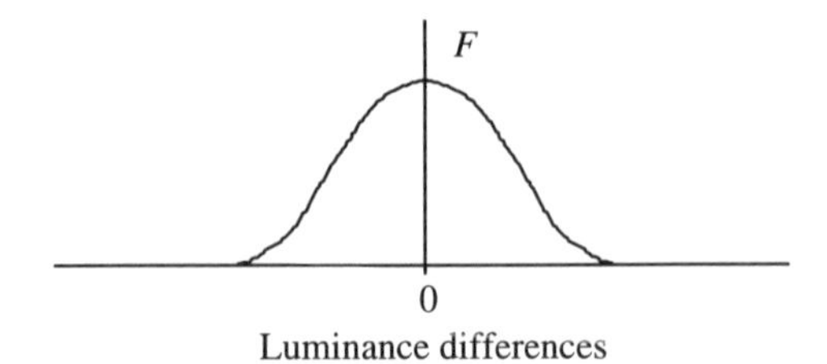

Figure 12.2: A possible choice for fuzzy set F.

12.3.1 FWM Filters Based on Fuzzy Sets

These filters adopt weights that are based on the luminance values $x_0, x_1, \ldots, x_N$ in the window. The adoption of weights aims at preserving fine details and textures during the smoothing process. The general structure of these filters is described by the following relationship:

$$y(\mathbf{n}) = \frac{\sum_{i=0}^{N} w_i(\mathbf{n}) x_i(\mathbf{n})}{\sum_{i=0}^{N} w_i(\mathbf{n})}. \tag{12.13}$$

The simplest filter of this group adopts one fuzzy set only. The filter operation is defined as follows:

$$y(\mathbf{n}) = \frac{\sum_{i=0}^{N} \mu_F(\Delta x_i)\, x_i(\mathbf{n})}{\sum_{i=0}^{N} \mu_F(\Delta x_i)}. \tag{12.14}$$

where F is a bell-shaped fuzzy set centered on zero. A possible choice for the membership function μ_F is shown in Fig. 12.2.

The fuzzy set shape aims at reducing the influence of pixels having large luminance differences with respect to the central one. The filter is effective for images degraded by Gaussian or uniform noise, and a membership function defined by two parameters suffices to obtain satisfactory results. Hybrid filters for mixed Gaussian and impulse noise removal can be obtained by resorting to median-based prefiltering to cancel outliers [Pen94, Pen95a,b,c].

Fuzzy cluster filters are another example of FWM filters because they typically adopt a weighted mean structure. The output is usually obtained by means of an iterative procedure:

$$y^{(k+1)}(\mathbf{n}) = \frac{\sum_{i=1}^{N} w_i^{(k)} x_i(\mathbf{n})}{\sum_{i=1}^{N} w_i^{(k)}}, \tag{12.15}$$

where $y^{(k+1)}$ is the candidate output (cluster center) at step $k+1$. Weights are evaluated by means of fuzzy membership functions and usually depend on the previous estimate of the cluster center [Dor96, Suc96].

An interesting class of fuzzy filters for color image processing is represented by fuzzy vector directional filters [Pla95]. These filters process multichannel image data by considering the weighted average of all of the vector-valued elements

$\mathbf{x}_0, \mathbf{x}_1, \ldots, \mathbf{x}_N$ inside the window. The evaluation of the weights is based on a distance criterion:

$$y(\mathbf{n}) = \frac{\sum_{i=0}^{N} \mu_{\text{sig}}(d_i)\mathbf{x}_i(\mathbf{n})}{\sum_{i=0}^{N} \mu_{\text{sig}}(d_i)}, \tag{12.16}$$

where

$$d_i = \sum_{j=0}^{N} A\left(\mathbf{x}_i, \mathbf{x}_j\right), \tag{12.17}$$

μ_{sig} is a sigmoid-shaped fuzzy set, d_i is the distance, and $A(\mathbf{x}_i, \mathbf{x}_j)$ denotes the angle between the vectors $\mathbf{x}_i$ and $\mathbf{x}_j$ ($0 \le A \le \pi$).

12.3.2 FWM Filters Based on Fuzzy Systems

A different class of FWM filters adopts fuzzy systems (i.e., fuzzy rules) to evaluate the weights $w_1, w_2, \ldots, w_N$ in Eq. (12.13). To achieve good detail-preserving behavior, fuzzy rules deal with local features that exploit some specific neighborhood information [Tag94]. A typical local feature for images corrupted by Gaussian noise is defined as follows:

$$K(\mathbf{n}) = \begin{cases} \frac{\sigma^2(\mathbf{n}) - \sigma_G^2}{\sigma^2(\mathbf{n})} & \text{if } \sigma^2(\mathbf{n}) \ge \sigma_G^2, \\ 0, & \text{otherwise,} \end{cases} \tag{12.18}$$

where $\sigma^2(\mathbf{n})$ is the variance of the data in the window and σ_G^2 is the variance of the noise. In uniform or gradually varying regions of the image, $\sigma(\mathbf{n})$ is similar to σ_G, so K is low. Conversely, in the presence of edges, $\sigma(\mathbf{n})$ is larger than σ_G, so K is high. Notice that the above considerations have been expressed in the form of fuzzy reasoning.

Another useful local feature deals with the difference between a pixel's luminance and the median x_{med} of the pixel values in the window:

$$E_i(\mathbf{n}) = \frac{|x_i(\mathbf{n}) - x_{\text{med}}(\mathbf{n})|}{\sigma_G}. \tag{12.19}$$

Finally, a third local feature is represented by the normalized distance $D_i(\mathbf{n})$ between any pixel and the central one. The goal is to reduce the importance of pixels that are far from the central element of the window.

The local features mentioned can be used as the inputs of a fuzzy system that processes them nonlinearly by means of fuzzy rules. A typical rule can be expressed as follows: IF K is large AND E_i is small AND D_i is small, THEN w_i is large. Good results have been obtained using a 27-rule fuzzy system for Gaussian noise cancellation [Tag95a]. The qth rule in this system ($q = 1, 2, \ldots, 27$) is formally defined as follows:

$$\text{Rule } q: \quad \text{IF } (K, F_{1,q}) \text{ AND } (E_i, F_{2,q}) \text{ AND } (D_i, F_{3,q}), \text{ THEN } (w_i, G_q), \tag{12.20}$$

where $F_{p,q}(p = 1, 2, 3)$ is a fuzzy set described by the triangular-shaped membership function

$$\mu_{F_{p,q}}(u_p) = \begin{cases} 1 - 2\frac{|u_p - a_{p,q}|}{b_{p,q}} & \text{for } -b_{p,q} \le u - a_{p,q} \le b_{p,q}, \\ 0, & \text{otherwise,} \end{cases} \tag{12.21}$$

and G_q is a fuzzy singleton defined on the interval $[0, 1]$ because a resulting weight ranging from zero to unity is required ($0 \le w_i \le 1$). According to the inference scheme described in the previous section, the output of the fuzzy system is determined using the following relation:

$$w_i = \frac{\sum_{q=1}^{27} v_q \lambda_q}{\sum_{q=1}^{27} \lambda_q}, \tag{12.22}$$

where λ_q is the degree of activation of the qth rule and v_q is the support point of G_q. The tuning of fuzzy sets, that is, the search for the optimal fuzzy set parameters, is performed by using a training image and a least mean-squares (LMS) algorithm. To derive a suitable expression for this purpose, we evaluate λ_q using the product instead of the minimum operator. A filter adopting a different inference scheme has also been proposed [Tag96].

Cancellation of mixed noise distributions (such as mixed Gaussian and impulse noise) can be performed by splitting the weight $w_i(\mathbf{n})$ into different components [Mun96]:

$$y(\mathbf{n}) = \frac{\sum_{i=0}^{N} w_i^{(I)}(\mathbf{n}) w_i^{(G)}(\mathbf{n}) x_i(\mathbf{n})}{\sum_{i=0}^{N} w_i^{(I)}(\mathbf{n}) w_i^{(G)}(\mathbf{n})}, \tag{12.23}$$

where $w_i^{(I)}(\mathbf{n})$ and $w_i^{(G)}(\mathbf{n})$ are weight components dealing with impulse and Gaussian noise, respectively. These components can be evaluated by fuzzy systems dealing with appropriate local features.

We finally observe that FWM filters can adopt weights to combine the outputs $y_1, y_2, \ldots, y_Q$ of different nonlinear techniques. The output of this class of filters (here called fuzzy combination filters) is given by the following relationship:

$$y(\mathbf{n}) = \frac{\sum_{k=1}^{Q} w_k(\mathbf{n}) y_k(\mathbf{n})}{\sum_{k=1}^{Q} w_k(\mathbf{n})}. \tag{12.24}$$

An adaptive filter for mixed noise removal combines the outputs of five ($Q = 5$) classical operators, such as midpoint, mean, median, and identity filters and a small-window median filter [Tag95b]. The weights are evaluated by a fuzzy system dealing with three local features. Another technique addresses image enhancement by combining the outputs of three ($Q = 3$) fuzzy filters designed to reduce Gaussian noise, remove outliers, and enhance edges [Cho97]. The different outputs are combined depending on the value of a local feature dealing with the luminance differences and spatial distances between the central point and the neighboring pixels.

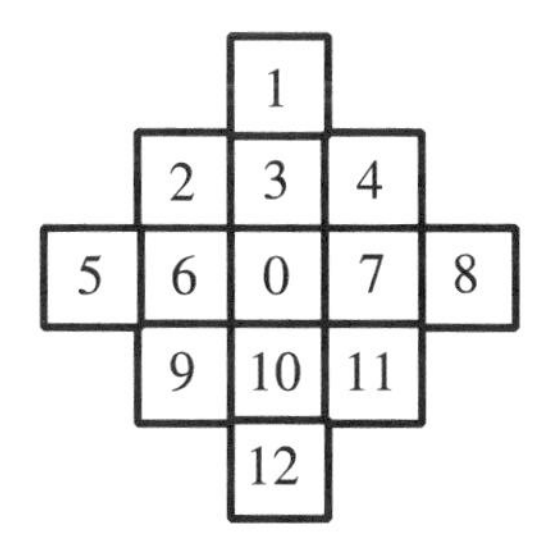

Figure 12.3: Neighborhood and pixel indexes.

12.4 FIRE Filters

FIRE (Fuzzy Inference Ruled by Else-action) filters are special fuzzy systems based on IF-THEN-ELSE reasoning [Rus92, Rus93]. Unlike FWM filters, FIRE filters directly yield the noise correction as the output of the inference process. Hence, FIRE filters belong to the class of direct approaches mentioned in Sec. 12.1. Typically, a FIRE filter adopts directional rules that deal with the luminance differences $\Delta x_1, \Delta x_2, \ldots, \Delta x_N$ between the central pixel and its neighbors. The fuzzy rule-base evaluates a positive or a negative correction Δy that aims at removing the noise (THEN action). If no rule is satisfied, the central pixel is left unchanged (ELSE action). Different noise statistics can be addressed by means of appropriate choices of the fuzzy sets, rules, and aggregation mechanisms [Rus95, Rus97].

12.4.1 FIRE Filters for Uniform Noise

The ΛΣ-FIRE filter is an example of a FIRE operator designed to deal with uniformly distributed noise [Rus96]. The filter adopts directional rules that aim at preserving the image edges during the noise removal. A repeated application of the operator does not increase the detail blur. As a result an effective noise cancellation can be obtained. The filter operates on the neighborhood depicted in Fig. 12.3.

As mentioned, the input variables are the luminance differences $\Delta x_i(\mathbf{n}) = x_i(\mathbf{n}) - x(\mathbf{n})$. By using fuzzy rules, the operator nonlinearly maps these inputs to the output variable $\Delta y(\mathbf{n})$. This represents the correction term that, when added to $x(\mathbf{n})$, yields the resulting luminance value $y(\mathbf{n}) = x(\mathbf{n}) + \Delta y(\mathbf{n})$. Fuzzy rules deal with two fuzzy sets labeled positive (PO) and negative (NE). The triangular-shaped membership function μ_{PO} is defined by the following relationship:

$$\mu_{\text{PO}}(u) = \begin{cases} 1 - \frac{1}{2}\frac{|u-c|}{c} & \text{for } -c \le u \le 3c, \\ 0, & \text{otherwise.} \end{cases} \tag{12.25}$$

The membership function μ_{NE} is symmetrically defined: $\mu_{\text{NE}}(u) = \mu_{\text{PO}}(-u)$. If the image is corrupted by uniformly distributed noise in the interval $[-A_n, A_n]$, the

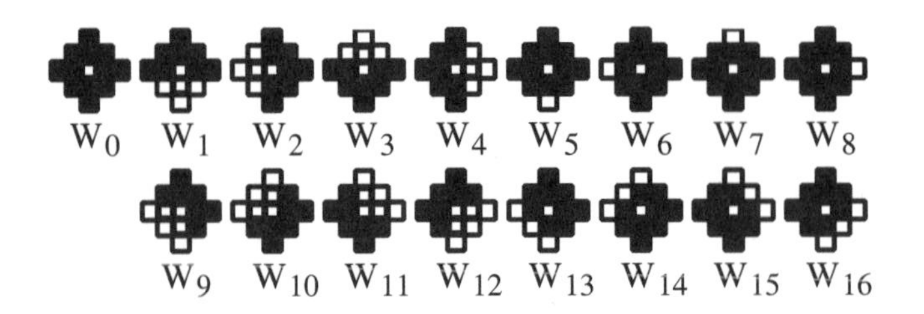

Figure 12.4: Pixel patterns.

parameter c is set as follows: $c = A_n$. The directional rules deal with the patterns of pixels W_i $(i = 0, \ldots, 16)$ represented in Fig. 12.4.

To perform a positive or a negative correction, we define two symmetrical smoothing rules for each pattern. For example, the following rules are associated with the pattern W_j:

$$\text{IF } (\Delta x_1, A_{1,j}) \text{ AND } \ldots \text{ AND } (\Delta x_N, A_{N,j}) \text{ THEN } (\Delta y, \text{PO}), \tag{12.26}$$

$$\text{IF } (\Delta x_1, A^*_{1,j}) \text{ AND } \ldots \text{ AND } (\Delta x_N, A^*_{N,j}) \text{ THEN } (\Delta y, \text{NE}), \tag{12.27}$$

where

$$A_{i,j} \equiv \begin{cases} \text{PO} & \text{for } x_i \in W_j, \\ \text{NE} & \text{for } x_i \notin W_j, \end{cases} \tag{12.28}$$

$$A^*_{i,j} \equiv \begin{cases} \text{NE} & \text{for } x_i \in W_j, \\ \text{PO} & \text{for } x_i \notin W_j. \end{cases} \tag{12.29}$$

The degrees of activation λ_j and λ^*_j of the above rules are respectively given by

$$\lambda_j = f_{\wedge\Sigma}\left(q, \mu_{A_{1,j}}(\Delta x_1), \ldots, \mu_{A_{N,j}}(\Delta x_N)\right), \tag{12.30}$$

$$\lambda^*_j = f_{\wedge\Sigma}\left(q, \mu_{A^*_{1,j}}(\Delta x_1), \ldots, \mu_{A^*_{N,j}}(\Delta x_N)\right), \tag{12.31}$$

where $f_{\wedge\Sigma}$ denotes a fuzzy aggregation function whose behavior is either the minimum operator or the arithmetic mean:

$$f_{\wedge\Sigma}(q, \mu_1, \mu_2, \ldots, \mu_N) = \begin{cases} \min_i \{\mu_i\} & \text{if } q = 1, \\ \frac{1}{N}\sum_{i=1}^{N} \mu_i & \text{if } q = 0. \end{cases} \tag{12.32}$$

The output Δy of the fuzzy operator is finally computed as

$$\Delta y = c\left(\max_j \{\lambda_j\} - \max_j \{\lambda^*_j\}\right). \tag{12.33}$$

Since the filter possesses a good detail-preserving behavior, it can be repeatedly applied to the image data to increase the noise cancellation (multipass filtering). In particular, an effective smoothing action can be obtained by varying the fuzzy aggregation scheme in Eq. (12.32) during the multipass process. The minimum operator ($q = 1$) defined by Eq. (12.32) can be used at the beginning of the multipass filtering, when a strong detail-preserving action is required. The arithmetic mean ($q = 0$) can be adopted at the end of the same process to smooth noisy pixels still present on the edges of the image.

1	2	3
4	0	5
6	7	8

Figure 12.5: Pixel indexes in the window.

12.4.2 FIRE Filters for Mixed Noise

To address mixed noise, a FIRE filter typically combines rules for different noise distributions by adopting a hierarchical structure. The filter operates on the neighborhood represented in Fig. 12.5. Input and output variables are defined as in the previous section. The filter adopts four piecewise linear fuzzy sets labeled large positive (LP), medium positive (MP), medium negative (MN), and large negative (LN). The membership functions m_{LP}, m_{MP}, m_{MN}, and m_{LN} of these sets are graphically represented in Fig. 12.6 ($0 < c < L - 1$; $0 < a < b$, $0 < b < L - 1$).

The filter operation can be described as follows. First, rules for impulse noise are activated [Rus99b]. These rules adopt fuzzy sets LP and LN, and deal with eight patterns identified by the sets of indexes $A_1 = \{2,4,5\}$, $A_2 = \{2,5,7\}$, $A_3 = \{4,5,7\}$, $A_4 = \{2,4,7\}$, $A_5 = \{1,2,5,8\}$, $A_6 = \{3,5,6,7\}$, $A_7 = \{1,4,7,8\}$, and $A_8 = \{2,3,4,6\}$. The overall activation of rules for impulse noise can be estimated by the following relation:

$$\alpha_{\max} = \max\{\alpha, \alpha^*\}, \tag{12.34}$$

where

$$\alpha = \max_j \left\{ \min_{i \in A_j} \{\mu_{\mathrm{LP}}(\Delta x_i)\} \right\}, \tag{12.35}$$

$$\alpha^* = \max_j \left\{ \min_{i \in A_j} \{\mu_{\mathrm{LN}}(\Delta x_i)\} \right\}. \tag{12.36}$$

If $\alpha_{\max} > \delta$ (typically $\delta = 0.1$), the ouput of the FIRE operator is given by

$$\Delta y = (L-1)(\alpha - \alpha^*). \tag{12.37}$$

Otherwise, a different group of rules is activated. Such rules adopt fuzzy sets MP and MN to address a different kind of noise, such as uniform and Gaussian noise. These rules deal with four pixel patterns defined by the following sets of indexes: $B_1 = \{1,2,3,4\}$, $B_2 = \{1,4,6,7\}$, $B_3 = \{5,6,7,8\}$, and $B_4 = \{2,3,5,8\}$. The output of the FIRE operator is then given by the following relation

$$\Delta y = c(\beta - \beta^*), \tag{12.38}$$

where

$$\beta = \max_j \left\{ \frac{1}{2}\left(\min_{i \in B_j} \{\mu_{\mathrm{MP}}(\Delta x_i)\} \right) + \frac{1}{2}\left(\frac{1}{N_j} \sum_{i \in B_j} \mu_{\mathrm{MP}}(\Delta x_i) \right) \right\}, \tag{12.39}$$

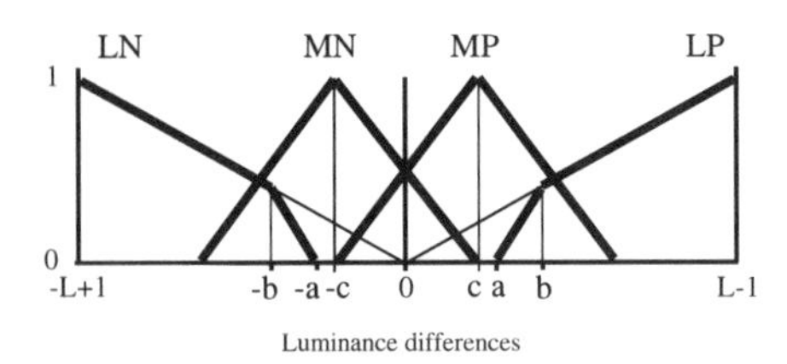

Figure 12.6: Possible choices for fuzzy sets LN, MN, MP, and LP.

$$\beta^* = \max_j \left\{ \frac{1}{2} \left(\min_{i \in B_j} \{\mu_{\mathrm{MN}}(\Delta x_i)\} \right) + \frac{1}{2} \left(\frac{1}{N_j} \sum_{i \in B_j} \mu_{\mathrm{MN}}(\Delta x_i) \right) \right\}, \quad (12.40)$$

where N_j is the number of elements of B_j. The filter is recursive; that is, the new value $y(\mathbf{n}) = x(\mathbf{n}) + \Delta y(\mathbf{n})$ is assigned to $x(\mathbf{n})$ at the end of the processing. It should be observed that the ability to detect groups of adjacent noise pulses is directly related to the choice of appropriate pixel patterns. As the probability of impulse noise increases, more patterns and more rules become necessary. On the other hand, the detail-preserving behavior is mainly related to the fuzzy set shapes. Fuzzy sets LP and LN are designed to perform full correction of large-amplitude noise pulses, partial correction of medium-amplitude noise pulses, and no correction of small-amplitude noise pulses.

Focusing on rules for nonimpulse noise, the detail-preservation mechanism is easily controlled by the parameter c, which automatically defines location and width of fuzzy sets MP and MN. A large value of the parameter c yields a strong smoothing effect, while a small value increases the detail-preserving capability. Finally, it should be observed that a heuristic search of the parameter values is not a very critical process because fuzzy rules and sets model human thinking and so represent easily understood concepts. It is clear, however, that a method for automatic generation of the fuzzy rulebase can represent a more effective approach. In particular, neuro-fuzzy architectures are very attractive to address this issue. In the next section we analyze in detail the key features of a neuro-fuzzy filter that can learn from examples.

12.5 Evolutionary Neural Fuzzy Filters: A Case Study

Evolutionary neural fuzzy (ENF) filters are a recently introduced class of nonlinear filters that combine the key advantages of fuzzy systems and artificial neural networks, that is, knowledge representation and knowledge acquisition [Rus99c]. The learning method is based on genetic algorithms (GAs), which search for the optimal solution of a problem by applying the mechanisms of natural selection and natural genetics to a population of potential solutions [Gol89, Ped97]. General advantages of the GAs are robustness and the ability to address hill-climbing problems. As a specific advantage, GAs do not require the adoption of differentiable functions in

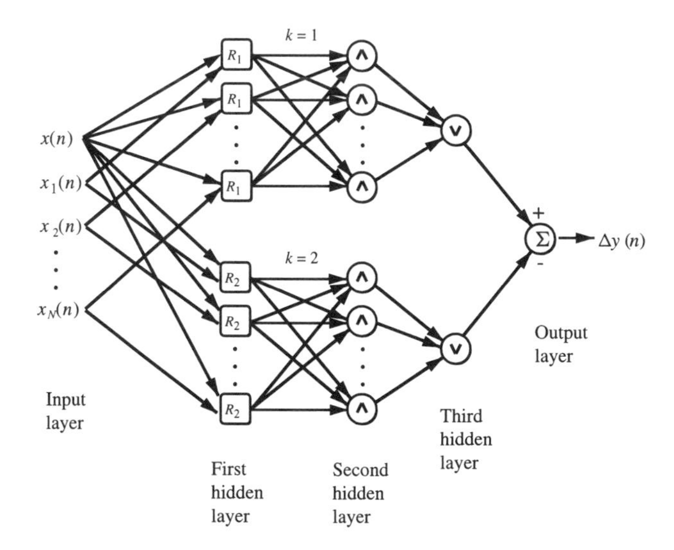

Figure 12.7: Network structure of the ENF filter.

the fuzzy inference scheme. Thus, fuzzy aggregation connectives such as the minimum and maximum operators can be adopted into the network structure without any limitation.

12.5.1 Network Structure

The structure of a filter for impulse noise removal is shown in Fig. 12.7. The network operates on a 3×3 window and includes two symmetrical subnetworks that deal with positive and negative noise pulses, respectively. The operation performed by the different layers is described as follows.

First Hidden Layer. Let $O_{k,i}^{(1)}$ be the output of the ith node in the kth subnetwork. The node function is expressed by

$$O_{k,i}^{(1)} = \mu_{R_k}(x_i, x); \quad k = 1, 2, \quad i = 1, \ldots, 8, \tag{12.41}$$

where $\mu_{R_1}(x_i, x)$ and $\mu_{R_2}(x_i, x)$ are 2-D membership functions that describe the following fuzzy relations

$$R_1 : x_i \text{ is much larger than } x,$$
$$R_2 : x_i \text{ is much smaller than } x.$$

The membership function μ_{R_1} is defined as follows:

$$\mu_{R_1}(x_i, x) = \begin{cases} \frac{x_i - x}{L-1} & \text{for } x_i - x \geq a_1, \\ \frac{a_1}{L-1}\frac{x_i - x - \alpha}{a_1 - \alpha} & \text{for } \alpha \leq x_i - x < a_1, \\ 0 & \text{for } x_i - x < \alpha, \end{cases} \tag{12.42}$$

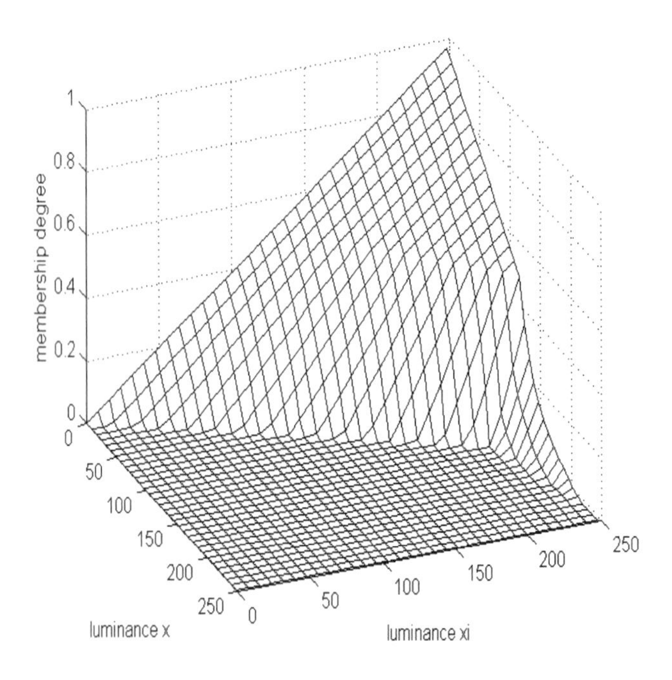

Figure 12.8: Graphical representation of fuzzy relation R_1.

where

$$\alpha(x) = \begin{cases} a_2 & \text{for } 0 \leq |x - L/2| < a_3, \\ \frac{a_2}{a_4 - a_3}(a_4 - |x - L/2|) & \text{for } a_3 \leq |x - L/2| < a_4, \\ 0 & \text{for } a_4 \leq |x - L/2| \leq \frac{L}{2}, \end{cases} \tag{12.43}$$

with $a_2 < a_1$. The membership function μ_{R_2} is symmetrically defined:

$$\mu_{R_2}(x_i, x) = \mu_{R_1}(x, x_i). \tag{12.44}$$

Fuzzy relations R_1 and R_2 are designed to avoid image blur during noise removal. Indeed, membership functions are weakly activated when the luminance of a neighboring element is close (or very close) to that of the central pixel. As a result, fine details are possibly preserved. The action produced by Eq. (12.43) aims at gradually disabling the detail-preserving mechanism as the luminance value of the central pixel approaches its minimum or maximum range. An example of graphical representation of the fuzzy relation R_1 is shown in Fig. 12.8.

Second Hidden Layer. Let $O_{k,j}^{(2)}$ be the output of the jth node in the kth subnetwork ($k = 1, 2$; $j = 1, \ldots, M$). The node function is defined by

$$O_{k,j}^{(2)} = \min_i \left\{ \left(O_{k,i}^{(1)} \right)^{w_{i,j}} \right\}, \tag{12.45}$$

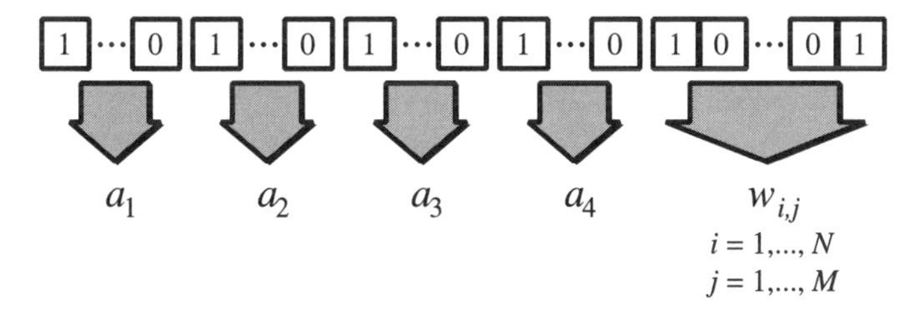

Figure 12.9: Parameter encoding.

where $w_{i,j}$ is a binary weight whose value is either unity or zero. The weight $w_{i,j}$ denotes the strength of the connection between the ith node in layer 1 and the jth node in layer 2. Subnetwork 2 adopts the same weight values as subnetwork 1.

Third Hidden Layer. Let $O_k^{(3)}$ be the output of the kth node. The node function is defined by

$$O_k^{(3)} = \max_j \left\{O_{k,j}^{(2)}\right\}. \tag{12.46}$$

Output Layer. Finally, the output Δy of the network is evaluated by the following relation:

$$\Delta y = (L - 1)\left(O_1^{(3)} - O_2^{(3)}\right). \tag{12.47}$$

The filter is recursive; that is, the new value $y(\mathbf{n}) = x(\mathbf{n}) + \Delta y(\mathbf{n})$ is assigned to $x(\mathbf{n})$ and is reused for computing the next outputs.

12.5.2 Genetic Learning

The parameters that define the ENF filter are the group of fuzzy set parameters (a_1, a_2, a_3, a_4) and the binary weights $w_{i,j}$ $(i = 1,\ldots,8,\ j = 1,\ldots,M)$. These parameters must be encoded as a binary string to activate the genetic training procedure (Fig. 12.9). The efficiency of the encoding scheme can be improved by considering the symmetry of the processing. It should be observed that each group of eight binary weights $\{w_{1,j}, w_{2,j},\ldots, w_{8,j}\}$ directly identifies a pixel pattern. Hence, we can adopt an 8-bit substring to encode this pattern and three related patterns obtained by 90°, 180° and 270° rotations.

The training procedure is based on a simple genetic algorithm and is briefly depicted in Fig. 12.10. The genetic learning starts with a randomly generated population of strings (individuals) and produces the subsequent populations by means of reproduction, crossover, and mutation operators [Gol89]. During reproduction, individual strings are copied depending on a performance index (fitness). Individuals having the best fitness have more chances to be reproduced. Then, the crossover, procedure may select a pair of newly reproduced strings (parents) to create two new individuals (children). This is performed by swapping some parent

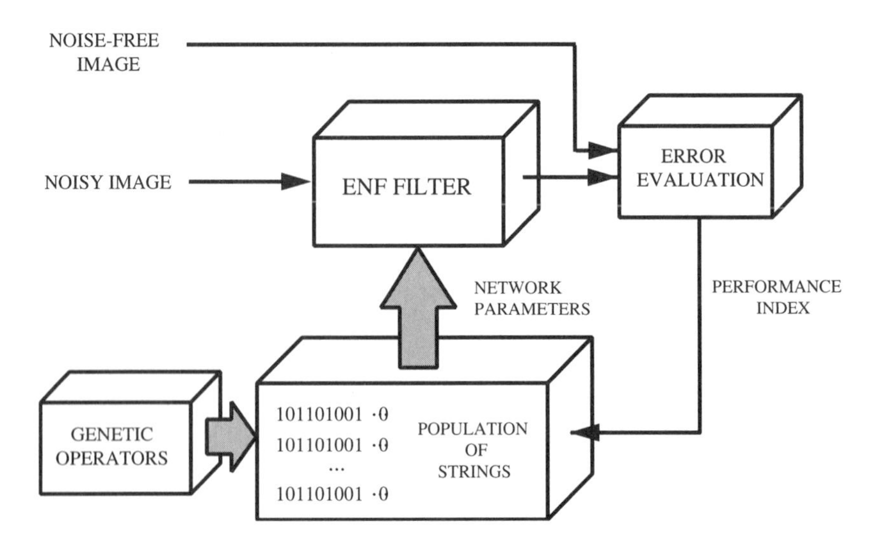

Figure 12.10: Genetic training procedure.

substrings. Finally, the mutation operator aims at randomly changing (with low probability) some bit values from 1 to 0 and vice versa.

The fitness measure is computed as follows: An image corrupted by impulse noise is used as input data and, for each string, the result generated by the corresponding ENF filter is considered. The fitness measure Φ is evaluated by considering the mean-square error (MSE) between the processed data and the original noise-free image:

$$\Phi = \frac{N_{\text{tot}}}{\sum_n (y(n) - s(n))^2}, \tag{12.48}$$

where N_{tot} is the number of processed pixels. The training of the network is usually stopped after an assigned number of generations. An example of results obtained is shown in Fig. 12.11. We chose an ENF filter having 24 nodes in each subnetwork ($M = 24$). The system was encoded by means of an 80-bit binary string. We chose a small population of 50 strings and set the parameters of the genetic learning as follows: crossover probability 1.0, mutation rate 0.01. The training data were generated by corrupting the 256×256 "Lena" image ($L = 256$) with salt-and-pepper impulse noise having probability 0.3 (Fig. 12.11b). The result obtained after 50 generations is shown in Fig. 12.11c.

12.5.3 Performance Analysis

The performance of the ENF filter is evaluated by using data different from those adopted during the training process. The well-known 256×256 "Pentagon" image was used for this purpose, and three noisy images were generated by superimposing impulse noise with probability 0.2, 0.3, and 0.4, respectively. The list of

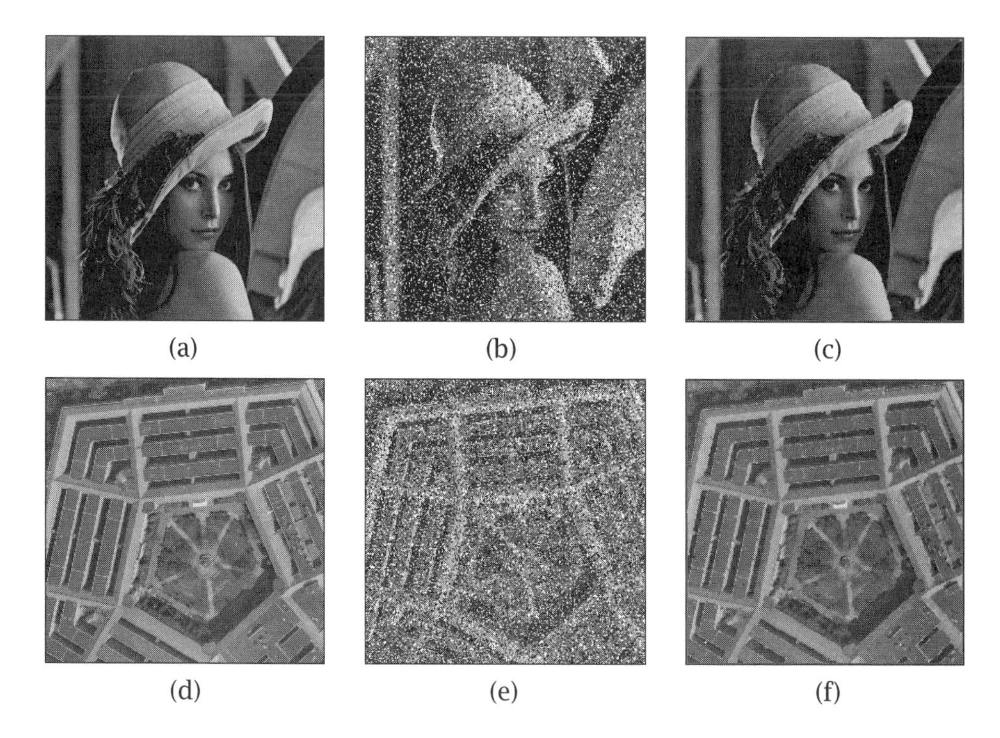

Figure 12.11: (a) Original Lena image, (b) training image corrupted by impulse noise with probability 0.3, and (c) result of the genetic training after 50 generations, (d) original Pentagon image, (e) test image corrupted by impulse noise with probability 0.3, and (f) result given by the ENF filter.

MSE values after processing by the ENF filter are reported in Table 12.1. The corresponding MSE values obtained by conventional 3×3 and 5×5 median operators are also included here for comparison.

It is known that the MSE value by itself is not sufficient to represent the visual quality of an image. However, a simple method can be adopted to highlight the different filtering behavior with respect to noise cancellation and detail preservation. Let the MSE be split into two different components $\mathrm{MSE_{nc}}$ and $\mathrm{MSE_{dp}}$ dealing with corrupted and uncorrupted pixels, respectively. Since the $\mathrm{MSE_{nc}}$ value deals with corrupted pixels only, it represents a good estimate of the ability to remove noise pulses (noise cancellation). Conversely, the $\mathrm{MSE_{dp}}$ addresses the class of uncorrupted pixels, that is, pixels still having the original luminance in the noisy image. Hence, the $\mathrm{MSE_{dp}}$ value represents a good estimate of the ability to preserve the original image structure (detail preservation). The $\mathrm{MSE_{nc}}$ and $\mathrm{MSE_{dp}}$ components are defined as follows:

$$\mathrm{MSE_{nc}} = \frac{\sum_{n \in I_1} (y(n) - s(n))^2}{N_1}, \tag{12.49}$$

Table 12.1: MSE values.

	Noise prob. 0.2			Noise prob. 0.3			Noise prob. 0.4		
	MSE	MSE$_{nc}$	MSE$_{dp}$	MSE	MSE$_{nc}$	MSE$_{dp}$	MSE	MSE$_{nc}$	MSE$_{dp}$
ENF filter	72.5	364.1	0.1	127.7	427.5	0.1	185.0	465.2	0.4
3x3 Median	349.7	514.6	308.8	572.9	924.1	423.5	1051.7	1577.0	705.4
5x5 Median	472.9	503.4	465.3	496.8	521.2	486.4	546.3	588.8	518.2

$$\mathrm{MSE_{dp}} = \frac{\sum_{n \in I_2} (y(n) - s(n))^2}{N_2}, \tag{12.50}$$

where I_1 is the set of coordinates that denotes the group of N_1 pixels corrupted by impulse noise $[x(\mathbf{n}) \neq s(\mathbf{n})]$ and I_2 is the set of coordinates that denotes the group of N_2 uncorrupted pixels $(x(\mathbf{n}) = s(\mathbf{n}))$ in the noisy image. The overall MSE is clearly given by

$$\mathrm{MSE} = \frac{N_1}{N_{\mathrm{tot}}}\mathrm{MSE_{nc}} + \frac{N_2}{N_{\mathrm{tot}}}\mathrm{MSE_{dp}}, \tag{12.51}$$

where $N_{\mathrm{tot}} = N_1 + N_2$. The list of $\mathrm{MSE_{nc}}$ and $\mathrm{MSE_{dp}}$ values for the mentioned test images are also reported in Table 12.1. The very good nonlinear behavior of the ENF filter is apparent. Another set of processed images is also shown in Figs. 12.11d–12.11f for visual evaluation. It should be observed that the ability to remove impulse noise directly depends on the quality of the training data. The learning process automatically acquires the necessary information about the size of detectable noise and its relationship with the size and shape of the objects in the image. Once this information has been obtained, it is used to process any further kind of data [Rus99c].

12.6 Concluding Remarks and Future Trends

Currently, fuzzy filters represent a well established technology for nonlinear image processing. In the past few years the number of different methods has increased rapidly and now fuzzy filters can deal with a variety of noise statistics, such as uniform noise, Gaussian noise, impulsive noise, and mixed noise distributions. Due to the intrinsic ability to address uncertainty, fuzzy filters are effective in removing noise and satisfactorily preserve fine details and textures. In particular, some recent techniques have been shown to perform significantly better than state-of-the-art nonlinear filters in the literature. It is worth pointing out that such techniques adopt fuzzy models in a less traditional way and take full advantage of the innovative paradigms of computational intelligence. It is expected that the combination of fuzzy, neural, and genetic paradigms will play a more and more significant role in the future of nonlinear filters.

References

[Cho97] Y. Choi and R. Krishnapuram. A robust approach to image enhancement based on fuzzy logic. *IEEE Trans. Image Process.* **6**(6), 808–825 (June 1997).

[Dor96] M. Doroodchi and A. M. Reza. Fuzzy cluster filter. In *Proc. IEEE Intl. Conf. on Image Processing*, pp. 939–942 (Lausanne, Switzerland, September 1996).

[Dyc84] H. Dyckoff and W. Pedrycz. Generalized means as a model of compensation connectives. *Fuzzy Sets and Syst.* **14**(2), 143–154 (1984).

[Gol89] D. E. Goldberg. *Genetic Algorithms in Search, Optimization and Machine Learning.* Addison-Wesley, Reading, MA (1989).

[Kos92] B. Kosko. *Neural Networks and Fuzzy Systems*, Prentice-Hall, Englewood Cliffs, NJ (1992).

[Mun96] M. Muneyasu, Y. Wada and T. Hinamoto. Edge-preserving smoothing by adaptive nonlinear filters based on fuzzy control laws. In *Proc. IEEE Intl. Conf. on Image Processing*, pp. 785–788 (Lausanne, Switzerland, September 1996).

[Ped97] W. Pedrycz. *Fuzzy Evolutionary Computation.* Kluwer, Boston, MA (1997).

[Pen94] S. Peng and L. Lucke. Fuzzy filtering for mixed noise removal during image processing. In *Proc. IEEE Intl. Conf. on Fuzzy Systems*, pp. 89–93 (Orlando, FL, June 1994).

[Pen95a] S. Peng and L. Lucke. Multi-level adaptive fuzzy filter for mixed noise removal. In *Proc. IEEE Intl. Symposium on Circuits and Systems* (Seattle, WA, April 1995).

[Pen95b] S. Peng and L. Lucke. An adaptive window hybrid filter. In *Proc. IEEE Workshop on Nonlinear Signal Processing*, pp. 313–316 (Neos Marmaras, Halkidiki, Greece, June 1995).

[Pen95c] S. Peng and L. Lucke. A hybrid filter for image enhancement. In *Proc. IEEE Intl. Conf. on Image Processing*, pp. 163–166 (Washington, DC, October 1995).

[Pla95] K. N. Plataniotis, D. Androutsos and A. N. Venetsanopoulos. Color image processing using fuzzy vector directional filters. In *Proc. IEEE Workshop on Nonlinear Signal Processing*, pp. 535–538 (Neos Marmaras, Halkidiki, Greece, June 1995).

[Rus92] F. Russo. A user-friendly research tool for image processing with fuzzy rules. In *Proc. First IEEE Intl. Conf. on Fuzzy Systems*, pp. 561–568 (San Diego, CA, March 1992).

[Rus93] F. Russo. A new class of fuzzy operators for image processing: design and implementation. In *Proc. Second IEEE Intl. Conf. on Fuzzy Systems*, pp. 815–820 (San Francisco, CA, March 1993).

[Rus95] F. Russo and G. Ramponi. A fuzzy operator for the enhancement of blurred and noisy images. *IEEE Trans. Image Process.* **4**(8), 1169–1174 (August 1995).

[Rus96] F. Russo. Fuzzy models for image-based measurement systems. In *Proc. IEEE Intl. Workshop on Emergent Technologies for Instrumentation and Measurements*, pp. 41–48 (Como, Italy, June 1996).

[Rus97] F. Russo. Fuzzy processing of image data using FIRE filters. *J. Intell. Fuzzy Syst.* **5**(4), 361–366 (1997).

[Rus99a] F. Russo. Fuzzy model fundamentals. In *Encyclopedia of Electrical and Electronics Engineering* (J. Webster, ed.), **8**, 158–166, Wiley (1999).

[Rus99b] F. Russo. FIRE operators for image processing. *Fuzzy Sets Syst.*, **103**(2), 265–275 (April 1999).

[Rus99c] F. Russo. Evolutionary neural fuzzy systems for noise cancellation in image data. *IEEE Trans. Instrum. Measure.* **48**(5), 915–920 (1999).

[Suc96] R. Sucher. A self-organizing nonlinear filter based on fuzzy clustering. In *Proc. IEEE Intl. Symposium on Circuits and Systems*, pp. 101–104 (Atlanta, GA, May 1996).

[Tag94] A. Taguchi, H. Takashima, and Y. Murata. Fuzzy filters for image smoothing. In *Nonlinear Image Processing V*, pp. 332–339, SPIE (1994).

[Tag95a] A. Taguchi, H. Takashima, and F. Russo. Data dependent filtering using the fuzzy inference. In *Proc. IEEE Instrumentation and Measurement Technology Conference*, pp. 752–756 (Waltham, MA, April 1995).

[Tag95b] A. Taguchi and M. Meguro. Adaptive *L*-filters based on fuzzy rules. In *Proc. IEEE Intl. Symp. on Circuits and Systems* (Seattle, WA, April 1995).

[Tag96] A. Taguchi and T. Kimura. Data-dependent filtering based on IF-THEN rules and ELSE-rule. In *Proc. VIII European Signal Processing Conference*, pp. 1713–1716 (Trieste, Italy, September 1996).

[Yag80] R. R. Yager. On a general class of fuzzy connectives. *Fuzzy Sets Syst.* **4**, 235–241 (1980).

[Zad65] L. A. Zadeh. Fuzzy sets. *Infor. Control*, **8**, 338–353 (1965).

13

Digital Halftoning

DANIEL L. LAU
GONZALO R. ARCE

Department of Electrical and Computer Engineering
University of Delaware
Newark, Delaware

13.1 Introduction

Since the introduction of photography in 1839, accurately reproducing an original continuous-tone photograph without loss of tonal value or detail has been a primary concern for the manufacturers of printing presses. At that time, the predominant form of printing was the letterpress, which was incapable of printing intermediate tones and could only produce black and white images. Continuous-tone, monochrome photographs could be reproduced only as line drawings created by highly skilled craftsmen, usually on scratchboard. In 1880, the halftoning process was invented, leading to a technical revolution in both photography and the printing industry.

In terms of photolithography (a process introduced in 1855 [Des94]), halftoning involved projecting light from the negative of a continuous-tone photograph, through a mesh screen, such as finely woven silk, onto a photosensitive plate. Bright light, as it passed through a pinhole opening in the silk screen, would form a large, round spot on the plate. Dim light would form a small spot. Light sensitive chemicals coating the plate would then form insoluble dots that varied in size according to the tones of the original photograph. After processing, the plate would have dots where ink was to be printed, raised slightly above the rest of the plate.

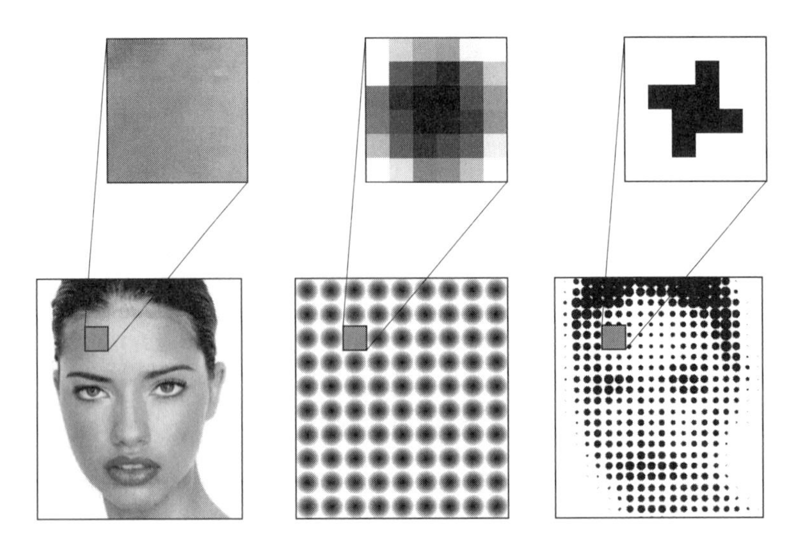

Figure 13.1: The AM halftoning process.

Later versions of the halftoning process employed screens made of glass that were coated on one side by an opaque substance. A mesh of parallel and equidistant lines was scratched in the opaque surface and then a second mesh was scratched perpendicular to them. Such screens would differ in the number of lines per inch that had been scratched into the opaque surface; while finer screens created better reproductions, the quality of the printing press would limit how fine a mesh could be employed.

Later still, the glass plate mesh was replaced altogether with a flexible piece of processed film, placed directly in contact with the unexposed lithographic film [Des94]. This contact screen had direct control of the dot structure. The screen controlled the screen frequency (the number of lines per inch), the dot shape (the shape of the dots as the size increased from light to dark), and the screen angle (the orientation of lines relative to the positive horizontal axis).

13.1.1 AM Halftoning

Today, printing is far more advanced with the introduction of nonimpact printing technologies and the emergence of desktop publishing. Brought on by advancements in the digital computer [Des94], the photomechanical screening process introduced in 1880 has been replaced, in many instances, by digital imagesetters. In some instances, printing is no longer binary since continuous-tone dye-sublimation printers are now readily available, but due to their speed and material requirements (special papers and inks), they have not reached the widespread acceptance that four color ink jet and electrophotographic (laser) printers have.

In these digital printers, the process of projecting a continuous-tone original through a halftone screen has been replaced with a raster image processor (RIP) that converts each pixel of the original image from an intermediate tone directly

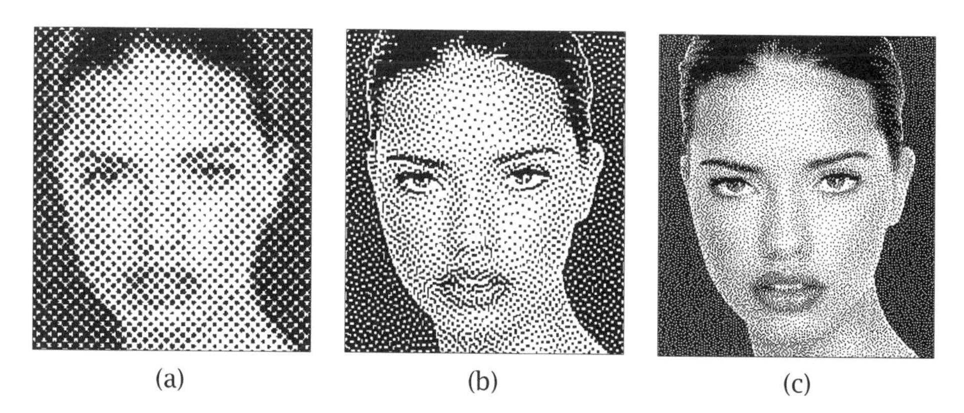

(a) (b) (c)

Figure 13.2: Grayscale images halftoned using (a) AM, (b) AM-FM, and (c) FM halftoning.

into a binary dot. When first introduced, the RIPs imitated the halftone patterns of contact screens by forming a regular grid of round dots that varied in size according to tone. This conversion, illustrated in Fig. 13.1 using the image "Adrian," was done using a dither array or mask in which pixels of the original image were compared with the pixels of a predefined array (the mask). Pixels of the original image that had intensities greater than their corresponding intensities in the mask were turned "on" (printed), while the remaining pixels were turned "off" (not printed). For large images, the mask was tiled end to end until the image space was completely covered by masks. Note how the mask is composed of a smaller mask that was tiled end to end.

For modulating the size of printed dots and not the number or frequency of the dots, these techniques are commonly referred to as amplitude modulated (AM) halftoning techniques. Shown in Fig. 13.2a is the image Adrian after applying AM halftoning. Like contact screens, the resulting patterns vary in their screen frequency, dot shape, and screen angle, defined in Fig. 13.3.

Screen Frequency

The screen frequency is the number of lines or rows of macrodots per inch of the resulting halftone pattern. Like the original glass plate screens, the finer screens create patterns with higher spatial resolutions. Depending on the resolution of the printer [measured in dots per inch (dpi)], screen frequency is limited by the number of unique gray levels the printer can represent. This relationship is defined as follows

$$\text{lines per inch} = \frac{\text{resolution of printer (in dpi)}}{\sqrt{\text{number of unique gray levels} - 1}}.$$

Dot Shape

Like the contact screen, the imagesetter controls the shape of the dots in the halftone pattern. These shapes are most clearly recognizable at gray level 1/2

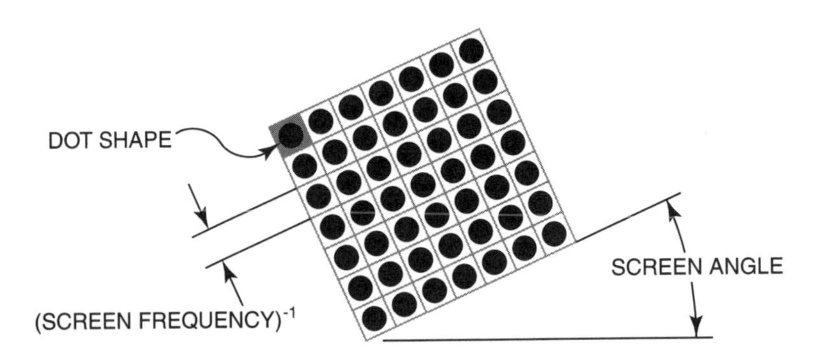

Figure 13.3: The dot shape, screen frequency, and screen angle of an AM halftone pattern.

(number of black dots equals the number of white dots). The most common dot shapes are round, square, and elliptical [Cou96], but special effect shapes have also been introduced [Bla93].

Screen Angle

The last parameter used to classify AM screens is the screen angle, or the orientation of screen lines relative to the horizontal axis. This parameter is a function of the human visual system, in which directional artifacts are least noticeable when oriented along the 45° diagonal [Cam96]. For black and white printing, this angle, therefore, should also be 45°.

13.1.2 FM Halftoning

As an alternative to AM screening, frequency modulated (FM) or stochastic screening techniques, in which dots are of constant size but variably spaced according to tone (Fig. 13.2c), are available to digital printers. Although relatively rare in commercial printing [Bla93], FM screening avoids the problems associated with AM methods [Rod94] such as Moiré pattern, the interference pattern created by superimposing two or more regular patterns. Moiré patterns, demonstrated in Fig. 13.4, are a problem associated with color printing. These interference patterns are minimized by commercial printers by offsetting the orientation of the halftone screen of each color, creating a desirable circular pattern called a *rosette* (Fig. 13.5; also see color insert). Great care must be taken when aligning these screens to avoid the creation of Moiré patterns [Ami94].

A further drawback of AM screening is image contouring, a visual banding effect created by an abruptly changing halftone texture. This artifact is minimized in AM patterns by increasing the maximum number of gray levels that a cluster can represent. However, to increase the number of achievable gray levels, we also must increase the total number of pixels that compose a cluster, reducing the spatial resolution of the resulting halftone image.

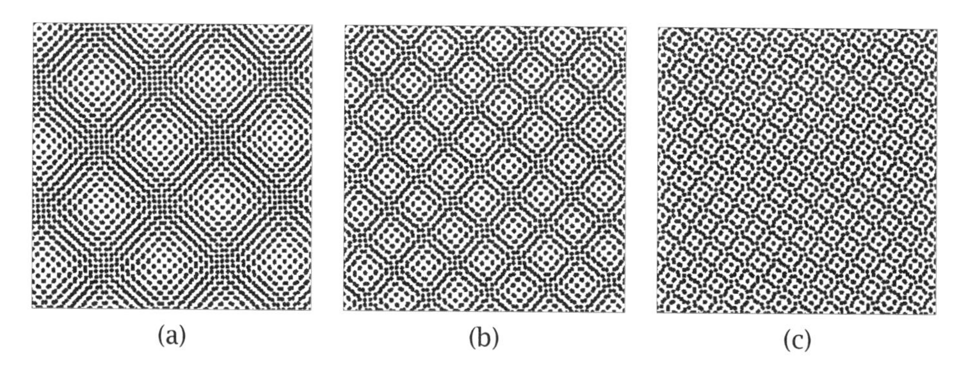

(a) (b) (c)

Figure 13.4: Moiré patterns created by offsetting two AM halftone patterns by (a) 5°, (b) 10°, and (c) 30°.

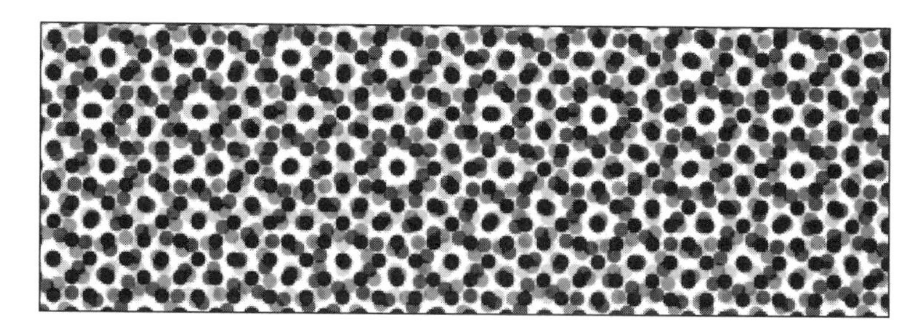

Figure 13.5: Rosette pattern created by setting the CMYK channels to screen angles 15°, 75°, 0°, and 45°, respectively.

While many techniques exist for converting a continuous-tone original into an FM halftone pattern, the best techniques are those that create halftone patterns, called blue-noise patterns, which are composed exclusively of high frequency spectral components. By isolating minority pixels in a homogeneous (stationary) and isotropic (radially symmetric) manner, blue-noise dither patterns minimize the visibility to the human eye of the individual dots and thereby minimize the addition of unwanted textures to the printed image. By distributing minority pixels in a random fashion, blue noise eliminates the need for screen angles and screen rulings, resulting in an image without artifacts and displaying higher spatial resolution.

The major drawback of blue-noise, and FM halftoning in general, is dot gain, the increase in size of the printed dot relative to the intended dot size of the original halftone film. When printing halftones, dot gain creates a reduction or compression in the printed tonal range, leading to a loss of definition and detail (contrast); furthermore, dot gain can lead to "plugged," or filled in, screens and cause a shift in color [Cou96] (Fig. 13.6).

Due to the nature of dot gain, FM screens typically suffer greater amounts of distortion than AM screens [Rod94] (Fig. 13.7), but it is not the amount of dot gain that limits the use of FM halftoning. Instead, it is the reproducibility of the printed dot. In a repeatable process in which the variation in dot size/shape from printed dot to printed dot is small, printers can rely on dot gain compensation

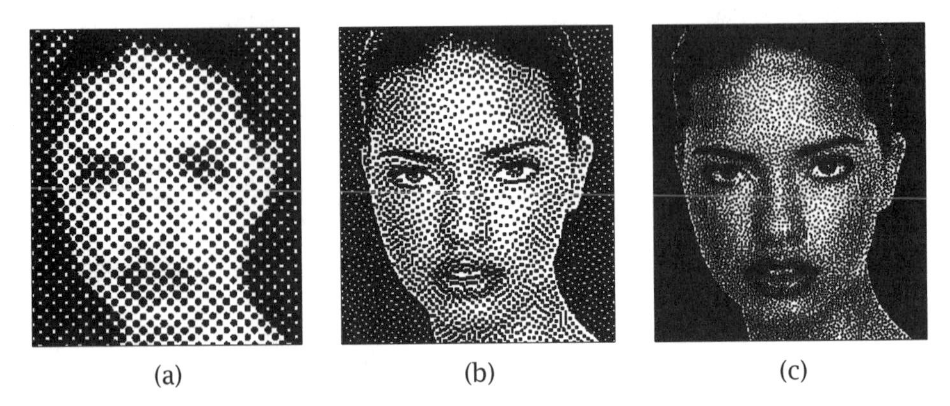

Figure 13.6: Grayscale images halftoned using (a) FM, (b) AM-FM, and (c) FM halftoning. Each is shown with dot/gain.

techniques to minimize the distortions introduced, but in a nonrepeatable process in which the variation in dot size/shape is high, pattern robustness (clustering) becomes a desired, and in many cases required, characteristic of the halftoning process [Uli87]. In summary, the choice between AM and FM screens is a function of the imaging system's ability to print individual pixels. If individual pixels can be reliably reproduced, the halftone can be based on individual pixels. Otherwise, the halftone must be composed of groups of pixels that, through clustering, form larger features that can be reliably reproduced [Fin92].

13.1.3 AM-FM Hybrids

Today, as printers are achieving higher and higher print resolutions (> 1200 dpi), the limits of FM halftoning are being reached. Researchers are beginning, therefore, to look at AM-FM hybrids that produce dot clusters that vary according to tone in both their size and spacing (Fig. 13.2b). In regard to the reproduction of monochrome images, AM-FM hybrids are capable, in general, of producing patterns with lower visibility (higher spatial resolution) compared to AM techniques and, if stochastic, do so without a periodic structure that adds an artificial texture to the printed image. With some amount of clustering, these halftones are easier to print reliably and with little variation in the resulting tone. In the reproduction of color, stochastic hybrids also maintain the same freedom from Moiré patterns associated with FM halftoning.

While various AM-FM halftoning techniques have been introduced, the best techniques are those that generate green-noise halftone patterns, which are binary dither patterns created exclusively of midfrequency spectral components. The goal of green-noise is to distribute minority pixel clusters in a homogeneous and isotropic fashion. The average size of these clusters can vary, with smaller clusters leading to halftoned images with higher spatial resolution and better edge detail and larger clusters leading to increased halftone robustness, the ability of the

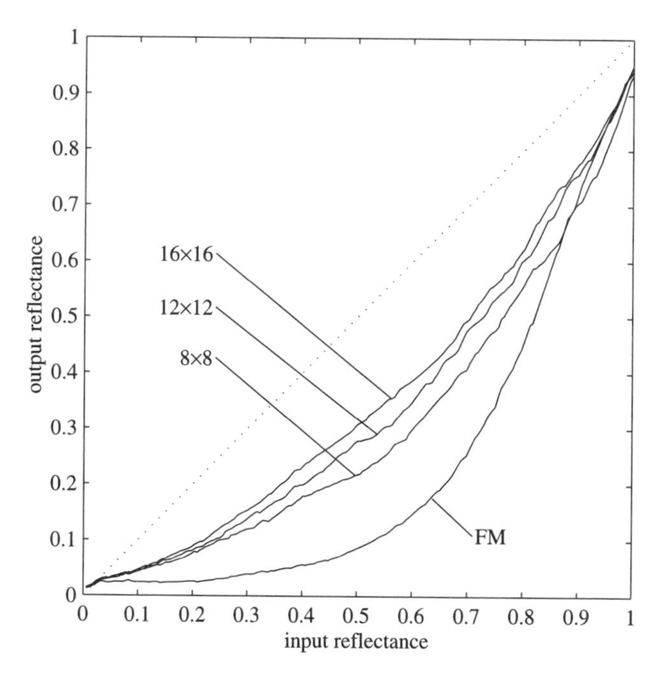

Figure 13.7: Measured input versus output reflectance curves for an ink jet printer using FM halftoning and using AM halftoning with cells of size 8×8, 12×12, and 16×16.

pattern to resist the distortions introduced during the printing process. As a tunable model, green-noise has, as a limiting case, blue noise, and by using green-noise, binary display devices, which were previously restricted to AM halftoning techniques, can now combine the maximum dispersion attributes of blue-noise with the clustering attributes of AM halftones [Lau98].

Note that the blue-noise model is well established and has had a profound impact on the printing industry since its introduction, paving the way for low-cost color output. The alternative green-noise model represents a new approach to stochastic halftoning that is just now starting to make inroads into the marketplace. It is, therefore, the purpose of this chapter to introduce the reader to both the blue- and green-noise halftoning models. Initially though, we introduce several metrics for characterizing stochastic dither patterns in both the spatial and spectral domains.

13.2 Halftone Statistics

Stochastic geometry is the area of mathematical research concerning complex geometrical patterns [Sto87]. Some of the problems addressed in this field include characterizing the growth of cells in a layer of organic tissue, calculating the average area covered by randomly placed disks of constant size and shape, and characterizing the location of individual trees within a forest [Gou95]. This last problem is an example of a *spatial point process* [Dig83], which is typically described using point process statistic metrics developed to describe the location of points in space. While many of the statistics have been developed for characterizing points in continuous space, they are perfectly suited to the study of digital halftone patterns [Lau98] such as those found in FM halftoning, in which minority pixels are randomly distributed.

13.2.1 Spatial Domain Statistics

For spatial domain analysis, the metrics developed for point process statistics are ideally suited, with ϕ being a sample of the stochastic point process Φ such that $\phi = \{x_i : i = 1, 2, 3, \ldots\}$ and $x_i \in \mathbf{R}^2$ (2D real space); furthermore, given $B \subset \mathbf{R}^2$, the scalar quantity $\phi(B)$ is the number of x_i's in the subset B. In terms of a digital halftone, $I_g(n)$ represents a binary pixel at sample index n of I_g, the halftone pattern created to represent a continuous-tone image of constant gray level g, with samples of the discrete-space image separated along the horizontal and vertical axis by a distance D. $\phi(n)$ is a pixel at sample index n of the point process Φ such that $\phi(n) = 1$ indicates that the pixel at sample index n is a minority pixel [$I_g(n) = 1$ for $0 \leq g < 1/2$ or $I_g(n) = 0$ for $1/2 \leq g < 1$] in I_g, while $\phi(n) = 0$ indicates that the pixel at sample index n is a majority pixel [$I_g(n) = 0$ for $0 \leq g < 1/2$ or $I_g(n) = 1$ for $1/2 \leq g < 1$].

A common metric used in stochastic geometry to measure the statistical properties of Φ is the *reduced second moment measure* $\mathcal{K}(n;m)$, defined as

$$\mathcal{K}(n;m) = \frac{\mathbf{E}\{\phi(n) | m \in \phi\}}{\mathbf{E}\{\phi(n)\}},$$

the ratio of the conditional expected value of $\phi(n)$ given that the sample at index m is a minority pixel ($\phi(m) = 1$) to the unconditional expected value of $\phi(n)$. For stationary Φ,

$$\begin{aligned} \mathbf{E}\{\phi(n)\} &= \mathbf{Pr}\{\phi(n) = 1\} \\ &= \begin{cases} g & \text{for } 0 \leq g < 1/2, \\ 1 - g & \text{for } 1/2 \leq g < 1, \end{cases} \end{aligned}$$

where $\mathbf{E}\{\phi(n)\}$ is commonly referred to as the *intensity* of ϕ and is labeled $\mathcal{I}$. Note that $\mathcal{I} \cdot \mathcal{K}(n;m)$ is, therefore, the probability that $\phi(n)$ is a minority pixel given that $\phi(m)$ is a minority pixel.

The reduced second moment measure has the useful property that $\mathcal{K}(n;m) > 1$ indicates a greater than average likelihood that a minority pixel exists at n given m, while $\mathcal{K}(n;m) < 1$ indicates a less than average likelihood. Figure 13.8b illustrates $\mathcal{K}(n;m)$, the reduced second moment measure, for the point process sample ϕ, given in 13.8a. In this figure, areas of $\mathcal{K}(n;m)$ shown in white are areas in which minority pixels are more likely to occur given a minority at the center position, while areas in black are regions in which minority pixels are less likely to occur.

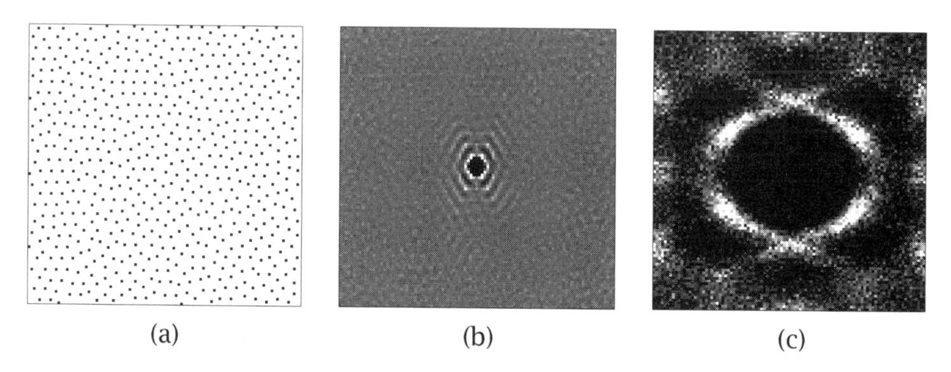

(a) (b) (c)

Figure 13.8: (a) Sample pattern ϕ and its corresponding (b) reduced second moment measure $\mathcal{K}(n;m)$ and (c) estimated power spectrum $\hat{P}(f)$.

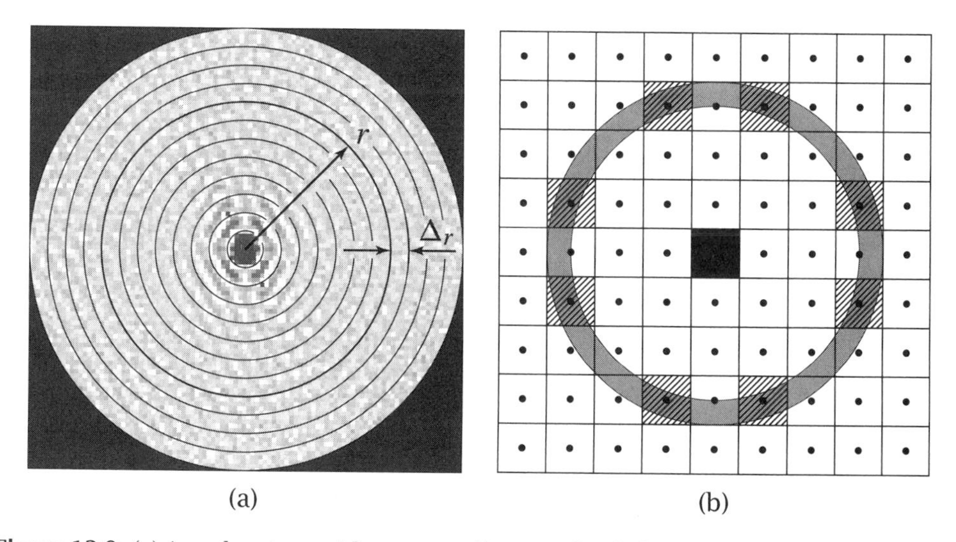

(a) (b)

Figure 13.9: (a) Annular rings with center radius r and radial width Δ_r, used to extract the pair correlation $\mathcal{R}(r)$ from the reduced second moment measure $\mathcal{K}(n;m)$; (b) the pixels, shown crosshatched, of $\mathcal{K}(n;m)$ that form the annular ring for $r = 3D$ and $\Delta_r = (1/2)D$.

For a stationary point process Φ, $\mathcal{K}(n;m) = \mathcal{K}(r,\theta)$, where r is the distance from m to n and θ is the direction. If Φ is also isotropic, $\mathcal{K}(r,\theta)$ is independent of θ and is commonly referred to as the *pair correlation* $\mathcal{R}(r)$, which is defined explicitly as

$$\mathcal{R}(r) = \frac{\mathbf{E}\left\{\phi\left(R_y(r)\right) \mid y \in \phi\right\}}{\mathbf{E}\left\{\phi\left(R_y(r)\right)\right\}},$$

the ratio of the conditional expected number of minority pixels located in the ring $R_m(r) = \{n : r \le |n - m| < r + \Delta_r\}$ (Fig. 13.9), given that a minority pixel exists at sample index m, to the unconditional expected number of minority pixels in $R_m(r)$. Figure 13.10 illustrates the pair correlation for the point process sample of Fig. 13.8a, with $\Delta_r = (1/2)D$.

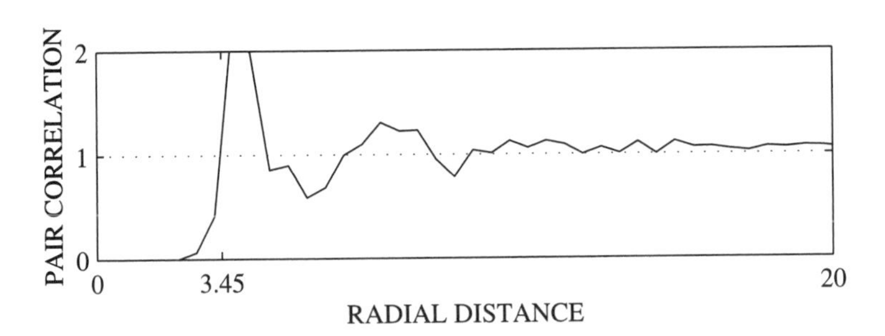

Figure 13.10: Pair correlation corresponding to the sample ϕ of Fig. 13.8a.

13.2.2 Spectral Statistics

Spectral analysis of a point process was first introduced by Bartlett to describe one-dimensional processes [Bar64a] and then again to study two-dimensional processes [Bar64b]. While statistics have been developed based on Bartlett's work [Mug96], statisticians have relied almost entirely on spatial statistics, such as the pair correlation, due to the intuitive characterizations they offer in describing spatial point processes. Within the signal processing community, though, spectral analysis is a tool commonly used for studying random processes.

Ulichney [Uli87], to characterize stochastic dither patterns created via error diffusion, first applied spectral analysis to stochastic dither patterns. To do so, Ulichney developed the radially averaged power spectra along with a measure of anisotropy. Both rely on estimating the power spectrum through Bartlett's method of averaging periodograms, the magnitude-square of the Fourier transform of the output pattern divided by the sample size, to produce the spectral estimate $\hat{P}(f)$. Although anisotropies of a dither pattern can be qualitatively observed by studying 3-D plots of $\hat{P}(f)$ such as in Fig. 13.8c, partitioning the spectral domain into a series of annular rings $R(f_\rho)$ of width Δ_ρ (Fig. 13.11) leads to a useful one-dimensional statistic.

The radially averaged power spectrum density (RAPSD) $P(f_\rho)$ is defined for discrete $\hat{P}(f)$ as the average power in the annular ring with center radius f_ρ such that

$$P(f_\rho) = \frac{1}{N\left(R(f_\rho)\right)} \sum_{f \in R(f_\rho)} \hat{P}(f),$$

where $N\left(R(f_\rho)\right)$ is the number of frequency samples in $R(f_\rho)$. The RAPSD maintains an intimate link with the spatial domain. In particular, a frequent occurrence of the interpoint distance r, indicated by maxima in $\mathcal{R}(r)$, implies a peak in $P(f_\rho)$ for radial frequency $f_\rho = r^{-1}$ proportional in magnitude to the peak in $R(r)$, meaning that a larger peak, in $P(f_\rho)$, leads to a larger peak in $\mathcal{R}(r)$. Figure 13.12 illustrates the RAPSD for the point process sample of Fig. 13.8a. Note that because the spectral composition of I_g along the horizontal (f_{h}) and vertical (f_{v}) axes is limited to discrete frequencies within the range $-(1/2)D^{-1} < f_{\mathrm{h,v}} < (1/2)D^{-1}$, $P(f_\rho)$ is defined only for $0 \le f_\rho < (\sqrt{2}/2)D^{-1}$ and the annular rings for which

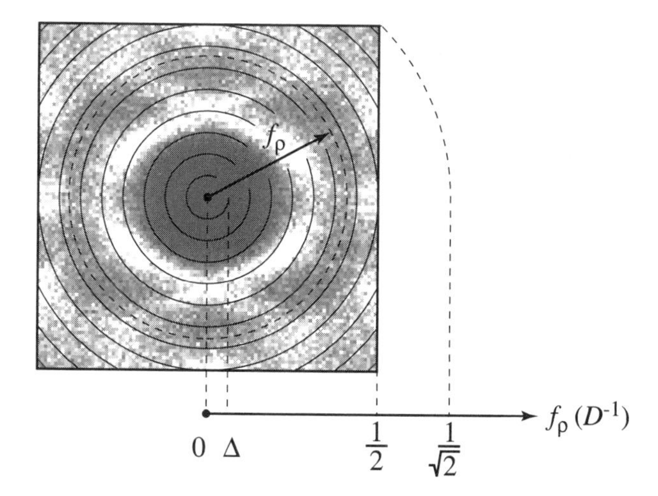

Figure 13.11: Annular rings used to divide the power spectrum $P(f)$ to form $P(f_\rho)$.

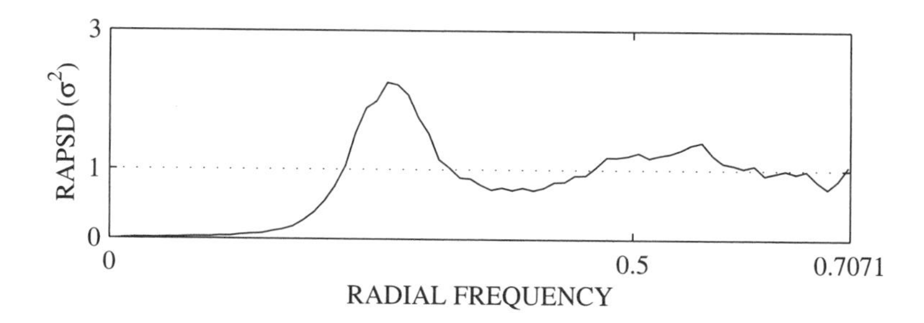

Figure 13.12: The radially averaged power spectral density corresponding to the sample ϕ of Fig. 13.8a.

$(1/2)D^{-1} \leq f_\rho < (\sqrt{2}/2)D^{-1}$ are incomplete. All plots of $P(f_\rho)$ therefore indicate the $f_\rho = (1/2)D^{-1}$ point along the horizontal axis.

13.3 Blue-Noise Dithering

Blue noise is a statistical model describing the ideal spatial and spectral characteristics of FM halftone patterns. The arrangement of minority pixels within a blue-noise halftone pattern is characterized by a distribution of binary pixels in which the minority pixels are spread as homogeneously as possible [Uli87]. Distributing pixels in this manner creates a pattern that is aperiodic, isotropic, and does not contain any low frequency spectral components. The result of halftoning a continuous-tone discrete-space monochrome image with blue noise is a pattern that is visually pleasing and does not impose an artificial texture onto an image.

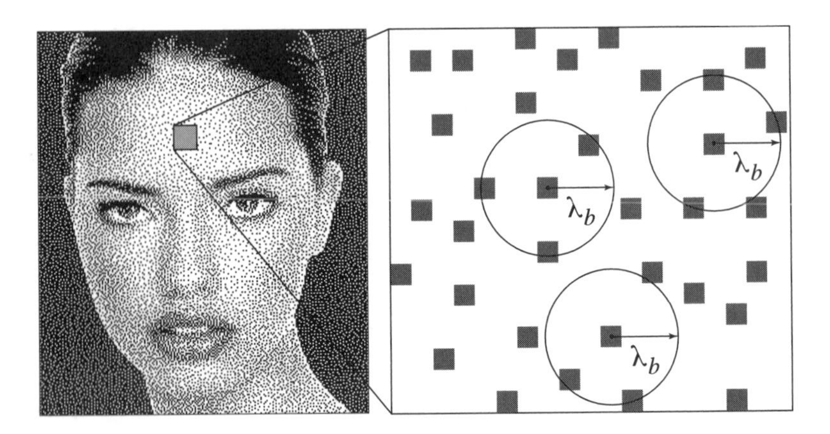

Figure 13.13: Minority pixels of blue noise separated by an average distance λ_b.

13.3.1 Spatial and Spectral Characteristics

Blue-noise, when applied to an image of constant gray level g, spreads the minority pixels of the resulting binary image such that they are separated by an average distance λ_b (Fig. 13.13), defined as

$$\lambda_b = \begin{cases} D/\sqrt{g} & \text{for } 0 < g \leq 1/2, \\ D/\sqrt{1-g} & \text{for } 1/2 < g \leq 1, \end{cases}$$

where D is the minimum distance between addressable points on the display. The parameter λ_b is referred to as the principal wavelength of blue noise, with its relationship to g justified by several intuitive properties:

1. As the gray value approaches perfect white ($g = 1$) or perfect black ($g = 0$), the principal wavelength approaches infinity.

2. The wavelength decreases symmetrically with equal deviations from black and white toward the middle gray ($g = 1/2$).

3. The square of the wavelength is inversely proportional to the number of minority pixels per unit area.

Again we note that the distribution of minority pixels is assumed to be stationary and isotropic.

Spatial Statistics

In light of the nature of blue noise to isolate minority pixels, blue-noise halftones are characterized in terms of the pair correlation $\mathcal{R}(r)$ by noting that

1. Few or no neighboring pixels lie within a radius of $r < \lambda_b$.

2. For $r > \lambda_b$, the expected number of minority pixels per unit area approaches 1 with increasing r.

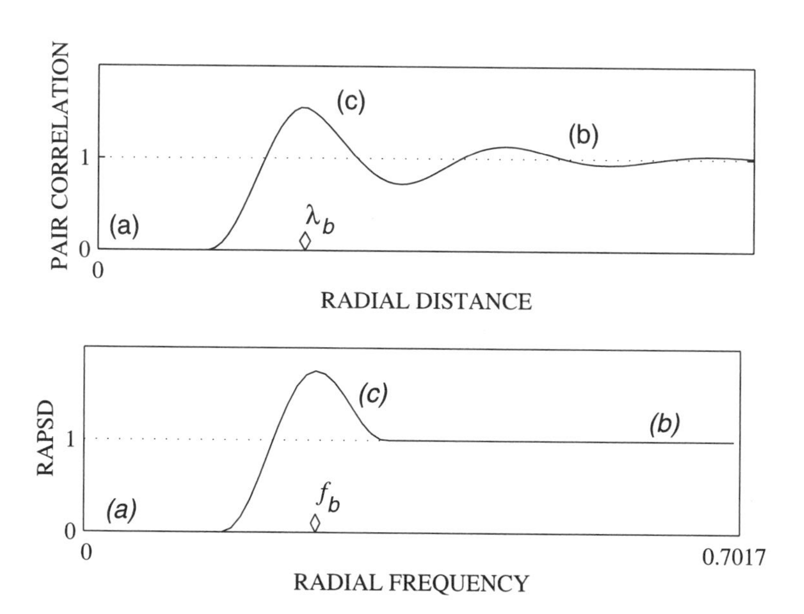

Figure 13.14: Top: pair correlation with principal wavelength λ_b. Bottom: RAPSD with principal frequency f_b of the ideal blue-noise pattern. (Reproduced with permission from [Lau98]. © 1998 IEEE.)

3. The average number of minority pixels within the radius r increases sharply near λ_b.

The resulting pair correlation for blue noise is therefore of the form indicated in Fig. 13.14 (top), in which $\mathcal{R}(r)$ shows a strong inhibition of minority pixels near $r = 0$, marked (a); decreasing correlation of minority pixels with increasing r [$\lim_{r\to\infty} \mathcal{R}(r) = 1$], marked (b); and a frequent occurrence of the interpoint distance λ_b, the principal wavelength, indicated by a series of peaks at integer multiples of λ_b, marked (c). The principal wavelength is indicated in the figure by a diamond located along the horizontal axis.

Spectral Statistics

Turning to the spectral domain, the spectral characteristics of blue noise in terms of $P(f_\rho)$ are shown in Fig. 13.14 (bottom) and can be described by three unique features: little or no low frequency spectral components, marked (a); a flat, high frequency (blue-noise) spectral region, marked (b); and a spectral peak at cutoff frequency f_b, the blue-noise principal frequency, marked (c), such that

$$f_b = \begin{cases} \sqrt{g}/D & \text{for } 0 < g \le 1/2, \\ \sqrt{1-g}/D & \text{for } 1/2 < g \le 1. \end{cases}$$

The principal frequency is indicated in the figure by a diamond located along the horizontal axis. Note that $P(f_\rho)$ is plotted in units of $\sigma^2 = g(1-g)$, the

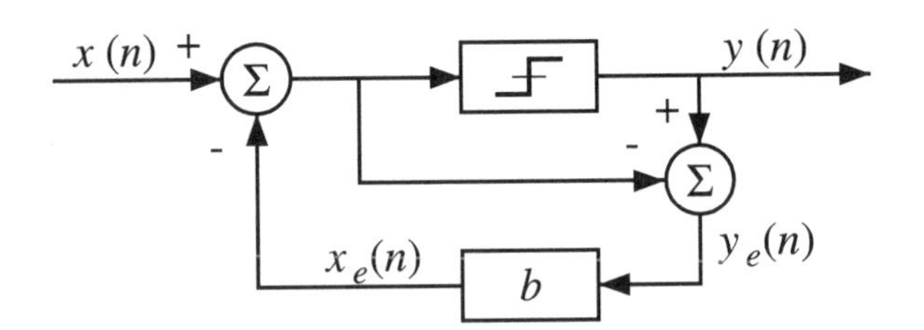

Figure 13.15: Error diffusion algorithm.

(a)

	•	7/16
3/16	5/16	1/16

(b)

		•	7/48	5/48
3/48	5/48	7/48	5/48	3/48
1/48	3/48	5/48	3/48	1/48

(c)

		•○	8/42	4/42
2/42	4/42	8/42	4/42	2/42
1/42	2/42	4/42	2/42	1/42

Figure 13.16: (a) Floyd and Steinberg, (b) Jarvis, Judice, and Ninke, and (c) Stucki error filters for error diffusion.

variance of an individual pixel in ϕ. The sharpness of the spectral peak at the blue-noise principal frequency is affected by the separation between minority pixels, which should have some variation. The wavelengths of this variation should not be significantly longer than λ_b since this adds low frequency spectral components to the corresponding dither pattern ϕ [Uli88], causing ϕ to appear more white than blue.

13.3.2 Error Diffusion

In error diffusion halftoning (Fig. 13.15), the output pixel $y(n)$ is determined by adjusting and thresholding the input pixel $x(n)$ such that

$$y(n) = \begin{cases} 1 & \text{for } [x(n) - x^e(n)] \geq 0, \\ 0 & \text{otherwise,} \end{cases}$$

where $x^e(n)$ is the diffused quantization error accumulated during previous iterations,

$$x^e(n) = \sum_{i=1}^{M} b_i \cdot y^e(n-i), \tag{13.1}$$

with $y^e(n) = y(n) - [x(n) - x^e(n)]$ and $\sum_{i=1}^{M} b_i = 1$. Using vector notation, Eq. (13.1) becomes

$$x^e(n) = \mathbf{b}^{\mathrm{T}} \mathbf{y}^e(n),$$

where $\mathbf{b} = [b_1, b_2, \ldots, b_M]^{\mathrm{T}}$ and $\mathbf{y}^e(n) = [y^e(n-1), y^e(n-2), \ldots, y^e(n-M)]^{\mathrm{T}}$.

First proposed by Floyd and Steinberg [Flo76], the original error weights, b_i, are shown in Fig. 13.16, along with those proposed later by Jarvis, Judice, and Ninke [Jar76] and by Stucki. The resulting halftoned images, produced by error

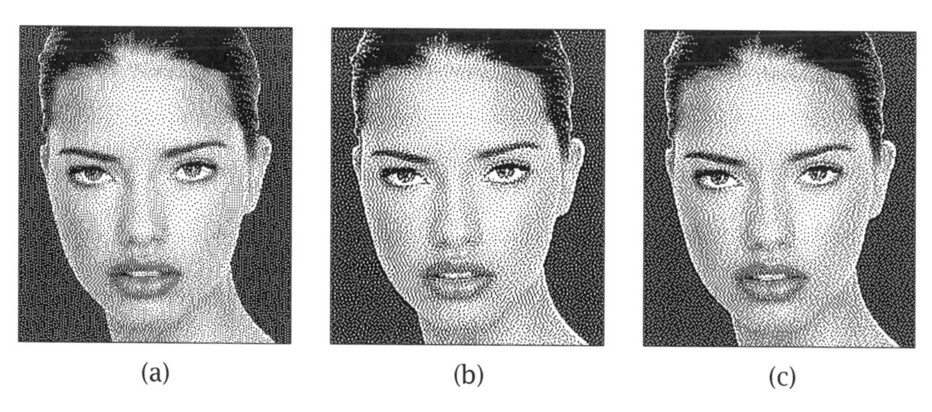

Figure 13.17: Grayscale image halftoned using error diffusion with (a) Floyd and Steinberg, (b) Jarvis, Judice and Ninke, and (c) Stucki filter weights employing a serpentine raster scan.

diffusion using these weights along with a serpentine left-to-right and then right-to-left raster scan, are shown in Fig. 13.17. Figure 13.18 illustrates the spatial and spectral characteristics of the three proposed error diffusion filters. While these spatial and spectral characteristics approximate the ideal blue-noise model, Ulichney [Uli87] showed that with some variation, error diffusion can be an improved generator of blue noise with resulting patterns having spatial and spectral characteristics much closer to those of the ideal blue-noise pattern.

Possible variations include using randomized weight positions and/or a perturbed threshold, a process that is equivalent to adding low variance white noise to the original input image [Kno92]. Another variation of particular significance involves perturbing the filter weights instead of the threshold. This perturbation of filter weights is accomplished by first pairing weights of comparable value, and then for each pair of weights, a scaled random value is added to one and subtracted from the other. To prevent negative values, we set the maximum noise amplitude (100%) to the value of the smaller weight in each pair. Using the Floyd and Steinberg filter weights, perturbing each of the pairs (7/16, 5/16) and (3/16, 1/16) creates a good blue-noise process, with the addition of 50% noise to each pair optimizing the trade-off between graininess and stable texture [Uli87].

13.3.3 Blue-Noise Dither Arrays

Since the concept of blue-noise was first introduced in 1987, many alternative techniques for generating visually pleasing blue-noise halftone patterns have been introduced. Two techniques of particular importance are the direct binary search algorithm [Ana92] and halftoning by means of blue-noise dither arrays [Sul91]. The direct binary search algorithm iteratively swaps binary pixels to minimize the difference between the original image and a low pass filtered version of the current binary pattern. While the computational complexity is much higher than that for error diffusion, the resulting images are far superior in quality.

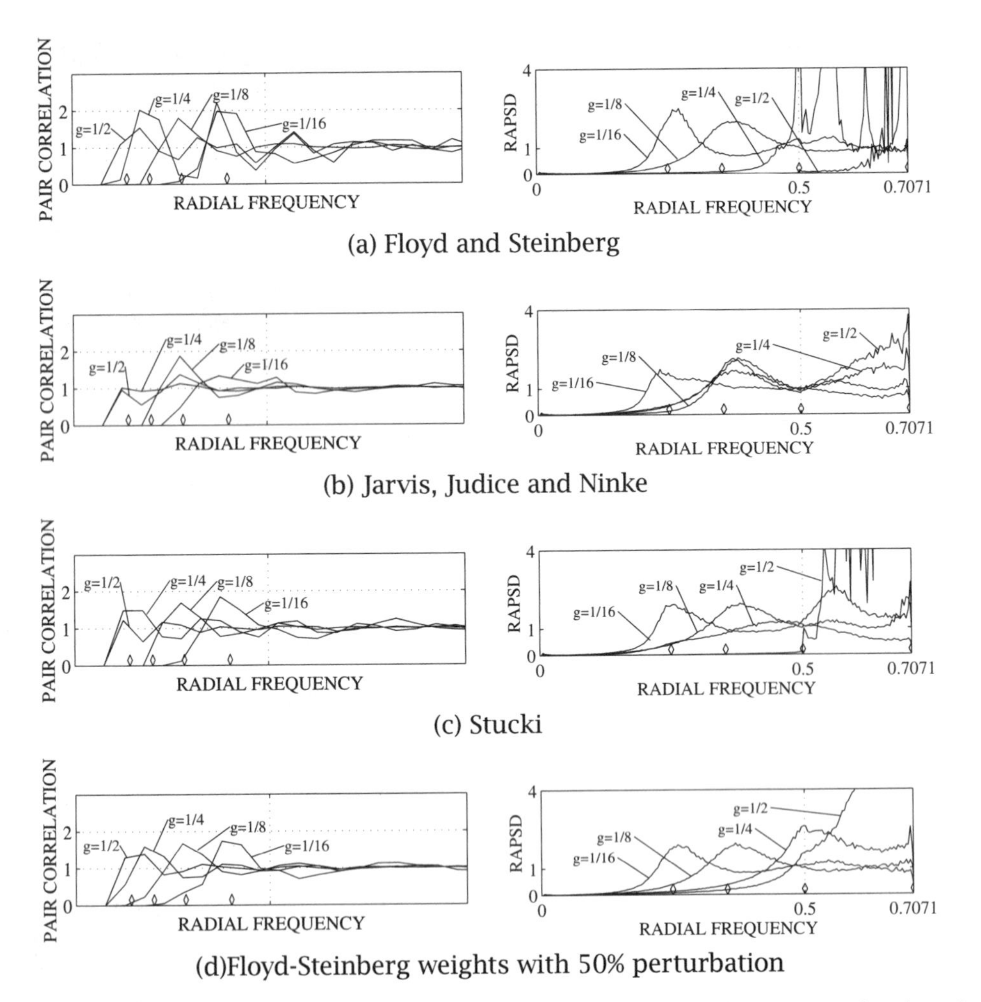

Figure 13.18: Spatial and spectral statistics using error diffusion for intensity levels $1/2$, $1/4$, $1/8$, and $1/16$.

Dither array halftoning is a technique in which the binary image is generated by comparing on a pixel-by-pixel basis the original continuous-tone image and a dither array or mask, as in AM halftoning. By designing these masks to have blue-noise characteristics, visually pleasing FM patterns can be generated with the absolute minimum in computational complexity. For a demonstration, Fig. 13.19 shows a blue-noise dither array and the corresponding halftoned image.

13.4 Green-Noise Dithering

Just as blue-noise is the high frequency component of white-noise, green-noise is the midfrequency component. Like blue, it benefits from aperiodic, uncorrelated structure without low frequency graininess, but unlike blue, green-noise patterns

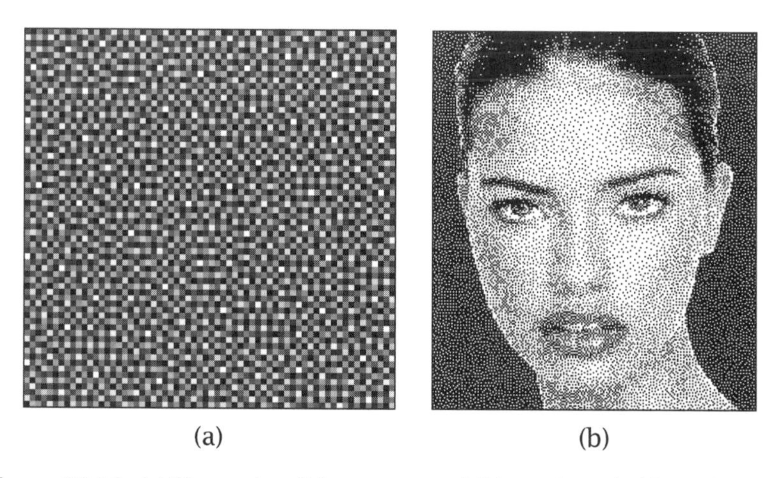

(a) (b)

Figure 13.19: (a) Blue-noise dither array and (b) resulting halftoned image.

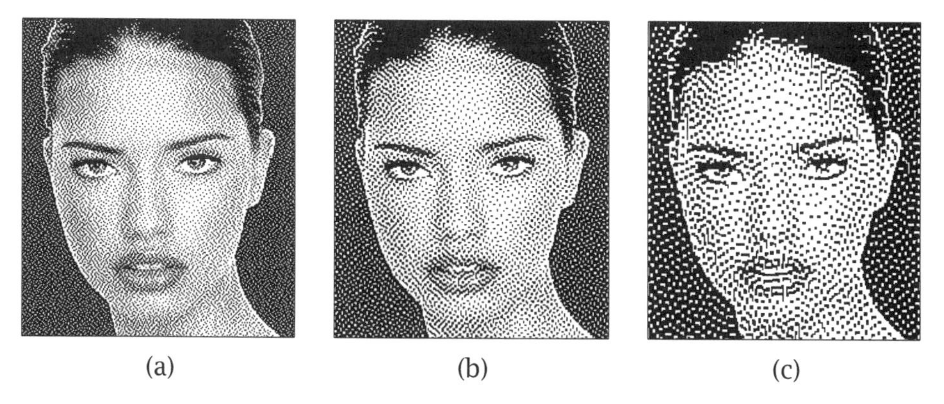

(a) (b) (c)

Figure 13.20: Grayscale image halftoned using green noise with increasing amounts of clustering, with (a) having the smallest clusters and (c) having the largest.

exhibit clustering. The result is a frequency content that lacks the high frequency component characteristic of blue-noise. Hence the term "green." The objective of using green-noise is to combine the maximum dispersion attributes of blue-noise with that of clustering of AM halftone patterns, and to do so at varying degrees, as shown in Fig. 13.20.

Noting Fig. 13.21, the motivation for green-noise is to offer varying levels of robustness, with the most robust patterns having tone reproduction curves close to those of AM patterns and the least robust having curves close to those of blue-noise. That is, green-noise can sacrifice spatial resolution for pattern robustness. Since different printers have different characteristics with respect to the printed dot, green-noise, therefore, can be tuned to offer the highest spatial resolution achievable for a given device.

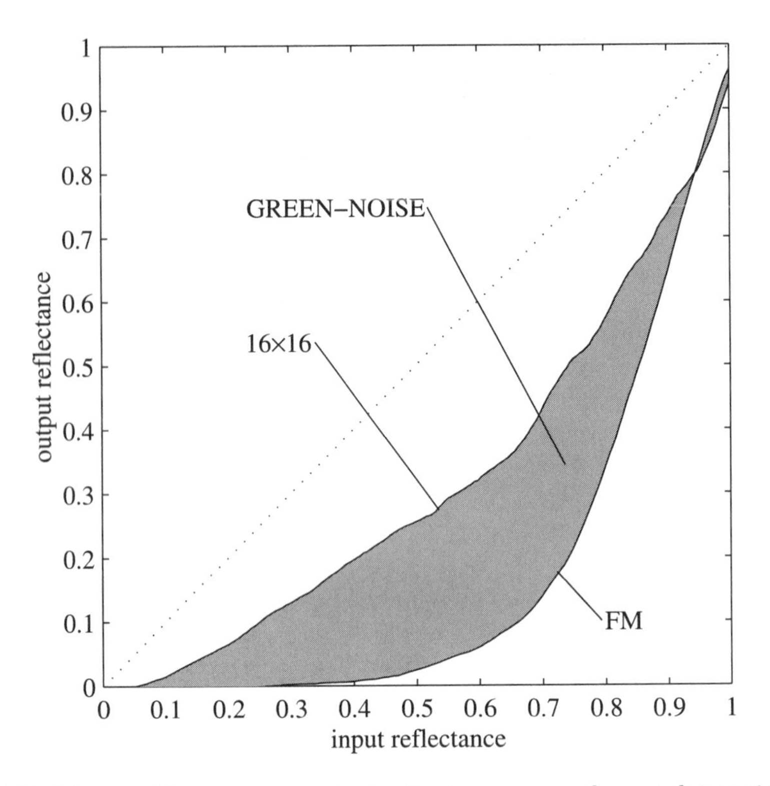

Figure 13.21: Measured input versus output reflectance curves for an ink jet printer using AM halftoning with cells of size 16×16, FM halftoning, and green-noise halftoning.

13.4.1 Spatial and Spectral Characteristics

Green-noise, when applied to an image of constant gray level g, arranges the minority pixels of the resulting binary image such that the minority pixels form clusters of average size $\overline{M}$ pixels separated by an average distance λ_g (Fig. 13.22). Point process statisticians have long described clustering processes such as those seen in green-noise by examining the cluster process in terms of two separate processes: (1) the *parent process* which describes the location of clusters,[1] and (2) the *daughter process*, which describes the shapes of clusters. In AM halftoning, clusters are placed along a regular lattice, and therefore variations in AM patterns occur in the cluster shape. In FM halftoning, the cluster shape is deterministic, a single pixel; it is the location of clusters that is of interest in characterizing FM patterns. Green-noise patterns, having variation in both cluster shape and cluster location, require an analysis that looks at both the parent and daughter processes.

Looking first at the parent process Φ_{p}, ϕ_{p} represents a single sample of the parent process such that $\phi_{\mathrm{p}} = \{x_i : i = 1, 2, \ldots, N_{\mathrm{c}}\}$, where N_{c} is the total number of clusters. For the daughter process Φ_{d}, ϕ_{d} represents a single sample cluster of Φ_{d} such that $\phi_{\mathrm{d}} = \{y_j : j = 1, 2, \ldots, M\}$, where M is the number of minority

[1]The location of a cluster refers to the centroid of all points within the cluster.

pixels in the cluster. We first define the translation, or shift in space, $T_x(B)$ of a set $B = \{y_i : i = 1, 2, 3, \ldots\}$ by x, relative to the origin, as

$$T_x(B) = \{y_i + x : i = 1, 2, 3, \ldots\},$$

and define ϕ_{d_i} as the ith sample cluster for $i = 1, 2, \ldots, N_{\mathrm{c}}$. A sample ϕ_{G} of the green-noise halftone process is defined as

$$\phi_{\mathrm{G}} = \sum_{x_i \in \phi_{\mathrm{p}}} T_{x_i}(\phi_{d_i}) = \sum_{x_i \in \phi_{\mathrm{p}}} \{y_{ji} - x_i : j = 1, 2, \ldots, M_i\},$$

the sum of N_{c} translated clusters. The overall operation is to replace each point of the parent sample ϕ_{p} of process Φ_{p} with its own cluster ϕ_{d} of process Φ_{d}.

To derive a relationship between the total number of clusters, the size of the clusters, and the gray level of a binary dither pattern, we define I_g as the binary dither pattern resulting from halftoning a continuous-tone discrete-space monochrome image of constant gray level g and $I_g[n]$ as the binary pixel of I_g with pixel index n. From the definition of $\phi(B)$ as the total number of points of ϕ in B, $\phi_{\mathrm{G}}(I_g)$ is the scalar quantity representing the total number of minority pixels in I_g, and $\phi_{\mathrm{p}}(I_g)$ is the total number of clusters in I_g such that $\phi_{\mathrm{p}}(I_g) = N_{\mathrm{c}}$. The intensity $\mathcal{I}$, being the expected number of minority pixels per unit area, can now be written as

$$\mathcal{I} = \frac{\phi_{\mathrm{G}}(I_g)}{N(I_g)} = \begin{cases} g & \text{for } 0 \le g < \frac{1}{2}, \\ 1 - g & \text{for } \frac{1}{2} \le g < 1, \end{cases} \tag{13.2}$$

the ratio of the total number of minority pixels in I_g to $N(I_g)$, the total number of pixels composing I_g. Given Eq. (13.2), $\overline{M}$, the average number of minority pixels per cluster in I_g, is

$$\overline{M} = \frac{\phi_{\mathrm{G}}(I_g)}{\phi_{\mathrm{p}}(I_g)} = \frac{\mathcal{I} \cdot N(I_g)}{\phi_{\mathrm{p}}(I_g)}, \tag{13.3}$$

the total number of minority pixels in I_g divided by the total number of clusters in I_g.

Although obvious, Eq. (13.3) shows the very important relationship between the total number of clusters, the average size of the clusters, and the intensity for I_g. AM halftoning is the limiting case when $\phi_{\mathrm{p}}(I_g)$ is held constant for varying $\mathcal{I}$, while FM halftoning is the limiting case when $\overline{M}$ is held constant. In addition, Eq. (13.3) says that the total number of clusters per unit area is proportional to $\mathcal{I}/\overline{M}$. For isolated minority pixels (blue noise), the square of the average separation (λ_b) between minority pixels is inversely proportional to $\mathcal{I}$, the average number of minority pixels per unit area [Uli87].

In green noise, it is the minority pixel clusters that are distributed as homogeneously as possible, leading to an average separation (centroid to centroid), (λ_g) between clusters whose square is inversely proportional to the average number of minority pixel clusters per unit area, $\mathcal{I}/\overline{M}$. Using the fact that $\lim_{\overline{M} \to 1} \lambda_g = \lambda_b$, the proportionality constant can be determined such that λ_g is defined as

$$\lambda_g = \begin{cases} D/\sqrt{g/\overline{M}} & \text{for } 0 < g \le 1/2, \\ D/\sqrt{(1-g)/\overline{M}} & \text{for } 1/2 < g \le 1, \end{cases}$$

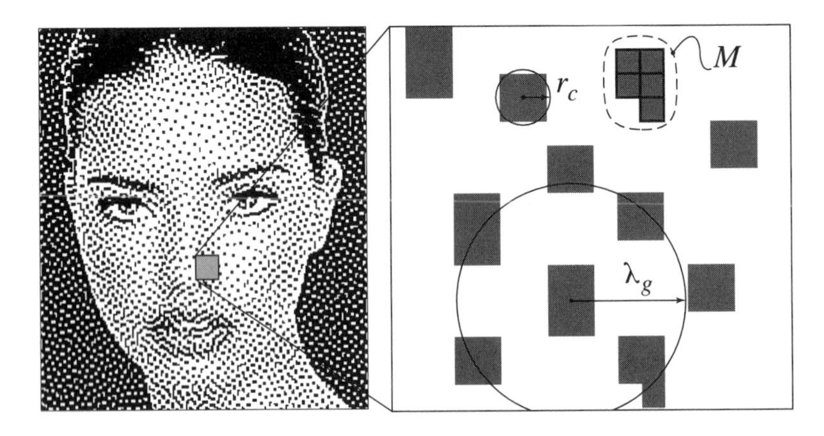

Figure 13.22: Minority pixel clusters of green-noise separated by an average distance λ_g.

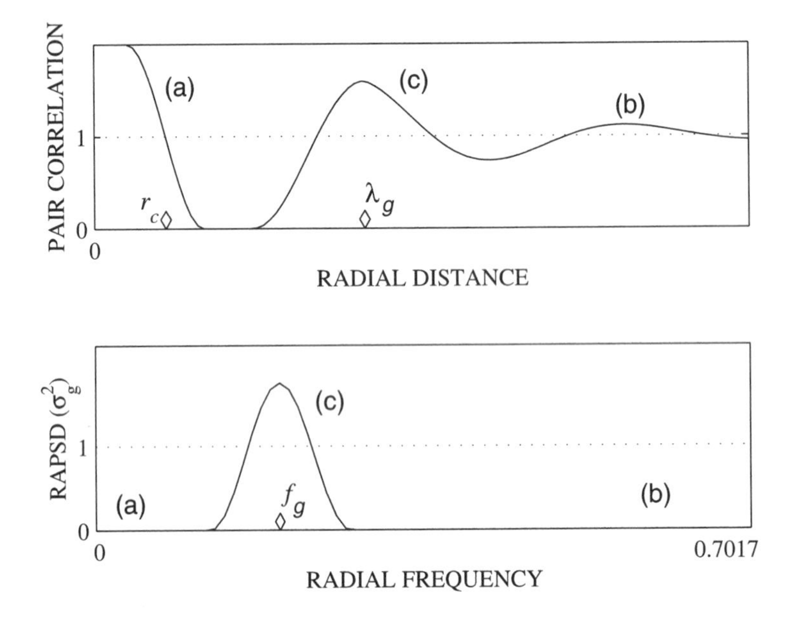

Figure 13.23: Top: pair correlation with principal wavelength λ_g and cluster radius r_c. Bottom: RAPSD with principal frequency f_g of the ideal green-noise pattern. (Reproduced with permission from [Lau98]. © 1998 IEEE.)

the *green-noise principal wavelength.* This implies that the binary dither pattern formed by replacing each cluster with a single minority pixel placed at the cluster's centroid is itself a blue-noise point process with intensity $1/\overline{M}$.

Spatial Statistics

If we assume a stationary and isotropic green-noise pattern, the pair correlation will have the form of Fig. 13.23 (top) given that:

1. Daughter pixels, on average, will fall within a circle of radius r_c centered around a parent point such that $\pi \cdot r_c^2 = \overline{M}$ (the area of the circle with radius r_c is equal to the average number of pixels forming a cluster).
2. Neighboring clusters are located at an average distance λ_g apart.
3. As r increases, the influence that clusters have on neighboring clusters decreases.

The result is a pair correlation that has (a) a nonzero component for $r < r_c$ due to clustering, (b) a decreasing influence as r increases, and (c) peaks at integer multiples of λ_g, indicating the average separation of pixel clusters. Note that the parameter r_c is also indicated by a diamond placed along the horizontal axis in Fig. 13.23 (top).

In the case of stationary and anisotropic green-noise patterns, the pair correlation will also be of the form of Fig. 13.23 (top), but because clusters are not radially symmetric, blurring occurs near the cluster radius r_c. In a similar fashion, because the separation between clusters will also vary with direction, blurring will occur at each peak in $\mathcal{R}(r)$ located at integer multiples of λ_g. This blurring will also occur as the result of variations in the cluster size–shape and in the variations in the separating distance between neighboring clusters. If this variation is too high, patterns tend to look "white" as the pair correlation begins to resemble that of a completely random, white-noise process, but too little variation leads to orderly, periodic textures that are also disturbing to the eye. So some variation is necessary for a visually pleasing, green-noise dither pattern. While radial symmetry is a desired characteristic of visually pleasing patterns [Uli88], it is also a desired characteristic for pattern robustness since elliptical clusters have a higher perimeter-to-area ratio of printed dots than radially symmetric clusters [Lau99].

Spectral Statistics

Assuming that the variation in cluster size is small for a given I_g, the placement of clusters apart leads to a strong spectral peak in $P(f_\rho)$ at $f_\rho = f_g$, the *green-noise principal frequency*, such that

$$f_g = \begin{cases} \sqrt{(g)/\overline{M}}/D & \text{for } 0 < g \le 1/2, \\ \sqrt{(1-g)/\overline{M}}/D & \text{for } 1/2 < g \le 1. \end{cases} \tag{13.4}$$

From Eq. (13.4), we make several intuitive observations: (1) as the average size of clusters increases, f_g approaches dc, and (2) as the size of clusters decreases, f_g approaches f_b. Figure 13.23 (bottom) illustrates the desired characteristics of $P(f_\rho)$ for ϕ_G, showing three distinct features: (a) little or no low frequency spectral components, (b) high-frequency spectral components that diminish with increased clustering, and (c) a spectral peak at f_g.

Like the pair correlation, the sharpness of the spectral peak in $P(f_\rho)$ at the green-noise principal frequency is affected by several factors. Consider first blue noise, in which the separation between minority pixels should have some variation. The wavelengths of this variation, in blue noise, should not be significantly longer than λ_b since this adds low frequency spectral components to the corresponding dither pattern I_g [Uli87], causing I_g to appear more white than blue. The same holds true for green noise, with large variations in cluster separation leading to a spectral peak at f_g that is not sharp but blurred since the variation in separation adds new spectral components to $P(f_\rho)$. This whitening effect on I_g is also created by increased variation in the size of clusters, with excessively large clusters leading to low frequency components and excessively small clusters leading to high. In summary, the sharpest spectral peak at f_g will be created when I_g is composed of round (isotropic) clusters whose variation in size is small and whose separation between nearest clusters is also isotropic with small variation.

13.4.2 Error Diffusion with Output-Dependent Feedback

Although error diffusion is a good generator of blue noise, the nature of green-noise to cluster pixels makes error diffusion inappropriate. As an alternative, error diffusion with output-dependent feedback, Fig. 13.24, has been proposed, in which a weighted sum of the previous output pixels is used to vary the threshold, making minority pixels more likely to occur in clusters. Furthermore, the amount of clustering is controlled through the scalar constant h, the hysteresis constant, with large values of h leading to large clusters and small values of h leading to small clusters.

Mathematically, the output pixel $y(n)$ is defined as

$$y(n) = \begin{cases} 1 & \text{for } \left[x(n) + x^e(n) + x^h(n)\right] \geq 0, \\ 0 & \text{otherwise,} \end{cases}$$

where $x^h(n)$ is the hysteresis or feedback term, defined as

$$x^h(n) = h \sum_{i=1}^{N} a_i \cdot y(n-i), \tag{13.5}$$

with $\sum_{i=0}^{N} a_i = 1$. The hysteresis constant h acts as a tuning parameter, with larger h leading to coarser output textures [Lev92] since h increases ($h > 0$) or decreases ($h < 0$) the likelihood of a minority pixel if the previous outputs were also minority pixels.

Equation (13.5) can also be written in vector notation as

$$x^h(n) = h\mathbf{a}^{\mathrm{T}}\mathbf{y}(n) \tag{13.6}$$

where $\mathbf{a} = [a_1, a_2, \ldots, a_N]^{\mathrm{T}}$ and $\mathbf{y}(n) = \left[y(n-1), y(n-2), \ldots, y(n-N)\right]^{\mathrm{T}}$. The calculation of the parameters $\mathbf{y}^e(n)$ and $x^e(n)$ remains unchanged from the original error diffusion algorithm. From Eqs. (13.5) and (13.6), calculation of the binary output pixel $y(n)$ can be summarized as

$$y(n) = \begin{cases} 1 & \text{for } [x(n) - \mathbf{b}^{\mathrm{T}}\mathbf{y}^e(n) + h\mathbf{a}^{\mathrm{T}}\mathbf{y}(n)] \geq 0, \\ 0 & \text{otherwise.} \end{cases}$$

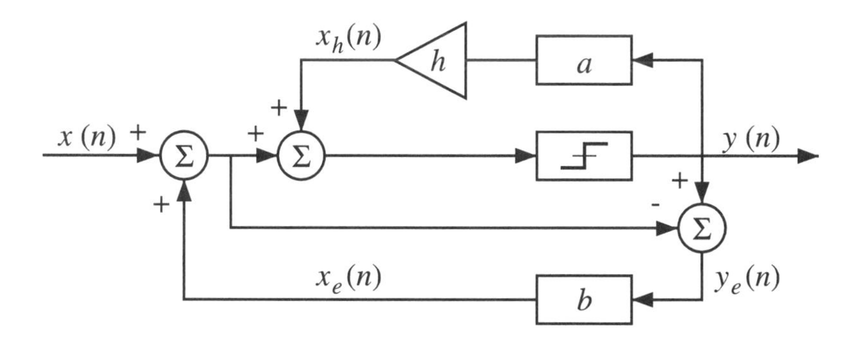

Figure 13.24: Error diffusion with the output-dependent feedback algorithm.

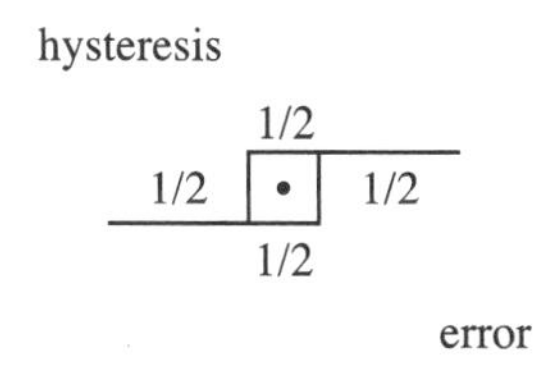

Figure 13.25: An arrangement of two hysteresis and two error diffusion coefficients for error diffusion with output-dependent feedback.

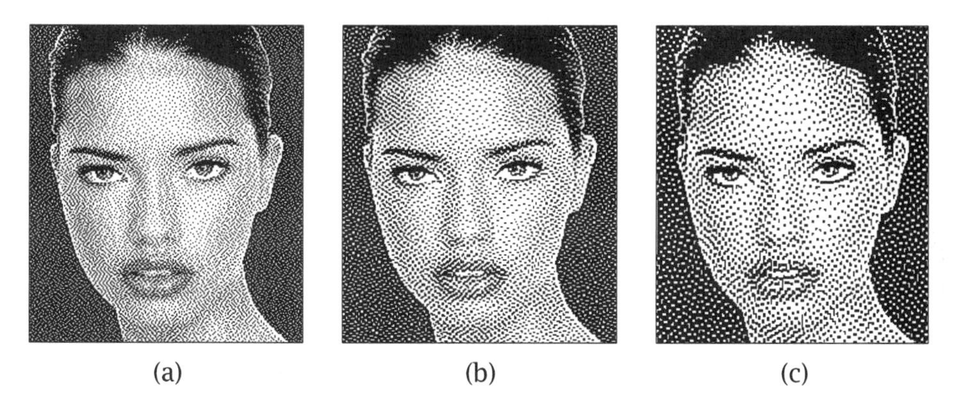

Figure 13.26: Grayscale images halftoned using error diffusion with output-dependent feedback with hysteresis parameter h equal to (a) 0.5, (b) 1.0, and (c) 1.5.

Using the arrangement of Fig. 13.25 for two hysteresis and two error filter coefficients, Fig. 13.26 shows the resulting halftoned images using hysteresis values $h = 0.5$, 1.0, and 1.5, respectively. The corresponding spatial and spectral characteristics are shown in Fig. 13.27. As with error diffusion, improved results can be achieved through alternative filter arrangements, perturbed filter weights, and perturbed thresholds.

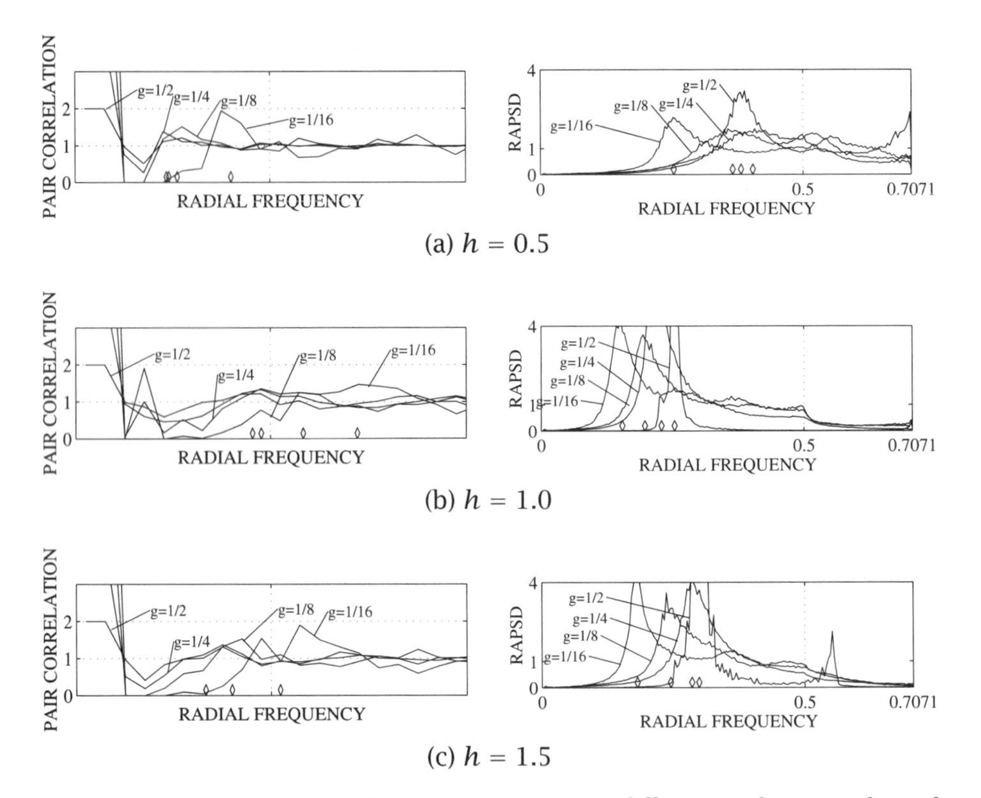

Figure 13.27: Spatial and spectral statistics using error diffusion with output-dependent feedback for gray levels 1/2, 1/4, 1/8, and 1/16.

13.4.3 Green-Noise Dither Arrays

While a direct binary search algorithm does not currently exist for generating green noise, it is possible to generate green-noise dither arrays [Lau99], which, just like blue-noise dither arrays, offer the absolute minimum in computational complexity leading to halftone patterns that are faster and less expensive to produce. Figure 13.28 shows a green-noise dither array and its corresponding halftoned image.

13.5 Conclusions

Invented in 1880, halftoning experienced the most significant step in its evolution with the introduction of error diffusion and FM halftoning in 1971, but it was not until 1990 (six years after desktop publishing was introduced through the Apple Macintosh computer in 1984) that FM halftoning was a viable alternative to AM for low cost ink jet printers. Only four years later, FM halftoning led to the introduction of affordable four color ink jet printers and an explosion of economic growth in the printing industry, with 2.1 million ink jet printers expected to be sold in 2002, compared to 0.6 million color laser printers [Pep98].

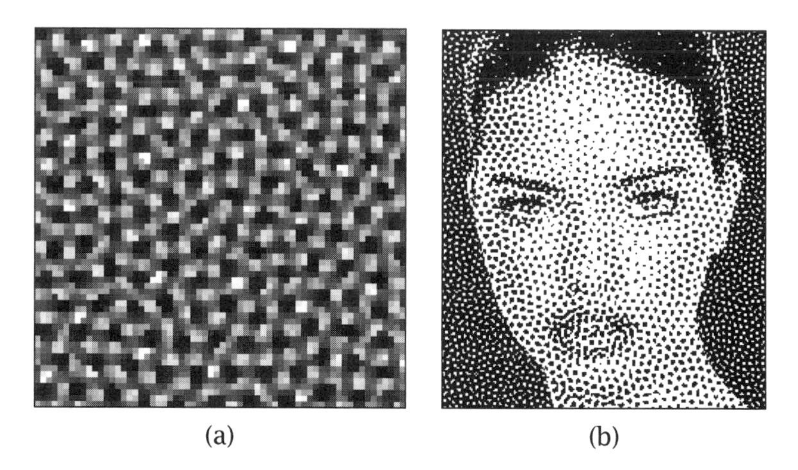

(a) (b)

Figure 13.28: (a) Green-noise dither array and (b) resulting halftoned image.

Today, we stand on the verge of a new technological revolution in printing as printers are achieving ultrahigh resolutions, but to take full advantage of the smaller dot sizes, a fundamental change must occur in the manner in which printed dots are arranged. Green noise represents that fundamental change, and as green noise is extended from theory to practice, the number of algorithms that generate it will dramatically increase since only a handful currently exist. Issues of particular importance include computational complexity and color reproduction.

References

[Ami94] I. Amidor, R. D. Hersch, and V. Ostromoukhov. Spectral analysis and minimization of Moiré patterns in color separation. *J. Electron. Imag.* **3**(7), 295–317 (July 1994).

[Ana92] M. Analoui and J. P. Allebach. Model based halftoning using direct binary search. In *Human Vision, Visual Processing, and Digital Display III*, Proc. SPIE Vol. 1666, pp. 96–108 (August 1992).

[Bar64a] M. S. Bartlett. The spectral analysis of a point process. *J. R. Statist. Soc. Ser. B* **25**(2), 264–280 (February 1964).

[Bar64b] M. S. Bartlett. The spectral analysis of two-dimensional point processes. *Biometrika* **51**(12), 299-311 (December 1964).

[Bla93] D. Blatner and S. Roth. *Real World Scanning and Halftones*. Addison-Wesley, Berkeley, CA (1993).

[Cam96] F. W. Campbell, J. J. Kulikowski, and J. Levinson. The effect of orientation on the visual resolution of gratings. *J. Physiol.* **187**(2), 427–436 (November 1996).

[Cou96] M. A. Coudray. Causes and corrections for dot gain on press. *Screen Printing: J. Technol. Manage.* **86**(3), 18–26 (August 1996).

[Des94] T. M. Destree (ed.). *The Lithographer's Manual.* Graphic Arts Technical Foundation, Pittsburgh, PA (1994).

[Dig83] P. J. Diggle. *Statistical Analysis of Spatial Point Patterns.* Academic Press, London (1983).

[Fin92] P. Fink. *PostScript Screening: Adobe Accurate Screens.* Adobe Press, Mountain View, CA (1992).

[Flo76] R. W. Floyd and L. Steinberg. An adaptive algorithm for spatial gray-scale. *Proc. SID* **17**(2), 75–79 (1976).

[Gou95] A. C. M. Goulard and L. Pages. Marked point process: Using correlation functions to explore a spatial data set. *Biomet. J.* **37**(7), 837–853 (July 1995).

[Jar76] J. F. Jarvis, C. N. Judice, and W. H. Ninke. A survey of techniques for the display of continuous-tone pictures on bilevel displays. *Comput. Graph. Image Process.* **5**(1), 13–40 (March 1976).

[Kno92] K. T. Knox. Error image in error diffusion. In *Image Processing Algorithms and Techniques III,* Proc. SPIE Vol. 1657, pp. 268–279 (May 1992).

[Lau98] D. L. Lau, G. R. Arce, and N. C. Gallagher. Green-noise digital halftoning. *Proc. IEEE* **86**(12), 2424–2444 (December 1998).

[Lau99] D. L. Lau, G. R. Arce, and N. C. Gallagher. Digital halftoning using green-noise masks. *J. Opt. Soc. Amer. A* **16**(7), 1575–1586 (July 1999).

[Lev92] R. Levien. Output dependent feedback in error diffusion halftoning. In *Proc. IS&T's Eighth Intl. Congress on Advances in Non-Impact Printing Technologies,* pp. 280–282 (1992).

[Mug96] M. A. Mugglestone and E. Renshaw. A practical guide to the spectral analysis of spatial point processes. *Comput. Statist. Data Anal.* **21**(1), 43 (January 1996).

[Pep98] M. Pepe. Inkjets moving into the enterprise. *Computer Reseller News* No. 810, p. 160 (October 1998).

[Rod94] M. Rodriguez. Graphic arts perspective of digital halftoning. In *Human Vision, Visual Processing, and Digital Display V.* Proc. SPIE Vol. 2179, pp. 144–149 (1994).

[Sto87] D. Stoyan, W. S. Kendall, and J. Mecke. *Stochastic Geometry and Its Applications.* Wiley, New York (1987).

[Sul91] J. Sullivan, L. Ray, and R. Miller. Design of minimum visual modulation halftone patterns. *IEEE Trans. Syst. Man Cybern.* **21**(1), 33–38 (January 1991).

[Uli87] R. A. Ulichney. *Digital Halftoning.* MIT Press, Cambridge, MA (1987).

[Uli88] R. A. Ulichney. Dithering with blue-noise. *Proc. IEEE* **76**, 56-79 (January 1988).

14

Intrinsic Dimensionality: Nonlinear Image Operators and Higher-Order Statistics

CHRISTOPH ZETZSCHE

Institut für Medizinische Psychologie
Ludwig-Maximilians-Universität München
München, Germany

GERHARD KRIEGER

Institut für Hochfrequenztechnik
Deutsches Zentrum für Luft- und Raumfahrt
Oberpfaffenhofen, Germany

14.1 Introduction

The classical approach in signal processing and communication engineering is based on linear systems theory and second-order statistics and has led to an impressive variety of tools for the processing of one- and multidimensional signals. Important applications in image processing are, for example, operators for the identification and localization of features (e.g., edge detectors), filters for noise removal, and block transforms for the exploitation of statistical redundancies (e.g., the cosine transform). In recent years, nonlinear operations, such as Volterra (or polynomial) filters, and higher-order statistics, such as cumulants and polyspectra, have received increasing interest [Boa95, Mat00, Nik93, Pin92, Pit90, Sch89, Sic92]. Meanwhile, there exist a number of impressive applications that demonstrate that these new methods indeed have a potential that exceeds that of the classical approach, but the specific gains they can yield in the various problem areas are still a subject of ongoing research. With respect to this potential for innovative applications, two directions can be distinguished. First of all, performance can be improved in those application areas that have already been tackled by the classical approaches in the past. Even more interesting, however, is the identification of such problems that, in principle, are unsolvable within the standard framework of linear systems theory.

For linear systems, the most basic operations can be readily described in the frequency domain since linear filtering can be regarded as the selective enhancement or blocking of certain frequency components of the input signal. Accordingly, basic types of linear filters are lowpass, highpass, and bandpass filters, and also bandstop filters. Multidimensional filters bring in a further elementary property, in being either isotropic or orientation specific. This raises two questions. First, do these elementary filtering properties also exist for nonlinear multidimensional operators? And second, are there additional basic properties that are *genuinely* nonlinear, i.e., cannot be provided by any linear system? The first question is readily answered since the basic filtering properties of linear multidimensional operators can all be generalized with some modifications to the nonlinear domain. With respect to the second question, the suggestion presented in this chapter is that the exploitation of the local *intrinsic dimensionality* of a signal can be an interesting candidate for a genuinely nonlinear property in multidimensional signal processing [Kri96, Zet90b].

Basically, the local intrinsic dimensionality determines how many of the degrees of freedom provided by a given domain are actually used by the local signal. The notion "intrinsic" is used in this context to enable this distinction between the dimension of the domain (e.g., two for images) and the actual variation of a specific local signal within this domain.[1] This type of signal property exists only in trivial form for one-dimensional signal processing and can only become nontrivial for two- or higher-dimensional signal domains. In the one-dimensional domain just one elementary distinction is possible: the signal can be constant or it can exhibit some variation. This distinction can be made within the standard linear framework by simply differentiating (highpass filtering) the signal. In the two-dimensional domain a new and nontrivial distinction comes into play: the distinction of intrinsically one-dimensional signals (i1D signals, e.g., straight edges, straight lines, and sinusoids) and intrinsically two-dimensional signals (i2D signals, e.g., corners, line ends, junctions, and spots). As we will see, this distinction cannot be obtained from a linear operator because of the logical OR gating that linear shift-invariant systems perform with respect to their intrinsically one-dimensional eigenfunctions. Accordingly, the detection of intrinsically two-dimensional signals requires the potential of an AND gating of complex exponentials [Zet90b], and this can be seen as just the basic operation being provided by polynomial operators.

[1]A formal definition of local intrinsic dimensionality is given in Sec. 14.2.1. A further reason for using the term "intrinsic" is the close relation to intrinsic surface properties in differential geometry (cf. Sec. 14.2.4).

The topics discussed in this chapter are relevant not only in the context of signal processing but are also of interest with respect to other scientific disciplines. In particular, there is a close relation to the information reduction strategies employed by biological vision systems [Zet93] and to the function of the "hypercomplex" or "end-stopped" neurons in the mammalian visual cortex [Hub65, Orb84]. There is also an interesting connection to the branch of mathematics being devoted to the analysis of the properties of surfaces and manifolds, which is differential geometry. And there exists a close relation to the non-Gaussian statistical properties of images, as revealed by higher-order statistics such as polyspectra.

The analysis and modeling of nonlinear visual neurons is relevant not only for a better understanding of biological vision systems, it is also of interest for engineering applications. Biological systems are the result of an extended evolutionary optimization process with respect to the specific structural properties of the terrestrial environment and hence are superior to the currently available engineering solutions in image processing, computer vision, and picture coding (at least where these have to deal with real world scenes). Biological systems hence can provide valuable hints for innovative research directions in these areas of engineering. A well known example for such a successful cross-fertilization is the wavelet approach (e.g., [Mac81, Mal89]). Of particular interest are also the close interdisciplinary relations with respect to statistical signal processing (e.g., [Zet93, Zet99a]). An interesting aspect in this context are the more recently realized connections to the issues of blind source separation and "independent component analysis" (e.g., [Bel97]). Our investigations hence may be regarded as an attempt to exploit this interdisciplinary potential and, in particular, to extend it beyond the limits of linear signal processing.

The chapter is organized as follows. In Sec. 14.2 we provide some general arguments regarding the relevance of the concept of intrinsic dimensionality. After a short formal definition we point out how intrinsic dimensionality is related to the statistical properties of natural images and to basic phenomena in visual neurophysiology and psychophysics. Related approaches in signal processing are discussed at the end of Sec. 14.2, with particular emphasis on the interpretation of images as surfaces. Section 14.3 contains the derivation of the conditions for the i2D selectivity of polynomial operators. We first show that i2D operators cannot be linear and consider then the logical gating function of polynomial operators with respect to the processing of intrinsic dimensionality. The derivation of the necessary and sufficient conditions for quadratic i2D operators are provided in the last part of Sec. 14.3. In Sec. 14.4 we present examples of different types of i2D operators. In particular, we present a frequency-domain design approach that is illustrated by the influence of stopband and passband modifications on the "curvature tuning." Furthermore, we provide isotropic and orientation-selective versions of i2D operators. In Sec. 14.5 we discuss the relationship between intrinsic dimensionality and higher-order statistics and provide some hints on possible further application areas. The chapter ends with a short discussion and a summary.

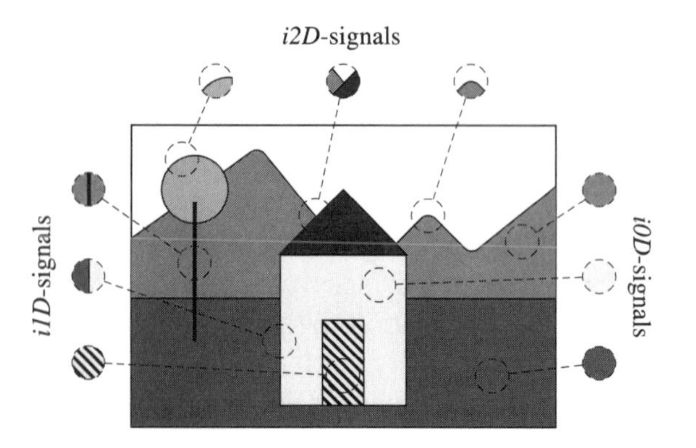

Figure 14.1: Typical examples for the three basic classes of intrinsic dimensionality in natural images. Each local region of the image can be classified according to its intrinsic dimensionality. Two aspects become intuitively apparent from this example: (1) There will be a gradual decrease of the probability of occurrence from i0D to i2D signals (i2D signals are "rare events"). (2) There is a hierarchy of intrinsic dimensionality in terms of redundancy, since knowledge of the i2D signals can be used to predict the more redundant i1D and i0D signals.

14.2 Transdisciplinary Relevance of Intrinsic Dimensionality

14.2.1 Definition

We are mainly interested in local signal properties and, accordingly, in operators with limited spatial support, but for convenience, we use signals with infinite spatial extent in all our definitions and derivations. For any operator with spatially limited support it is simply irrelevant whether the input signal extends outside the support region.

Intrinsic dimensionality is directly related to the degrees of freedom being used by an actual image signal [Zet90b]. A signal $u(x,y)$ can be classified as a member of one of the classes i0D, i1D, or i2D according to the following rule:

$$u(x,y) \in \begin{cases} \{\text{i0D}\} & \text{for } u(x,y) = \text{const}, \\ \{\text{i1D}\} & \text{for } u(x,y) = \text{func}(ax+by), (a \neq 0) \vee (b \neq 0), \\ \{\text{i2D}\} & \text{otherwise.} \end{cases} \quad (14.1)$$

The first class is given by all signals that are constant, that is, do not depend on (x,y). The second class comprises all signals that can be written as a function of one variable in an appropriately rotated coordinate system, where the constants a and b are the coefficients of the corresponding affine transform. Signals that cannot be classified as either i0D or i1D belong to the third class.

Typical local image configurations with different types of intrinsic dimensionality are illustrated in Fig. 14.1, which shows a simplified view of a natural image. Some basic features, such as the substantially differing probabilities or the different degrees of predictability associated with the three types of local signals, are already evident from this simple example. A detailed statistical evaluation is given in the next section. Most important for the current context is the fact that i2D signals are the most significant ones.

Since we make extensive use of the spectral representation of signals and operators in subsequent derivations, we shall also give the corresponding Fourier-domain classification rules for the intrinsic dimensionality. i0D signals may be represented as $U(f_x, f_y) = \mathrm{K} \cdot \delta(f_x, f_y)$ where K is a constant and δ denotes the Dirac delta function. To derive a spectral representation of i1D signals, we start from the subset of i1D signals that can be written as

$$u(x, y) = \psi(x), \qquad \forall x, y \in \mathbf{R}, \tag{14.2}$$

where $\psi(x)$ is an arbitrary function of one variable x. $\psi(x)$ is independent of y and therefore constant for all sections parallel to the y-axis. Equation (14.2) can be expressed in the spatial frequency domain as

$$U(f_x, f_y) = \int_{-\infty}^{\infty} \int_{-\infty}^{\infty} \psi(x) \mathrm{e}^{-j2\pi(f_x x + f_y y)} \, \mathrm{d}x \, \mathrm{d}y = \Psi(f_x) \cdot \delta(f_y), \tag{14.3}$$

where $\Psi(f_x)$ is the 1-D Fourier transform of $\psi(x)$. Hence, the i1D signal of Eq. (14.2) corresponds to a modulated Dirac line in the frequency domain.

Applying an affine coordinate transform to $u(x, y)$ in Eq. (14.2) (which includes rotation as a special case) and using the Fourier correspondence [Bam89]

$$u(\mathbf{Ax}) \Longleftrightarrow \frac{1}{\det(\mathbf{A})} \cdot U\left((\mathbf{A}^{-1})^t \mathbf{f}\right), \qquad \det(\mathbf{A}) \neq 0, \tag{14.4}$$

where

$$\mathbf{A} = \begin{pmatrix} a_{11} & a_{12} \\ a_{21} & a_{22} \end{pmatrix}, \qquad \mathbf{x} = \begin{pmatrix} x \\ y \end{pmatrix}, \qquad \mathbf{f} = \begin{pmatrix} f_x \\ f_y \end{pmatrix}, \tag{14.5}$$

we obtain

$$\psi(a_{11}x + a_{12}y) \Longleftrightarrow \frac{1}{|\det(\mathbf{A})|} \cdot \Psi\left(\frac{a_{22}f_x - a_{21}f_y}{\det(\mathbf{A})}\right) \cdot \delta\left(\frac{a_{11}f_y - a_{12}f_x}{\det(\mathbf{A})}\right). \tag{14.6}$$

Hence, every signal which can be written as $u(x, y) = \psi(a_{11}x + a_{12}y)$, i.e., which is constant along the direction perpendicular to the vector $\binom{a_{11}}{a_{12}}$, has as its Fourier transform a modulated Dirac delta line through the origin. This permits us to give a definition of intrinsic dimensionality in the frequency domain:

$$U(f_x, f_y) \in \begin{cases} \{\text{i0D}\} & \text{for } U(f_x, f_y) = \mathrm{K} \cdot \delta(f_x, f_y), \\ \{\text{i1D}\} & \text{for } U(f_x, f_y) = U(f_x, f_y) \cdot \delta(af_x + bf_y), (a \neq 0 \vee b \neq 0), \\ \{\text{i2D}\} & \text{otherwise.} \end{cases} \tag{14.7}$$

In the following sections we provide some general arguments supporting the relevance of the concept of intrinsic dimensionality for communication engineering and biological vision systems. We further provide a short discussion of earlier approaches related to the processing of i2D signals.

14.2.2 Intrinsic Dimensionality and Image Statistics

There exists a close relationship between the concept of intrinsic dimensionality and the statistical properties of natural images [Bar93, Zet90c, Zet93]. This is already evident from Fig. 14.1, with its simplified view of a typical natural scene. Two basic aspects of the statistics can be deduced from this figure. First, the different local image features show a clear order in their probability of occurrence, and this order is determined by their intrinsic dimensionality. i0D signals (regions with low variation of luminance) have the highest probability. They correspond to the homogeneous "interior" of objects. The next common type of image features are i1D signals. They correspond to the object borders, which, from a statistical point of view, are more or less straight in most cases. i2D-signals are the rarest events in natural images, corresponding to the less frequently occurring abrupt changes in contour orientation or to occlusions, that is, to local configurations in which intrinsically 1-D contours cover each other.

Although Fig. 14.1 shows only a highly simplified version of the real world, it correctly reflects the basic statistical properties of actual natural scenes. As in this figure, each local region of a natural image can be classified into one of the three types of intrinsic dimensionality, though the class borders may be more fuzzy. This hierarchy of intrinsic dimensionality has been systematically investigated by a statistical analysis of a set of natural images [Weg90a, Zet90a, Zet93, Zet99a], and the related statistical properties of image sequences have been demonstrated [Weg92]. The results of this statistical analysis are illustrated in Fig. 14.2, and they fully confirm the above considerations, derived from the idealistic model image, that the three classes of intrinsic dimensionality occur with decaying probability. From the viewpoint of information theory this implies that the amount of information carried by a local image feature increases systematically with the feature's rank of intrinsic dimensionality. It is important to note that this type of statistical hierarchy is a specific property of natural images. It will not occur with a Gaussian-noise image, for example.

Another point that is intuitively apparent from Fig. 14.1 is the high predictive potential of i2D features [Zet90a, Zet90c]. Knowing only the intrinsically two-dimensional regions of an image does restrict the set of possible images that might be compatible with this information. For simple configurations, for example, polygons, the i2D signals obviously do provide the complete information about the configuration, but for more realistic images, it is an open question whether the predictive power of the i2D information is sufficient to specify the image, since the analysis of this problem corresponds to the invertability of nonlinear differential equations. However, using relaxation methods, it has been shown that a substantial part of the image information can be recovered from mere knowledge of its i2D regions [Bar93]. As a counterpart to their predictive power, i2D signals can be seen as the most nonpredictable information in images [Zet90c, Zet93]. A related interpretation of neural end-stopping in terms of such nonpredictable residual information has recently been suggested [Rao99].

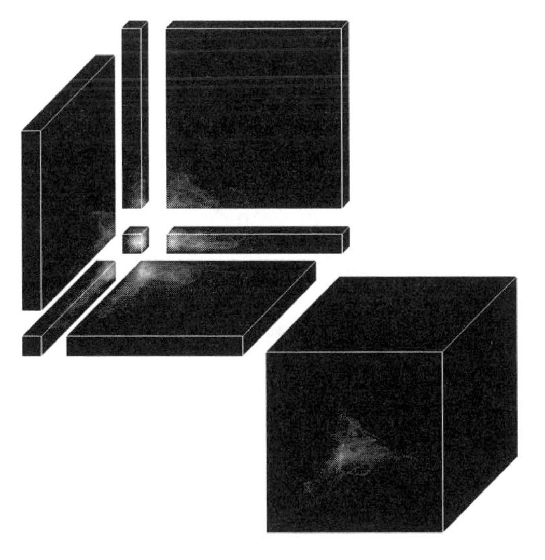

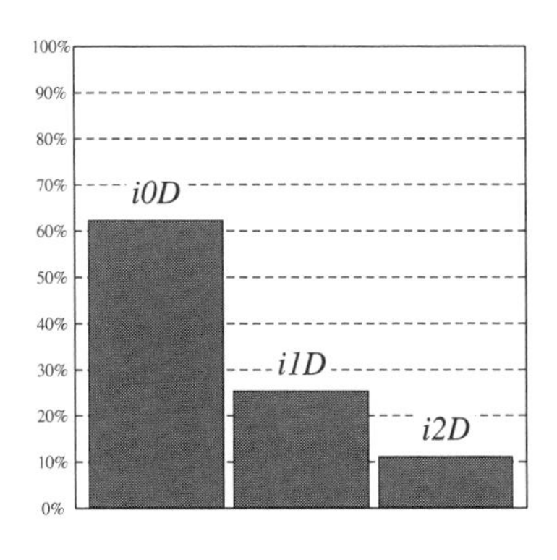

Figure 14.2: Statistical distribution of signals with different ranks of intrinsic dimensionality. The left image shows a 3-D visualization of the multivariate probability density function of natural images [Weg90b]. In this visualization i0D signals are located at the origin of the (hyper)cube, i1D signals mainly on the axes, and i2D signals mainly in the residual space (for details, see [Weg90b, Weg96]). A quantitative statistical evaluation confirms the visual impression: i0D signals are most common, and i2D signals are quite rare events in natural images (right figure).

A detailed analysis of the importance of intrinsic dimensionality with respect to image statistics and image coding and a discussion of the specific potentials for image compression provided by linear and nonlinear coding methods can be found elsewhere [Weg96, Zet93, Zet99a]. An analysis of the relation between the higher-order statistics of natural images, as measured by polyspectra, and the processing properties of i2D-selective nonlinear operators is provided in Sec. 14.5.

14.2.3 Neurophysiology and Psychology of Vision

The relevance of the concept of intrinsic dimensionality with respect to the processing of images is also supported by a variety of results from neurophysiological and psychophysical research on biological vision systems. Recent investigations have revealed a close relationship between biological and technical image processing in the form of an intimate connection between the wavelet transform and the processing characteristics of simple cells in the visual cortex of higher vertebrates [Dau85, Mal89, Mar80]. This relationship is further supported by the analysis of biological vision in terms of signal statistics and information theory (e.g., [Ati90, Fie87, Fie94, Rud94, Zet87, Zet93, Zet99a]). In particular, wavelets with a close resemblance to biological receptive fields have emerged in several variants of independent component analysis (ICA) of natural images [Bel97, Ols96, van98]. However, the operations performed by the wavelet transform and by cortical simple cells are basically linear, and the signals involved are mainly i1D (as will become clear in Sec. 14.5). Therefore, these simple cells are limited in their processing potential, in particular with respect to the exploitation of higher-order statistical redundancies [Zet93]. But a substantial part of the neural machinery of the mammalian visual cortex cannot be described in a satisfying manner by linear filter models, and most of these nonlinear operations can be interpreted as further steps toward a more efficient exploitation of the statistical redundancies of the natural environment [Weg90b, Zet99a, Zet99b]. The "cortical gain control" [Car94], for example, is a ubiquitous nonlinearity, influencing even the apparently linear simple cells and making them nonlinear devices, which separates the encoding of contrast from their phase tuning properties. Another example is the phase invariance of complex cells, which makes them "energy detectors" (see Sec. 14.4.4).

These nonlinearities, however, can still be seen as "moderate" modifications of the underlying linear wavelet transform, whereas there exists also a peculiar sort of neuron that cannot be described as some modified version of a linear system but can only be adequately treated within a genuinely nonlinear framework. Such highly nonlinear neurons are found not only at the cortical level (e.g., "hypercomplex" [Hub65], "end-stopped" [Orb84], or "dot-responsive" [Sai88] cells) but also in the retina of lower vertebrates (e.g., the "bug-detector" in frogs [Let59]). A complete characterization of the processing properties of this class of cells does not yet exist. End-stopped neurons are usually characterized by their tuning function with respect to the length of a short line segment, which decreases with increasing line length (see Sec. 14.4.6, Fig. 14.15). Many end-stopped cells are also selective for the orientation, but some neurons in cortical area 19 show a clear spatially isotropic behavior while exhibiting the same vanishing response to extended straight lines. Bug detectors in the frog retina are also isotropic but are even more restricted in their selectivity, reacting preferentially only to isolated dark spots on a brighter background. But aside from their differences they share all one essential property: none of them reacts to straight lines, straight edges, sinusoidal gratings, or any other straight pattern. Instead, they respond to spots, corners, line ends, curved edges, and similar patterns. This means that they are natural examples of operators that block all i0D as well as i1D signals and only respond to certain i2D signals [Zet90b].

Although these cells have been known for more than three decades, the first attempts to model them are more recent [Dob87, Koe87, Koe88, Zet87, Zet88, Zet90b], which is presumably due to the fact that appropriate models have to be highly nonlinear (see Sec. 14.3.1 and [Zet90b]). Some of the early models were developed by heuristic strategies and lack the formal basis required to be safe against any unpredictable reaction to untested signal configurations [Dob87, Hei92]; others are subject to the inherent limitations of the derivative operators used in differential geometry [Koe88]. To avoid such problems, we developed a formal framework for the construction of nonlinear i2D-selective $\Sigma\Pi$ networks [Zet90b, Zet90c], which is extended here to the more general Volterra-Wiener theory [Kri96].

The importance of i2D signals in terms of statistical redundancy is also reflected in psychophysical and psychological results. Attneave [Att54] demon-

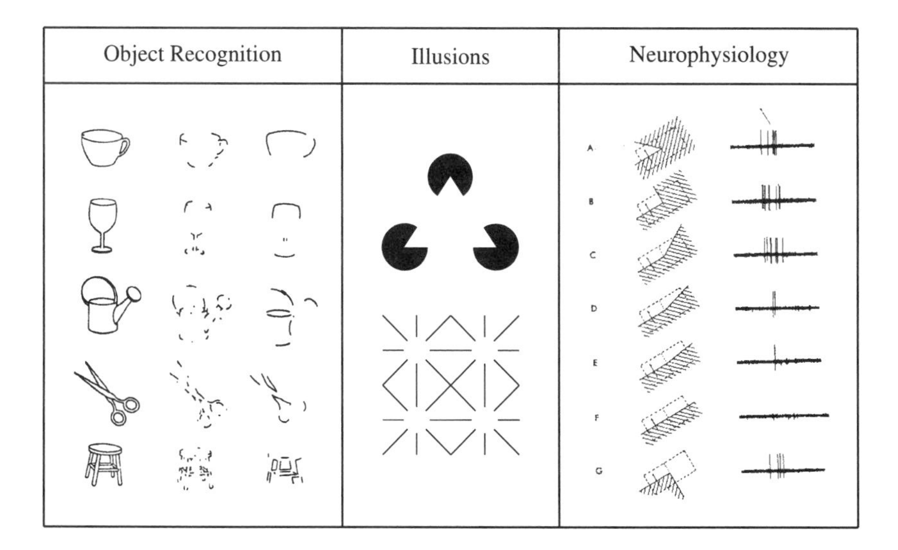

Figure 14.3: Examples from various scientific contexts that provide empirical evidence for the special importance of i2D signals in vision. Left: Psychological experiments demonstrate that such signals are most crucial for the recognition of objects ("recognition by components", reproduced with permission from [Bie85], © 1985 Academic Press; see also [Att54]). Middle: Many visual illusions that evoke the sensation of physically nonexistent i0D regions and i1D signals are triggered by the physical presence of local signals with the least redundancy, that is, i2D signals. Right: A substantial number of cells in the mammalian visual cortex are selectively sensitive to the presence of i2D signals (e.g., the "hypercomplex cell" reproduced with permission from [Hub65], © 1965 The American Physiological Society).

strated this importance with the evaluation of subjective ratings of the relevant points of line drawings and with his famous "reconstruction" of the drawing of a cat from i2D signals. Further support comes from the "recognition by components" theory [Bie85] and from the observation that i2D signals are involved in the control of saccadic eye movements [Kri00a, Mac67, Zet98].

Some examples for the relevance of i2D signals in biological vision are shown in Fig. 14.3. These empirical findings from vision research, together with the considerations on the statistics of natural images, provide strong evidence for the relevance of intrinsic dimensionality with respect to the efficient processing of pictorial information.

14.2.4 Differential Geometry

The need for the development of operators capable of detecting certain i2D signals was recognized early in image processing [Bea78], and there exist, in particular, a number of suggestions for the construction of "corner detectors" and for methods to evaluate "geometrical" properties of images (e.g., [Bea78, Bes86, Kit82, Koe88, Wat85]). Although many approaches have employed heuristic design strategies,

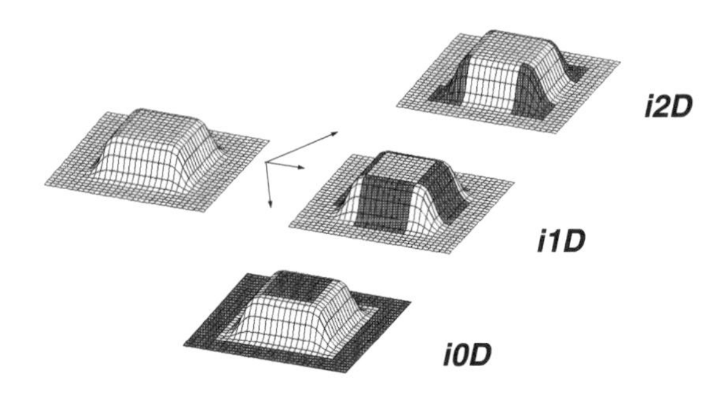

Figure 14.4: Surface types corresponding to different ranks of intrinsic dimensionality (from [Zet93]). To apply concepts from differential geometry to image processing, we have to interpret the 2-D image-intensity function as a surface in 3-D space. The left side shows the surface representation of a band-limited bright square on dark background. The decomposition into different surface types is illustrated on the right side (from bottom to top): i0D signals correspond to planar surface regions, i1D signals correspond to parabolic regions, and i2D signals correspond to elliptic–hyperbolic regions.

there is one general concept which is used, often implicitly, by most of them. This common idea is the reinterpretation of the two-dimensional image signal as a surface in three-dimensional space. Once this is done, conceptualizations of surface properties such as "peak," "pit," "ridge," etc. may be applied [Har83, Pat75]. Even more important is the access to an elaborated formal theory for the determination of such properties, namely, differential geometry (e.g., [Doc76]). The majority of the existing approaches for corner detectors and for other types of procedures that evaluate topographic properties of images [Har83] can be interpreted in terms of differential geometry and surface predicates.

The interpretation of the image-intensity signal as a surface is illustrated in Fig. 14.4. Planar surface regions correspond to i0D signals, parabolic surface regions to i1D signals, and elliptic–hyperbolic surface regions to i2D signals (see [Bar93, Zet90b, Zet90c, Zet93] for a detailed discussion).

The discrimination between planar and parabolic surfaces (or between i0D and i1D signals) is easy within a linear signal processing framework because it can basically be obtained by some sort of differentiation or high-pass filtering or by an edge-detector. More complicated is the separation of elliptic–hyperbolic regions from planar and parabolic regions. This can be obtained by evaluation of the "Gaussian curvature" of the surface regions. Hence, the geometrical concept of Gaussian curvature is closely related to the proposed concept of *intrinsic dimensionality* [Bar93, Zet90b]. In short, the image-intensity function $l(x,y)$ can be regarded as a Monge patch, defined by

$$\mathbf{S}(x,y) = \{x, y, l(x,y)\} \ . \tag{14.8}$$

The Gaussian curvature is defined as

$$K = \frac{\det(\mathbf{B})}{\det(\mathbf{G})}, \tag{14.9}$$

where the first and second fundamental form coefficients for Monge patches [Eq. (14.8)] are

$$\mathbf{G} = \begin{bmatrix} 1 + l_x^2 & l_x l_y \\ l_x l_y & 1 + l_y^2 \end{bmatrix} \tag{14.10}$$

and

$$\mathbf{B} = \frac{1}{\sqrt{1 + l_x^2 + l_y^2}} \cdot \begin{bmatrix} l_{xx} & l_{xy} \\ l_{xy} & l_{yy} \end{bmatrix} = \frac{1}{\sqrt{1 + l_x^2 + l_y^2}} \cdot \mathbf{H} \ . \tag{14.11}$$

Here $\mathbf{H}$ is the Hessian and l_x, l_{xy}, etc. are partial derivatives in the direction(s) indicated by the subscript(s). The classification properties of Gaussian curvature are completely determined by

$$\det(\mathbf{H}) = l_{xx} \cdot l_{yy} - l_{xy}^2 \ . \tag{14.12}$$

This equation exhibits the basic operations required for the construction of an i2D operator. The decision on the intrinsic dimensionality is made on the basis of a nonlinear, AND-like combination $(l_{xx} \cdot l_{yy})$ of the results of two elementary directional measurements of signal variation (l_{xx} and l_{yy}). The elementary measurements can be obtained by linear operations. The measurements may not be independent, which becomes apparent if we are not operating in the principal axes. In these cases, the results of the AND combination may comprise a subset of false contributions from i1D signals, which need to be cancelled out by the squared mixed derivative of Eq. (14.12).

Although it is possible to use the above analysis of the basic ingredients of one special i2D operator as a starting point for the development of a more general class of operators, we do not make use of the concepts of differential geometry in the following derivations. It turns out, however, that the essential underlying principle, namely, the AND-like combination of orientation-specific measurements [Zet90b], is also the basic operation provided by i2D-specific Volterra operators and that Eq. (14.12) can be interpreted as a prominent example out of this class of quadratic Volterra filters.

14.3 i2D-Selective Nonlinear Operators

14.3.1 A Fundamental Limitation of Multidimensional Linear Systems

Linear filters can easily be tailored to match certain prescribed signal characteristics (matched filtering, e.g., [Ros82]), and they can also be used to selectively block certain signals (e.g., high or low frequencies). Hence, it may be suggestive to assume that it should be possible to construct a linear filter that is selectively tuned to i2D signals, that is, that blocks i0D and i1D signals. However, a short look at the basic "logical" structure of the operations that can be performed with linear filters shows that this task is just a prototypical example for the inherent limitations of linear systems [Zet90b].

Consider a linear, shift-invariant system with an input $u_1(x, y)$ and an output $u_2(x, y)$. Such a system is completely determined by its impulse response (point spread function) $h(x, y)$ or, equivalently, by the Fourier transform $H(f_x, f_y)$ of the impulse response, the frequency transfer function. If we denote the Fourier transform of the input signal $u_1(x, y)$ as $U_1(f_x, f_y)$, the output can be written as

$$u_2(x, y) = \int_{-\infty}^{\infty} \int_{-\infty}^{\infty} H(f_x, f_y) \cdot \left\{ U_1(f_x, f_y) \exp\left[j2\pi(f_x x + f_y y) \right] \right\} \mathrm{d}f_x \, \mathrm{d}f_y. \tag{14.13}$$

From Eq. (14.13) we can see that the basic logical operation is a weighted superposition, that is, an OR-like combination of complex exponentials, the eigenfunctions of linear systems, into which any input signal can be decomposed (Fig. 14.5, left). Note that, according to Eq. (14.1), these complex exponentials represent i1D signals if $f_x \neq 0$ or $f_y \neq 0$ and i0D signals if $f_x = 0$ and $f_y = 0$. With an OR gate it is impossible to block the gate inputs in a nontrivial manner. This becomes visible if we choose a single complex exponential as the input, that is,

$$u_1(x, y) = \exp[j2\pi(f_x x + f_y y)]. \tag{14.14}$$

In this case the resulting output is

$$u_2(x, y) = H(f_x, f_y) \exp[j2\pi(f_x x + f_y y)]. \tag{14.15}$$

Hence, if we are interested in building a system that does not respond to i0D and i1D signals, we must choose the trivial solution:

$$H(f_x, f_y) = 0, \qquad \forall f_x, f_y. \tag{14.16}$$

A system with this transfer function can be of only limited interest.

These simple considerations tell us that linear systems are definitely unsuitable to build an i2D operator. But knowing that the system has to be nonlinear means only that it must be one representative of the remainder of all possible systems. It thus becomes necessary to identify minimal structural requirements for i2D operators.

It has been argued that the critical shortcoming of linear operators is their restriction to OR gatings of their intrinsically one-dimensional eigenfunctions and that the essential prerequisite for the construction of i2D operators is a potential for AND combinations of the sinusoidal components of the signal [Zet90b]. This conjecture was initially derived from a structural analysis of the concept of Gaussian curvature and then developed toward a general framework for the design of i2D operators based on ΣΠ networks of linear subunits [Bar93, Zet90c, Zet93]. In the following the previous results are extended to the formalism of Volterra filters, which is one of the most developed theoretical frameworks for nonlinear operators (see Chapters 6 and 7 and [Mat00]).

We start from the definition of intrinsic dimensionality and derive the necessary and sufficient conditions for a quadratic Volterra operator to respond exclusively to i2D signals. Furthermore, we argue that this potential of Volterra operators can be regarded as a characteristic and elementary property of multidimensional polynomial filters and that it may be seen as one of the basic extensions beyond the potential of linear filtering techniques in multidimensional signal processing.

14.3.2 Volterra Expansion of Nonlinear Operators

The Volterra formalism is based on a power series expansion of nonlinear functionals which relates the N-dimensional input $u_1(\mathbf{x})$ of a nonlinear, shift-invariant system to its N-dimensional output $u_2(\mathbf{x})$ as follows (see Chapters 6 and 7 and [Pit90, Rug81, Sch89, Sic92]):

$$\begin{aligned} u_2(\mathbf{x}) &= h_0 + \int_{-\infty}^{\infty} h_1(\mathbf{x}_1) \cdot u_1(\mathbf{x}-\mathbf{x}_1) \cdot d\mathbf{x}_1 \\ &+ \int_{-\infty}^{\infty}\int_{-\infty}^{\infty} h_2(\mathbf{x}_1,\mathbf{x}_2) \cdot u_1(\mathbf{x}-\mathbf{x}_1) \cdot u_1(\mathbf{x}-\mathbf{x}_2) \cdot u_1(\mathbf{x}-\mathbf{x}_2) \cdot d\mathbf{x}_1\, d\mathbf{x}_2 \\ &+ \cdots . \end{aligned} \tag{14.17}$$

The functions $h_0, h_1(\mathbf{x}_1)$ and $h_2(\mathbf{x}_1,\mathbf{x}_2)$ are the Volterra kernels, the knowledge of which completely determines the behavior of the nonlinear operator. Note that for a linear system only $h_1(\mathbf{x}_1)$ would be different from zero. According to Eq. (14.17), Volterra systems can be seen as multilinear operators that act on ordered tuples of input values. For example, the second-order Volterra kernel $h_2(\mathbf{x}_1,\mathbf{x}_2)$ can be seen as a linear weighting function being applied to pairwise products of pixel values. This operation can be conveniently interpreted (e.g., [Sch89]) as a multidimensional linear convolution of $h_2(\mathbf{x}_1,\mathbf{x}_2)$ with the "expanded" input $\tilde{u}_1(\mathbf{x}_1,\mathbf{x}_2) := u_1(\mathbf{x}_1) \cdot u_1(\mathbf{x}_2)$ evaluated at the coordinates $\mathbf{x}_1 = \mathbf{x}_2 = \mathbf{x}$, that is,

$$u_2(\mathbf{x}) = \tilde{u}_2(\mathbf{x}_1,\mathbf{x}_2)|_{\mathbf{x}_1=\mathbf{x}_2=\mathbf{x}} , \tag{14.18}$$

where $\tilde{u}_2(\mathbf{x}_1,\mathbf{x}_2) := h_2(\mathbf{x}_1,\mathbf{x}_2) \overset{\mathbf{x}_1}{*} \overset{\mathbf{x}_2}{*} \tilde{u}_1(\mathbf{x}_1,\mathbf{x}_2)$ denotes the 2N-dimensional convolution result in the expanded space.

This transition from the expanded space to the original signal space, and its counterpart in the frequency domain, can be easily interpreted by use of the central slice theorem since Eq. (14.18) can be regarded as a "slice" through the expanded convolution. This becomes immediately apparent if we rewrite the expanded convolution in suitably scaled diagonal coordinates (see Fig. 14.13, center column) as $\hat{u}_2(\mathbf{x}, \mathbf{x}_0) = \tilde{u}_2(\mathbf{x}_1, \mathbf{x}_2)$ with

$$\begin{bmatrix} \mathbf{x} \\ \mathbf{x}_0 \end{bmatrix} = \mathbf{A} \begin{bmatrix} \mathbf{x}_1 \\ \mathbf{x}_2 \end{bmatrix}, \qquad \mathbf{A} = 1/2 \begin{pmatrix} \mathbf{1} & \mathbf{1} \\ -\mathbf{1} & \mathbf{1} \end{pmatrix}. \tag{14.19}$$

Here $\mathbf{1}$ denotes the identity matrix with size equal to the dimension of the input signal. In the new coordinates, the output can simply be written as

$$u_2(\mathbf{x}) = \hat{u}_2(\mathbf{x}, \mathbf{0}) \,, \tag{14.20}$$

and we can apply the Fourier projection slice theorem (the central slice theorem, e.g., [Bra86]) to obtain its Fourier transform:

$$U_2(\mathbf{f}) = \mathcal{F}\{\hat{u}_2(\mathbf{x}, \mathbf{0})\} = \int_{-\infty}^{\infty} \hat{U}_2(\mathbf{f}, \mathbf{f}_0)\, d\mathbf{f}_0. \tag{14.21}$$

Making use of the similarity theorem (see Eq. 14.4), this can be expressed as

$$U_2(\mathbf{f}) = \det(\mathbf{A}) \int_{-\infty}^{\infty} \tilde{U}_2\left(\frac{\mathbf{f} - \mathbf{f}_0}{2}, \frac{\mathbf{f} + \mathbf{f}_0}{2}\right) d\mathbf{f}_0 = \det(\mathbf{A}) \int_{-\infty}^{\infty} \tilde{U}_2(\mathbf{f}_1, \mathbf{f}_2)\, d\mathbf{f}_0, \tag{14.22}$$

where

$$\tilde{U}_2(\mathbf{f}_1, \mathbf{f}_2) := \mathcal{F}\{\tilde{u}_2(\mathbf{x}_1, \mathbf{x}_2)\} = H_2(\mathbf{f}_1, \mathbf{f}_2) \cdot U_1(\mathbf{f}_1) \cdot U_1(\mathbf{f}_2) \tag{14.23}$$

is the expanded output spectrum and $H_2(\mathbf{f}_1, \mathbf{f}_2)$ is the multidimensional Fourier transform of $h_2(\mathbf{x}_1, \mathbf{x}_2)$. The "contraction" can thus be interpreted as a projection of the expanded output spectrum $\tilde{U}_2(\mathbf{f}_1, \mathbf{f}_2)$ along the secondary diagonal axes $\mathbf{f}_0 = -\mathbf{f}_1 + \mathbf{f}_2$. We make use of such an interpretation in Sec. 14.4.4, where we also provide a graphical illustration. A similar interpretation of the contraction in terms of an integration along diagonal lines is also discussed in detail in Chapter 6. Note that by substitution of $\mathbf{f}_0 = -2\mathbf{f}_a + \mathbf{f}$ into Eq. (14.22) we can also derive the classical form of the output spectrum [Sch89]:

$$U_2(\mathbf{f}) = \int_{-\infty}^{\infty} \tilde{U}_2(\mathbf{f}_a, \mathbf{f} - \mathbf{f}_a)\, d\mathbf{f}_a. \tag{14.24}$$

14.3.3 The Logical Gating Structure of Nonlinear Operators

An important advantage for the understanding of the nonlinear operators in the context of intrinsic dimensionality is a consideration of their processing properties in the frequency domain [Kri96, Zet90c]. An analysis of certain characteristics of Volterra filters by use of a frequency-domain interperpretation has also proven successful in the design of the nonlinear operators discussed in Chapters 6 and 7. In the following, we concentrate on the processing of two-dimensional images with the spatial and spectral dimensions $\mathbf{x} = [x, y]$ and $\mathbf{f} = \left[f_x, f_y\right]$, respectively. The concepts proposed can be extended to arbitrary dimensions, however.

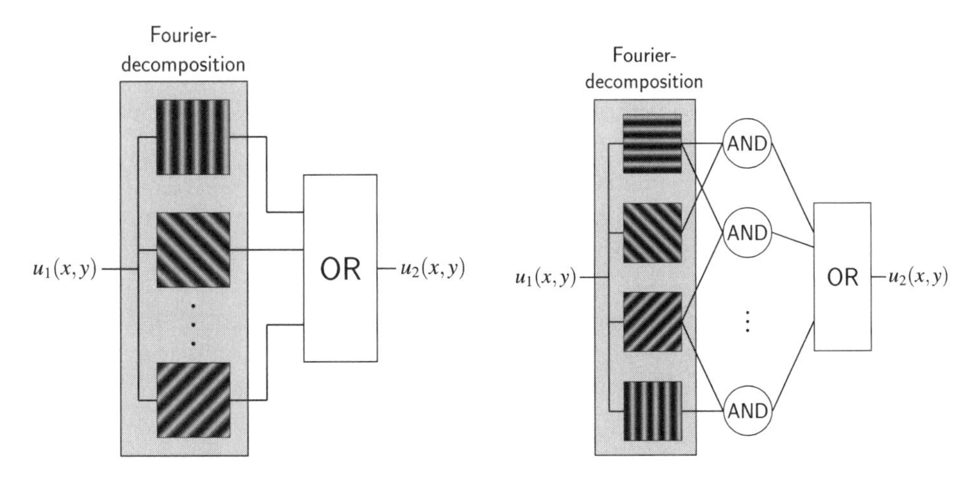

Figure 14.5: Schematic description of the logical gating function of linear operators (left) and quadratic Volterra operators (right) with regard to the processing of intrinsic dimensionality in two-dimensional images. As illustrated on the left, linear systems can be seen to perform an OR-gating of their intrinsically one-dimensional sinusoidal eigenfunctions. Therefore, it is impossible to find a linear operator that responds to i2D signals but blocks all i1D and i0D signals. As shown on the right,with nonlinear Volterra operators the input can be thought of as being decomposed into its frequency components in the first stage. The Volterra operator can then be interpreted as acting with AND gates on pairs of intrinsically one-dimensional exponentials. The results of these pairwise combinations are then fed into a final OR gate.

With respect to intrinsic dimensionality, a key feature of nonlinear Volterra operators is the provision of a specific type of logical gating with respect to the input frequency components. This becomes apparent if we rearrange the terms in the inverse Fourier transform of Eq. (14.24) to obtain [Ben90]

$$
\begin{aligned}
u_2(x,y) = & \int_{-\infty}^{\infty}\int_{-\infty}^{\infty}\int_{-\infty}^{\infty}\int_{-\infty}^{\infty} H_2(f_{x1}, f_{y1}, f_{x2}, f_{y2}) \cdot \\
& \left[\left(U_1(f_{x1}, f_{y1}) \mathrm{e}^{j2\pi(f_{x1}x + f_{y1}y)} \right) \cdot \left(U_1(f_{x2}, f_{y2}) \mathrm{e}^{j2\pi(f_{x2}x + f_{y2}y)} \right) \right] \\
& \mathrm{d}f_{x1}\,\mathrm{d}f_{y1}\,\mathrm{d}f_{x2}\,\mathrm{d}f_{y2}.
\end{aligned}
\tag{14.25}
$$

The structure of Eq. (14.25) suggests that the basic logical operation of a bilinear operator with respect to the intrinsic dimensionality of the input signal can be described as a two-stage process. In a first stage, all possible pair-wise AND combinations (complex products) of the intrinsically one-dimensional exponentials of the input signal are computed. In a second stage, these products are fed into an OR gate [that is, they are weighted with $H_2(f_{x1}, f_{y1}, f_{x2}, f_{y2})$ and superimposed]. These operations are schematically illustrated on the right side of Fig. 14.5.

Compared to the classical techniques of linear signal processing, the processing of products of frequencies (instead of the weighting of the frequency components per se) leads to a novel type of "filtering" with two main new features. First, the frequency products lead to the emergence of new frequencies at the output of the nonlinear system. The new high-frequency terms may be used, for example, to improve the sharpness of images (Chapter 7), while the newly generated low-frequency terms may be used for demodulation, as in the local energy mechanism of Sec. 14.4.4 or in Teager energy operators (Chapter 6). Even more important for the current context, however, is the second consequence of the potential for a processing of frequency combinations: the creation of a new type of passbands and stopbands. While for linear filters these bands are defined only by the frequencies per se (e.g., all frequencies beyond a certain bandlimit f_s, or all frequencies between $f_{\min}$ and $f_{\max}$) the bands for nonlinear filters can be defined in terms of *relations* between frequencies. Only specific frequency combinations may be allowed to influence the output, for example.

Note that this view requires a modified interpretation of the concept of filtering. Neither complex exponentials nor their products are eigenfunctions of the nonlinear system since new frequency terms are generated that are not present in the input. Thus, "filtering" in the modified sense controls only whether certain input signals can contribute to the output, without specifying these contributions as a weighted superposition. With nonlinear filters, the term "stopband" means only that the input signals falling into this band will not contribute to the output, and "passband" indicates only their potential to contribute.

This novel feature of nonlinear operators, the potential to act on relations between frequencies, raises the general question of which types of relations should be considered as elementary. The current context suggests that relations associated with the intrinsic dimensionality may have such an elementary status. Intrinsic dimensionality is no trivial type of relationship because an unspecific AND gating of frequency components is not sufficient for the construction of i2D operators. This can easily be seen from the fact that any linear filter followed by a pointwise squaring operation (a "Wiener system") is a quadratic Volterra filter, that is, it acts on AND combinations of pairs of frequencies, but is clearly no i2D operator. Evidently, it is necessary to impose specific restrictions on the frequency relations; that is, we have to define which AND combinations have to be blocked and which may be allowed to pass. Stated in filtering terms, we have to define passbands and stopbands for i2D operators.

14.3.4 i2D Selectivity: Necessary and Sufficient Conditions

Using the frequency-domain description of the quadratic part of the Volterra series we can derive simple rules for the interpretation of the kernels with respect to their selectivity for the intrinsic dimensionality of input signals. In the following, we give a short sketch of the derivation of necessary and sufficient conditions for quadratic i2D operators for two-dimensional images. A more detailed derivation can be found elsewhere [Kri96].

Regarding necessity, we consider an i1D signal consisting of two exponentials with equal orientation φ and radial frequencies f_a and f_b. In this case, the output of a quadratic Volterra system consists of three equally oriented frequency components at $2f_a$, $2f_b$, and $f_a + f_b$. The last term is due to the nonlinear interaction between the two frequency components comprising the input of the quadratic system. Its contribution to the output spectrum $U_2(f_x, f_y)$ is given by

$$\begin{aligned} U_2(f_x, f_y) &= 2H_2^{\text{sym}}(f_a \cos\varphi, f_a \sin\varphi, f_b \cos\varphi, f_b \sin\varphi) \\ &\cdot \delta(f_x - (f_a + f_b)\cos\varphi, f_y - (f_a + f_b)\sin\varphi), \end{aligned} \tag{14.26}$$

where H_2^{sym} is the symmetric frequency-domain Volterra kernel

$$\begin{aligned} H_2^{\text{sym}}(f_{x1}, f_{y1}, f_{x2}, f_{y2}) &= \frac{1}{2}\big[H_2(f_{x1}, f_{y1}, f_{x2}, f_{y2}) \\ &+ H_2(f_{x2}, f_{y2}, f_{x1}, f_{y1})\big]. \end{aligned} \tag{14.27}$$

Hence, if we require the output of a quadratic Volterra system to vanish for all i1D inputs consisting of two exponentials with equal orientation, the following condition is necessary:

$$\begin{aligned} & H_2^{\text{sym}}(f_{x1}, f_{y1}, f_{x2}, f_{y2}) = 0, \\ & \forall \begin{bmatrix} (f_{x1}, f_{y1}) \\ (f_{x2}, f_{y2}) \end{bmatrix} = \begin{bmatrix} (f_a \cos\varphi, f_a \sin\varphi) \\ (f_b \cos\varphi, f_b \sin\varphi) \end{bmatrix}. \end{aligned} \tag{14.28}$$

From the right-hand side of this equation it becomes apparent that the operator must block all combinations of frequency pairs with a joint orientation φ (see [Zet90b]). The right-hand side of Eq. (14.28), defining a parametric surface in a four-dimensional space, may also be rewritten by the implicit equation $f_{x1} f_{y2} = f_{x2} f_{y1}$. Hence, a necessary condition for a quadratic Volterra operator to be insensitive to i1D as well as i0D signals is

$$H_2^{\text{sym}}(f_{x1}, f_{y1}, f_{x2}, f_{y2}) = 0, \qquad \forall f_{x1} \cdot f_{y2} = f_{x2} \cdot f_{y1}. \tag{14.29}$$

In the following, we show that Eq. (14.29) is also sufficient for a quadratic system's insensitivity with respect to i1D and i0D signals. For i0D signals, it is obviously sufficient to require $H_2^{\text{sym}}(0,0,0,0) = 0$. To derive a condition for i1D signals, we note that according to Eq. (14.6) the Fourier transform of every i1D signal can be written as a modulated Dirac line passing through the origin. For such an i1D input the expanded output spectrum [$\tilde{U}_2(f_{x1}, f_{y1}, f_{x2}, f_{y2})$; cf. Eq. (14.23)] of a quadratic Volterra system is given by

$$\begin{aligned} \tilde{U}_2(f_{x1}, f_{y1}, f_{x2}, f_{y2}) &= \frac{1}{|\det(\mathbf{A})|^2} \cdot H_2^{\text{sym}}(f_{x1}, f_{y1}, f_{x2}, f_{y2}) \\ &\cdot \Xi\left(\frac{a_{22} f_{x1} - a_{21} f_{y1}}{\det(\mathbf{A})}\right) \cdot \delta\left(\frac{a_{11} f_{y1} - a_{12} f_{x1}}{\det(\mathbf{A})}\right) \\ &\cdot \Xi\left(\frac{a_{22} f_{x2} - a_{21} f_{y2}}{\det(\mathbf{A})}\right) \cdot \delta\left(\frac{a_{11} f_{y2} - a_{12} f_{x2}}{\det(\mathbf{A})}\right), \end{aligned} \tag{14.30}$$

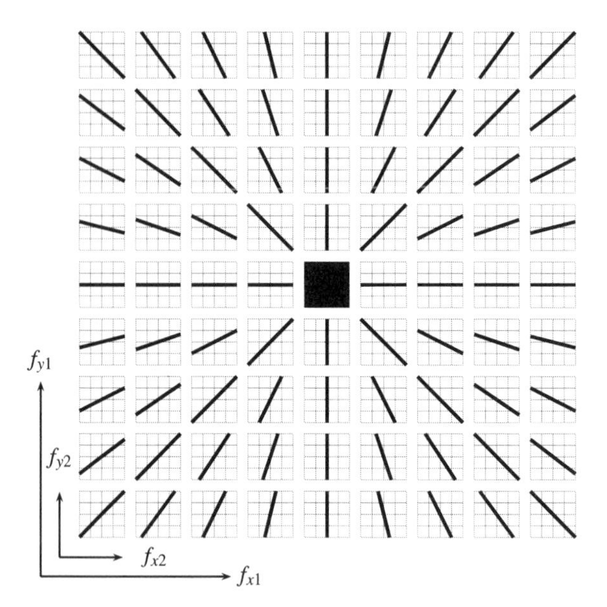

Figure 14.6: "Forbidden zones," or minimum stopbands, wherein the kernel has to vanish to yield i2D specificity. The figure shows schematic sections of the frequency-domain Volterra kernel $H_2(f_{x1}, f_{y1}, f_{x2}, f_{y2})$, or nonlinear "filter" function, of a quadratic system. Each section (subpicture) is indexed by the first frequency coordinates (f_{x1}, f_{y1}). Within each section, every position is indexed by (f_{x2}, f_{y2}). The black bars indicate the "forbidden zones" which must be subregions of the stopbands of any i2D operator. With respect to the four-dimensional space, the forbidden zones represent one connected three-dimensional subspace.

where Ξ is an arbitrary function of one argument and $\mathbf{A}$ is defined as in Eq. (14.5). Thus, if the kernel transform H_2^{sym} is zero for all $(f_{x1}, f_{y1}, f_{x2}, f_{y2})$ for which the arguments of the delta functions are zero, that is, for

$$\left.\begin{array}{l} a_{11} f_{y1} = a_{12} f_{x1} \\ a_{12} f_{x2} = a_{11} f_{y2} \end{array}\right\} \Longleftrightarrow f_{x1} \cdot f_{y2} = f_{x2} \cdot f_{y1}, \tag{14.31}$$

we can be sure that $\tilde{U}_2(f_{x1}, f_{y1}, f_{x2}, f_{y2}) = 0$, and therefore $U_2(f_x, f_y) = 0$. It follows that Eq. (14.29) is a necessary and sufficient condition for a genuine i2D operator.

The corresponding "forbidden zones," or minimum stopband regions, of a bilinear frequency-domain kernel are illustrated in Fig. 14.6. The forbidden zones define a connected three-dimensional subspace that appears as an oriented line within the two-dimensional sections of the four-dimensional kernel.[2] According to the right-hand side of Eq. (14.28), the orientation of these lines with respect to (f_{x2}, f_{y2}), that is, their orientation within each section, is determined by the position (f_{x1}, f_{y1}) of the respective section since $\varphi = \arctan(f_{y1}/f_{x1})$.

[2]Referring to the decomposition of the kernels into 2-D sections, we may use the plural form "stopbands" in spite of the connectedness of this region in the 4-D kernel space.

14.4 Frequency Design Methods for i2D Operators

In the following we provide several examples of quadratic Volterra operators belonging to the class of i2D operators. For these operators we employ frequency design methods since both the minimal requirements of i2D operators [Eq. (14.29)] and the specification of their additional signal processing properties can be obtained more conveniently in the frequency domain. Although we do not consider space-domain design strategies here we also describe a typical space-domain kernel of an i2D operator.

The design of i2D operators not only has to address the i2D selectivity but also involves other design dimensions. The response behavior of an i2D operator will typically be the result of a specific combination of these basic properties. The design properties considered in this section are

- i2D sensitivity ("curvature tuning")
- spatial isotropy
- orientation selectivity
- phase invariance ("local energy")

Further extensions of the design space, such as an increased order of the Volterra operators or the generalization from products to other elementary nonlinearities, are considered in Sec. 14.6.

14.4.1 A Basic Decomposition into Multiplicative Combinations of Linear Filters

Before going into the detailed design considerations, we shortly discuss an implementation issue. Although it is possible to realize a Volterra operator in a direct fashion by weighted products of the signal values, it is often more convenient to consider an alternative method, a linear filter based decomposition. The implementation structures derived by such a decomposition consist of a set of parallel filters, subsequent multiplication (or squaring) operations, and a final summation stage (Fig. 14.7). This type of structure is a quite useful basis for the design of a variety of i2D operators (see [Bar93, Zet90b, Zet90c, Zet93]). Furthermore, it provides a connection to neural network models with $\Sigma\Pi$ architectures [Dur89].

The simplest version, a single linear filter followed by a squaring operation (a "Wiener model"), is not suitable for the construction of i2D operators. As we have already shown, a linear filter itself cannot block intrinsically one-dimensional signals due to its logical OR gating. Obviously, this deficit cannot be corrected by a subsequent squaring. Thus the simplest possible realization of an i2D operator is the multiplicative combination of the outputs of two linear filters with different orientation preferences. However, with this architecture there is a strong constraint on the passbands of the two filters in that their orientation tuning functions are not allowed to overlap [Zet90b, Zet90c]. This is because all i1D signals can be thought of as being composed of frequencies with joint orientation. Such frequency pairs

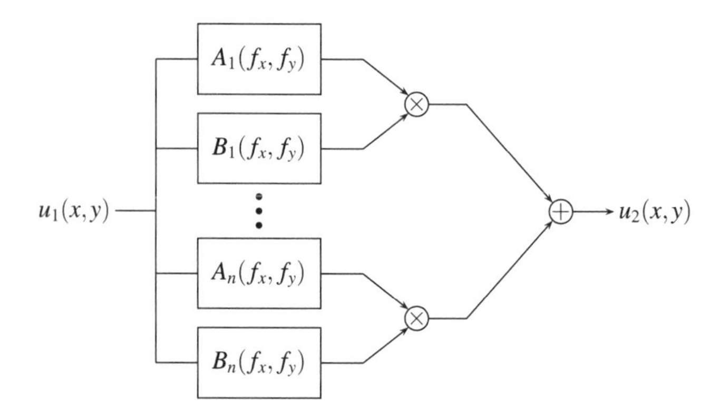

Figure 14.7: A general ΣΠ architecture for the realization of i2D operators. For $n = 2$, $A_1(f_x, f_y) = (j2\pi f_x)^2$, $B_1(f_x, f_y) = (j2\pi f_y)^2$, and $A_2(f_x, f_y) = B_2(f_x, f_y) = 4\pi^2 f_x f_y$ we obtain, as a special case, an operator closely related to Gaussian curvature.

are hence not allowed to pass an i2D system [cf. Eq. (14.29)]. While such a simple two-filter decomposition can provide the i2D selectivity, it restricts substantially the adaptability of the resulting operator.

A substantially larger class of i2D operators can be obtained with decompositions that consist of several branches, with each branch combining the outputs of a set of linear filters (two in the case of quadratic operators) in a multiplicative fashion (see [Bar93, Zet90b, Zet90c, Zet93]). With respect to the i2D selectivity, this structure allows an interpretation, which can often be helpful in the design process. For example, we can now allow the overlap of the orientation tuning functions in one branch, provided the false responses to i1D signals caused by this overlap are cancelled by appropriate contributions from other branches [this is expressed in the "compensation equation" [Zet90c], a ΣΠ counterpart to Eq. (14.29)]. We have found that one useful design strategy is to organize the branches into two groups, an "enhancing" group used to maximize the operator's response to the desired features and a "suppressing" group used to avoid false responses to i1D signals [Zet90c]. However, any choice that yields the appropriate cancellation effect will lead to a valid i2D operator.

14.4.2 Isotropic "Gaussian" i2D Operator

We start with a simple example that illustrates two important aspects. First, it illustrates which conditions must be met to guarantee the spatially isotropic response behavior of a Volterra operator, and second, it provides the link between i2D operators and the concept of Gaussian curvature in differential geometry.

In the spectral domain, the general condition for the isotropy of a second-order operator can be expressed as

$$\begin{aligned} H_2^{\text{sym}}(f_{x1}, f_{y1}, f_{x2}, f_{y2}) = H_2^{\text{sym}}(&f_{x1}\cos\varphi + f_{y1}\sin\varphi, \\ &-f_{x1}\sin\varphi + f_{y1}\cos\varphi, f_{x2}\cos\varphi + f_{y2}\sin\varphi, \\ &-f_{x2}\sin\varphi + f_{y2}\cos\varphi), \quad \forall\varphi. \end{aligned}$$

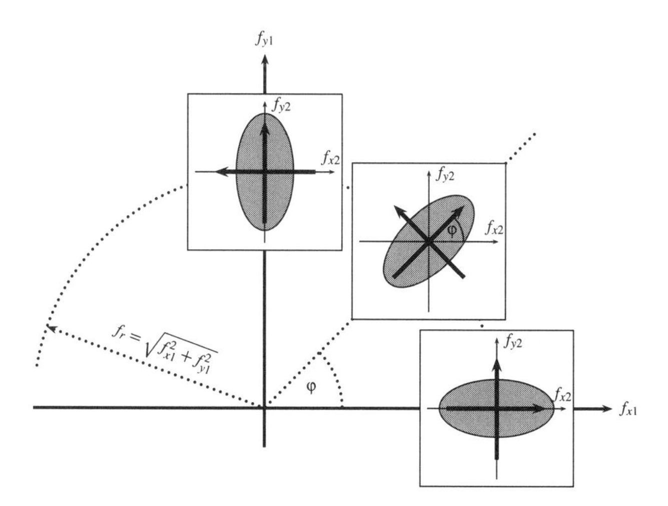

Figure 14.8: Conditions for the isotropy of a quadratic operator. The drawing shows how each two-dimensional section of the symmetric four-dimensional frequency-domain kernel must reappear in rotated form in other sections with the same $f_r = (f_{x1}^2 + f_{y1}^2)^{1/2}$. If the index of a section is rotated with respect to another section by an angle of φ, the function within the section must also be rotated by the same angle φ to ensure isotropy.

This constraint on the form of the frequency-domain kernel is schematically illustrated in Fig. 14.8. Basically, it states that any two-dimensional section of the four-dimensional kernel must reappear, in rotated form, in other sections with the same $f_r = (f_{x1}^2 + f_{y1}^2)^{1/2}$ and that the relative rotation of the functions within two such sections must be equivalent to the rotation angle φ between the indexing coordinates (f_{x1a}, f_{y1a}) and (f_{x1b}, f_{y1b}) of the two sections. By the Fourier transform rules, the same constraints are also valid for the corresponding space-domain kernel.

Let us now consider the class of isotropic i2D operators that can be designed with $\Sigma\Pi$ architectures with Π-order two. To simplify the design task, we assume polar separability of the filter functions, that is, $A_i = A_{\rho,i} \cdot A_{\phi,i}$, and identical radial tuning functions for all filters, that is, $A_{\rho,i} = A_{\rho,j}$, $\forall i, j$. Furthermore, we use only the squared version of the $\Sigma\Pi$ system; that is, we assume that the two filters in each branch are identical, $A_i = B_i$. To obtain the desired isotropy, we can then start by choosing for the first branch a simple isotropic filter, a Laplace operator, and square its output ($A_{\phi,1} = A_{\phi,2} = \text{const}$). This branch is of course not i2D selective. Now two conditions must be met for the remaining branches. First, they must together represent an isotropic kernel, and second, they must cancel the undesired responses to i1D signals from the first branch. Since the radial tuning is fixed, only the angular tuning functions have to be determined. A straightforward solution is to choose two branches with "complementary" anisotropic kernels such that the two add up to an isotropic kernel. With a sinusoidal basis for the angular tuning functions, a straightforward realization of such a structure is given by the

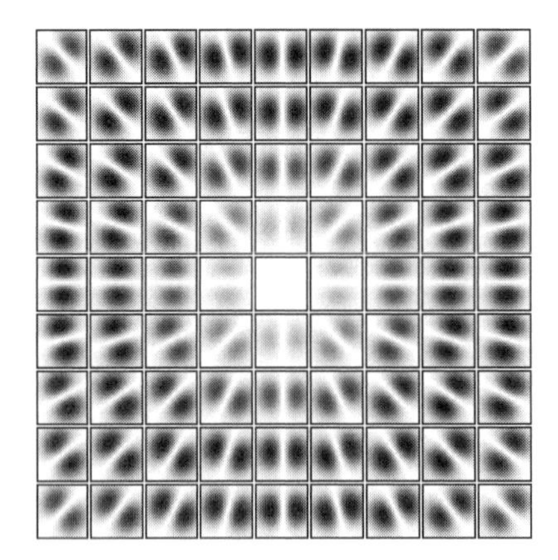
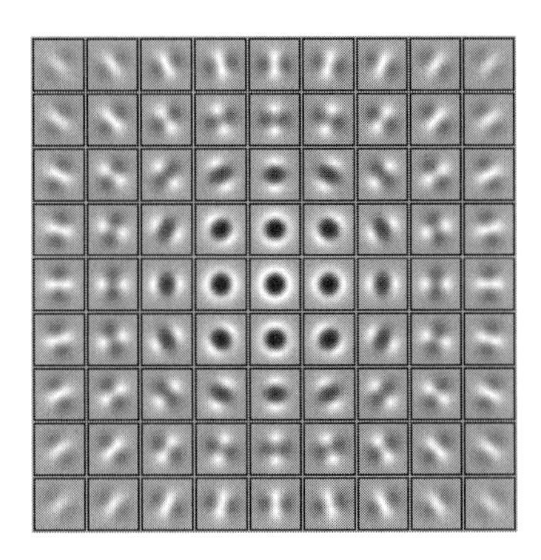

Figure 14.9: An isotropic i2D operator as characterized by its nonlinear filter functions in the frequency (left) and spatial (right) domains. The arrangement of sections of the 4-D Volterra kernel is the same as in Fig. 14.6. The selectivity for intrinsically 2-D signals may easily be seen from the frequency-domain kernel: since the minimum stopbands according to the black forbidden zones of Fig. 14.6 are taken into account, the operator will not respond to i0D and i1D signals. From a comparison with the corresponding space-domain kernel, it should be evident that a design strategy that tries to maintain the i2D specificity by use of appropriate space-domain integrals would be much more complicated than the simple preservation of appropriate frequency-domain stopbands. The isotropy of the operator (i.e., the independence of its response behavior with respect to rotations of the two-dimensional image coordinates) is reflected in the fact that each two-dimensional section of the four-dimensional nonlinear filter function reappears in rotated form in other sections with $f_r = (f_{x1}^2 + f_{y1}^2)^{1/2}$. If the indexing coordinates of two sections are related by an angle φ, the functions within the section must also appear rotated with respect to each other by the same angle φ (see Fig. 14.8).

trigonometric identity $\sin^2(\phi) + \cos^2(\phi) = 1$. The choice of a sinusoidal basis is interesting because the filters can then also be expressed as partial derivatives, with $\sin(\phi)$ and $\cos(\phi)$ corresponding to l_{xy} and l_{uv}, where u and v represent the diagonal directions $u = (1/\sqrt{2})(x + y)$ and $v = (1/\sqrt{2})(x - y)$. The whole operator can then be written as $(l_{xx} + l_{yy})^2 - l_{xy}^2 - l_{uv}^2$, and this can be further simplified to $l_{xx} \cdot l_{yy} - l_{xy}^2$, which corresponds exactly to the essential determinant in the definition of the Gaussian curvature of the local image intensity function interpreted as a surface in 3-D space [Eq. (14.12)].

The frequency-domain Volterra kernel of the band-limited operator is depicted on the left side of Fig. 14.9. From the structure of the stopbands, which appear as white regions, it is easy to see that the kernel meets the basic requirements for i2D operators illustrated in Figure 14.6. The corresponding space-domain kernel is shown on the right of Figure 14.9.

The effect of applying the "Gaussian" i2D operator to two test images is shown in Fig. 14.11(b). It can be seen that the operator is sensitive only to local i2D signals with a high degree of curvature; that is, it detects spots, sharp corners, line ends, etc. but not the less curved circles. Such a restricted operation range may be suited for specific applications, but in general it will not be desirable that curved circles are classified as straight lines (i1D signals). To avoid this effect and increase the curvature sensitivity of the operator, we must get rid of the strong inherent limitations imposed by the partial derivatives in the definition of Gaussian curvature and other surface-based predicates (see [Bar93, Zet90b, Zet90c]).

Given the present framework, it is no problem to derive more general i2D operators. For example, we can consider a system that may be regarded as a direct derivation from the requirement that H_2^{sym} has to vanish for $f_{x1}f_{y2} = f_{x2}f_{y1}$, as shown in Eq. (14.29). Such a system can be expressed as a quadratic Volterra filter with the frequency-domain kernel given by

$$H_2(f_{x1}, f_{y1}, f_{x2}, f_{y2}) = (j2\pi)^2 \cdot (f_{x1}f_{y2} - f_{y1}f_{x2}) \cdot A(f_{x1}, f_{y1}) \cdot B(f_{x2}, f_{y2}). \quad (14.32)$$

Substituting $f_{x1}f_{y2} = f_{y1}f_{x2}$, it is easy to verify that the symmetric kernel of Eq. (14.32) satisfies Eq. (14.29). Hence we can use arbitrary filter functions $A(f_x, f_y)$ and $B(f_x, f_y)$, and Eq. (14.32) will always yield the kernel of a genuine i2D operator. The Gaussian curvature operator then becomes just a special case if we choose in Eq. (14.32) the filter functions $A(f_x, f_y)$ and $B(f_x, f_y)$ as

$$A(f_x, f_y) = j2\pi f_x \quad , \quad B(f_x, f_y) = j2\pi f_y. \quad (14.33)$$

A variety of other generalizations in the context of ΣΠ architectures has been derived ([Zet90c]). The design freedom offered by a more general framework can be fully exploited, however, only if it can be used in a systematic fashion for the generation of desired response properties of the operators. In the following section, we hence consider how the tuning properties of i2D operators are influenced by the spectral shape of the Volterra kernel.

14.4.3 The Influence of Passbands and Stopbands on the "Curvature Tuning" of i2D Operators

An important feature offered by the framework of linear signal processing is the possibility to adapt the filtering properties of linear operators to the specific demands of the signal processing task. A simple example for such an adjustment is the determination of suitable passbands and stopbands. Thus, it is an important question, whether the signal processing properties of nonlinear i2D operators can also be adapted by use of a nonlinear analog to the stopbands and passbands of linear filters.

Figure 14.10 shows several examples of possible arrangements for the passbands and stopbands of i2D operators. The kernels themselves are shown in the top row, and a schematic version of a section of each kernel is depicted in the bottom row.

Figures 14.10a and 14.10b illustrate the posssible variations of the widths of the stopbands and passbands for isotropic i2D operators. Basically, the width of the stopband determines the minimum curvature of the image features that

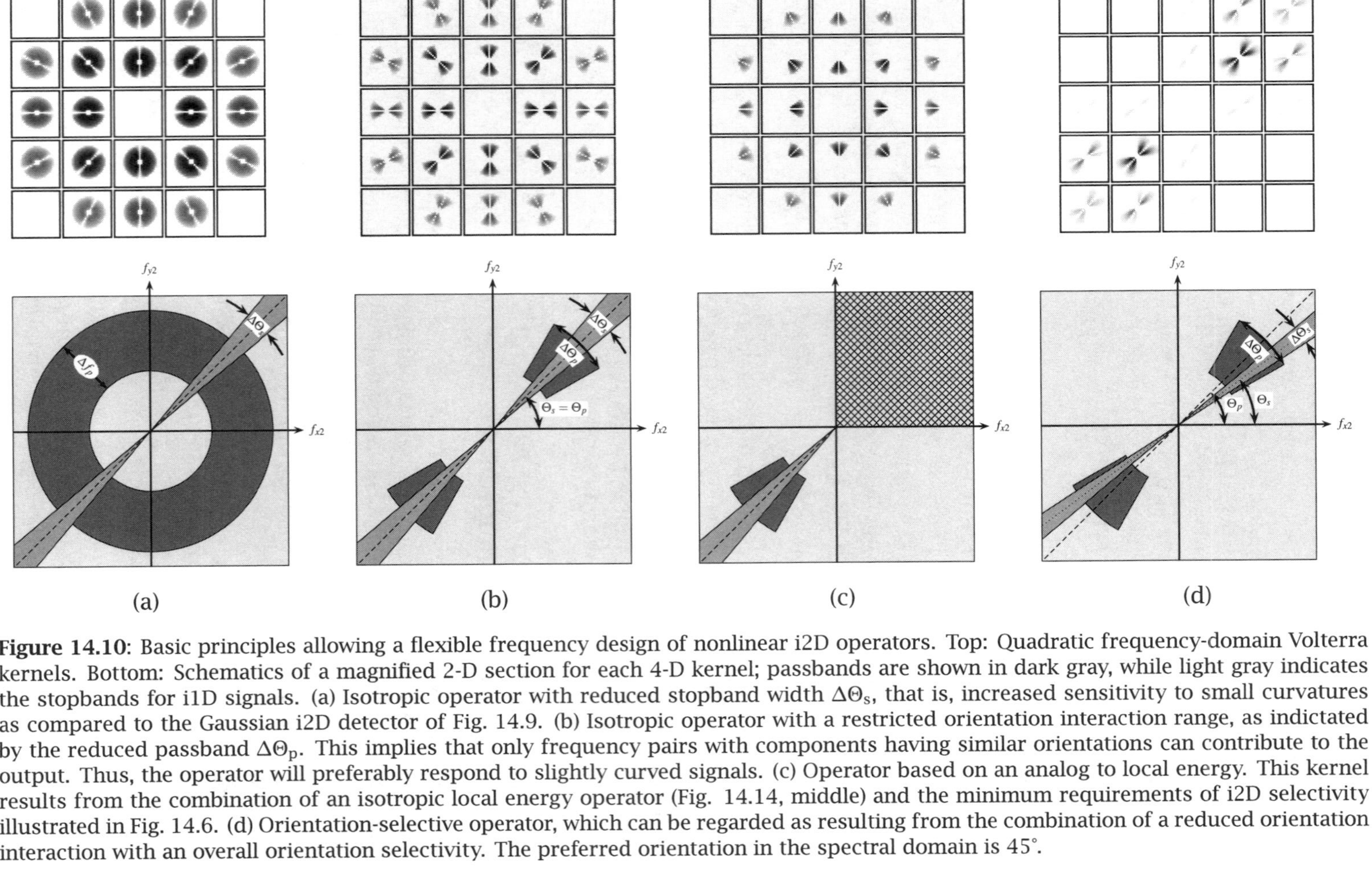

Figure 14.10: Basic principles allowing a flexible frequency design of nonlinear i2D operators. Top: Quadratic frequency-domain Volterra kernels. Bottom: Schematics of a magnified 2-D section for each 4-D kernel; passbands are shown in dark gray, while light gray indicates the stopbands for i1D signals. (a) Isotropic operator with reduced stopband width $\Delta\Theta_s$, that is, increased sensitivity to small curvatures as compared to the Gaussian i2D detector of Fig. 14.9. (b) Isotropic operator with a restricted orientation interaction range, as indictated by the reduced passband $\Delta\Theta_p$. This implies that only frequency pairs with components having similar orientations can contribute to the output. Thus, the operator will preferably respond to slightly curved signals. (c) Operator based on an analog to local energy. This kernel results from the combination of an isotropic local energy operator (Fig. 14.14, middle) and the minimum requirements of i2D selectivity illustrated in Fig. 14.6. (d) Orientation-selective operator, which can be regarded as resulting from the combination of a reduced orientation interaction with an overall orientation selectivity. The preferred orientation in the spectral domain is 45°.

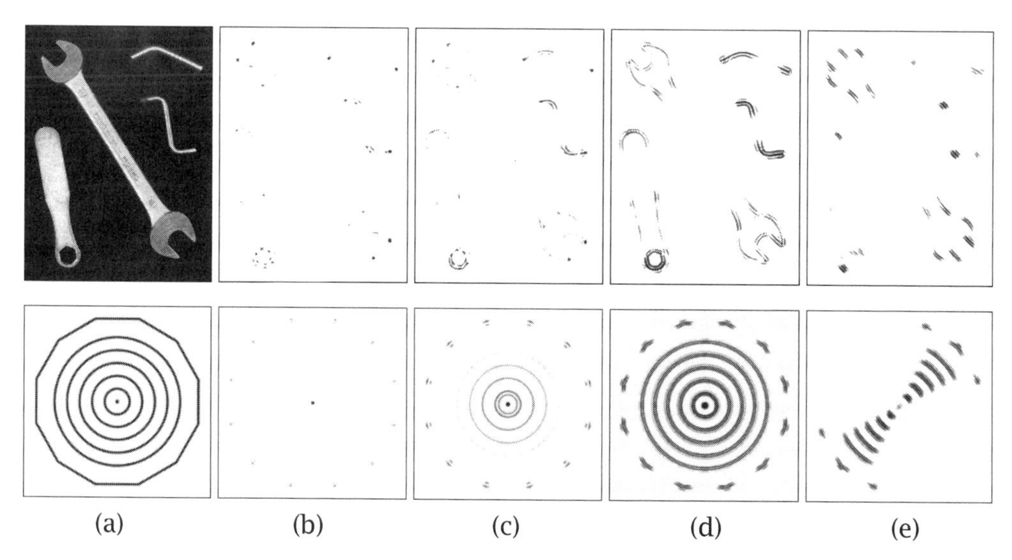

Figure 14.11: (a) Two test images and their responses from different types of i2D operators. (b) "Gaussian," (c) isotropic with a reduced width of the stopband for i1D signals, (d) isotropic with both reduced stopband width and restricted range of orientation interaction in the passband and (e) anisotropic, orientation-selective. For all operators only positive, square-rooted responses were used, to ensure proportionality to image intensity. The general tendency is an increase of curvature sensitivity from (b) through (d). The Gaussian operator (b) responded only to highly curved features (spots, line ends, etc.). The operator with reduced stopband (c) responded additionally to the less curved signals (e.g., inner circles). The one with reduced orientation-interaction range (d) preferred even the less curved signals. The responses of the orientation-selective operator (e) can be regarded as a subset of (d).

can be detected by the operator. This influence can be understood by taking into account that all frequency pairs with angular differences $\Delta\varphi = \arctan(f_{y1}/f_{x1}) - \arctan(f_{y2}/f_{x2})$ that are smaller than the angular bandwidth $\Delta\Theta_s/2$ of the stopband are blocked by the operator. A narrow stopband $\Delta\Theta_s$ will thus block only those signals that are very close to i1D signals, that is, are nearly perfectly straight, so that already signals with small curvature are allowed to pass. A wider stopband, such as in the Gaussian curvature operator of Fig. 14.9, will block those signals with small curvature and allow only responses to pronounced i2D signals such as sharp corners, which contain frequency components with widely separated orientations.

Figure 14.10a shows a frequency-domain kernel of an i2D operator with a more sensitive curvature tuning than that of the Gaussian i2D operator, which becomes evident by comparison of its stopbands with those of the Gaussian curvature operator (Fig. 14.9).

The influence of this modification of the nonlinear filter function can be seen in Fig. 14.11c. Compared to the responses of the Gaussian i2D operator in the second column, additional responses appear at locations with less sharply curved features, for example, at the locations of the inner circles. A further general ten-

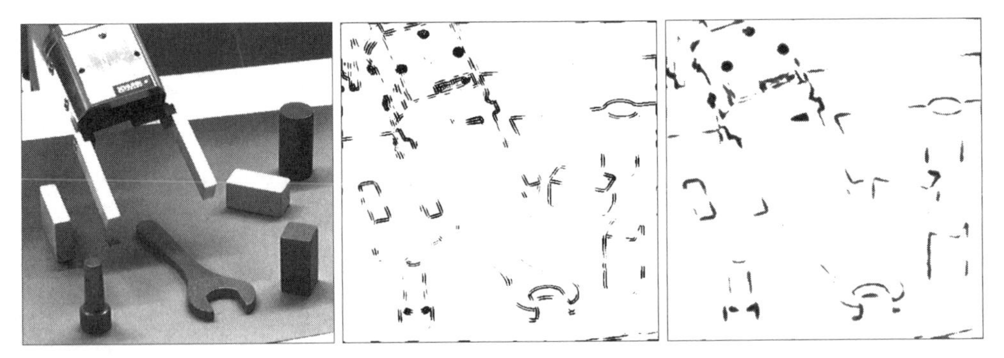

Figure 14.12: Responses of an i2D operator with reduced passband (middle; see Fig. 14.10b) and of an isotropic *i2D* energy operator (right, see Fig. 14.10c).

dency is that the form of the responses becomes more elongated. The strongest responses, however, do not differ substantially from those of the Gaussian operator. This has to be expected since the basic form of the passband has not changed substantially.

Indeed, a further design variable is given by the width of the passband $\Delta\Theta_p$ of an i2D operator. This alters the relation between the responses to optimum signals and the responses to features with minimum curvature. For the first two operators, there has been little restriction on the basic passbands, except for a general bandlimit determining the highest and lowest frequencies that are allowed as components. Thus signals with high curvature are passed by both i2D operators. To shift the optimum response toward lower curvature, we have to restrict the admissible orientation interactions by allowing only frequency pairs in the passbands for which the orientation of the components does not differ by more than a given limiting angle. The frequency-domain kernel and a schematic characterization of the resulting kernel sections are shown in Figure 14.10b. The main consequence of the restriction of orientation interactions is that the basic shape of the passbands will no longer be circular but elliptical and more concentrated in the immediate surroundings of the forbidden zones. A point to note is that the isotropy of a Volterra operator (with respect to its response) does not exclude the possibility of a strong anisotropy of the individual sections of the corresponding four-dimensional kernel.

The responses of this operator to the test images are shown in Fig. 14.11d. It can be seen that the relation between the responses to strongly curved signals (such as isolated points) on the one hand and to less curved signals (such as the outer circle) on the other, has substantially changed. In general, there is an increased tendency to respond to signals with low curvatures.

14.4.4 Local Amplitude-Energy for Nonlinear Operators: Complex Cells

A further dimension for the design of nonlinear operators is given by the local phase properties. Since two frequency components are combined, there are also two phase values that influence the operator response. In principle, this enables a further differentiation of the input signals, for example, a distinction between line ends and corners, but the issue is complicated by the additional new frequency terms that are generated by the nonlinear interaction. These terms lead to a complex, multilobed pattern of the local operator responses. In many cases, it would thus be more desirable to get rid of the phase dependencies. For such a phase invariance there already exists a suitable signal processing concept, the "analytic signal" [Pap62], which enables the definition of the local amplitude (or energy) $a(x)$ and the local phase $\phi(x)$ of a bandpass signal. This concept has found some interest in the context of biological image processing (e.g., [Ade85, Bur92, Weg90b]), since the local energy component is closely related to the nonlinear operation of cortical complex cells, which exhibit a phase-invariant, demodulating response behavior [DeV88]. There is also a close relationship to the characteristic statistical dependencies that occur between the wavelet coefficients of natural images [Weg90b, Zet99b]. Since we are interested in using such a local phase invariance for i2D operators, we provide first an analysis of the basic mechanism of the local energy computation in terms of Volterra filters (see [Kri92]). A related extensive discussion of the Volterra representation of Teager energy operators can be found in Chapter 6.

The interpretation of the local energy computation in terms of Volterra filters is possible because it is defined in terms of squaring nonlinearities,

$$u_2(x) = a^2(x) = [u_1(x) \star h_1^{\text{even}}(x)]^2 + [u_1(x) \star h_1^{\text{odd}}(x)]^2 \ . \qquad (14.34)$$

Here, $h_1^{\text{even}}(x)$ and $h_1^{\text{odd}}(x)$ are the impulse responses of even- and odd-symmetric linear filters with identical magnitude transfer functions, that is, $|H_1^{\text{even}}(f)| = |H_1^{\text{odd}}(f)|$. In the frequency domain, the corresponding second-order Volterra kernel is given by

$$H_2(f_1, f_2) = H_1^{\text{even}}(f_1) \cdot H_1^{\text{even}}(f_2) + H_1^{\text{odd}}(f_1) \cdot H_1^{\text{odd}}(f_2). \qquad (14.35)$$

The properties of the local energy computation are thus caused by a specific arrangement of passbands and stopbands of the spectral Volterra kernel, which is reflected in a mirror symmetry of the kernel (Fig. 14.13). The consequences of this structure can be easily understood by considering the processing of two sinusoids with frequencies f_a,f_b. Due to the complementary structure of the quadrature filter pathways, the frequency-domain kernel H_2 vanishes in the quadrants in which f_1 and f_2 have the same sign. Since the interpretation in terms of the central slice theorem (Sec. 14.3.2) suggests that we view the contraction as a projection of the expanded spectrum $\tilde{U}_2(f_1, f_2)$ onto the line $f = f_1 + f_2$, it is readily apparent that the expanded spectral frequency components in these stopband quadrants are just the components that can generate the high-frequency sum terms at the output of the system ($2f_a$, $f_a + f_b$, $2f_b$). These terms are blocked by the energy operator, and only the low- and zero-frequency difference terms ($f_a - f_b$, $f_a - f_a = f_b - f_b = 0$) are allowed to pass. This selective blocking corresponds

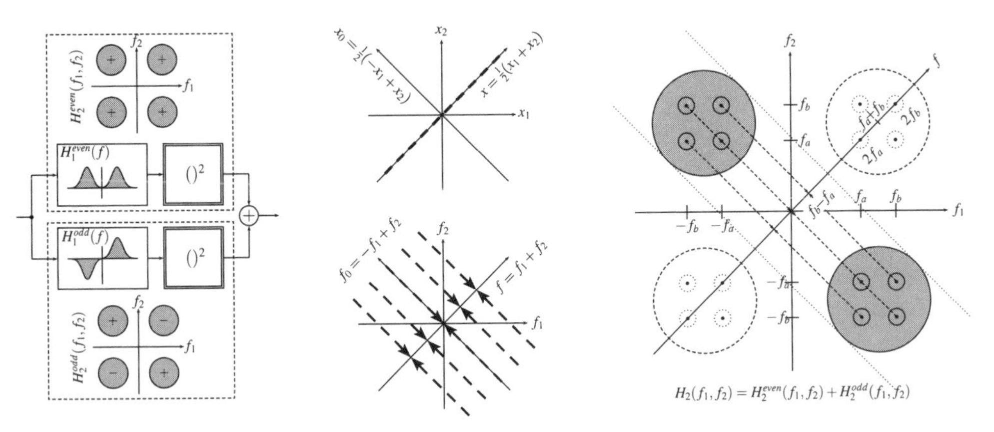

Figure 14.13: Volterra kernel of a 1-D operator for the computation of the local energy of bandpass signals. This kernel results from the summation of the partial kernels for the even- and odd-symmetric filter paths, as shown on the left side. The essential demodulation property is due to the cancellation of the kernel in the quadrants in which both frequencies have equal signs, which results in a specific mirror symmetry of the kernel. According to the interpretation by means of the central slice theorem (middle figure; see Sec. 14.3.2), the output spectrum results from the diagonal projection onto the axis $f = f_1 + f_2$. The vanishing of the kernel in the lower left and upper right quadrants thus guarantees that only frequency components with opposite signs are combined, which yields the low-frequency terms required for the computation of the local energy.

to a local phase invariance, or to a demodulation property, of the local energy operator.

There is a close relationship to the Teager energy filters that are analyzed in detail in Chapter 6. Our condition of a vanishing transfer function within the quadrants in which the frequency terms have equal signs is just a stronger restriction than the condition $H_2(f, f) = 0$, that is, the vanishing of the transfer function only at the main diagonal, which is used in the design of the Teager filters. A difference between the two types of energy operators is thus that the local energy operators block the high-frequency contributions from all sum terms, including the mixed terms $f_a + f_b$.

This principle for the construction of local energy operators can be extended for the design of i2D operators. But before going into this, let us introduce one last design dimension: orientation selectivity. After we have the list of design dimensions completed, we show in subsequent section how these dimensions can be combined to derive i2D operators with specific properties.

14.4.5 An Anisotropic, Orientation-Selective Operator

Isotropic operators are desirable in many applications, but in certain cases this type of response invariance can also be disadvantageous. For example, an operator may be required that indicates both the presence of some noncircular i2D feature and its main orientation axis at the same time. Orientation selectivity may also be important for achieving high S/N ratios in environments with isotropic Gaussian noise if i2D signals that become close to i1D signals should be detected. Edges with low curvature, for example, have their spectral energy concentrated at frequency pairs that differ only by a small angle. An orientation-specific operator will then provide a better S/N ratio than an isotropic one. To combine this higher S/N ratio with an isotropic behavior we first apply a suitable threshold to each of a set of orientation-selective branches and then combine these thresholded outputs.

Such advantages seem also to be relevant for biological vision. The visual systems of mammals pursue a strategy that is closely related to a processing by orientation-selective i2D operators. The classical "hypercomplex" cells of Hubel and Wiesel [Hub65] and also end-stopped simple cells exhibit a combined selectivity with respect to both orientation and intrinsic dimensionality [Zet90b, Zet90c]. Thus it is interesting from both an engineering and a biological perspective to provide operators that combine these two properties. The kernel of an orientation-selective i2D operator and one schematic section of it are illustrated in Fig. 14.10d. The orientation selectivity of the operator is immediately visible from the global shape of the frequency-domain kernel since the kernel vanishes for all sections (f_{x1}, f_{y1}) with angles $\varphi = \arctan(f_{y1}/f_{x1})$ outside the preferred orientation range. The orientation selectivity also implicitly restricts the range of possible orientation interactions such that orientation-selective i2D operators can, in principle, be considered as components of the isotropic i2D operator with restricted orientation interaction, as described in Sec. 14.4.3.

The responses of the orientation-selective i2D operator to the test images are shown in Fig. 14.11e. As expected, the responses can be considered as orientation-specific subsets of the responses of the isotropic i2D operator with restricted orientation interactions.

14.4.6 i2D Operators Based on an Analog to Local Energy

The analytic signal is a one-dimensional concept. Before we derive the kernels that combine i2D selectivity and energy computation we first show how the local energy computation as such can be extended to the two-dimensional domain. The simplest possibility for this is to use oriented two-dimensional bandpass filters. Such filters exhibit a bandpass characteristic in one direction and a low-pass characteristic in the orthogonal direction. The energy computation is then simply performed with respect to the bandpass direction. An isotropic energy computation can then be thought of as consisting of a superposition of a whole set of oriented energy detectors that cover the complete range of orientations. The corresponding frequency-domain Volterra kernels for these two schemes are shown on the left and middle of Fig. 14.14.

Shown on the right in the figure is the spatial-domain kernel corresponding to the orientation-selective energy operator. As mentioned, cortical complex cells can be assumed to correspond to the oriented version of a two-dimensional energy computation. The second-order kernels of cortical complex cells have been

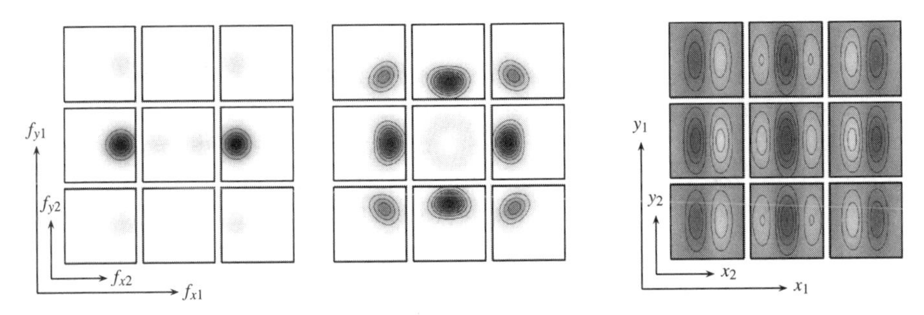

Figure 14.14: Frequency-domain Volterra kernels of an orientation-selective (left) and an isotropic (middle) local energy operator. The spatial-domain kernel corresponding to the orientation-selective local energy operator is shown on the right.

measured [Szu90], and they indeed show the basic structure of the right kernel in Fig. 14.14.

Now we can proceed to the desired combination of i2D selectivity and energy computation. This can be obtained from a combination of the respective kernel structures, that is, from the combination of forbidden zones with the blocking of frequency combinations with equal signs [Kri99]. In good approximation this can already be realized with a second-order kernel An example of the second-order kernel of an isotropic i2D energy operator is shown in Fig. 14.10c. The response of this operator is illustrated in Fig. 14.12.

It should be noted that the straightforward combination of the two kernel properties for i2D selectivity and for energy detection cannot provide an ideal type of i2D-selective energy operator if they are combined in a second-order kernel. The desired strict nonnegativity and the envelope demodulation property cannot be achieved exactly within the constraints given by a second-order operator because there is a conflict between the demand to avoid combinations of identical frequency components, as required for i2D selectivity, and the fact that the mean of the output of a second-order bandpass system is due to the zero frequency that results from the difference terms from frequencies with opposite sign. However, a true combination of both properties is possible with higher-order operators (see Sec. 14.6).

As mentioned, the combination of energy detection with both i2D and orientation selectivity is also realized in biological vision systems. Figure 14.15 shows a comparison of end-stopped cells of type "simple," which combines orientation and i2D selectivity, and "complex" (also known as hypercomplex), which combines all three properties. The simulations by appropriate Volterra operators are in good agreement with the measured physiological data.

14.5 i2D Operators and Higher-Order Statistics

Higher-order statistics (HOS) is a rapidly evolving tool for the analysis of non-Gaussian signals, and past years have witnessed successful applications in many

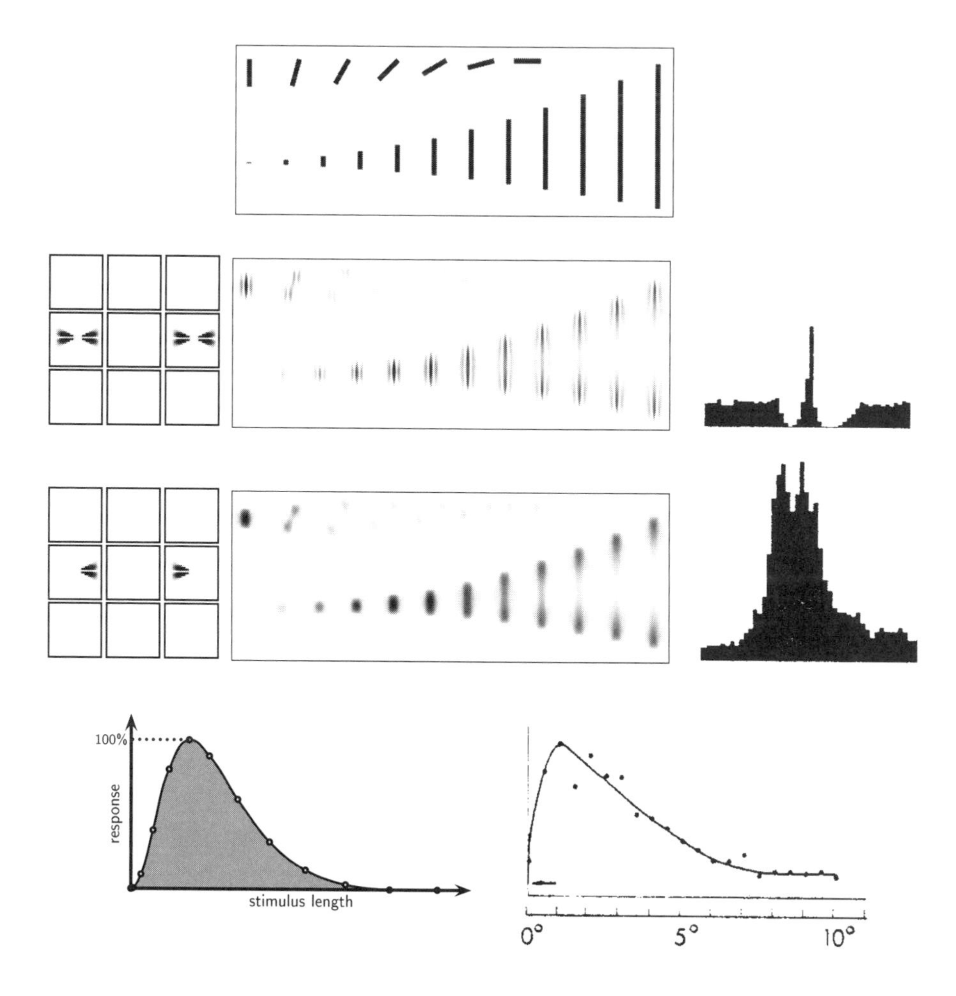

Figure 14.15: Modeling of end-stopped cells in the visual cortex. Top: Bars differing in length and orientation, which are often used to investigate neural responses in physiological experiments. Second and third rows: Kernels of orientation-selective i2D operators (left), which may serve as models for end-stopped cells of simple and complex type; responses of these i2D operators (middle); and measured activity profiles (right), which were obtained by shifting a bar of optimum length and orientation perpendicular to its principal axis. Simple cells have inhibitory sidelobes, while complex cells give an unmodulated low-pass response [DeV88]. This is in good agreement with the simulation results in the middle column. Bottom: Length tuning profiles for the simulated (left) and measured (right) cell (physiological data from [Kat78], © 1978 The American Physiological Society).

different signal processing areas, such as the analysis of speech signals, the classification of textures, and the detection of signals in noisy environments. The investigation of the relevance of HOS for the description of natural scenes was started by an analysis of oriented image features in terms of higher-order cumulants [Zet93] and then extended to a more detailed analysis of two-dimensional images in terms of polyspectra [Kri95, Kri97, Kri00b, Zet99a, Zet00]. A one-dimensional analysis of image polyspectra with respect to radial frequency terms can be found elsewhere [Tho97].

The analysis of the polyspectra of natural images reveals two important points. First, it can provide an explanation for the "sparse coding" that can be achieved by linear decompositions with orientation-selective filters [Dau88, Fie87, Wat87, Zet87]. Second, and this is the more relevant issue for the present context, it shows how and why nonlinear i2D-selective operators can exploit the typical higher-order dependencies in natural images.

We start with a short definition of the cumulants and polyspectra of a stationary random process $\{u(\mathbf{x})\}$. The cumulants are defined in terms of a power series expansion of the joint multivariate probability density function $p[u(\mathbf{x}_1),\ldots,u(\mathbf{x}_n)]$ via the second characteristic function [Nik93]. For zero mean signals, the second-order cumulant $\mathbf{c}_2^{\mathbf{u}}(\mathbf{x}_1) = \mathsf{E}[u(\mathbf{x}) \cdot u(\mathbf{x}+\mathbf{x}_1)]$ is well known as the autocorrelation function, and the third-order cumulant is given by

$$\mathbf{c}_3^{\mathbf{u}}(\mathbf{x}_1,\mathbf{x}_2) = \mathsf{E}[u(\mathbf{x}) \cdot u(\mathbf{x}+\mathbf{x}_1) \cdot u(\mathbf{x}+\mathbf{x}_2)], \tag{14.36}$$

where E denotes the expectation operator. Likewise, it can be shown that the fourth-order cumulant is given by

$$\begin{aligned} \mathbf{c}_4^{\mathbf{u}}(\mathbf{x}_1,\mathbf{x}_2,\mathbf{x}_3) &= \mathsf{E}[u(\mathbf{x}) \cdot u(\mathbf{x}+\mathbf{x}_1) \cdot u(\mathbf{x}+\mathbf{x}_2) \cdot u(\mathbf{x}+\mathbf{x}_3)] \\ &\quad -\mathbf{c}_2^{\mathbf{u}}(\mathbf{x}_1) \cdot \mathbf{c}_2^{\mathbf{u}}(\mathbf{x}_2-\mathbf{x}_3) - \mathbf{c}_2^{\mathbf{u}}(\mathbf{x}_2) \cdot \mathbf{c}_2^{\mathbf{u}}(\mathbf{x}_1-\mathbf{x}_3) \\ &\quad -\mathbf{c}_2^{\mathbf{u}}(\mathbf{x}_3) \cdot \mathbf{c}_2^{\mathbf{u}}(\mathbf{x}_1-\mathbf{x}_2). \end{aligned} \tag{14.37}$$

An important property of the cumulants of order higher than two is that, as opposed to the moments, they vanish for Gaussian signals. Especially the third- and fourth-order cumulants may be interpreted as the difference between the moments of a given process and the moments of its (second-order) equivalent Gaussian representation. The cumulants are thus very well suited to describe the statistical properties of non-Gaussian signals (such as images from realistic environments).

Similar to the second-order statistics, it is often more convenient to investigate the higher-order statistical dependencies in the frequency domain. The Fourier transforms of the third-order cumulant $\mathbf{c}_3^{\mathbf{u}}(\mathbf{x}_1,\mathbf{x}_2)$ and the fourth-order cumulant $\mathbf{c}_4^{\mathbf{u}}(\mathbf{x}_1,\mathbf{x}_2,\mathbf{x}_3)$ are known as the bispectrum $\mathbf{C}_3^{\mathbf{U}}(\mathbf{f}_1,\mathbf{f}_2)$ and the trispectrum $\mathbf{C}_4^{\mathbf{U}}(\mathbf{f}_1,\mathbf{f}_2,\mathbf{f}_3)$, respectively. A more direct interpretation of these polyspectra may be given if we express $\{u(\mathbf{x})\}$ in terms of the frequency components $dU(\mathbf{f})$ of the Fourier-Stieltjes representation of a stochastic process [Ros85]. Using this spectral representation, the power spectral density is given by $\mathbf{C}_2^{\mathbf{U}}(\mathbf{f}) = \mathsf{E}[dU(\mathbf{f}) \cdot dU^*(\mathbf{f})]$, and for the bispectrum we obtain

$$\mathsf{E}[dU(\mathbf{f}_1) \cdot dU(\mathbf{f}_2) \cdot dU^*(\mathbf{f}_3)] = \begin{cases} \mathbf{C}_3^{\mathbf{U}}(\mathbf{f}_1,\mathbf{f}_2) \cdot d\mathbf{f}_1 d\mathbf{f}_2 & \text{for } \mathbf{f}_3 = \mathbf{f}_1 + \mathbf{f}_2, \\ 0 & \text{otherwise.} \end{cases} \tag{14.38}$$

From this equation it becomes apparent that the bispectrum is given by the expectation of three Fourier components, the sum of whose frequencies equals zero. A similar equation may be derived for the trispectrum. Hence, as opposed to the

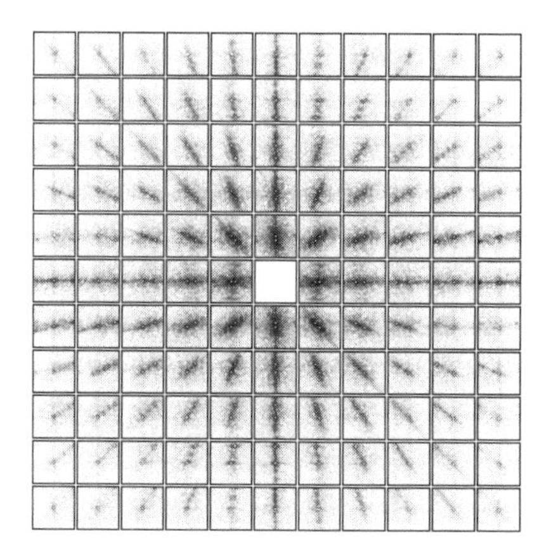

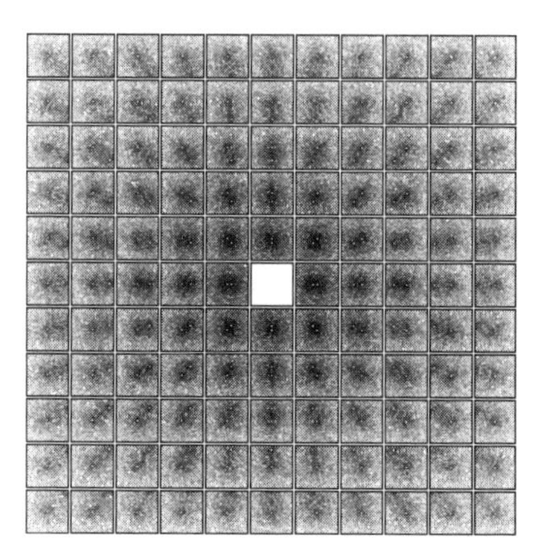

Figure 14.16: Bispectra of a set of natural images (left) and a non-Gaussian noise image (right) with first- and second-order statistics approximately equivalent to natural scenes. Shown are several slices of the four-dimensional bispectrum with f_{x1} = const and f_{y1} = const. Note that only for the natural images does there exist an elongated concentration of bispectral "energy" in those regions in which the frequency components are aligned to each other with respect to orientation. This implies substantial statistical dependency between frequency components of equal orientation.

power spectrum, both the bispectrum and the trispectrum measure the statistical dependencies between frequency components. Equation (14.38) is also the starting point for a direct computation of the polyspectra in the frequency domain [Nik93], on which we base our analysis of the higher-order statistics of natural images. An example for the bispectrum derived from a set of natural scenes is given in Fig. 14.16 (left).

Both types of polyspectra show a structurally similar concentration of the polyspectral "energy" in those regions in which the frequency components are aligned to each other, that is, have the same orientation. This uneven polyspectral energy distribution can be interpreted as statistical redundancy, similar to the well known $1/f^2$ decay of the power spectrum of natural images. As with second-order statistics, we can consider two basic strategies for the exploitation of this redundancy, a predictive coding strategy and a transform coding strategy (that is, some sort of multichannel decomposition of the signal).

Let us begin with the latter. As mentioned, one of the most basic recent findings with respect to the efficient encoding of natural images in both biological systems and image compression schemes has been the observation of the "sparse coding" that results from the application of a wavelet-like size- and orientation-selective filter decomposition [Fie87, Wat87, Zet87]. Interest in this topic has further increased due to the recent discovery of a strong relation between this wavelet-like image decomposition and the concepts of "independent component analysis" and blind source separation [Bel97, Com94, Ols96, van98]. However, a complete understanding of the relationship between the structure of the multivariate probability density function (pdf) and these sparse coding effects has yet to be achieved. It has been shown that the standard second-order approach is inherently blind to the presence of local orientations in natural images and hence cannot provide a reasonable explanation for why an orientation-selective decomposition should exhibit this advantageous behavior [Zet93, Zet97]. The analysis of the structure of the multivariate pdf by means of wavelet statistics has revealed that the pdf can be approximated by a compound of low-dimensional orthogonal subspaces [Weg90a, Zet93, Zet99a]. The advantage of the wavelet-like decompositions is thus due to the fact that for each wavelet coefficient, only a smaller part of these subspaces is aligned to the coordinate that corresponds to the coefficient, while the larger part is more or less orthogonal to it. This enables the sparse coding with reduced interunit dependencies.

More recently we have shown that this sparse coding effect can also be deduced from the structure of the polyspectra [Kri98, Kri00b, Zet00, Zet99a]. Without going into the details here, we just mention that this explanation is based on the fact that a useful measure for the sparseness of a filter decomposition can be the kurtosis γ,[3] which can be seen as the combinded result of the fourth-order dependencies, as measured by the trispectrum, and the specific weighting by the filter functions $H(\mathbf{f})$. Since $\gamma = \mathrm{c}_4^{\mathrm{u}}(\mathbf{0},\mathbf{0},\mathbf{0})$, we obtain

$$\gamma = \iiint \mathrm{C}_4^{\mathrm{U}}(\mathbf{f}_1,\mathbf{f}_2,\mathbf{f}_3) \cdot H(\mathbf{f}_1) \cdot H(\mathbf{f}_2) \cdot H(\mathbf{f}_3) \cdot H^\star(\mathbf{f}_1+\mathbf{f}_2+\mathbf{f}_3)\, d\mathbf{f}_1\, d\mathbf{f}_2\, d\mathbf{f}_3. \tag{14.39}$$

The maximum kurtosis can thus be obtained if the shape of the filter function is matched in a suitable way to the structure of the trispectrum. Since the trispectrum, like the bispectrum, also exhibits a concentration of the spectral energy at those frequency combinations at which the components are aligned to each other, the optimal linear filters have to be confined to those frequencies that are more or less aligned to each other, and thus they have to be orientation selective.

A signal decomposition by oriented filters is hence optimum if we are restricted to linear operations. But what about nonlinear schemes? The second coding strategy for the exploitation of the uneven distribution of the polyspectral energy is suggested by the analogy to the predictive, or "whitening," strategy of the standard second-order approach. The transfer function of a linear predictive scheme is inversely related to the shape of the power spectrum (hence the resulting whitening of the output spectrum). A white power spectrum means that the autocorrelation function is zero peaked, which in turn implies that there are no more spatial second-order correlations present in the output signal. The same type of reasoning can now be applied to the polyspectra. If the trispectrum is constant, this implies that the cumulant function is zero peaked, that is, that there are no longer any spatial fourth-order correlations in the output signal. A suitable strategy for this higher-order whitening can thus be a kind of "inverse filtering," which has to act in a suppressive fashion in those regions of the polyspectrum with a high energy concentration, but in an enhancing fashion in those regions in which the

[3] More precisely, the normalized kurtosis is defined as $\gamma_n = \gamma/\sigma^4 = \frac{\mathrm{E}[(u-\overline{u})^4]-3\mathrm{E}[(u-\overline{u})^2]^2}{\mathrm{E}[(u-\overline{u})^2]^2}$, where $\overline{u} = \mathrm{E}[u]$.

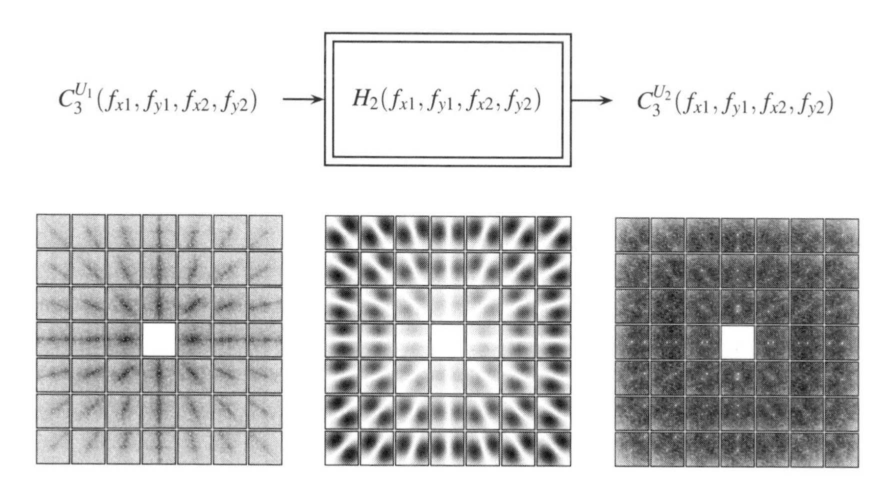

Figure 14.17: Higher-order whitening: There exists an inverse relationship between the polyspectral energy distribution, which shows its maximum concentration in the regions corresponding to i1D signals (left), and the Volterra-Wiener kernels of i2D-selective neurons, which spare out these forbidden i1D zones (right). This is analogous to the inverse relationship between a decaying power spectrum, with its concentration of spectral energy at low frequencies, and the linear high-pass filter, which is required for its whitening.

polyspectral energy concentration is low. However, this type of operation cannot be achieved by a linear filter since such filters cannot act selectively on arbitrary combinations of frequency terms. The potential for such a defined action on AND combinations of frequency terms is exactly what distinguishes nonlinear operators from linear operators, as we showed in Sec. 14.3.3 (see Fig. 14.5). In addition, since a nonlinear operator suitable for higher-order whitening has to block the redundant terms of the polyspectrum (frequency tuples with aligned orientations, i.e., i1D signals) and enhance the nonredundant terms (frequency tuples with heterogeneous orientations, i.e., i2D signals), it will have a structure identical to the i2D operators discussed in the previous section. The higher-order whitening of natural images can thus be achieved by nonlinear i2D-selective operators, as illustrated in Fig. 14.17. Again, this is a result of interest not only for communication engineering but also for our understanding of biological vision systems. In the latter context, the nonlinear processing properties of end-stopped and dot-responsive neurons can be interpreted as the result of a higher-order whitening strategy of the visual cortex.

14.6 Discussion

In this chapter we have discussed the relevance of intrinsically two-dimensional signals in image processing and provided a framework for the design of appropriate i2D-selective nonlinear operators. However, the issues discussed here can also be seen as parts of a larger and more general territory of future research. In the following, we shortly discuss some possible extensions.

Extension to Higher Dimensions. Here we have concentrated on the processing of two-dimensional images, but the basic principle of intrinsic dimensionality can well be extended to higher-dimensional domains, for example, to the processing of image sequences or of volume images. For image sequences, we have also found a natural hierarchy of local signal redundancy in terms of intrinsic dimensionality [Weg92]. All signals that move with approximately constant velocity, for example, are i1D or i2D signals. Only if there is a sudden change of velocity, or an occlusion effect, can an i3D signal result. A nonlinear operator for the detection of intrinsically three-dimensional signals (an i3D operator) can thus be used to identify significant motion discontinuities, and it does not require the commonly assumed necessity of a computationally demanding calculation of the complete flow field [Zet91].

Extension to Higher Orders. Using the same arguments as presented in Sec. 14.3 the suggested formalism can be extended to higher-dimensional signal domains as well as to Volterra operators of any order [Kri92]. Third-order kernels, for example, open the interesting possibility of using two frequency components in the "energy mode," and using this energy as a "modulator" for the third component, which preserves the "original" sign and frequency content of the input image. With fourth-order operators, it is possible to come even closer to an ideal i2D-selective energy operator. This can be seen from the simple example of the multiplicative combination of two oriented "local energy" terms according to Eq. (14.35) (see [Zet90c]). If the basic filters have no orientation overlap, the resulting operator is obviously i2D selective. This principle may be generalized to a fourth-order kernel which then enables the mathematically strict combination of the principles of i2D selectivity, energy computation, and, if required, isotropy within one nonlinear operator.

Extension to Other Nonlinearities. A point often criticized in the Volterra approach is the tendency toward instability, which arises from the multiplicative interactions, or seen from the other side, the problems with the approximation of saturation nonlinearities. Thus, the question arises of whether the principles discussed here can also be transferred to systems based on nonmultiplicative nonlinear interactions. Some examples for such an extension, which are derived from a generalization of the ΣΠ approach, are given elsewhere [Zet90c]. We could also show that the basic principle can be extended to the usage of even-symmetric power functions with arbitrary real valued exponents [$NLIN(x) = |x|^p$]. Of particular interest with respect to robustness is the combination of (1) filter functions with high orientation selectivity (which is not possible with spatial derivatives), (2) a nonlinearity with exponent $p = 1.0$, that is, a semilinear behavior, and (3) the ability to have an internal gain factor that controls the sensitivity for the distinction between i1D and i2D signals. Furthermore, some information about the "polarity" of the signal (e.g., peak or pit) should be provided. These are the features of a specific i2D operator, the "clipped eigenvalue operator" [Zet90c], that is based on

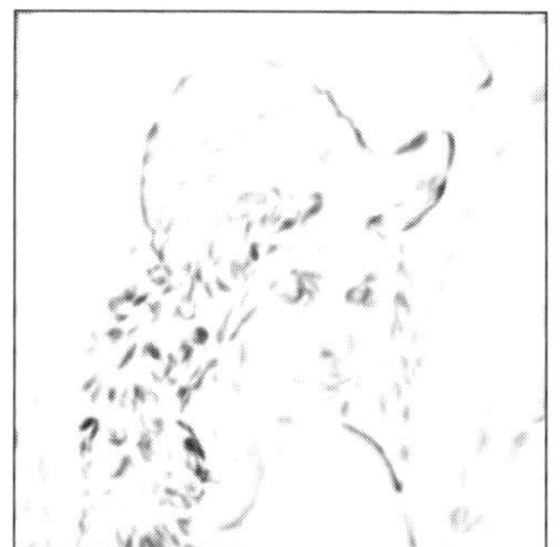

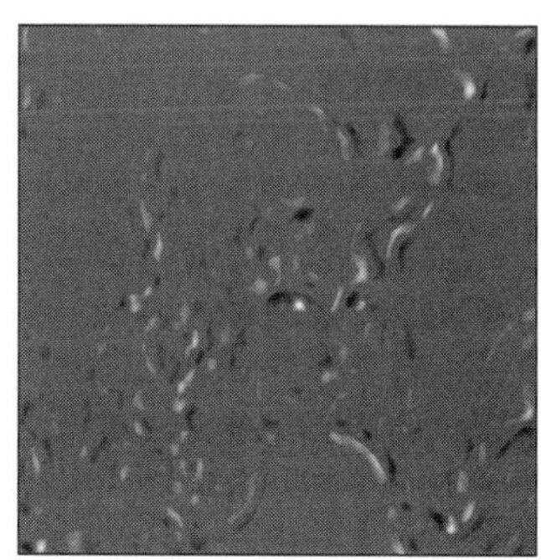

Figure 14.18: Generalization of i2D operators. Left: Input image. Middle: Output of an isotropic i2D energy operator (see Fig. 14.10c). Right: Output of an i2D-selective "clipped eigenvalues" operator (see [Bar93, Zet90c]).

an appropriate generalization of the eigenvalues of the Hessian. Its application to a natural image is shown in Fig. 14.18.

Filter-Nonlinearity-Filter (FNF) Systems. We have seen that a Wiener system, that is, a linear spatial filter, followed by a point nonlinearity is insufficient for the construction of i2D operators. But what about the addition of a further filtering stage, that is, an FNF system? This subclass of nonlinear systems can indeed be i2D selective. The simplest example is an FNF system based on two oriented bandpass filters with orthogonal orientation. For i1D signals, only a restricted range of orientations can pass the first filter, and the nonlinearity only can generate frequencies with the same orientation. If the second filter can block these orientations, no i1D signals can pass the system. For i2D signals, however, the nonlinearity can generate combination frequency terms that extend outside the orientation range of the first filter and can pass the second filter. Another possibility is the usage of appropriate radial bandpass filters. If we use a squaring nonlinearity and the first filter has an octave band, for example, the sum and difference frequency terms from i1D signals will be higher than the upper frequency limit or lower than the lower frequency limit, respectively. If the second bandpass is restricted to the same range as the first one, it can thus be passed only by frequency terms that are due to i2D signals. It should be noted, however, that such FNF systems will always be much more restricted in their tuning properties than the more general schemes presented in this chapter.

Relation to Adaptive Filtering. An interesting point to note is the close relationship between stationary nonlinear systems and adaptive linear systems. Formally, any adaptive linear system is equivalent to a static nonlinear system with memory. The Gaussian curvature operator, for example, can be seen to compute the product of the "principal curvatures." These are two orthogonal derivatives, which are computed in a specific coordinate frame. Since they are equivalent to the eigenvalues of the Hessian, the output of the "clipped eigenvalue" operator is a linear measure (a derivative), which is automatically computed in an optimal coordinate system; that is, it is an example for a specific sort of adaptive linear filter.

Range and Volume Images. A further interesting application area for i2D operators is the analysis of range and volume images. Typical schemes for the classification and segmentation of range images of solid shapes use a direct implementation of the formulas provided by differential geometry [Bes88]. With respect to the i2D parts, this corresponds to the evaluation of Gaussian curvature of the corresponding surfaces; that is, it is related to the application of the "Gaussian" i2D operator. However, the resulting operators, while being formally correct, are not easily adaptable to the specific signal processing requirements, in which a greater discriminability between shallow i2D signals and true i1D or i0D signals is often desired and where noise has to be taken into account. Similar problems arise for volume images, as they appear in CT, MRI, and other medical image processing contexts. These problems can be handled more flexibly by using the class of tunable operators proposed in this chapter (see also [Bar93, Zet90c]).

Image Compression. The close relationship between intrinsic dimensionality and the statistical redundancies in images suggests an application of the concepts in picture coding. Here two directions can be distinguished. First, intrinsic dimensionality can be used in the context of vector quantization. A wavelet-based feature-specific vector quantizer has been developed that takes into account knowledge about both the image statistics and neural processing mechanisms [Weg96]. In addition to the statistical advantages, the approach also provides an implicit quality metric that avoids the well known disadvantages of the classical MSE metric [Zet89]. Since the perceptual metric is determined by neural mechanisms, a compression scheme that is closely related to the biological structure can be more easily optimized for the subjective criteria. A second direction to explore is the direct exploitation of the predictive potential of i2D signals. This more complex issue corresponds to the inversion of a nonlinear signal representation, for which we currently have only approximate and time-consuming solutions. Nevertheless, simulations have indicated that an inversion from i2D signals seems possible in principle [Bar93].

Taxonomy and Classification of Visual Neurons. The close relationship between i2D operators and certain classes of visual neurons has already been mentioned. In this context it is important to note that i2D-selective neurons appear not only in the well known form of orientation-selective hypercomplex cells [Hub65] but also in a variety of versions, including, for example, isotropic "dot-responsive" cells [Sai88] and "bug detectors" [Let59]. Although the basic phenomenon of i2D selectivity has been known for four decades in neurophysiology, the modeling of those cells has only recently started (e.g., [Dob87, Koe87, Zet87, Zet90b]). In addition to providing a framework for this modeling, our proposed concept also offers a strategy for the systematic neurophysiological measurement of the nonlinear processing properties of i2D-selective neurons by the well established methods for the identification of Volterra-Wiener systems [Kor88, Mar78] . This, in turn, may lead to improved criteria for the taxonomy and classification of the variety of nonlinear visual neurons.

14.6.1 Summary

The evaluation of the statistical properties of natural images and the analysis of their relationship to basic neurophysiological and psychophysical aspects of biological image processing have led to the concept of local intrinsic dimensionality [Bar93, Zet90b, Zet90c, Zet93, Zet99a]. Here we have suggested that the processing of intrinsic dimensionality may be regarded as an elementary application of nonlinear Volterra operators.

Linear filters are inherently restricted to an OR gating of their sinusoidal eigenfunctions. They cannot change the form of these frequency components, and they can control only the passage of the components per se, without being able to address specific relations. Polynomial operators, in contrast, provide an AND gating of frequency components with specific relations, and in addition, they change the form of the components. Quadratic operators, for example, enable the selective filtering, that is, passing or blocking, of pairwise AND combinations, and they allow the selective or joint generation of higher (sum) and lower (difference) frequency terms at the output of the system.

Since complex exponentials are intrinsically one-dimensional functions, the potential for an AND gating of these frequency components is the essential prerequisite for the design of i2D operators. By use of the Volterra formalism, we have derived necessary and sufficient conditions for the frequency-domain kernels to ensure this i2D selectivity. We have then introduced a generalized nonlinear filtering concept for the frequency-domain design of nonlinear operators that is based on a new type of stopband and passbands, which are defined in terms of relations between frequency components. A basic requirement for i2D selectivity is that all frequency pairs whose components are jointly oriented have to be blocked completely. Modification of the arrangement and the extension of the passbands and stopbands then enables the control of the "curvature tuning" and of other processing properties, such as isotropy versus orientation selectivity, or the combination with local energy detection. The basic mechanism for local energy detection (local phase invariance, demodulation) can also be interpreted in terms of the generalized nonlinear filtering concept because it requires the blocking of products in which the frequency components have equal signs, while the products of components with different sign are allowed to pass.

The separate processing of constant (that is, low-frequency) and of varying (that is, high-frequency) signals is one of the most basic capabilities of linear systems. In analogy, the distinction between i1D and i2D signals can be regarded as a basic potential of two-dimensional polynomial operators. The resulting i2D operators are capable of exploiting the elementary higher-order redundancies of multidimensional signals.

Acknowledgment

This work has been supported by Deutsche Forschungsgemeinschaft SFB 462/B5 and Re337/10.

References

[Ade85] E. H. Adelson and J. R. Bergen. Spatiotemporal energy models for the perception of motion. *J. Opt. Soc. Amer. A* **2**, 284–299 (1985).

[Ati90] J. J. Atick and A. N. Redlich. Towards a theory of early visual processing. *Neural Comput.* **2**, 308–320 (1990).

[Att54] F. Attneave. Some informational aspects of visual perception. *Psychol. Review* **61**, 183–193 (1954).

[Bam89] R. Bamler. *Mehrdimensionale lineare Systeme.* Springer, Berlin (1989).

[Bar93] E. Barth, T. Caelli, and C. Zetzsche. Image encoding, labelling and reconstruction from differential geometry. *CVGIP: Graph. Models Image Process.* **55**(6), 428–446 (1993).

[Bea78] P. R. Beaudet. Rotationally invariant image operators. In *Proc. 4th Intl. Joint Conf. on Pattern Recognition*, pp. 578–583. IEEE Press (1978).

[Bel97] A. Bell and T. Sejnowski. The "independent components" of natural scenes are edge filters. *Vision Research* **37**, 3327–3338 (1997).

[Ben90] J. S. Bendat. *Nonlinear System Analysis and Identification.* Wiley, New York (1990).

[Bes86] P. J. Besl and R. C. Jain. Invariant surface characteristics for 3D object recognition in range images. *Comput. Vision Graph. Image Process.* **33**, 33–80 (1986).

[Bes88] P. J. Besl and R. C. Jain. Segmentation through variable-order surface fitting. *IEEE Trans. Pattern Anal. Machine Intell.* **10**(2), 167–192 (1988).

[Bie85] I. Biedermann. Human image understanding: recent research and a theory. *Comput. Vision Graph. Image Process.* **32**,29–73 (1985).

[Boa95] B. Boashash, E. J. Powers, and A. M. Zoubir, eds. *Higher-Order Statistical Signal Processing.* Longman and Wiley, Melbourne and New York (1995).

[Bra86] R. Bracewell. *The Fourier Transform and Its Applications.* McGraw-Hill, New York (1986).

[Bur92] D. C. Burr and M. C. Morrone. A nonlinear model of feature detection. In *Nonlinear Vision: Determination of Neural Receptive Fields, Function, and Networks*, B. Pinter and B. Nabet, eds., pp. 309–327. CRC Press, Boca Raton, FL (1992).

[Car94] M. Carandini and D. G. Heeger. Summation and division by neurons in primate visual cortex. *Science* **264**, 1333–1336 (1994).

[Com94] P. Comon. Independent component analysis, a new concept? *Signal Process.* **36**, 287–314 (1994).

[Dau85] J. G. Daugman. Uncertainty relation for resolution in space, spatial frequency, and orientation optimized by two-dimensional visual cortical filters. *J. Opt. Soc. Amer.* A**2**(7), 1160–1169 (1985).

[Dau88] J. G. Daugman. Complete discrete 2-D Gabor transforms by neural networks for image analysis and compression. *IEEE Trans. Acoust. Speech and Signal Processing* ASSP-**36**, 1169–1179 (1988).

[Doc76] M. P. DoCarmo. *Differential Geometry of Curves and Surfaces.* Prentice-Hall, Englewood Cliffs, NJ (1976).

[DeV88] R. L. DeValois and K. K. DeValois. *Spatial Vision.* Oxford University Press, New York (1988).

[Dob87] A. Dobbins, S. W. Zucker, and M. S. Cynader. Endstopped neurons in the visual cortex as a substrate for calculating curvature. *Nature* **329**, 438–441 (October 1987).

[Dur89] R. Durbin and D. Rumelhart. Product units: A computationally powerful and biologically plausible extension to backpropagation networks. *Neural Comput.* **1**, 133–142 (1989).

[Fie87] D. J. Field. Relations between the statistics of natural images and the response properties of cortical cells. *J. Opt. Soc. Amer.* A**4**, 2379–239 (1987).

[Fie94] D.J. Field. What is the goal of sensory coding? *Neural Comput.* **6**, 559–601 (1994).

[Har83] R. M. Haralick, L. T. Watson, and T. J. Laffey. The topographic primal sketch. *Intl. J. Robotic Research* **2**(1), 50–72 (1983).

[Hei92] F. Heitger, L. Rosenthaler, R. von der Heydt, E. Peterhans, and O. Kübler. Simulation of neural contour mechanisms: From simple to end-stopped cells. *Vision Research* **32**(5), 63–981 (1992).

[Hub65] D. H. Hubel and T. N. Wiesel. Receptive fields and functional architecture in two nonstriate visual areas (18 and 19) of the cat. *J. of Neurophysiol.* **28**, 229–289 (1965).

[Kat78] H. Kato, P. O. Bishop, and G. A. Orban. Hypercomplex and simple/complex cell classifications in cat striate cortex. *J. Neurophysiol.* **41**, 1071–1095 (1978).

[Kit82] L. Kitchen and A. Rosenfeld. Gray-level corner detection. *Pattern Recog. Lett.* **1**, 95–102 (1982).

[Koe87] J. J. Koenderink and A. J. van Doorn. Representation of local geometry in the visual system. *Biol. Cybern.* **55**, 367–375 (1987).

[Koe88] J. J. Koenderink and W. Richards. Two-dimensional curvature operators. *J. Opt. Soc. Amer. A* **5**(7), 1136–1141 (1988).

[Kor88] M. J. Korenberg. Exact orthogonal kernel estimation from finite data records: Extending Wiener's identification of nonlinear systems. *Ann. Biomed. Eng.* **16**, 201–214 (1988).

[Kri92] G. Krieger. *Analyse und Beschreibung nichtlinearer Systeme mit Volterra-Reihen.* Diploma Thesis, Lehrstuhl für Nachrichtentechnik, TU-München (1992).

[Kri95] G. Krieger, C. Zetzsche, and E. Barth. Nonlinear image operators for the detection of local intrinsic dimensionality. In *Proc. IEEE Workshop on Nonlinear Signal and Image Processing*, pp. 182–185 (1995).

[Kri96] G. Krieger and C. Zetzsche. Nonlinear image operators for the evaluation of local intrinsic dimensionality. *IEEE Trans. Image Process.* Sepcial Issue on Nonlinear Image Processing. **5**(6), 1026–1042 (1996).

[Kri97] G. Krieger, C. Zetzsche, and E. Barth. Higher-order statistics of natural images and their exploitation by operators selective to intrinsic dimensionality. In *Proc. IEEE Signal Processing Workshop on Higher-Order Statistics*, Vol. PR08005, pp. 147–151. IEEE Computer Society, Los Alamitos, CA (1997).

[Kri98] G. Krieger and C. Zetzsche. Higher-order redundancies of natural scenes and their relation to biological vision. *Perception*, **27**, 11–12 (1998).

[Kri99] G. Krieger. *Nichtlineare Informationsverarbeitung in biologischen und technischen Sehsystemen: Eine Analyse mit Volterra-Reihen und Statistiken höherer Ordnung.* Ph.D. thesis, Lehrstuhl für Nachrichtentechnik, Technische Universität München (1999).

[Kri00a] G. Krieger, I. Rentschler, G. Hauske, and C. Zetzsche. Object and scene analysis by saccadic eye-movements: An investigation with higher-order statistics. *Spatial Vision*, in press (2000).

[Kri00b] G. Krieger and C. Zetzsche. The higher-order statistics of natural images: I. Structural properties. In preparation (2000).

[Let59] J. Y. Lettvin, H. R. Maturana, W. S. McCulloch, and W. H. Pitts. What the frog's eye tells the frog's brain. *Proc. IRE* **47**, 1940–1951 (1959).

[Mac67] N. H. Mackworth and A. J. Morandi. The gaze selects informative details within pictures. *Percep. Psychophys.* **2**, 547–552 (1967).

[Mac81] D. M. MacKay. Strife over visual cortical function. *Nature*, **289**, 117-118 (1981).

[Mal89] S. G. Mallat. A theory for multiresolution signal decomposition: the wavelet representation. *IEEE Trans. Pattern Anal. Machine Intell.*, **11**(7), 674-693 (1989).

[Mar78] P. Z. Marmarelis and V. Z. Marmarelis. *Analysis of Physiological Systems—The White-Noise Approach*, Vol. 1. Plenum Press, New York (1978).

[Mar80] S. Marcelja. Mathematical description of the responses of simple cortical cells. *J. Opt. Soc. Amer. A* **70**(11), 1297–1300 (1980).

[Mat00] V. J. Mathews and G. L. Sicuranza. *Polynomial Signal Processing.* Wiley, New York (2000).

[Nik93] C. L. Nikias and A. P. Petropulu. *Higher-Order Spectral Analysis: A Nonlinear Signal Processing Framework.* Prentice-Hall, Englewood Cliffs, NJ (1993).

[Ols96] B. A. Olshausen and D. J. Field. Wavelet-like receptive fields emerge from a network that learns sparse codes for natural images. *Nature*, **381**, 607-609 (1996).

[Orb84] G. A. Orban. *Neuronal Operations in the Visual Cortex.* Springer, Heidelberg (1984).

[Pap62] A. Papoulis. *The Fourier Integral and Its Applications.* McGraw-Hill, New York (1962).

[Pat75] K. Paton. Picture description using Legendere polynomials. *Comput. Graph. Image Process.* **4**, 40–54 (1975).

[Pin92] R. B. Pinter and B. Nabet, eds. *Nonlinear Vision.* CRC Press, Boca Raton, FL (1992).

[Pit90] I. Pitas and A. N. Venetsanopoulos. *Nonlinear Digital Filters: Principles and Applications.* Kluwer Academic Publishers, Boston (1990).

[Rao99] R. Rao and D. Ballard. Predictive coding in the visual cortex: A functional interpretation of some extra-classical receptive-field effects. *Nature Neurosci.* **2**, 79–87 (1999).

[Ros82] A. Rosenfeld and A. C. Kak. *Digital Picture Processing.* Academic Press, Orlando, FL (1982).

[Ros85] M. Rosenblatt. *Stationary Sequences and Random Fields.* Birkhäuser, Boston (1985).

[Rud94] D. Ruderman. The statistics of natural images. *Network*, **5**, 517–548 (1994).

[Rug81] W. J. Rugh. *Nonlinear System Theory*. Johns Hopkins University Press, Baltimore, MD (1981).

[Sai88] H. Saito, K. Tanaka, Y. Fukada, and H. Oyamada. Analysis of discontinuity in visual contours in area 19 of the cat. *J. Neurosci.* **8**, 1131–1143 (1988).

[Sch89] M. Schetzen. *The Volterra and Wiener Theories of Nonlinear Systems*. Krieger's, Malabar, FL, updated edition (1989).

[Sic92] G. L. Sicuranza. Quadratic filters for signal processing. *Proc. IEEE* **80**(8), 1263–1285 (1992).

[Szu90] R. G. Szulborski and L. A. Palmer. The two-dimensional spatial structure of nonlinear subunits in the receptive fields of complex cells. *Vision Research*, **30**, 249–254 (1990).

[Tho97] M. G. Thompson and D. H. Foster. Role of second- and third-order statistics in the discriminability of natural images. *J. Opt. Soc. Amer. A* **14**, 2081–2090 (1997).

[van98] J. van Hateren and A. van der Schaaf. Independent component filters of natural images compared with simple cells in primary visual cortex. *Proc. Royal Soc. London B* **265**, 359–366 (1998).

[Wat85] L. T. Watson, T. J. Laffey, and R. M. Haralick. Topographic classification of digital image intensity surfaces using generalized splines and the discrete cosine transformation. *Comput. Vision Graph. Image Process.* **29**, 143–167 (1985).

[Wat87] A. B. Watson. Efficiency of a model human image code. *J. Opt. Soc. Amer. A* **4**, 2401–2417 (1987).

[Weg90a] B. Wegmann and C. Zetzsche. Statistical dependence between orientation filter outputs used in a human vision based image code. In *Visual Communication and Image Processing* (M. Kunt, ed.), Proc. SPIE, Vol. 1360, pp. 909–923 (1990).

[Weg90b] B. Wegmann and C. Zetzsche. Visual system based polar quantization of local amplitude and local phase of orientation filter outputs. In *Human Vision and Electronic Imaging: Models, Methods, and Applications* (B. Rogowitz, ed.), Proc. SPIE, Vol. 1249, pp. 306–317 (1990).

[Weg92] B. Wegmann and C. Zetzsche. Efficient image sequence coding by vector quantization of spatiotemporal bandpass outputs. In *Visual Communications and Image Processing '92* (P. Maragos, ed.) Proc. SPIE, Vol 1818, pp. 1146–1154 (1992).

[Weg96] B. Wegmann and C. Zetzsche. Feature-specific vector quantization of images. Special Issue: Vector Quantization, *IEEE Trans. Image Process.* **5**, 274–288 (1996).

[Zet87] C. Zetzsche and W. Schönecker. Orientation selective filters lead to entropy reduction in the processing of natural images. *Perception,* **16**, 229 (1987).

[Zet88] C. Zetzsche. Statistical properties of the representation of natural images at different levels in the visual system. *Perception,* **17**, 359 (1988).

[Zet89] C. Zetzsche and G. Hauske. Multiple channel model for the prediction of subjective image quality. In *Human Vision, Visual Processing and Visual Display* (B. Rogowitz, ed.), Proc. SPIE, Vol. 1077, pp. 209–216 (1989).

[Zet90a] C. Zetzsche. Sparse coding: the link between low level vision and associative memory. In *Parallel Processing in Neural Systems and Computers* (R. Eckmiller, G. Hartmann, and G. Hauske, eds.), pp. 273–276, Elsevier, Amsterdam (1990).

[Zet90b] C. Zetzsche and E. Barth. Fundamental limits of linear filters in the visual processing of two-dimensional signals. *Vision Research,* **30**, 1111–1117 (1990).

[Zet90c] C. Zetzsche and E. Barth. Image surface predicates and the neural encoding of two-dimensional signal variation. In *Human Vision and Electronic Imaging: Models, Methods, and Applications* (B. Rogowitz, ed.), Proc. SPIE, Vol. 1249, pp. 160–177 (1990).

[Zet91] C. Zetzsche and E. Barth. Direct detection of flow discontinuities by 3D-curvature operators. *Pattern Recog. Lett.* **12**, 771–779 (1991).

[Zet93] C. Zetzsche, E. Barth, and B. Wegmann. The importance of intrinsically two-dimensional image features in biological vision and picture coding. In *Digital Images and Human Vision* (A. Watson, ed.), pp. 109–138, MIT Press, Cambridge, MA (1993).

[Zet97] C. Zetzsche, E. Barth, G. Krieger, and B. Wegmann. Neural network models and the visual cortex: The missing link between cortical orientation selectivity and the natural environment. *Neurosci. Lett.* **228**(3), 155–158 (1997).

[Zet98] C. Zetzsche, K. Schill, H. Deubel, G. Krieger, E. Umkehrer, and S. Beinlich. Investigation of a sensorimotor system for saccadic scene analysis: an integrated approach. In *Proc. 5th Intl. Conf. Soc. Adaptive Behavior* (R. Pfeifer, B. Blumenberg, J. Meyer, and S. Wilson, eds.). Volume 5, pp. 120–126, MIT Press, Cambridge, MA (1998).

[Zet99a] C. Zetzsche and G. Krieger. Nonlinear neurons and higher-order statistics: New approaches to human vision and electronic image processing. In *Human Vision and Electronic Image Processing* (B. Rogowitz and T. Pappas, eds.). Proc. SPIE, Vol. 3644, pp. 2–33 (1999).

[Zet99b] C. Zetzsche, G. Krieger, and B. Wegmann. The atoms of vision: Cartesian or polar? *J. Opt. Soc. Amer. A* **16**(7), 1554–1565 (1999).

[Zet00] C. Zetzsche and G. Krieger. The higher-order statistics of natural images: II. Exploitation by linear and nonlinear signal transforms. In preparation (2000).

Index